GaInAsP

ALLOY SEMICONDUCTORS

GaInAsP
ALLOY SEMICONDUCTORS

Edited by
T. P. Pearsall
Optical Electronics Research Department
Bell Laboratories, Murray Hill,
New Jersey, USA

A Wiley–Interscience Publication

JOHN WILEY & SONS
Chichester · New York · Brisbane · Toronto · Singapore

Library of Congress Cataloging in Publication Data:

Main entry under title:

GaInAsP alloy semiconductors.

'A Wiley–Interscience publication.'
Includes bibliographical references and index.
1. Gallium arsenide—Congresses. 2. Semicon ductors—Congresses. I. Pearsall, T. P.
TK7871. 15. G3G34 621. 3815′2 81–15922
AACR2

ISBN 0 471 10119 2

British Library Cataloguing in Publication Data:

GaInAsP alloy semiconductors.
1. Gallium arsenide
2. Semiconductors
I. Pearsall, T. P.
537.6′22 QC611.8.G3

ISBN 0 471 10119 2

Typeset by Macmillan India Ltd., Bangalore.
Printed at Vail-Ballou Press Inc., New York.

Contents

Contributing Authors vii

Editor's Preface xi

I INTRODUCTION

GaInAsP Alloy Semiconductors: Introduction *G. A. Antypas* 3

II EPITAXIAL CRYSTAL GROWTH

1 **Vapour-phase Epitaxy of GaInAsP** 11
G. H. Olsen

2 **Liquid-phase Epitaxy** 43
K. Nakajima

3 **$Ga_{0.47}$ $In_{0.53}$ As/InP and GaInAsP/InP Double Heterostructures Grown by Low-pressure Metal-organic Vapour-phase Epitaxy** 61
J. P. Hirtz, M. Razeghi, M. Bonnet, and J. P. Duchemin

4 **III–V Alloy Growth by Molecular-beam Epitaxy** 87
C. E. C. Wood

5 **Ion Implantation** 107
F. H. Eisen and L. R. Tomasetta

6 **High-purity Material** 121
G. E. Stillman, L. W. Cook, T. J. Roth, T. S. Low, and B. J. Skromme

III GaInAsP MATERIALS PROPERTIES

7 **Defect Motion and Growth of Extended Non-radiative Defect Structures in GaInAsP** 169
W. D. Johnston, Jr

8 **Low-field Carrier Mobility** 189
J. R. Hayes, A. R. Adams, and P. D. Greene

9 **Low-field Transport Calculations** 213
Y. Takeda

10 **Hot Electron Transport in n-Type $Ga_{1-x}In_xAs_yP_{1-y}$ Alloys Lattice-matched to InP** 243
M. A. Littlejohn, T. H. Glisson, and J. R. Hauser

11 **High-field Transport Measurements** 275
R. F. Leheny

12 **Electronic Structure of $Ga_xIn_{1-x}As_yP_{1-y}$ Alloys Lattice-matched to InP** 295
T. P. Pearsall

13 **Photoluminescence and Optical Gain of GaInAsP** 313
E. O. Göbel

IV GaInAsP DEVICE TECHNOLOGY AND PERFORMANCE

14 **Double-heterostructure Lasers** 341
Y. Suematsu, K. Iga, and K. Kishino

15 **Temperature Dependence of Laser Threshold Current** 379
Y. Horikoshi

16 **Photodetectors** 413
Y. Matsushima and K. Sakai

17 **Field-effect Transistors** 437
H. Ohno and J. Barnard

Appendix: Some Physical Properties 456

Index 459

Contributing Authors

A. R. Adams	*Department of Physics, University of Surrey, Guildford, Surrey, UK*
G. A. Antypas	*CrystaComm Inc., 486 Ellis St, Mt View, California, USA*
J. Barnard	*School of Electrical Engineering Cornell University, Ithaca, NY 14853, USA*
M. Bonnet	*Thomson-CSF, LCR, Domaine de Corbeville, BP No. 10, 91401 Orsay, France*
L. W. Cook	*Electrical Engineering Research Laboratory, University of Illinois at Urbana-Champaign, Urbana, IL 61801, USA*
J. P. Duchemin	*Thomson-CSF, LCR, Domaine de Corbeville, BP No. 10, 91401 Orsay, France*
F. H. Eisen	*Rockwell International Microelectronics Research and Development Center, Thousand Oaks, CA 91360, USA*
T. H. Glisson	*Department of Electrical Engineering, North Carolina State University, Raleigh, NC 27650, USA*
E. Göbel	*Max-Planck-Institut für Festkörperforschung, Heisenbergstrasse 1, D-7000 Stuttgart 80, FRG*
P. D. Greene	*Standard Telecommunication Laboratories Ltd, London Road, Harlow, Essex, UK*
J. R. Hauser	*Department of Electrical Engineering, North Carolina State University, Raleigh, NC 27650, USA*
J. R. Hayes	*Department of Physics, University of Surrey, Guildford, Surrey, UK*
J. P. Hirtz	*Thomson-CSF, LCR, Domaine de Corbeville, BP No. 10, 91401 Orsay, France*

Y. Horikoshi — *Musashino Electrical Communication Laboratories, Nippon Telegraph and Telephone Public Corporation, Musashino-shi, Tokyo 180, Japan*

K. Iga — *Research Laboratories of Precision Machinery and Electronics, 4259 Nagatsuta, Midori-ku, Yokohama 227, Japan*

W. D. Johnston, Jr — *Bell Laboratories, Murray Hill, NJ 07974, USA*

K. Kishino — *Tokyo Institute of Technology, Department of Physical Electronics, 2-12-1 O-okayama, Meguro-ku, Tokyo 152, Japan*

R. F. Leheny — *Bell Laboratories Inc., Holmdel, NJ 07733, USA*

M. A. Littlejohn — *Department of Electrical Engineering, North Carolina State University, Raleigh, NC 27650, USA*

T. S. Low — *Electrical Engineering Research Laboratory, University of Illinois at Urbana-Champaign, Urbana, IL 61801, USA*

Y. Matsushima — *KDD Research and Development Laboratories, 2-1-23 Nakameguro, Meguro-ku, Tokyo 153, Japan*

K. Nakajima — *Fujitsu Laboratories Ltd, Semiconductor Materials Laboratory, 1015 Kamikodanaka, Nakahara-ku, Kawasaki 211, Japan*

H. Ohno — *Dept. of Electrical Engineering, Hokkaido University, Sapporo, 060, Japan.*

G. H. Olsen — *RCA Laboratories, Princeton, NJ 08540, USA*

T. P. Pearsall — *Bell Laboratories, Murray Hill, NJ 07974, USA*

M. Razeghi — *Thomson-CSF, LCR, Domaine de Corbeville, BP No. 10, 91401 Orsay, France*

T. J. Roth — *Electrical Engineering Research Laboratory, University of Illinois at Urbana-Champaign, Urbana, IL 61801, USA*

K. Sakai — *KDD Research and Development Laboratories, 2-1-23 Nakameguro, Meguro-ku, Tokyo 153, Japan*

B. J. Skromme — *Electrical Engineering Research Laboratory, University of Illinois at Urbana-Champaign, Urbana, IL 61801, USA*

G. E. Stillman — *Electrical Engineering Research Laboratory, University of Illionois at Urbana-Champaign, Urbana, IL 61801, USA*

Y. Suematsu — *Tokyo Institute of Technology, Department of Physical Electronics, 2-12-1 O-okayama, Meguro-ku, Tokyo 152, Japan*

Y. Takeda — *Department of Electrical Engineering, North Carolina State University, Raleigh, NC 27650, USA*

L. R. Tomasetta — *Rockwell International Microelectronics Research and Development Center, Thousand Oaks, CA., 91630 U.S.A.*

C. E. C. Wood — *School of Electrical Engineering, Cornell University, Ithaca, NY 14853, USA*

Editor's Preface

Not quite a decade has passed since Antypas and colleagues at Varian proposed the growth of $Ga_xIn_{1-x}As_yP_{1-y}$ Lattice-matched to InP substrates. During the first five years following this announcement, initial device results were greeted with some enthusiasm but mostly incredulity that anything practical could ever be developed from what appeared to be a hopelessly complex semiconductor alloy. In 1976 the announcement of high efficiency light emitters whose spectral output matched the region of low attenuation for optical fibre transmission ignited an explosive research programme around the world to develop devices from this semiconductor material. It would be fitting to acknowledge here those in the programme office of the Night Vision Laboratory, Fort Belvoir, Virginia, who had the courage and vision to support this work in its early stages of development.

Today, only five years later, the performance of GaInAsP emitters and detectors is such that they have relegated their counterparts made from GaAs and Si to a secondary role in optical fibre telecommunications systems. Many of the scientists who have made fundamental contributions to this remarkable development are authors of chapters in this book. I would like to express my thanks to them for their cooperative effort in producing this work. I would also like to express my deep gratitude to the referees of each of the chapters whose prompt and conscientious work has made a significant contribution to this volume.

Murray Hill, New Jersey
2 November 1981

I INTRODUCTION

GaInAsP Alloy Semiconductors
Edited by T. P. Pearsall

GaInAsP Alloy Semiconductors: Introduction

G. A. Antypas

Crysta Comm Inc.,
486 Ellis St, Mt View,
California, USA

1. HISTORICAL BACKGROUND

Almost concurrently with the discovery that binary III–V compounds form a class of semiconductors, it was realized that solid solutions of these compounds could provide the basis for the practical engineering of a specific semiconductor to meet certain device requirements. The III–V semiconductors have energy bandgaps that span the range 0.18–2.42 eV. In order to study the properties of this class of alloy semiconductors, considerable effort was expended in the late 1950s and early 1960s to develop growth processes for these new materials. The principal methods used were vapour-phase epitaxy, liquid-phase epitaxy, Bridgman growth, zone levelling, gradient freeze and modified Czochralski, among others. These efforts were modestly successful in generating samples of sufficient quality for characterization of the basic properties of these new alloy semiconductors, e.g. the composition dependence of band structure and lattice parameter. While GaAs devices began to move out of the research laboratory in the early 1970s, ternary compounds, with the exception of $GaAs_xP_{1-x}$ for visible light-emitting diodes (LEDs), remained in the laboratory. It was the communications revolution brought on by the development of low-loss optical fibre that provided the motivation to develop, from alloy semiconductors, practical optoelectronic devices whose properties were optimized with respect to the fibre transmission characteristics.

Even though the basic heterojunction device concepts were well understood at this time, practical device development was stymied by the lack of a suitable semiconductor system in which these devices could be realized. Theoretically, alloy III–V semiconductors could provide this vehicle; however, the main drawback for efficient device development can be seen in Figure 1 which shows the variation of lattice constant as a function of bandgap for the whole family of

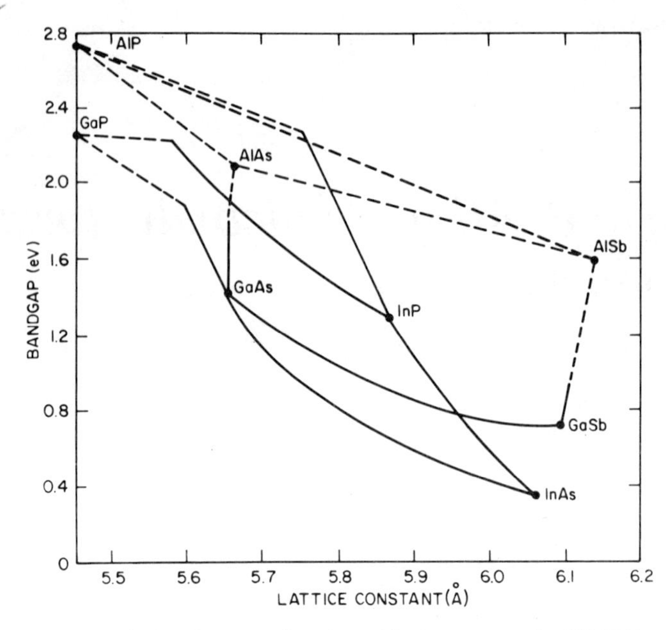

Figure 1 Variation of the bandgap as a function of lattice constant for III–V binary and alloy semiconductors

tenary III–V semiconductors. The one-to-one relationship between bandgap and lattice constant is a major obstacle to device-quality crystal growth because only a few alloy compositions could be grown nearly lattice-matched on available substrates. For many years, this difficulty was the major impediment to the development of efficient heterojunction lasers and LEDs. Two factors are necessary in developing efficient heterojunction devices: (a) single-crystal substrates available on a commercial scale, and (b) a semiconductor system that provides an energy bandgap range at constant lattice parameter. From Figure 1, it can be seen that the group III (Ga,Al)–group V (P, As, Sb) systems can yield alloy semiconductors of varying bandgaps at almost constant lattice parameter. Since the GaAs technology was better developed than other possible III–V semiconductors, GaAs–AlAs emerged first as the vehicle for the development of a new class of efficient semiconductor heterojunction devices. The most prominent of these devices was the GaAlAs heterojunction laser, a key element in the rapidly emerging fibre-optic communication industry.

At the same time as the development of semiconductor devices based on GaAs–AlAs, a number of applications in the field of night vision required efficient heterojunction detectors which were sensitive to light at longer wavelengths, in the range of 1.0–1.2 μm where GaAs could not be used because of its larger bandgap. Since the $Al_xGa_{1-x}As$ alloy system could not meet the requirement for these heterojunction devices, considerable effort was invested in minimizing the lattice-mismatch effects in other ternary III–V alloys which were photosensitive at wavelengths longer than 1.0 μm. Such alloys were GaAs–InAs, GaAs–GaSb, and InAs–InP. In order to overcome the lattice-mismatch limitation, compositional grading between the substrate and the device material was achieved in order to minimize the incorporation of a high concentration of line defects that would adversely affect device performance. By comparing rapid progress that was being made on heterojunction devices based on the GaAs–AlAs system with the difficulties facing device development in mismatched III–V alloy systems, it soon became evident that lattice matching across a heterojunction interface is essential for the development of efficient and reliable devices.

2. THE GaInAsP–InP LATTICE-MATCHED QUATERNARY ALLOY

The restrictions that impose a one-to-one relationship between lattice constant and bandgap in III–V alloys can be relaxed by incorporating a fourth component, thus adding an extra degree of freedom so that both bandgap and lattice constant can be adjusted independently. In Figure 1, binary compounds are represented by points, ternary compounds are represented by lines, and quaternary compounds by areas bounded by ternary alloy lines. The GaInAsP quaternary system shown in Figure 1 is bounded by the InAsP, GaInAs, and GaAsP ternaries. The key advantage of this system is that two commercially available substrates, GaAs and InP, can be employed for the epitaxial growth of lattice-matched quaternary layers. When GaAs is used as a substrate, the lattice-matched GaInAsP alloy spans the energy range of 1.42–1.91 eV. This system has received little attention, primarily because it almost reproduces the energy range spanned by the AlGaAs system. When InP is used as a substrate, lattice-matched InGaAsP alloys can be prepared whose bandgaps span the energy range between 0.75 and 1.35 eV.

We first reported the preparation of lattice-matched GaInAsP/InP heterojunctions at Varian Associates in 1972.[1] Our research was motivated by efforts to develop a negative-electron-affinity photocathode detector operating at 1.06–1.1 μm. The first such commercial device which incorporated a GaInAsP photocathode was introduced by Varian in 1973. This high-speed photomultiplier was sensitive to light whose wavelength extended to 1.1 μm, with a quantum efficiency far superior to that of the best S-1 photomultipliers. While the group at Varian continued its research effort on photoemission from GaInAsP, Bogatov and coworkers[2] in 1975 and Hsieh[3] in 1976 reported the fabrication of heterojunction laser devices operating efficiently at room temperature. These developments, along with reports that fibre transmission properties are optimized at longer wavelengths with respect to both attenuation and wavelength dispersion, initiated an explosive effort in studying the GaInAsP system. This effect can best be seen in the bibliography compilation made by A. R. Clawson[4]

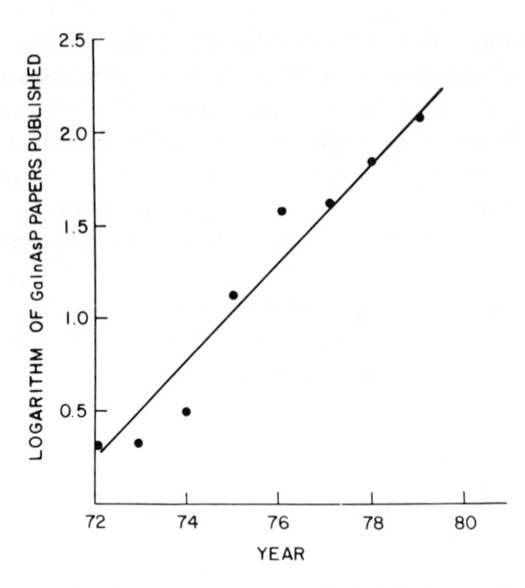

Figure 2 Publication rate of papers on GaInAsP alloy semiconductors

(see Figure 2) which shows that the number of papers published on GaInAsP every year since 1975 has increased at an exponential rate.

As will be described in the chapters that follow, GaInAsP/InP heterojunctions can be prepared epitaxially by every major crystal growth technique: liquid-phase epitaxy, vapour-phase epitaxy, organometallic chemical vapour-phase epitaxy, and molecular-beam epitaxy. The principal orientation in device development is towards fibre-optic communications using light sources such as lasers and LEDs operating at wavelengths from 1.06 to 1.55 μm and light detectors such as p–i–n and avalanche photodiodes sensitive in the same wavelength range. The performance and lifetime characteristics of these devices at the present time rival those made from $Al_xGa_{1-x}As$. GaInAsP/InP semiconductor devices now appear to be suitable for development in commercial fibre-optic communication networks. In addition to the application of the GaInAsP/InP heterojunction devices to fibre-optic technology, band structure and transport studies indicate that these alloys could be used to improve the performance of high-frequency microwave discrete devices and integrated circuits.

The foundation of GaInAsP device technology is the quality of the InP substrate. While efforts towards the improvement of InP substrates represent a small fraction of the time and money invested in the related device technology, present evaluation of InP single-crystal ingots shows that the quality of the InP is at least in a state adequate to support commercial device development. The work of Seki and coworkers[5] on the growth of low-defect-density, heavily doped n-type and p-type InP has been reproduced in many laboratories. Now, InP with an almost zero dislocation density is commercially available. While further progress is expected, the future of the GaInAsP/InP technology is assured with clearly recognized applications in fibre-optic communications and potential improvements in microwave and integrated-circuit technologies.

REFERENCES

1. G. A. Antypas, R. L. Moon, L. W. James, J. Edgecumbe, and R. L. Bell, 'III–V quaternary alloys', *Gallium Arsenide and Related Compounds, 1972*, Conf. Ser. no. 17, Institute of Physics, London, 1973, p. 48.
2. A. P. Bogatov, L. M. Dolginov, L. V. Druzhinina, P. G. Eliseev, B. N. Sverdlov, and E. G. Shevchenko. 'Heterojunction lasers made of GaInAsP and AlGaSbAs solid solutions', *Sov. J. Quantum Electron.*, **4**, 1281, 1975.
3. J. J. Hsieh, 'Room-temperature operation of GaInAsP/InP double-heterostructure diode lasers emitting at 1.1 μm', *Appl. Phys. Lett.*, **28**, 283, 1976.
4. A. R. Clawson, 'Bibliography on the quaternary III–V semiconductor InGaAsP', *Technical Note* 830, Naval Ocean Systems Center, San Diego, CA, 1980.
5. Y. Seki, J. Matsui, and H. Watanabe, 'Impurity effect on the growth of dislocation-free InP single crystals', *J. Appl. Phys.*, **47**, 3374, 1975.

II EPITAXIAL CRYSTAL GROWTH

GaInAsP Alloy Semiconductors
Edited by T. P. Pearsall

Chapter 1

Vapour-phase Epitaxy of GaInAsP

G. H. OLSEN

RCA Laboratories,
Princeton, NJ 08540, USA

1.1 INTRODUCTION

GaInAsP alloys can be grown by liquid-phase epitaxy (LPE), metalorganic chemical vapour deposition (MOCVD), molecular-beam epitaxy (MBE), and conventional vapour-phase epitaxy (VPE)—each discussed in an appropriate chapter of this book. The LPE process has the advantages that it can be a 'near-equilibrium' thermodynamic process, is relatively simple and inexpensive, Al and Sb alloy growth is no problem, and the best device results achieved to date have been with LPE material. Its drawbacks include growth problems with GaInAsP alloys for $\lambda > 1.4\ \mu m$ and potential non-uniform growth as well as melt carry-over and terrace formation. MOCVD has all of the advantages discussed below for conventional VPE as well as the ability to grow Al and Sb alloys. Its drawbacks include the formation of addition compounds with InP-bearing alloys and the possibility of organic contamination (carbon). MBE had the advantage of a clean growth environment together with the capability for single-atomic-layer growth and excellent uniformity. However, it is a very expensive and complex process, and difficulties have been reported with p-type doping and with the growth of phosphorus-bearing alloys (phosphorus clogs the vacuum pump).

The conventional (i.e. non-metalorganic) VPE process[1-3] will be the subject of this chapter. This process has demonstrated good thickness and compositional uniformity, flexible control of alloy composition (e.g. compositional grading), and the ability to be scaled up for mass production. Its drawbacks include difficulties with the growth of Al and Sb compounds, potential for hillock and haze formation, and interfacial decomposition during the 'preheat' stage.

The VPE process can be performed by (1) the 'trichloride' method whereby $AsCl_3$ or PCl_3 is passed over either elemental In or Ga—to form metal chlorides—or passed over binary source wafers such as GaAs and InP; or (2) by the 'hydride' method whereby metal chlorides are formed by passing HCl gas (either from a tank of HCl or cracked $PCl_3/AsCl_3$) over hot In or Ga metal and combined with cracked hydrides of arsenic (AsH_3) and/or phosphorus (PH_3).

Although the purest (lowest background doping levels) InP and GaAs have been synthesized via the trichloride method—which is therefore most often applied to microwave devices—the majority of electro-optic devices which have been synthesized by VPE were done by the hydride method. Most VPE work on GaInAsP has been done via the hydride method and this technique will receive most of the attention in this chapter. Results with the trichloride technique will be discussed where applicable.

1.2 VPE REACTOR DESIGN

The VPE process in its present form is largely an evolution of the method first demonstrated by Tietjen and Amick[1] for the growth of Ga(As,P) alloys using the hydrides arsine (AsH_3) and phosphine (PH_3). Therefore, the present-day RCA hydride VPE system for the growth of (Ga,In)(As,P) alloys will be described in some detail.[2,3] Other systems will then be described with differences pointed out. At the time of this writing, reports on the hydride/chloride VPE growth of (Ga,In)(As,P) alloys include at least Beuchet and coworkers[4] at Thomson-CSF (France), Enda and coworkers[5] at Nippon Telegraph and Telephone (NTT) (Japan), Enstrom and coworkers[6] at RCA (USA), Hyder and coworkers[7] at Varian (USA), Johnston and coworkers[8] at Bell Laboratories (USA), Kanbe and coworkers[9] at NTT, Mizutani and coworkers[10] at Nippon Electric Co. (NEC) (Japan), Nagai and coworkers[11] at NTT, Olsen and coworkers[12] at RCA, Saxena and coworkers[13] at Varian, Seki and coworkers[14] at Tokyo University of Agriculture and Technology (Japan), Sugiyama and coworkers[15] at NTT, Susa and coworkers[16] at NTT, Vohl[17] at Lincoln Laboratories (USA), and Zinkiewicz and coworkers[18] at the University of Illinois (USA). Growth details for these systems are summarized in Table 1.1.

The RCA VPE growth system is shown schematically in Figure 1.1. The system embodies extensive modifications upon the one developed by Tietjen and Amick.[1] The reactor tube (~ 25 mm internal diameter) is made of quartz except for areas that are not heated to high temperatures which are made of Pyrex. Heat

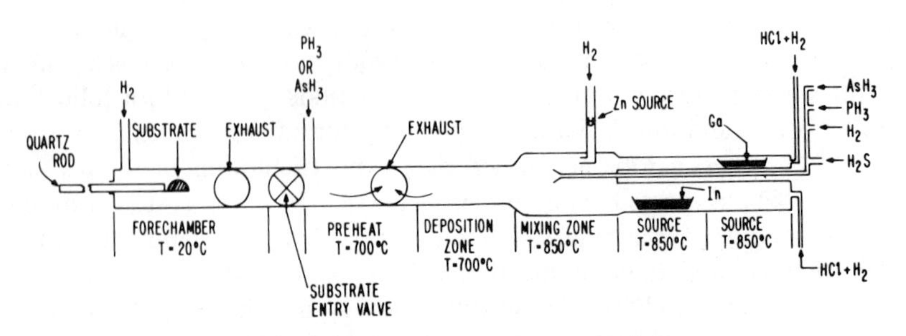

Figure 1.1 Sketch of RCA VPE growth system

Table 1.1 Summary of growth techniques

	Enda[5]	Sugiyama[15]	Johnston[8]	Susa[16]	Kanbe[9]	Seki[14]	Zinkiewicz[18]	Hyder[7]	Olsen[12]	Mizutani[10]	Vohl[17]	Beuchet[4]
λ (μm)	1.38	0.6–1.0	1.5	1.6–1.7	1.6–1.7		1,3–1.7	1.0–1.7	1.3–1.7	1.3	1.3	1.3–1.67
HCl source	HCl/AsH_3	$AsCl_3/AsH_3$	HCl/AsH_3	$AsCl_3/AsH$		HCl/AsH_3	HCl/AsH_3	HCl/AsH_3		HCl/AsH_3	$PCl_3/AsCl_3$	HCl/AsH_3
% AsH, PH_3	5	5	2, 4	5	5		10		10			pure
T_{source} (°C)	830	830	840	800	800		750–800	850	850	800	750–820	750
T_{mix} (°C)	900	900		825	825		750		850	850	830	800
T_{grow} (°C)	660	650–725	700	730	650–750	650–700	650	690	700	650–700	700	700
ΔT (°C cm^{-1})	14	10			1		1	5	0.1	±0.5		10
Preheat	As/PH_3 chamber	none	none	sliding quartz boat	AsH_3 atmos.	near baffle	preheat zone	sliding quartz	AsH_3/PH_3 preheat		preheat zone	none (etching chamber)
H_2 (total) (cm^3 min^{-1})		665	5000		500	400	500–900	1100	5000	1800		1–2.5
Growth rate (μm h^{-1})	5–8	5–15	~15		6–12	5	3–120		~25		1–3	
In area (cm^2)							14	16	100			20
Comments		40 mm, quartz, stoich-iometric Ga	diethyl-zinc, computer, diethyl-telluride, 5% HCl 2% HCl	40 mm diam.	45 mm diam.	single flat temper-ature	magnetic rotator, vertical, indium buckets	40 mm diam., slider boat, 3:1 metal rich,			(110) growth, PCl_3, $AsCl_3$, GaAs, InP, InAs	Four chamber, 30 mm diam., pure HCl for In, 10% HCl for Ga

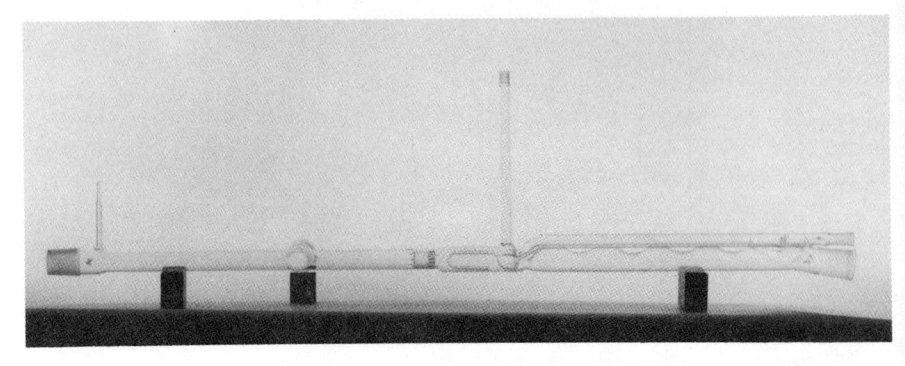

Figure 1.2 Photograph of quartz tube used in RCA VPE system

Figure 1.3 Photograph of automated RCA VPE system

is provided by the use of 'clamshell' resistance furnaces which surround the tube. Photographs of the quartz growth tube and of the complete system are shown in Figures 1.2 and 1.3. The furnaces are left on at all times (except for disassembly or cleaning) and hydrogen flows through the tube constantly. Deposition is initiated by passing HCl gas over the indium and/or gallium metal (which is held at 850–900°C) in order to form metal chlorides. The area of indium metal held in quartz boats is considerably larger ($\sim 100\,cm^2$) than that of gallium metal ($\sim 25\,cm^2$). Arsine and/or phosphine (10% in H_2) are brought in through a separate tube and then mixed with the metal chlorides in a mixing zone. P-type doping is accomplished by heating a zinc bucket in a hydrogen atmosphere in order to obtain elemental zinc vapour which is then transported by H_2. N-type doping is accomplished by adding about 100 ppm H_2S gas to the group V line. All input reactant flows are controlled by electronic mass-flow controllers. Meanwhile, a polished substrate is held on a rotatable sample holder with a quartz spring (shown in Figure 1.4) which is attached to the end of the quartz rod and inserted into the forechamber which is then flushed out with hydrogen while the input gas flows are being equilibrated. The substrate is sealed off from the growth chamber by a large Pyrex stopcock (substrate entry valve). After the forechamber is flushed out ($\sim$ 15 min at $2000\,cm^3$ min of H_2), the stopcock is opened and the quartz rod (which is supported by a close-tolerance 'truebore' gas

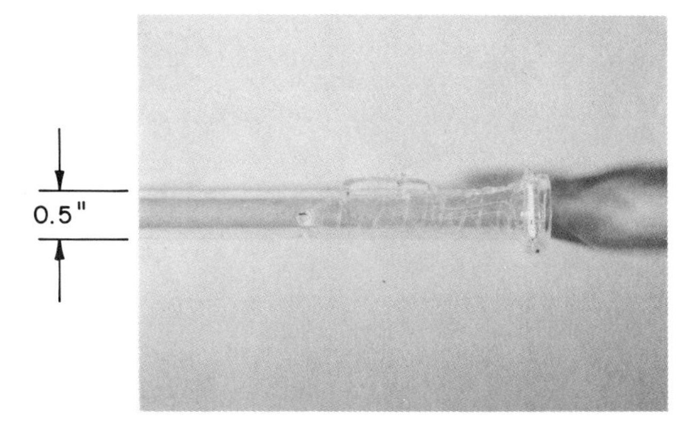

Figure 1.4 Photograph of the rotatable quartz substrate holder used to improve the uniformity of VPE growths

bearing) is pushed in so that the substrate is moved to the preheat zone. This zone has an atmosphere of arsine and/or phosphine, corresponding to the substrate group V constituent, in order to minimize decomposition effects. This decomposition[2,19] is a result of preferential evaporation of the group V constituent at elevated temperatures and is especially pronounced for InP at temperatures above 400°C. Figure 1.5 contains an optical photograph of an InP surface which had been heated to 700°C in a H_2 ambient. Pronounced surface damage had occurred. Figure 1.6 contains cross-sectional transmission electron micrographs from III–V compound heterointerfaces which were grown with and without group V atmospheres.[3] The 'loops' shown in Figure 1.6 are dislocation loops, formed from coalescence of vacancies. Such interfacial defects might well impair the proper operation of a device. The substrate is then heated to near the growth temperature and inserted into the growth zone where deposition takes place. The substrate is rotated ($\sim$ 10 rpm) during growth in order to smooth out any non-uniformities in temperature or gas flow. The temperature over the substrate is constant to within ± 0.1°C. Thickness uniformities of ± 5 % and compositional uniformities of ± 0.1 mol % have been measured with $In_{0.5}Ga_{0.5}P$ grown on GaAs. Growth is ended by withdrawing the substrate to the forechamber (before altering any of the flows) where it cools to room temperature in a hydrogen ambient. If the substrate is to be removed, the stopcock must first be closed. However, if subsequent layers are to be grown, the substrate is held in the forechamber while the reactant flows are changed and equilibrated (typically 15–30 min). The above process is then repeated. If compositional grading is desired, the control voltage to the appropriate mass-

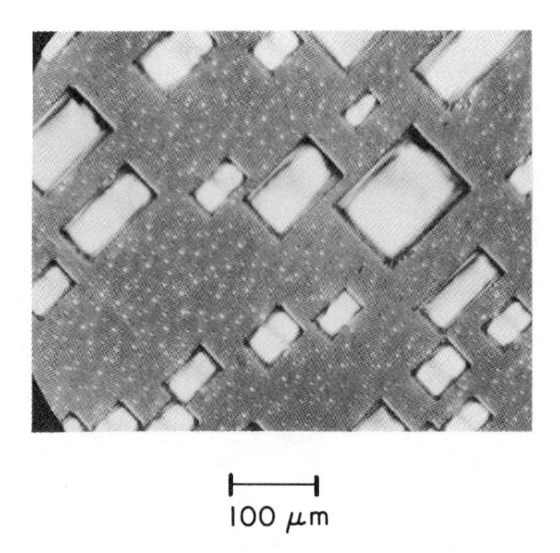

Figure 1.5 Optical photograph of an InP surface heated in H_2 at 670°C for 1 h

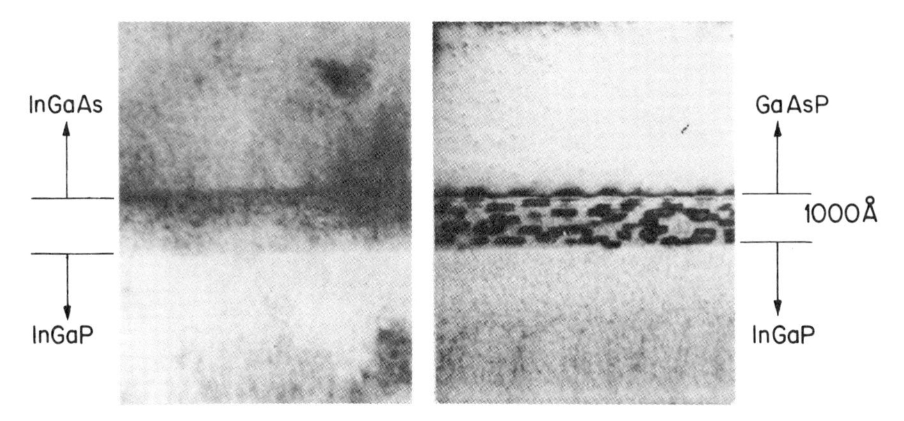

Figure 1.6 Cross-sectional {011} TEM micrographs of InGaAs and GaAsP vapour grown on InGaP. The GaAsP sample was not preheated in PH_3 and contains dislocation loops which have vacancy character

flow controller can be varied either abruptly (in discrete steps) or smoothly in a continuous fashion.[19]

The Varian system [7] (see Figure 1.7) is somewhat similar to the RCA system in design although no provisions are made for a separate preheat zone. The substrate is 'preheated' by using a quartz 'slider-boat' (see Figure 1.8) in which the substrate can be left in the growth zone and protected from gas ambients until growth begins by pulling back the slider top. A thermocouple is also embedded in the holder. Total H_2 flow quoted is limited to 1100 cm^3 min^{-1}. A two-zone heat-pipe furnace is used for heating.

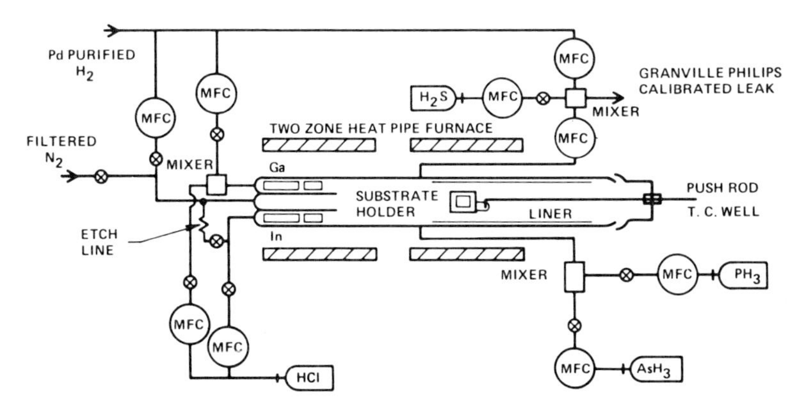

Figure 1.7 Sketch of VPE growth system described by Hyder. (Reproduced by permission of S. B. Hyder.)

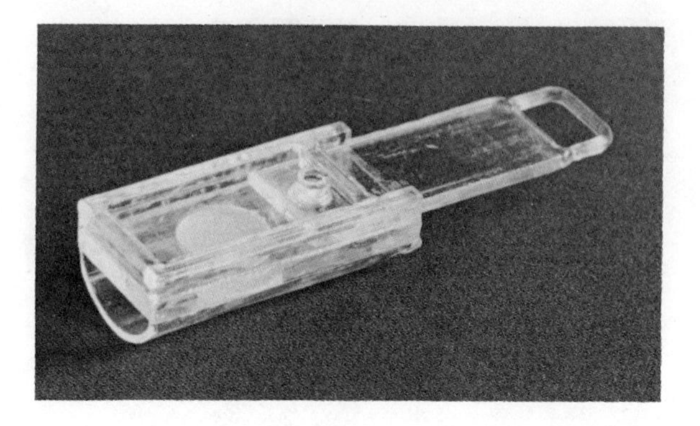

Figure 1.8 Photograph of Varian 'slider boat' substrate holder. (Reproduced by permission of S. B. Hyder.)

The University of Illinois system[18] is also similar to the RCA system except that the growth tube is vertical and total H_2 flow is 500–900 sccm (standard $cm^3\,min^{-1}$). A novel system of suspended indium boats (see Figure 1.9) is used to ensure intimate contact of HCl gas (with the metal sources). A metal surface area of only $14\,cm^2$ is employed. A magnetic feedthrough is used to lift and rotate the substrate.

The system of Johnston and Strege[8] is a horizontal one which uses 5000 sccm total H_2 flow. A sketch is shown in Figure 1.10. No preheat zone is provided. Note that 2 % and 5 % HCl concentrations in H_2 are used to generate GaCl and InCl. Diethyl-telluride (n-type) and diethyl-zinc (p-type) were the preferred dopant sources. The apparatus consists of a quartz reaction vessel with four inlet tubes connected to a single larger growth tube. All gas flows are controlled at the inlet end by electronic mass-flow controllers. AsH_3 and PH_3 flow directly into the growth tube while the HCl is passed over graphite boats containing liquid indium and liquid gallium. Dopants are added through the fourth tube. Mixing of the metal chlorides and the hydrides is delayed by passing the hydride gases through a nozzle. This reduces wall deposits upstream from the substrate, which may cause changes in growth rate and composition. The nozzle tip is perforated to enhance active gas mixing just before flowing over the substrate. The substrate is pushed in from the exhaust end to a point about 5 cm downstream from the nozzle. The reactor is mounted in a horizontal furnace and heated to a temperature of 840°C at the source zone and 700°C at the growth zone.

The system of Sugiyama *et al.*[15] is shown in Figure 1.11. $AsCl_3$ is used as an HCl source. A residence time of over 4 s is claimed between the metals and HCl gas. No provision for p-type doping is specified.

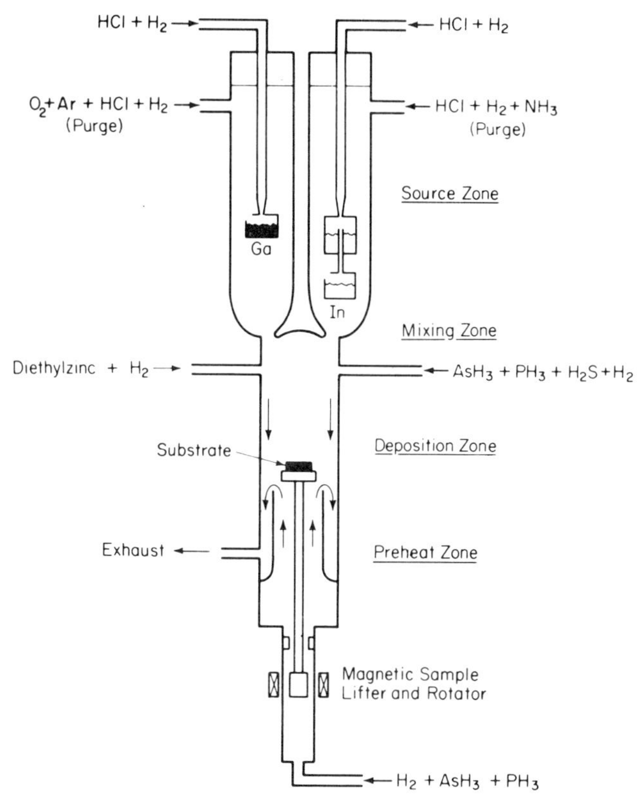

Figure 1.9 Sketch of GaInAsP VPE growth system described by Zinkiewicz. (Reproduced by permission of L. M. Zinkiewicz.)

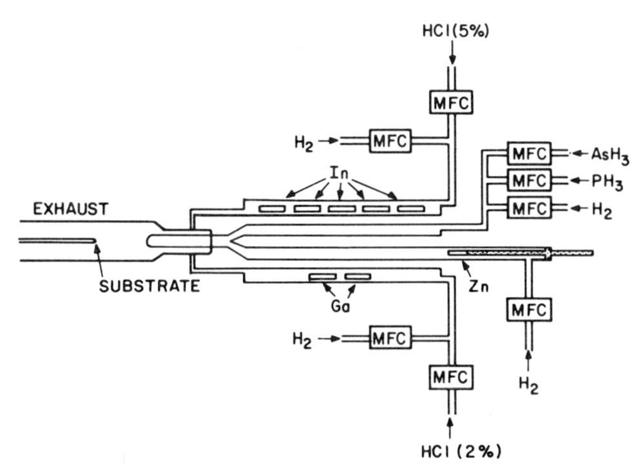

Figure 1.10 VPE reactor of Johnston and Strege. (Reproduced by permission of W. D. Johnston, Jr, and Bell Telephone Laboratories, Inc.)

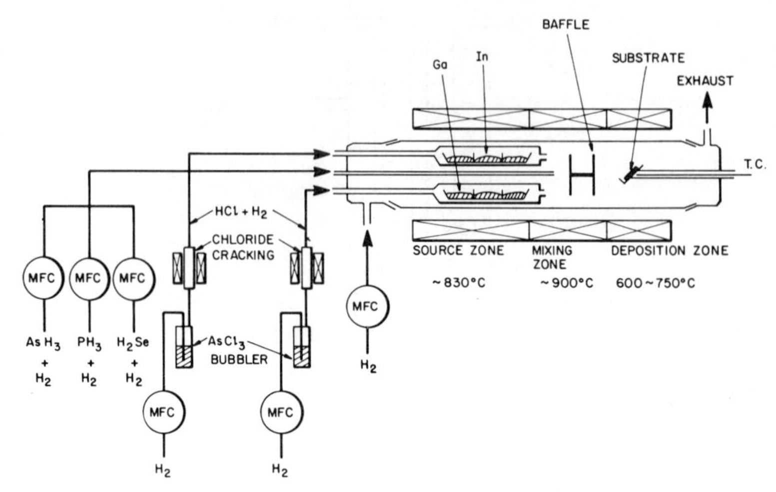

Figure 1.11 Sketch of a VPE reactor according to Sugiyama *et al.*[15] (Reproduced by permission of K. Sugiyama and the *Japanese Journal of Applied Physics.*)

The system of Enda[5] employs a sealed preheat chamber in which the substrate temperature can be brought up to growth temperature in a group V hydride ambient (see Figure 1.12).

The system of Susa,[16] primarily for InP and InGaAs growth, uses $AsCl_3$ as an HCl source and also uses a sliding quartz boat to protect substrates from decomposition. Improved crystalline quality was attributed to this boat.

Vohl[17] has grown alloys by the trichloride method which uses PCl_3 and $AsCl_3$ as sources for P and As. Three compound sources, InP, InAs, and GaAs—each in a separate tube—are reacted with a gas mixture containing the same group V element. Each reaction is independently controlled. The deposition of the alloy is

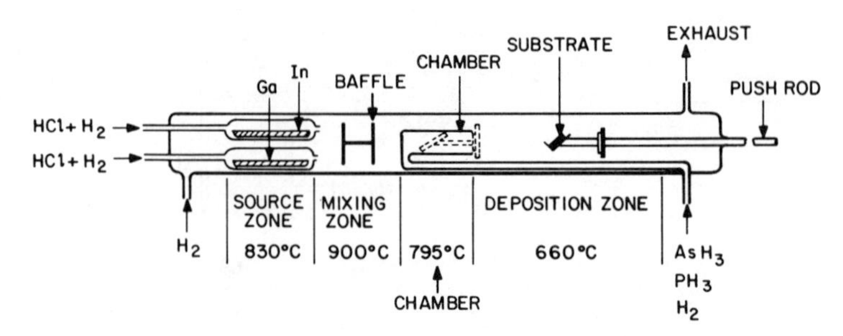

Figure 1.12 VPE system described by Enda.[5] (Reproduced by permission of H. Enda and the *Japanese Journal of Applied Physics.*)

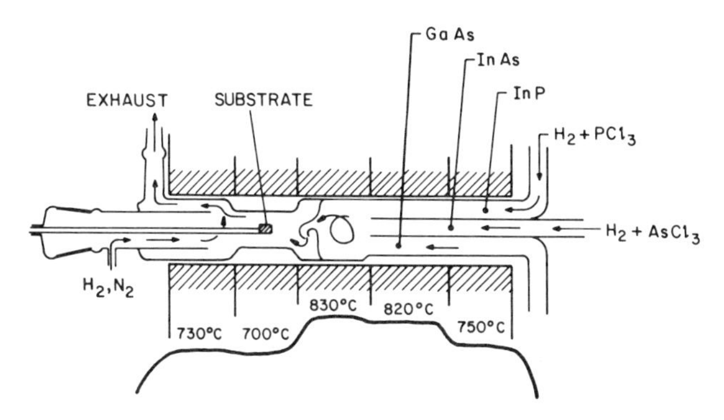

Figure 1.13 VPE system described by Vohl. (Reproduced by permission of P. Vohl.)

regarded as the deposition of a mixture of GaAs, InP, and InAs. The alloy composition is obtained by adjusting the relative amounts of the three compounds that are deposited on the substrate. A sketch of the furnace is shown in Figure 1.13.

1.3 AUTOMATED VPE REACTORS

Reactors of the type discussed in section 1.2 can be automated to improve control and reproducibility. Perhaps the most sophisticated automated growth system is that of Johnston.[20] There are eight gas flows which must be controlled: arsine, phosphine, hydrogen chloride for the gallium, hydrogen chloride for the indium, plus complementary hydrogen flows for each of the four inlet tubes. Since it becomes very difficult for an operator to handle eight mass-flow controllers during the growth, a computer has been interfaced to the growth system to carry out the grading (see Figure 1.14). The computer directs the digital-to-analogue converter to output a voltage signal to each of the eight mass-flow controllers. The flow controllers in turn send a voltage signal to the analogue-to-digital converters which describe the actual gas flows passing through the controllers. This incoming flow information is compared with the control signals sent to the flow controllers to assure proper gas flows. In addition the computer stores both the input and output information for later analysis. Since the compositional grading is under program control, changes in the grading procedure can be easily made by changing the control program.

In order to improve the reproducibility of vapour-grown devices, the RCA VPE system has also been automated. Figure 1.3 shows a photograph of an automated VPE system. The unit is controlled by a Tylan Process Controller, which can store 16 independent SPDT relay positions for up to 96 different timing steps. All reactant gas valves are actuated by air-operated bellows valves, which in turn are actuated by remote electrical solenoids (to eliminate the danger

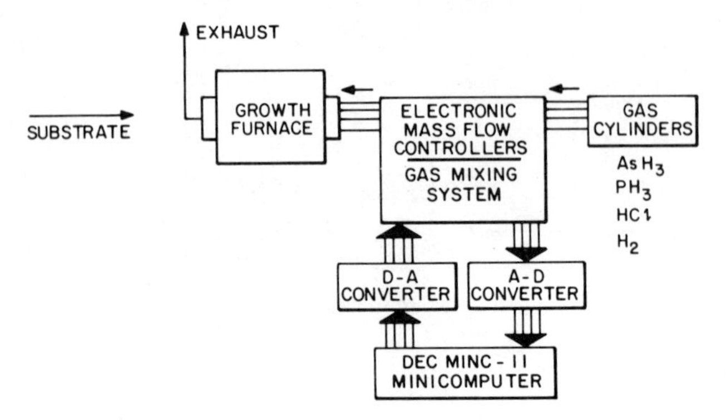

Figure 1.14 Flow diagram of an automated computer-controlled VPE reactor described by Johnston and Strege. (Reproduced by permission of W. D. Johnston, Jr, and Bell Telephone Laboratories, Inc.)

of sparking near flammable gases). All reactant gases are controlled by electrically operated mass-flow controllers. Insertion and removal of a zinc bucket (for p-type doping) are accomplished via a magnetic feedthrough which is raised or lowered by an electric motor. Substrate position is controlled by a motorized belt drive, which has three stop positions (forechamber, preheat, and growth). The substrate holder can also be rotated during growth. Control of each layer of a heterojunction device structure is accomplished via five potentiometers (for the HCl_{In}, HCl_{Ga}, PH_3, AsH_3 and H_2S mass-flow controller settings). Up to five separate layer settings can be programmed. Even more layers can be grown if only doping changes are required. In order to carry out a run, all steps (relay positions and times) are first programmed into the process controller. Once stored, these data can be kept indefinitely or changed at will. Thereafter, once a fresh substrate is loaded into the forechamber, an operator merely activates the start button, and returns two hours later to receive the fully grown heterostructure.

Zinkiewicz[21] has also described an automated VPE system. Electronic mass-flow controllers and air-operated bellows valves were employed throughout. They could be controlled by an H-8 computer. Provision was also made for the introduction of NH_3, N_2, and O_3 gases.

1.4 MULTIBARREL REACTORS

A dual-growth-chamber GaAs VPE reactor was proposed by Watanabe *et al.*[22] in 1977 to produce abrupt changes in doping profile. This system was upgraded into an GaInAsP dual-growth-chamber reactor by Mizutani *et al.*[10] Other types of multibarrel reactors were independently conceived to grow (Ga,In)(As,P) by Olsen and Zamerowski[3] (the 'double-barrel' reactor) and by Beuchet *et al.*[4] (the

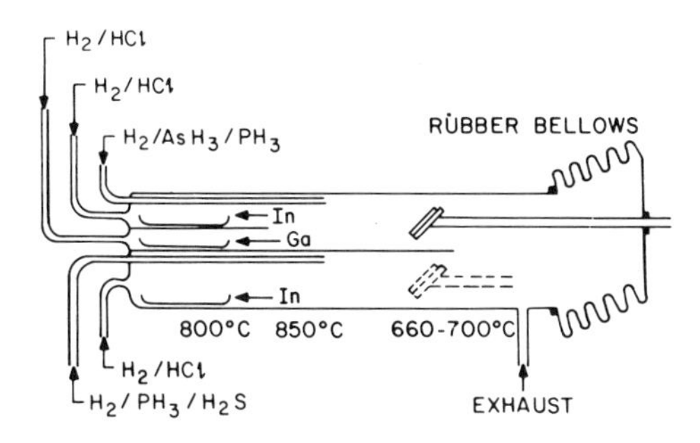

Figure 1.15 'Dual-growth-chamber' VPE reactor described by Mizutani *et al.*[10] (Reproduced by permission T. Mizutani and the *Japanese Journal of Applied Physics.*)

four-barrel 'multichamber' reactor). The three types of reactors are shown in Figures 1.15, 1.16, and 1.17.

The dual-growth-chamber reactor of Mizutani consists of one horizontal chamber to grow InP and another parallel chamber to grow alloys of (Ga,In)(As,P). The substrate is supported by a rod mounted on a flexible rubber bellows so that transfer between chambers is accomplished by withdrawing and shifting the rod. Transfer can be effected within 2 s and heterointerface transition widths less than 50–60 Å are claimed. The chemistry and flow settings are similar to those described in section 1.2. Further details are listed in Table 1.1. Although (in its present form) gallium-bearing alloys can only be grown in one of the chambers, and no provisions have been made for p-type doping, high-quality

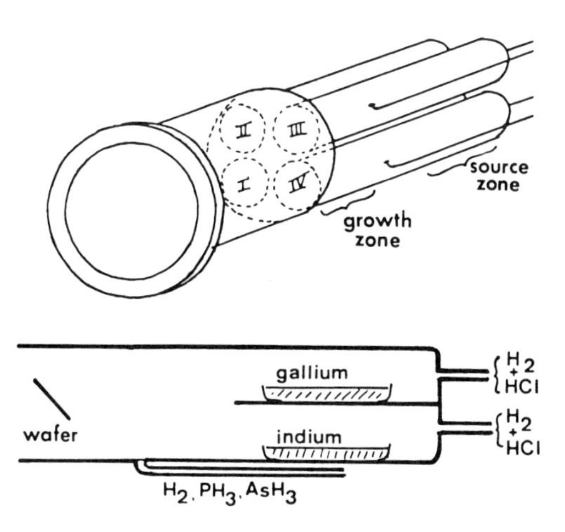

Figure 1.16 Multichamber VPE reactor described by Beuchet *et al.* (Reproduced by permission of G. Beuchet.)

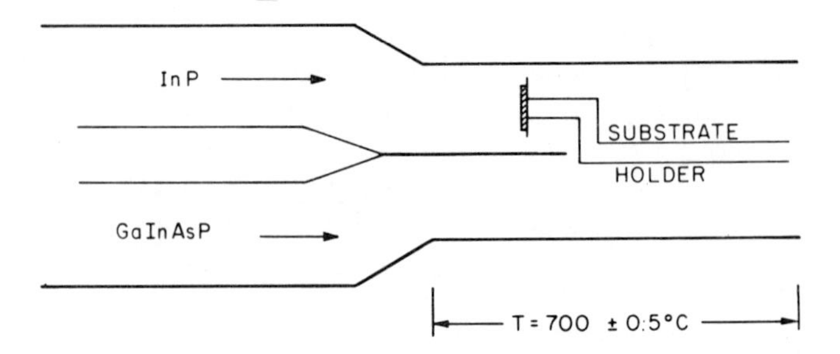

Figure 1.17 Double-barrel VPE reactor described by Olsen and Zamerowski

1.3 μm InP/InGaAsP/InP continuous-wave (CW) lasers have been synthesized in the system (Cd was diffused in after growth for the p contact).

The 'double-barrel' reactor of Olsen and Zamerowski is sketched in Figure 1.17. The concept involves the use of two conventional RCA VPE systems (see section 1.2) placed in parallel, feeding into a single growth chamber. With this construction, different gases can be run through each tube, so that double-heterostructure lasers (e.g. InP/GaInAsP/InP/GaInAsP) can be prepared simply by switching the substrate from one tube to the other (in less than 1 s) thus removing the need for preheat cycles which may limit device performance by the introduction of interfacial defects. A photograph of the quartz tube is shown in Figure 1.18. A heat pipe surrounds the deposition zone (in order to maintain a uniform growth temperature) and obscures its view. A frontal photograph of this zone may be seen in Figure 1.19. Here, the two separate circular flow tubes can be seen along with the flat quartz plate which separates them and leads into the single-barrel growth zone. A sketch of the growth zone is shown in Figure 1.17. The substrate can either be held just in front of each tube or be inserted *into* the tube if any mixing problems occur. Conversely, by holding the substrate at the end of the quartz plate and rotating it, extremely thin multilayers might

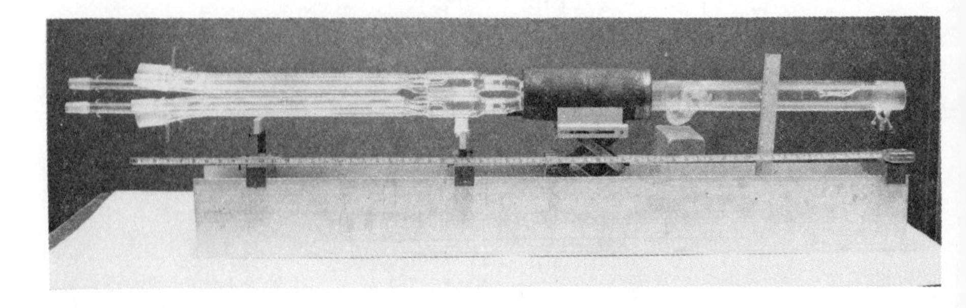

Figure 1.18 Photograph of the quartz tube in the RCA 'double-barrel' VPE reactor

Figure 1.19 Frontal photograph of the quartz tube in the RCA 'double-barrel' VPE reactor

be grown. It is estimated that by rotating the substrate at 250 rpm under normal VPE growth conditions, *single atomic layers of each material could be deposited*. This offers the possibility of preparing structures that exhibit quantum size effects.[23] A 50-layer InAsP/InP structure with thicknesses of about 200 Å prepared in such a manner is shown in Figure 1.20.

Although exciting fundamental studies of this nature may be performed with structures prepared in the new reactor, its main practical advantage is a saving in growth time and materials as well as the possible attainment of superior interfacial properties. Growth of a typical laser structure usually takes about two hours and involves four perheat cycles, including one at each of the two cavity interfaces. InP is particularly susceptible to decomposition effects at elevated temperatures ($> 400°C$). With the new system, the total growth time is reduced to about 30 min, and all preheat cycles are eliminated (except for the original one, which affects only the original InP/InP interface which lies 5 to 10 μm below the laser cavity). The reduced growth time would enable 15 or more double-heterostructure laser wafers to be grown in a single day. The elimination of the preheat cycles yields cleaner interfaces which should provide lasers with lower threshold current densities, higher efficiencies, and better reliability. Improved interfaces should similarly enhance the performance of avalanche photodiodes.

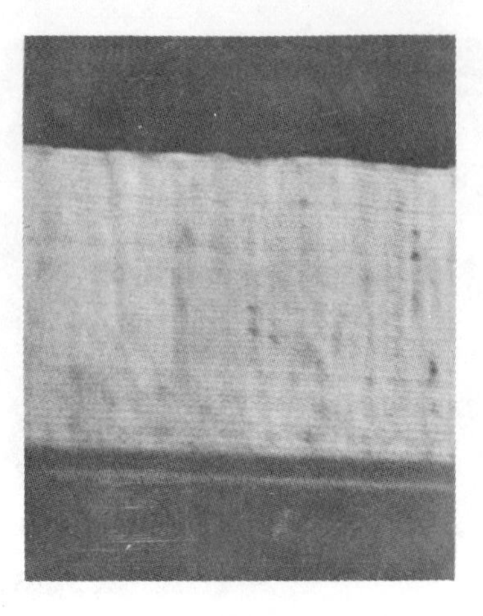

50 LAYER VPE STRUCTURE
≃ 200 Å EACH
GROWN IN DOUBLE-BARREL REACTOR

Figure 1.20 SEM micrograph of a stained cleavage edge of a 50 layer InAsP/InP multilayer heterostructure grown in the RCA 'double-barrel' VPE reactor. Each layer is about 200 Å thick

The 'multichamber' reactor of Beuchet *et al.*[4] (Figure 1.16) consists of four parallel tubes, each with a specific function. The system has been automated so that interchamber transfer can be effected within 1 s. The purpose of each chamber is as follows:

first: preheat and substrate etch
second: growth of n-InP
third: growth of n-GaInAs
fourth: growth of p-InP.

Metallic zinc is used for p-type doping while H_2S is the n-type dopant. Other variants of the above procedures are possible.

1.5 THERMODYNAMICS AND KINETICS

1.5.1 Thermodynamics

The chemical reactions that occur in the deposition region are:

$$2GaCl + \tfrac{1}{2}As_4 + H_2 = 2GaAs\ (alloy) + 2HCl \qquad (1.1)$$

$$2InCl + \tfrac{1}{2}As_4 + H_2 = 2InAs\ (alloy) + 2HCl \qquad (1.2)$$

$$2GaCl + \tfrac{1}{2}P_4 + H_2 = 2GaP\ (\text{alloy}) + 2HCl \tag{1.3}$$

$$2InCl + \tfrac{1}{2}P_4 + H_2 = 2InP\ (\text{alloy}) + 2HCl \tag{1.4}$$

$$As_4 = 2As_2 \tag{1.5}$$

$$P_4 = 2P_2 \tag{1.6}$$

$$AsH_3 = \tfrac{1}{2}As_2 + \tfrac{3}{2}H_2 \tag{1.7}$$

$$PH_3 = \tfrac{1}{2}P_2 + \tfrac{3}{2}H_2 \tag{1.8}$$

Assuming the system to be at chemical equilibrium (a conditional assumption), equilibrium thermodynamics[24] can be applied to calculate the solid alloy composition of (Ga,In)(As,P) given input vapour flows and fundamental thermodynamic data. This is a difficult calculation to complete in closed form. Nagai[25] has made a simplified calculation by neglecting the presence of P_2 and As_2 (and higher chlorides) species, Koukitu and Seki[26] performed a similar analysis whereby the presence of P_2 and As_2 was not neglected. A computer solution was employed. Both calculations give reasonable agreement with the experimental data of Sugiyama *et al.*[15] Figure 1.21 shows a comparison of the experimental data of Olsen with a calculation performed by Koukitu and Seki for alloys ranging from 60 to 100 % indium and 0 to 80 % phosphorus. In general, the calculation is within 10 mol % of the experimental data, which is reasonable, considering the possible non-equilibrium effect of substrate position, vapour flow patterns, etc.

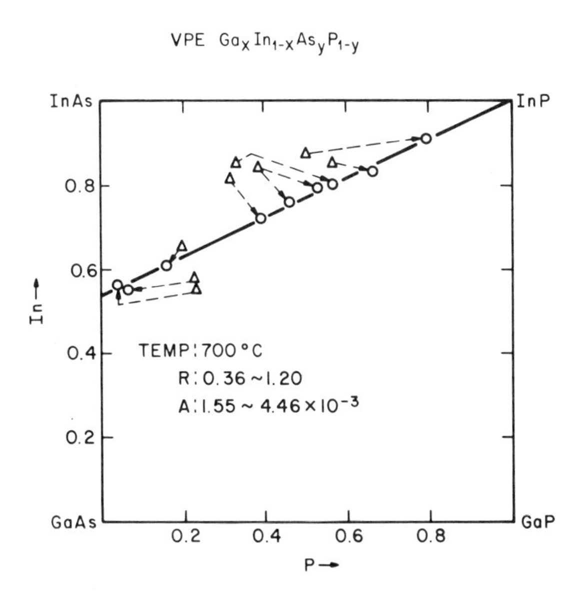

Figure 1.21 Comparison of the experimental data (O) of Olsen and Zamerowski with the thermodynamic calculations(Δ) of Koukitu and Seki (K_1 and K_3 from Ref. 24)

1.5.2 Kinetics

The kinetics of VPE are influenced by such effects as mass flow and chemical reaction rates, which can often cause considerable deviations from thermodynamic equilibrium. For example, the growth rate and composition of layers grown together under identical conditions upon substrates with differing crystallographic orientation can differ significantly. Kanbe *et al.*[9] found that the growth rate of InGaAs on InP at 700°C varied from 9.2 $\mu m\,h^{-1}$ for the (111)A face to 1.2 $\mu m\,h^{-1}$ for the (111)B face. Similarly the indium content varied from 58 % for the (111)A to 64 % for the (111)B face. A simplified view of how the VPE growth of a (Ga,In)(As,P) alloy might take place is shown in Figure 1.22. These steps include: (1) transport of reactants to the surface; (2) adsorption of reactants at the surface; (3) surface events, such as chemical reactions, surface transport, lattice incorporation, etc.; (4) desorption of products from the surface; and finally (5) transport of products from the surface. The above steps are consecutive and the slowest one is the rate-determining step.

The deposition rate is generally limited by one of three regimes: (1) mass-transport-limited; (2) mass-transfer- (or diffusion-) limited; and (3) reaction-limited. *Mass-transport*-limited simply means that not enough reacting gas is being supplied or that the reaction products are not being conducted away fast enough by the vapour stream. *Mass-transfer*-limited means that not enough mass is being transported between the gas stream and the substrate, usually due to limited diffusion of the reactant species through the vapour stream due to a stagnant boundary layer which exists near the substrate. Finally, *reaction*- or *kinetically* limited implies that the rate is being limited by the speed of the chemical reaction or the adsorption of the species on the surface. Mass-transport-limited processes generally have low temperature and substrate-orientation dependence but high flow-rate dependence. Kinetically limited processes, on the other hand, have high temperature and orientation-dependence but low flow-rate dependence. These considerations can sometimes dictate the required precision of experimental equipment.

1.6 SUBSTRATE PREPARATION

Lattice-matched GaInAsP alloys have been grown on GaAs ($0.6 < \lambda < 0.9\,\mu m$) and InP ($0.95 < \lambda < 1.7\,\mu m$) substrates. More complete details on substrate preparation can be found elsewhere.[27] Good practice calls for degreasing (methanol–acetone–hot trichloroethylene–acetone–methanol) followed by a non-preferential chemical etch *immediately* before growth. A generally accepted etch for GaAs is *Caros* acid (5:1:1 H_2SO_4:H_2O_2:H_2O) for about 15 min at room temperature. A number of etches have been suggested for InP substrates. Good results have been obtained with 0.5 % bromine in methanol (BM) and with 17:1:1

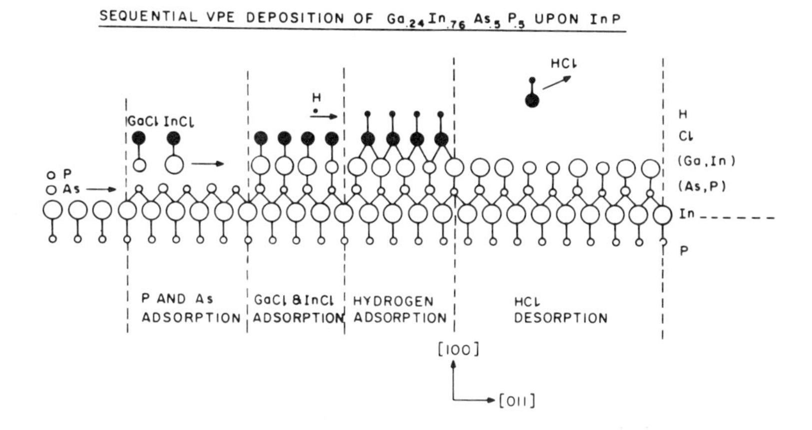

Figure 1.22 Proposed nucleation of GaInAsP via the VPE growth process

HBr:Br:H_2O. *Caros* acid is often used before and/or after bromine etches are used to obtain a smooth, haze-free surface.

1.7 GROWTH CONDITIONS

As mentioned previously it is difficult to predict alloy composition from vapour flows owing to the complexity of the VPE process together with deviations from equilibrium. Similarly, it is difficult—and sometimes misleading—to extrapolate relationships between alloy composition and vapour flows from one growth system to another owing to the vast differences in reactor design and experimental procedures. Differences in properties such as growth tube diameter, carrier gas (H_2) flow rates and substrate positioning can greatly affect alloy composition even if reactant gas partial pressures and growth temperatures are identical. This behaviour illustrates how differences in *kinetic* behaviour of the system can overwhelm similarities in *thermodynamics*. This author wishes to *emphasize* that proper gas flows for particular alloys will *always* have to be determined empirically by trial and error on the growth system in use. Thermodynamic calculations and other system results should be used, however, for qualitative guidance.

Table 1.1 contains a summary of growth conditions for VPE GaInAsP found in the literature. The only parameters for which reasonable agreement *might* be said to exist would be:

$$T_{\text{source}}: \ 800\text{–}850^\circ\text{C}$$
$$T_{\text{mix}}: \ 825\text{–}900^\circ\text{C}$$
$$T_{\text{grow}}: \ 650\text{–}725^\circ\text{C}$$

Other parameters generally vary too widely for serious comparison. The range of total H_2 flows reported (500–5000 $cm^3 min^{-1}$) could very well change the vapour stream from a slow-moving laminar flow to a rapid turbulent flow.

A comparison of the lattice parameter mismatch ($\Delta a/a$) and bandgap resulting from growth conditions at two different laboratories 2, 3, 7, 13 is shown in Table 1.2. The alloy compositions were *estimated* with an extrapolation technique.[28] (It must be emphasized that errors can result from such extrapolations unless lattice mismatch is properly considered. These data are also plotted in Figure 1.23 as the ratio of R_5 (= $PH_3/(PH_3+AsH_3)$) to R_3 (= $GaCl/(GaCl+InCl)$) vs bandgap wavelength (λ_g) in the finished alloy. Reasonable qualitative agreement exists, considering the big difference in hydrogen carrier flow (1100 vs 5000 $cm^3 min^{-1}$).

The presence of lattice mismatch can sometimes be identified by the appearance of a surface cross-hatch pattern as shown in Figure 1.24. The cross-hatch pattern is a growth perturbation caused by the presence of interfacial misfit dislocations.[19] However, the lack of cross-hatch in no way guarantees the absence of lattice mismatch. InP/GaInAsP alloys seem to accommodate higher amounts of elastic strain without misfit dislocation formation—probably due to the smaller elastic constants of InP compared with those of GaAs. Misfit-dislocation-free GaInAsP/InP interfaces with lattice mismatch values as high as 0.64 %[29] and 0.5 %[30] have been reported. See references 29–31 for more complete discussion of this topic.

Most authors who have published on the VPE growth of GaInAsP (see Table 1.1) have included *some* information which relates alloy composition to vapour flows. Consult the references for full details.

Table 1.2 Comparison of partial pressures (p) to obtain alloys with lattice mismatch ($\Delta a/a$) and bandgap wavelengths (λ_g)

	p_{GaCl}	p_{InCl}	p_{AsH_3}	p_{PH_3}	$\Delta a/a$ (%)	λ_g (μm)
	2.0	86.0	3.6	46.8	+0.04	1.04
	3.3	86.0	5.2	46.8	+0.09	1.11
RCA	2.6	84.2	8.0	22.8	<0.03	1.23
(H_2 = 5000 $cm^3 min^{-1}$,	2.6	83.4	8.4	22.8	<0.03	1.28
T_g = 700°C)	3.6	83.4	17.0	22.8	<0.03	1.33
	4.4	46.0	55.0	6.0	+0.03	1.55
	4.8	46.0	55.0	0	+0.07	1.66
	5.5	176.4	40.0	142.0	<0.03	1.17
	2.7	85.5	36.4	64.5	<0.03	1.28
Varian	2.7	88.2	28.2	62.7	<0.03	1.30
(H_2 = 1100 $cm^3 min^{-1}$,	5.5	103.6	40.9	68.2	<0.03	1.45
T_g = 688°C)	2.7	55.4	34.5	22.7	<0.03	1.48
	5.5	75.5	45.5	33.6	<0.03	1.53
	5.5	58.6	63.6	0	<0.03	1.67
	2.8	40.9	45.5	0	<0.03	1.67

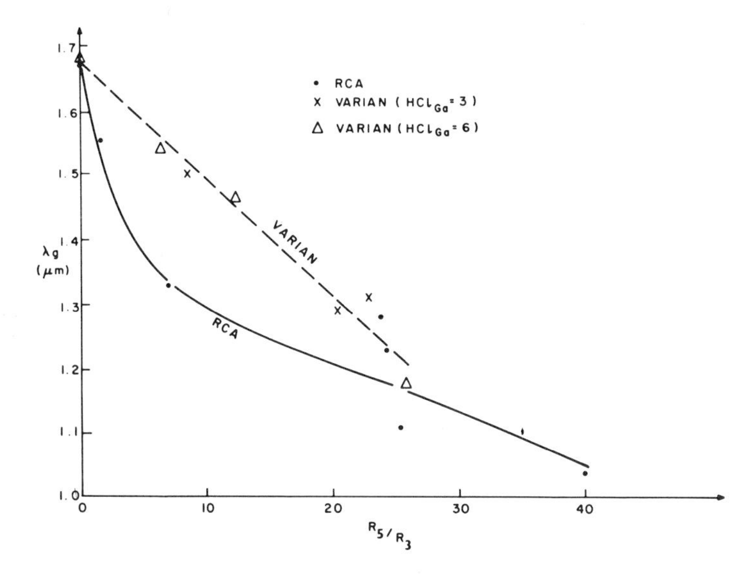

Figure 1.23 Comparison of RCA and Varian VPE growth conditions: ratio of R_5: $PH_3/(PH_3 + AsH_3)$ to R_3: $(GaCl/(InCl + GaCl))$ vs bandgap wavelength of solid GaInAsP

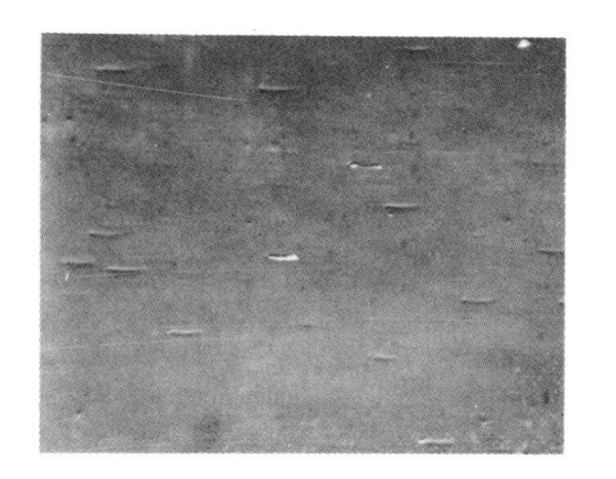

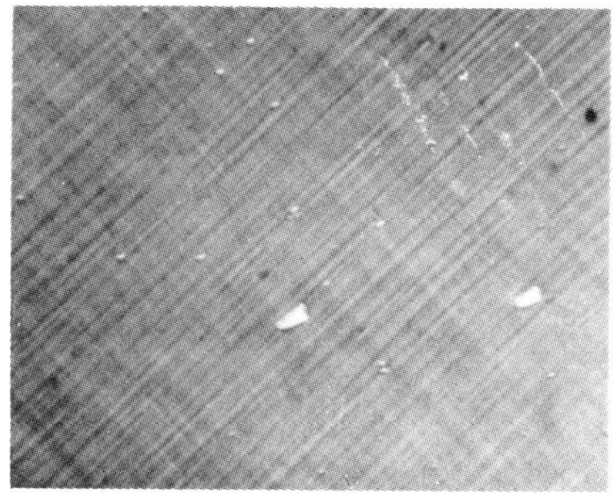

Figure 1.24 Optical micrographs of top surfaces of lattice-matched and lattice-mismatched VPE GaInAsP/InP double-heterostructure lasers

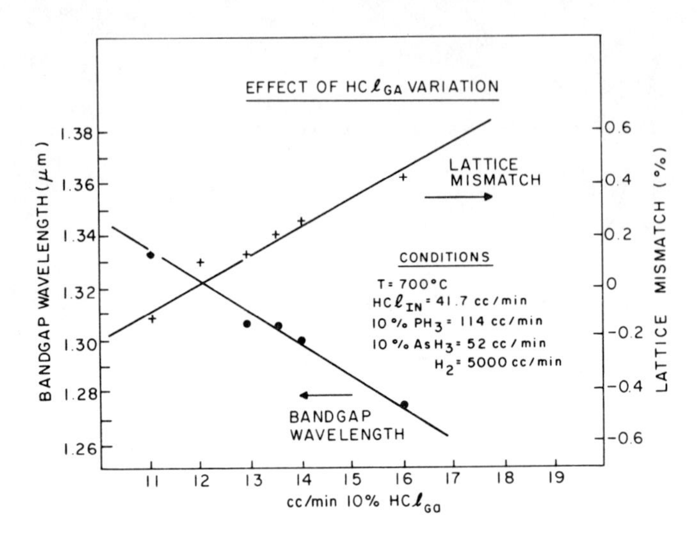

Figure 1.25 Lattice mismatch (+) and bandgap (•) vs HCl_{Ga} flow for VPE GaInAsP alloys grown on InP substrate

The growth conditions of Olsen and Zamerowski[2] are contained in Figures 1.25–1.29. The dependence of λ_g and $\Delta a/a$ on HCl_{Ga} flow and AsH_3 flow are found in Figures 1.25 and 1.26. The variation of In and P content with growth temperature are shown in Figure 1.27. Figure 1.28 and 1.29 show the variation in n-and p-type doping levels with H_2S flow and Zn bucket temperature. Finally, the

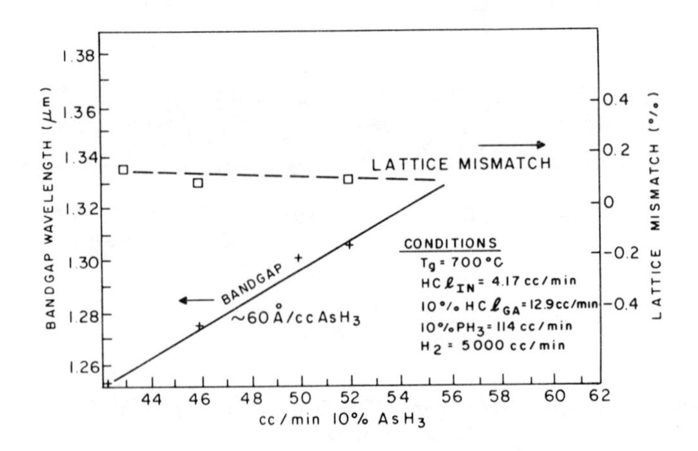

Figure 1.26 Lattice mismatch (□) and bandgap (+) vs AsH_3 flow for VPE GaInAsP alloys grown on InP substrates

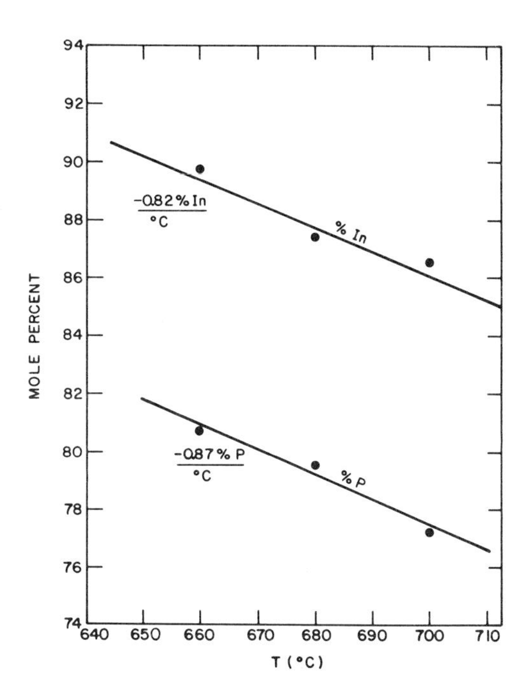

Figure 1.27 In and P composition of VPE $Ga_{1-x}In_xAs_yP_{1-y}$ alloys vs growth temperature

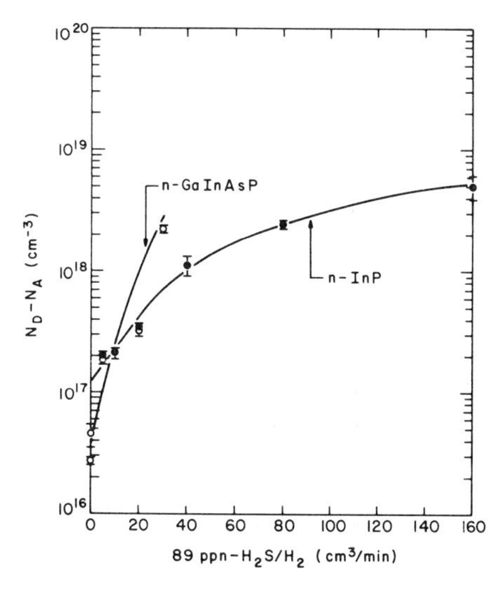

Figure 1.28 *N*-type doping level (carrier concentration) vs H_2S dopant flow for VPE GaInAsP and InP alloys

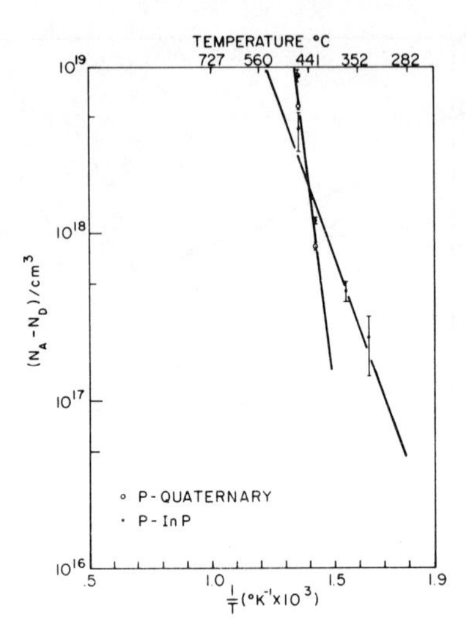

Figure 1.29 *P*-type doping level vs zinc dopant bucket temperature for VPE GaInAsP and InP alloys

reproducibility of the VPE process is amply demonstrated by the data in Table 1.3 for eleven consecutive runs.

Growth conditions for GaInAsP alloys on GaAs substrates with bandgap wavelengths in the range of 0.6–0.9 μm have been published by Sugiyama *et al.*[15]

Table 1.3 Compositional reproducibility of GaInAsP alloys grown on InP substrates

Sample no.	$\lambda_{PL\,(\mu m)}$	$\Delta a/a$ (%)	% In	% As
1	1.270	+0.05	74.5	53.8
2	1.270	≤0.03	76.1	52.4
3	1.270	≤0.03	76.1	52.4
4	1.270	≤0.03	76.1	52.4
5	1.270	≤0.03	76.1	52.4
6	1.265	≤0.03	76.4	52.2
7	1.270	≤0.03	76.1	52.4
8	1.280	−0.09	77.2	52.2
9	1.270	≤0.03	76.1	52.4
10	1.280	−0.07	77.0	52.4
11	1.265	≤0.03	76.4	52.2
	$\lambda = 1.27 \pm 0.005$		76.2 ± 0.7	52.5 ± 0.4

1.8 MATERIAL PURITY AND ELECTRICAL PROPERTIES

The effect of growth parameters upon purity and background carrier concentration in III–V compounds has been discussed elsewhere.[18] The main impurity in these materials is thought to be silicon. Silicon is introduced into the grown crystal via the 'mole-fraction' effect whereby chlorosilanes are generated by the reaction of HCl, H_2 and the quartzware (SiO_2) generally used for the growth apparatus.[32] The incorporation of silicon would thus be inversely proportional to the mole fraction of the reactants (since $HCl + H_2 + SiO_2 \rightarrow SiCl + \ldots$ and $K_{Si} = p_{SiCl}/(p_{HCl} + p_{H_2})$).

Starting material impurity is also a major concern. Gas purities (H_2, AsH_3, PH_3, HCl) are at least 99.999 %. Indium and gallium purities are usually at least 99.9999 %. Silicon is one of the largest impurities in indium. Zinkiewicz[18] has pointed out that while commercial-grade quartz has a metallic impurity concentration of 30 ppm, synthetic quartz has only about 1 ppm. Other

Table 1.4 Electronic properties of p-type InP and InGaAsP

Composition	E_g (eV)	N_A (cm^{-3})	L (μm)	S ($cm\,s^{-1}$)	D ($cm^2\,s^{-1}$)	μ ($cm^2\,V^{-1}\,s^{-1}$)
InP	1.35	2×10^{19}	1.0	5×10^5	30	1200
$Ga_{0.19}In_{0.81}As_{0.42}P_{0.58}$	1.01	5×10^{18}	0.5	—	50	2000
$Ga_{0.22}In_{0.78}As_{0.48}P_{0.52}$	1.04	2×10^{18}	1.75	3×10^4	57	2300

E_g = energy bandgap.
N_A = hole density.
L = diffusion length.
S = recombination velocity.
D = diffusion coefficient.
μ = mobility.

Table 1.5 Best[a] reported VPE electrical properties

Reference	Material	N_{300} (cm^{-3})	N_{77} (cm^{-3})	μ_{300}	μ_{77}
Olsen[2,3]	InP	1.6×10^{16}	1.1×10^{16}	3 600	20 000
Zinkiewicz[21]	InP	3.8×10^{14}	2.3×10^{14}	3 600	57 500
Clark[35]	InP	1.6×10^{14}	1×10^{14}	5 300	90 000
Fairhurst[36]	InP	1.1×10^{14}	8.8×10^{13}	6 000	121 050
Fairman[37]	InP	1.2×10^{14}	1.3×10^{14}	6 060	140 220
Zinkiewicz[21]	$Ga_{0.47}In_{0.53}As$	6.3×10^{15}	4.4×10^{15}	6 780	18 170
Hyder[7]	$Ga_{0.47}In_{0.53}As$	8×10^{14}	3.6×10^{15}	5 826	20 200
Olsen[2,3]	$Ga_{0.47}In_{0.53}As$	2.8×10^{15}	2.6×10^{15}	9 900	33 600
Susa[16]	$Ga_{0.47}In_{0.53}As$	1×10^{15}	1×10^{15}	10 050	35 400
Oliver[38] (LPE)	$Ga_{0.47}In_{0.53}As$		3.5×10^{14}	13 800	77 000
Enda[5]	$Ga_{0.69}In_{0.31}As_{0.72}P_{0.28}$	6.0×10^{15}	5.2×10^{15}	6 600	16 000
Olsen[2,3]	$Ga_{0.39}In_{0.61}As_{0.85}P_{0.15}$	3.2×10^{15}	2.5×10^{15}	7 900	19 000
Beuchet[4]	$Ga_{0.47}In_{0.53}As$	1×10^{15}	6.5×10^{14}	10 000	28 000

[a] Best *individual* values—samples with highest mobilities did not necessarily have *lowest* carrier concentration.

factors that affect purity include leaks, substrate and source temperatures, and substrate orientation.

Very little work has been done on the measurement of diffusion lengths in VPE GaInAsP. Ettenberg[33, 34] has measured an electron diffusion length of 1.75 μm in p-type $Ga_{0.22}In_{0.78}As_{0.48}P_{0.52}$. Further data are contained in Table 1.4.

A summary of the best reported VPE mobilities and background carrier concentrations is contained in Table 1.5.

1.9 NOVEL GROWTH PHENOMENA

The VPE growth of GaInAsP alloys on (100) InP substrates tends to be flat and planar—even on etched 'dovetail' channels as illustrated in Figure 1.30. This tendency can be overcome by etching mesas on oxide-defined (110) InP substrates and then rounding off the mesa with further etching after oxide removal. Figure 1.31 shows a photograph of a double-heterostructure grown in such a manner.

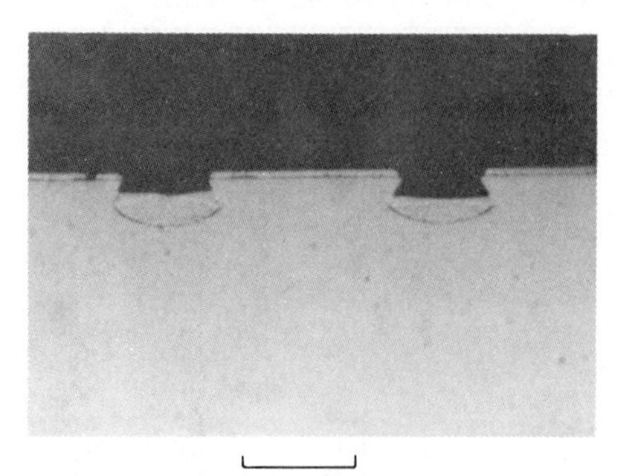

Figure 1.30 Optical micrographs of stained {011} cleavage edges of VPE InP grown over 'dovetail' grooves in InP substrates. Note the tendency for flat, planar growths associated with the VPE process

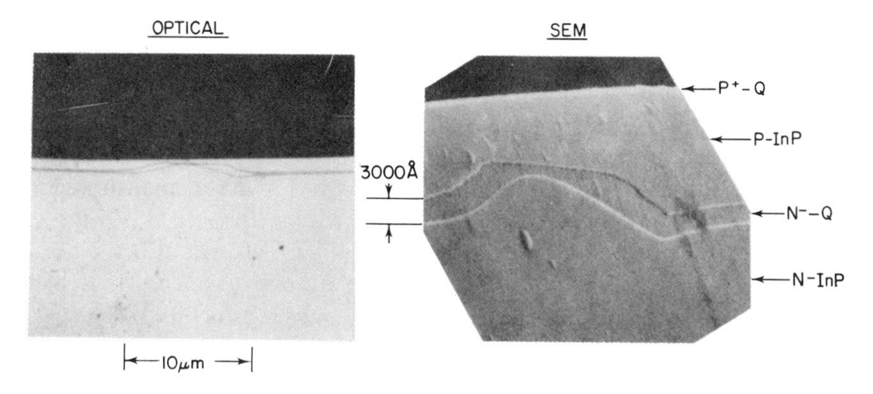

Figure 1.31 SEM micrograph of a stained cleavage edge from a non-planar GaInAsP/InP heterostructure vapour grown over etched 'humps' in (110) InP substrates

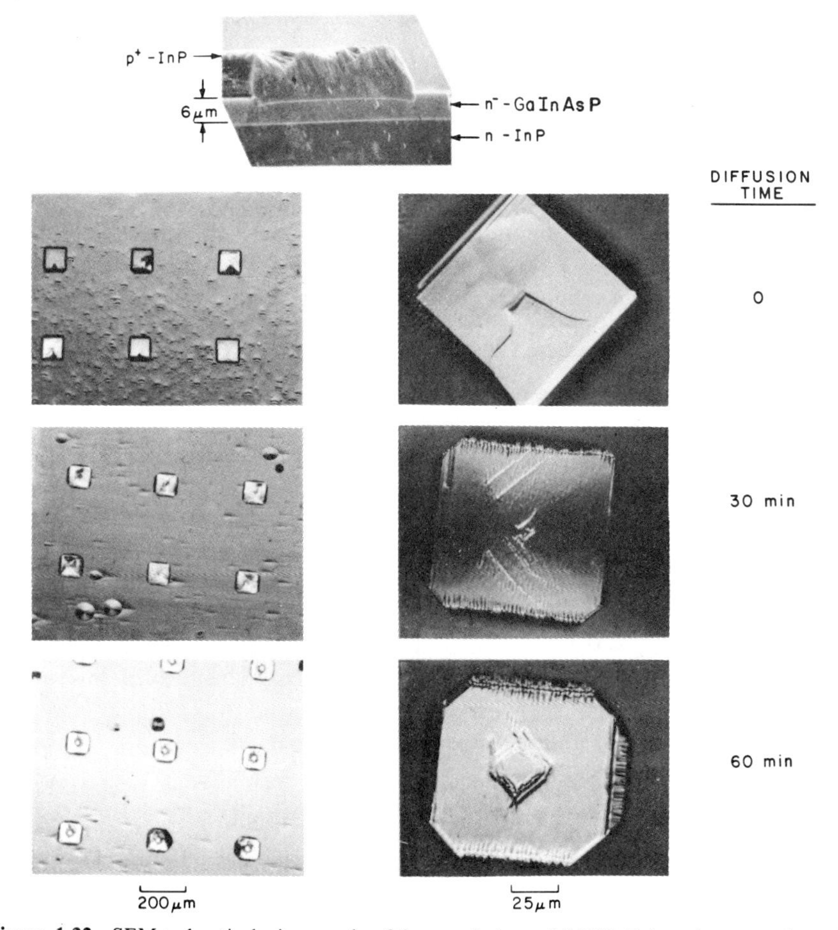

Figure 1.32 SEM and optical micrographs of the morphology of (100) InP deposits grown through 150 μm circular holes in SiO_2 onto a GaInAsP layer

Notice how the cavity thickness varies from about 1000 Å to cover 3000 Å, depending on the position of the etched 'hump'. Growth on (110) substrates has become popular during the last year, perhaps due to superior growth properties offered by a perfectly stoichiometric surface. Hawrylo[39] has demonstrated low-threshold current density and CW lasing action with (110) GaInAsP/InP 1.3 μm lasers. Vohl[17] has demonstrated extended lateral VPE growth of InP on oxide-defined (100) and (110) InP substrates. The author has recently achieved exceptionally high quality VPE growth of InP and GaInAs on (311) and (511) InP substrates.

Figure 1.32 illustrates the VPE growth of InP in circular holes defined in SiO_2-covered (100) GaInAsP alloys. Notice that, as the growth proceeds out of the holes, it tends to grow as a square, with edges defined by (111) facets.

1.10 DEVICE RESULTS WITH VPE GaInAsP

Perhaps the best endorsement of VPE process is the impressive array of outstanding device results obtained to date.

Olsen and coworkers[12,40] have demonstrated CW room-temperature lasing action at 1.25, 1.55, and 1.65 μm with GaInAsP/InP double-heterostructure lasers. Single lateral and longitudinal mode behaviour, low threshold currents (85 mA), and high reliability (> 30,000 h CW at 22°C) have been obtained with 1.25 μm devices. High-quality 1.3 μm avalanche photodiodes[41] and low leakage (10 nA), high quantum efficiency (QE ~ 70 %), high speed (< 0.5 ns), and high reliability (> 10,000 h stable at 60°C) have also been demonstrated[42] with 1.0–1.7 μm GaInAs/InP p–i–n detectors.

Johnston and Strege[8] have made 1.52 μm GaInAsP/InP VPE lasers with pulsed threshold current densities below 1200 A cm^{-2}. Mizutani *et al.*[10] have demonstrated CW lasing action near 1.3 μm with VPE GaInAsP/InP lasers grown with their new 'dual-growth-chamber' reactor. Susa *et al.*[43] and Ando *et al.*[44] have demonstrated impressive detector results with 1.0–1.7 μm $Ga_{0.47}In_{0.53}As$/InP VPE structures including a SAM (separate absorption and multiplication) device whereby light is absorbed in the low-bandgap $Ga_{0.47}In_{0.53}As$ ($E_g \sim 0.74$ eV) but multiplied via avalanche gain in the high-bandgap InP ($E_g \sim 1.35$ eV) region. Dark current densities as low as 7.8×10^{-4} A cm^{-2} at avalanche gains of 10 were reported. A maximum gain of 60 and a pulse risetime of about 100 ps were also observed.

1.11 FUTURE DIRECTIONS

As long-wavelength (1.0–1.7 μm) GaInAsP electro-optical devices[45] become more widely used, motivated by (a) low fibre absorption and dispersion; (b) high transmission through water and smoke; and (c) greatly enhanced eye safety at wavelengths greater than 1.4 μm, the VPE process should enjoy a renaissance not seen since the 1960s. Vapour-phase epitaxy offers its usual advantages of smooth, uniform surfaces, uniformly low background doping density, and economy of scale for large-area devices (e.g. large detectors). However, a particular advantage exists for the VPE growth of devices for wavelengths above 1.4 μm compared to liquid-phase epitaxy (LPE). The LPE growth of InP onto GaInAsP alloys which have $\lambda_g > 1.4\,\mu$m is difficult because of dissolution of the alloy by the InP melt. Although this problem can be circumvented by special growth techniques, it does limit the flexibility of the LPE technique since the usual growth techniques cannot be used here. No such problems exist for VPE, and multilayer heterostructures with bandgap wavelengths in excess of 2.0 μm can be fabricated using the same techniques as for 1.3 μm devices. As long-wavelength GaInAsP lasers, LEDs, and photodetectors begin to replace GaAs emitters and Si detectors for low-loss fibre applications and as new applications arise in the 1.0–1.7 μm regime (e.g. GaInAs MISFET devices[46]), look for the VPE process to play an increasingly important role in the technology of electronic devices.

REFERENCES

1. J. J. Tietjen and J. A. Amick, 'The preparation and properties of vapor-deposited epitaxial GaAsP using arsine and phosphine', *J. Electrochem. Soc.*, **113**, 724–728, 1966.
2. G. H. Olsen and T. J. Zamerowski, 'Vapor-phase growth of (In,Ga)(As,P) quaternary alloys', *IEEE J. Quantum Electron.*, **QE-17**, 128–138, 1981.
3. G. H. Olsen and T. J. Zamerowski, 'Crystal growth and properties of binary, ternary and quaternary (In,Ga)(As,P) alloys grown by the hydride vapor phase epitaxy technique', *Progress in Crystal Growth and Characterization*, vol. II, ed. B. R. Pamplin, London, Pergamon, 1979, pp. 309–375.
4. G. Beuchet, M. Bonnet, and J. P. Duchemin, 'New type of reactor using the chloride transport method for growing GaInAs/InP heterostructures', *Proc. 1980 NATO Conf. on InP, Rome Air Development Center Tech. Memo.* RADC-TM-80-07, Hanscom Air Force Base, MA, 1980, p. 303.
5. H. Enda, 'Preparation and properties of VPE-grown GaInAsP', *Jpn. J. Appl. Phys.*, **18**, 2167–2168, 1979.
6. R. E. Enstrom, C. J. Nuese, J. R. Appert, and N. J. DiGiuseppe, 'Vapor phase growth of InGaAsP alloys for 1.4 μm heterojunction lasers', *Electrochem. Soc. Extended Abstracts*, **78-1**, 504–506, 1978.
7. S. B. Hyder, R. R. Saxena, and C. C. Cooper, 'Vapor-phase epitaxial growth of quaternary $In_{1-x}Ga_xAs_yP_{1-y}$ in the 0.75–1.35 eV band-gap range', *Appl. Phys. Lett.*, **34**, 584–586, 1979.
   S. B. Hyder, R. R. Saxena, S. H. Chiao, and R. Yeats, 'VPE growth of InGaAs lattice-matched to (100) InP for photodiode application', *Appl. Phys. Lett.*, **35**, 787–789, 1979.
   S. B. Hyder, 'Vapor phase epitaxial growth of InP-based compound semiconductor materials', *J. Cryst. Growth*, **54**, 109–116, 1981.
8. W. D. Johnstron, Jr, and K. E. Strege, 'Uniformly low-threshold diode lasers at 1.5–1.55 μm form VPE InGaAsP material', *38th Annual IEEE Device Research Conf. Abstracts, Cornell University*, vol. IVB-3, June 1980.
9. H. Kanbe, Y. Yamauchi, and N. Susa, 'VPE InGaAs on (100), (111)A and (111)B InP substrates', *Appl. Phys. Lett.*, **35**, 603–605, 1979.
10. T. Mizutani, M. Yoshida, A. Usai, H. Watanabe, T. Yuasa, and I. Hayashi, 'Vapor phase growth of InGaAsP/InP DH structures by the dual-growth-chamber method', *Jpn. J. Appl. Phys.*, **19**, L113–L116, 1980.

11. H. Nagai, 'A simple analysis of VPE growth; citing an instance of GaInAs', *J. Electrochem. Soc.*, **126**, 1400–1403, 1979.
12. G. H. Olsen, C. J. Nuese, and M. Ettenberg, 'Low-threshold 1.25 μm vapor-grown InGaAsP CW lasers', *Appl. Phys. Lett.*, **34**, 262–264, 1979.
13. R. R. Saxena, S. B. Hyder, P. E. Gregory, and J. S. Escher, 'Vapor phase epitaxial growth of indium gallium arsenide/indium arsenide phosphide heterojunctions for long wavelength transferred electron photocathodes', *J. Cryst. Growth*, **50**, 481–484, 1980.
R. R. Saxena, S. B. Hyder, P. E. Gregory, and J. S. Escher, 'VPE growth of indium gallium phosphide/indium gallium arsenide structures for transferred-electron photocathodes', *J. Electrochem. Soc.*, **127**, 733–737, 1980.
14. H. Seki, A. Koukitu, and M. Matsumura, 'The VPE growth of InGaAsP alloys by the single flat temperature zone', *J. Cryst. Growth*, **54**, 615–617, 1981.
15. K. Sugiyama, H. Kojima, H. Enda, and M. Shibata, 'Vapor phase epitaxial growth and characterization of $Ga_{1-y}In_yAs_{1-x}P_x$ quaternary alloys', *Jpn. J. Appl. Phys.*, **16**, 2197–2203, 1977.
16. N. Susa, Y. Yamauchi, H. Ando, and H. Kanbe, 'Vapor-phase epitaxial growth of indium gallium arsenide on (100) indium phosphide substrate', *Jpn. J. Appl. Phys.*, **19**, L17–L20, 1980.
17. P. Vohl, 'Vapor phase epitaxy of GaInAsP and InP', *J. Cryst. Growth*, **54**, 101–108, 1981.
18. L. M. Zinkiewicz, T. R. Lepkowski, T. J. Roth, and G. E. Stillman, 'Vapor phase growth of InP and InGaAs by the hydride (In-Ga-AsH_3-HCl-H_2) technique', Ref. 4, *ibid.*, pp. 349–374.
19. G. H. Olsen and M. Ettenberg, 'Growth effects in the hetero-epitaxy of III–V compounds', in *Crystal Growth: Theory and Technique*, vol. II, ed. C. H. L. Goodman, New York, Plenum Press, 1978, pp. 1–56.
20. W. D. Johnston, Jr (private communication).
21. L. M. Zinkiewicz, 'Vapor phase epitaxial growth of GaAs, InP and InGaAs by the hydride technique', *PhD Thesis*, University of Illinois, 1981.
22. H. Watanabe, M. Yoshida, and Y. Seki, 'Vapor growth of GaAs multi-layer with abrupt carrier concentration profile', *Electrochem. Soc. Extended Abstracts*, 151*st Meeting, Philadelphia*, pp. 255–256, May 1977.
23. N. Holonyak, Jr, R. M. Kolbas, R. D. Dupuis, and P. D. Dapkus 'Quantum-well heterostructure lasers', *IEEE J. Quantum Electron.*, **QE-16**, 170–186, 1980.
24. V. S. Ban, 'Mass spectrometric and thermodynamic studies of the CVD of some III–V compounds', *J. Cryst. Growth*, **17**, 19–30, 1972.
25. H. Nagai, 'Thermodynamic analysis of $Ga_xIn_{1-x}As_yP_{1-y}$ CVD: Ga-In-As-P-H system', *J. Cryst. Growth*, **48**, 359–362, 1980.
26. A. Koukitu and H. Seki, 'Thermodynamic analysis for InGaAsP epitaxial growth by the chloride–CVD process', *J. Crystl. Growth*, **49**, 325–333, 1980.
27. A. R. Clawson, D. A. Collins, D. I. Elder, and J. J. Monroe, 'Laboratory procedures for etching and polishing InP semiconductor', *Tech. Note* NOSC 592, Naval Ocean Systems Center, San Diego, CA, December 1978.
28. G. H. Olsen, 'InGaAsP quaternary alloys: composition, refractive index and lattice mismatch', *J. Electron. Mater.*, **9**, 977–987, 1980.
29. K. Oe, Y. Shinoda, and K. Sugiyama, 'Lattice deformations and misfit dislocations in GaInAsP/InP double-heterostructure layers', *Appl. Phys. Lett.*, **33**, 962–964, 1978.
30. K. Nakajima, S. Yamazaki, S. Komiya, and K. Akita, 'Misfit dislocation-free $In_{1-x}Ga_xAs_{1-y}P_y$ heterostructure wafers grown by LPE', *J. Appl. Phys.*, **52**, 4575–4582, 1981.

31. G. H. Olsen, 'Interfacial lattice-mismatch effects in III–V compounds', *J. Cryst. Growth*, **31**, 223–239, 1975.
32. M. E. Weiner, 'Si contamination in open flow quartz systems for the growth of GaAs and GaP', *J. Electrochem. Soc.*, **119**, 496–504, 1972.
33. M. Ettenberg, G. H. Olsen, and C. J. Nuese, 'The effect of gas-phase stoichiometry on the minority carrier diffusion length in vapor-grown GaAs', *Appl. Phys. Lett.*, **29**, 141–142, 1976.
34. G. H. Olsen, C. J. Nuese, R. E. Enstrom, M. Ettenberg, and R. T. Smith, 'III–V heterojunction structures for long-wavelength injection lasers', *Annual Report* US Army/DARPA Contract no. DAAB07-76-C-0872, Ft. Monmouth, NJ, September 1978.
35. R. C. Clark, 'Indium phosphide vapor epitaxy: a review', *J. Cryst. Growth*, **54**, 88–100, 1981.
36. K. Fairhurst, D. Lee, D. S. Robertson, H. T. Parfitt, and W. H. E. Wilgoss, 'A study of the vapor phase epitaxy of indium phosphide', Ref. 4, *ibid.*, pp. 313–347.
37. R. D. Fairman, M. Omuri, and F. B. Fauk, 'Recent progress in the control of high-purity VPE InP by the PCl/In/H technique', *Gallium Arsenide and Related Compounds* (*St Louis*), *1976*, Conf. Ser. no. 33b, Institute of Physics, London, 1977, pp. 45–54.
38. J. D. Oliver, 'LPE $Ga_xIn_{1-x}As$/InP for high frequency device applications', *J. Cryst. Growth*, **54**, 64–68. 1981.
39. F. Z. Hawrylo, 'LPE growth of 1.3 μm InGaAsP CW lasers on (110) InP substrates', *Electron. Lett.*, **17**, 282–283, 1981.
40. G. H. Olsen, T. J. Zamerowski, and N. J. DiGiuseppe, '1.5–1.7 μm VPE indium gallium arsenide phosphide (InGaAsP)/InP CW lasers', *Electron. Lett.*, **16**, 516–518, 1980.
41. G. H. Olsen and H. Kressel, 'Vapor-grown 1.3 μm InGaAsP/InP avalanche photodiodes', *Electron. Lett.*, **15**, 141–142, 1979.
42. G. H. Olsen, 'Low-leakage, high efficiency, reliable VPE InGaAs 1.0–1.7 μm photodiodes', *IEEE Electron. Dev. Lett.*, **EDL-2**, 217–219, 1981.
43. N. Susa, H. Nakagome, H. Anod, and H. Kanbe, 'Characteristics in InGaAs/InP avalanche photodiodes with separated absorption and multiplication regions', *IEEE J. Quantum Electron.*, **QE-17**, 243–249, 1981.
44. H. Ando, Y. Yamauchi, H. Nakagome, N. Susa, and H. Kanbe, 'InGaAs/InP separated absorption and multiplication regions APD using LPE and VPE', *IEEE J. Quantum Electron.*, **QE-17**, 250–254, 1981.
45. G. H. Olsen, 'Laser diodes for the 1.5 μm–2.0 μm wavelength range', *J. Opt. Commun.*, **2**, 11–19, 1981.
46. P. O'Connor, T. P. Pearsall, K. Y. Cheng, A. Y. Cho, J. C. M. Hwang, and K. Alavi, "$In_{0.53}Ga_{0.47}As$ FETs with insulator-assisted Schottky gates", *Electron. Dev. Lett.*, **EDL-3**, 64–66, 1982.

GaInAsP Alloy Semiconductors
Edited by T. P. Pearsall

Chapter 2

Liquid-phase Epitaxy

KAZUO NAKAJIMA

Fujitsu Laboratories Ltd, Semiconductor Materials Laboratory, 1015 Kamikodanaka, Nakahara-ku, Kawasaki 211, Japan

2.1 INTRODUCTION

The GaInAsP/InP alloy systems have received considerable attention and have been studied extensively as materials for optical devices because they have many advantages relating to their growth and material properties, and because of the interest in 0.9–1.6 μm wavelength emitters for optical fibre communication. The greatest advantages are that, in this quaternary system, single-phase solid solutions exist throughout the whole range of compositions[1] and that the alloy can be epitaxially grown lattice-matched on InP substrates over a wide range of bandgaps from 1.34 eV (0.92 μm)[2] to 0.74 eV (1.68 μm)[3,4] at room temperature. Therefore, compositional grading layers[5] to relieve lattice-mismatch strain are unnecessary. As a substrate for the growth of GaInAsP, InP of good quality and with low dislocation density is available. In addition the temperature of fusion of InP is high (1070° C)[6] enough for convenient epitaxial growth upon it. Moreover, smooth surfaces of InP can be obtained reproducibly by effective chemical etchants.[3,7] With the regard to the stability of the GaInAsP crystals, they are not deliquescent and are chemically stable in the atmosphere and water. Surface oxidation is also a lesser problem than that for crystals containing Al. Therefore, complicated structures can easily be prepared by a two-step LPE growth and/or selective etching techniques. The GaInAsP/InP alloy system has many such advantages, but on the other hand there are several special problems in the growth of the alloy systems.

In this chapter, the features of liquid-phase epitaxial (LPE) growth which present special problems to and advantages for the GaInAsP/InP systems are described. The Ga–In–As–P phase diagram, LPE growth conditions for the growth of lattice-matched GaInAsP layers on InP, the composition dependence of bandgaps, and the growth conditions of misfit–dislocation–free layers are also described in detail.

2.2 PHASE DIAGRAM

To produce GaInAsP/InP heterostructures by LPE growth, an accurate Ga–In–As–P equilibrium phase diagram is required. Several early reports[8,9] on the quaternary phase diagram did not supply sufficient information for the growth of lattice-matched layers. Nakajima *et al.*[10,11] determined the equilibrium phase diagram experimentally to facilitate the preparation of closely lattice-matched GaInAsP/InP heterostructures at 600 and 650°C. Temperatures between 600 and 650° C are commonly used for the growth because the thermal etching of InP surfaces increases rapidly above 650° C[12], and the composition variation of epitaxial layers grown below 600° C becomes a serious problem as a result of the depletion of P in the growth solution.[11] The experimental phase diagram at 650° C is described in this section.

Liquidus data are generally determined by the seed dissolution technique[13] or the liquid observation method.[14] The InP seed dissolution technique can be used to determine the Ga–In–As–P liquidus isotherms at 650° C.[10,11] In the experiments, the ternary undersaturated Ga + In + As solution is saturated with P at 650° C by bringing the solution into contact with an InP seed. The As concentration in the ternary solution was always below the solubility limit at 650° C. The P solubility is calculated from the weight loss of the seed after removal of the solution. Although equilibrium cannot be established between a Ga + In + As + P solution and an InP seed because the solid in equilibrium with such a solution must be $Ga_xIn_{1-x}As_{1-y}P_y$, a steady state is reached by the formation of a very thin film on the seed once the liquidus composition finishes by dissolution of the InP seed. Fortunately, the film and the solidified solution

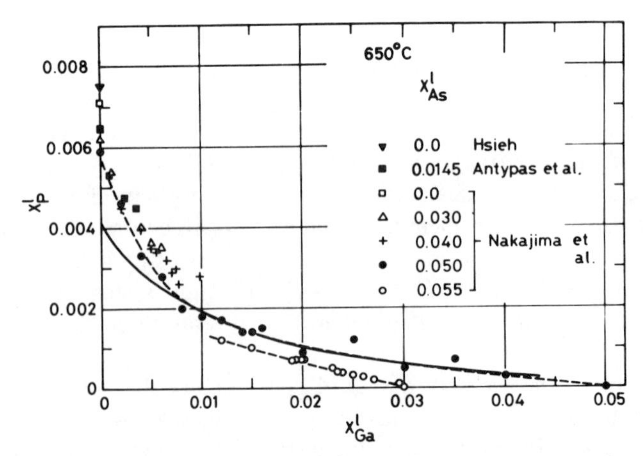

Figure 2.1 650°C liquidus isotherms in the Ga–In–As–P system at several X^l_{As}. Liquidus data include values from Nakajima *et al.*,[10,11] Antypas *et al.*,[8] and Hsieh[15]. The full curve represents the calculated liquidus at $X^l_{As} = 0.050$. Broken curves are the experimental liquidus isotherms at $X^l_{As} = 0.050$ and 0.055

adhered to it can be completely etched off in nitric acid without any dissolution of the InP seed, permitting accurate measurement of the weight loss. Figure 2.1 shows the experimental 650° C liquidus isotherms. The main effect of the addition of Ga to the quaternary solutions is appreciably to decrease the solubility of P in the solutions, while the presence of As has a less pronounced effect in this range.

The Ga + In + As ternary liquidus isotherm can also be determined accurately by the InP seed dissolution technique,[4] but accurate liquidus data cannot be obtained by using an InAs or GaAs seed because a very rough ternary polycrystalline film is always grown on the seed and the solution cannot be completely wiped off the surface of the seed. An InP seed is used as a medium for determining the As solubility in the ternary Ga + In + As solution. Figure 2.2 shows the P solubility in solutions at 650° C as a function of X^{l}_{As} at several X^{l}_{Ga}, where X^{l}_{i} represents the atomic fraction of element *i* in the solution. At constant X^{l}_{Ga}, X^{l}_{P} decreases with increasing X^{l}_{As}. If the As concentration in the solution is just at or above the solubility limit, P cannot dissolve from the InP seed, and a weight loss cannot be detected. Therefore, X^{l}_{P} becomes zero at a certain value of X^{l}_{As}, and X^{l}_{P} remains zero when X^{l}_{As} is above that value. The value of X^{l}_{As} at which X^{l}_{P} becomes zero is the As solubility in the ternary Ga–In–As solution just saturated at 650° C. Figure 2.2 suggests an important fact that GaInAs epitaxial layers can be grown on InP substrates from just saturated solutions without notable melting back of InP. From the value shown in Figure 2.2 the liquidus isotherm at 650° C is determined as shown by the full curve in Figure 2.3. The broken curve is the 620° C liquidus isotherm calculated using parameters reported by Pearsall and Hopson[20].

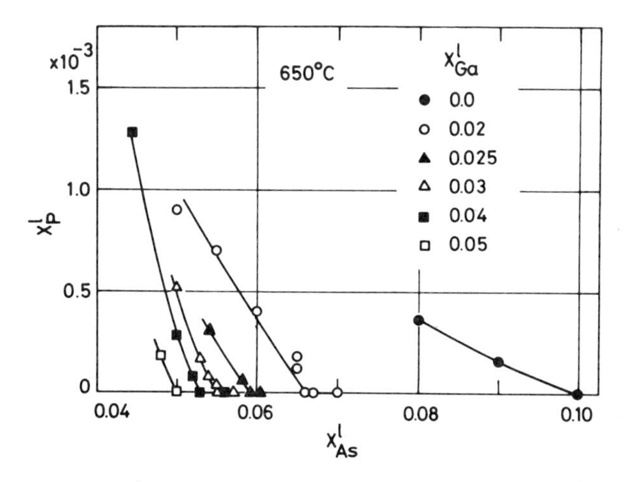

Figure 2.2 P solubility, X^{l}_{P} at 650°C in Ga–In–As–P melts as a function of X^{l}_{As} at several values of X^{l}_{Ga}, as determined by the seed dissolution technique

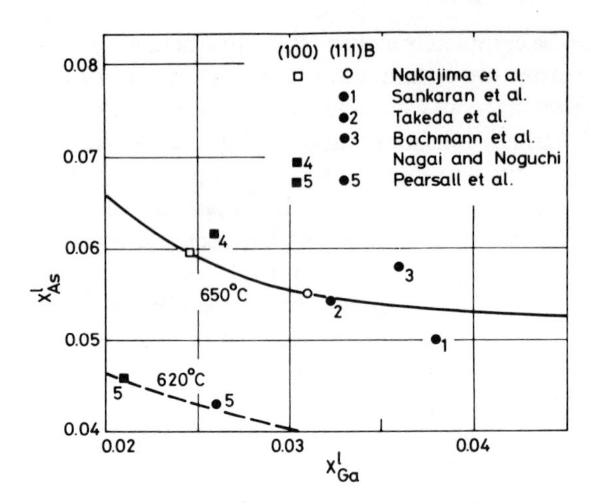

Figure 2.3 Experimentally determined liquidus isotherm at 650°C, and melt compositions for the growth of lattice-matched $Ga_{0.47}In_{0.53}As$ alloys on (100) and (111)B InP substrates from Nakajima *et al.*,[4] Sankaran *et al.*,[16] Takeda *et al.*,[17] Bachmann and Shay,[18] Nagai and Noguchi,[19] and Pearsall *et al.*[20,21] The broken curve is the calculated liquidus isotherm at 620°C

Solidus data are determined by electron-probe microanalysis performed on surfaces of epitaxial layers. Figures 2.4 and 2.5 show the solid solubility isotherms for Ga and P into $Ga_xIn_{1-x}As_{1-y}P_y$ alloys at 650° C. The quaternary epitaxial layers are grown on InP (111)B substrates under the equilibrium conditions. Prior to the growth, the undersaturated Ga + In + As solution is held on an InP

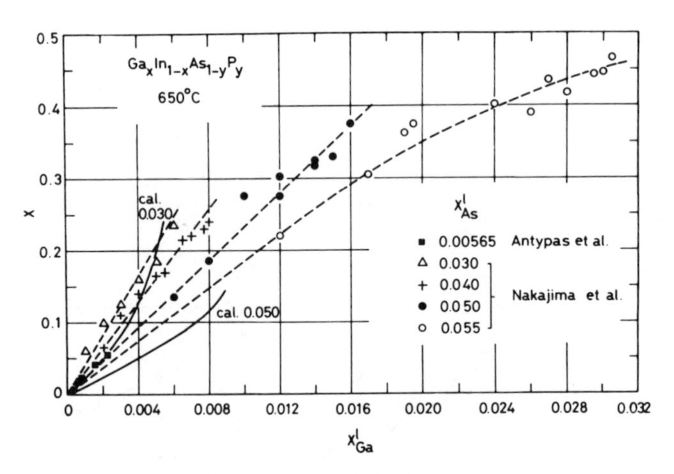

Figure 2.4 650°C solid solubility isotherms for Ga into $Ga_xIn_{1-x}As_{1-y}P_y$ alloys at several X^l_{As}, including points from Nakajima *et al.*,[10,11] and Antypas *et al.*[8] Full curves represent the calculated solidus isotherms at $X^l_{As} = 0.030$ and 0.050

seed for 1 h at 650° C to achieve saturation. The distribution coefficient for Ga increases with decreasing X^{l}_{As}, and the P concentration in alloys decreases remarkably with increasing X^{l}_{Ga} at constant X^{l}_{As}.

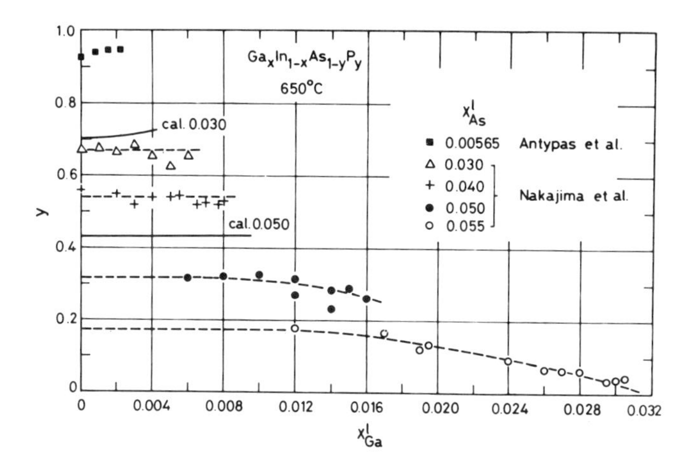

Figure 2.5 650°C solid solubility isotherms for P into $Ga_xIn_{1-x}As_{1-y}P_y$ alloys at several X^{l}_{As}, including points from Nakajima *et al.*[10,11] and Antypas *et al.*[8] Full curves represent the calculated solidus isotherms at $X^{l}_{As} = 0.030$ and 0.050

Several calculation methods[22–24] have been derived to predict the III_A–III_B–V_C–V_D type phase diagram. Jordan and Ilegems[24] have presented the most rigorous thermodynamic treatment of solid–liquid equilibrium in this type of quaternary system. The full curves shown in Figures 2.1, 2.3 and 2.5 are the calculated phase diagram obtained by treating the quaternary liquid and solid as simple solutions[25] in Jordan and Ilegems' model. The interaction parameters necessary for the calculation are derived from reference 25. The calculated results do not agree well with the experimental ones. One of the reasons for the disagreement is the unsuitability of the interaction parameters. These parameters have been derived from curve fitting with the binary or ternary experimental phase diagrams, but most of them have not been determined at such a low temperature as 650° C. Therefore, it is difficult to fit the calculated results entirely to the experimental data. To obtain reasonable agreement, values of several parameters need to be changed. That is, the interaction parameters are functions of temperature and composition in the strict sense because of the departure from regularity in the thermodynamic properties of III–V systems.[26, 27] An example of a modified calculation of the Ga–In–As–P phase diagram was reported by Perea and Fonstad.[28]

2.3 GROWTH CONDITIONS FOR LATTICE MATCHING

In order to grow high-quality $Ga_xIn_{1-x}As_{1-y}P_y$ on InP, the determination of the correct solution compositions for lattice matching is most important. The solution compositions for equilibrium LPE growth of lattice-matched layers are found from the results of the phase diagram and lattice constant measurements of these layers grown from solutions with compositions on the liquidus isotherms. Lattice constants are measured by the double-crystal x-ray diffraction technique. The precise diffraction angles of the layers are determined by using the substrate reflection as an internal standard.

The distribution coefficients of the constituent elements in the Ga–In–As–P system are strongly affected by the crystallographic orientation of the substrate.[29,30] Therefore, different solution compositions are required to grow lattice-matched layers on both (100) and (111)B substrates. These values were systematically determined at 650° C by Nakajima *et al.*[10,11] and Sankaran *et al.*[16] for the (111)B face and by Nagai and Noguchi[19] and Feng *et al.*[31] for the (100) face. Pollack *et al.*[32] also determined these values entirely at 620° C for the (100) face by the two-phase solution method. Figure 2.6 shows quaternary solution compositions, X^l_{As}, X^l_P, and X^l_{Ga}, required for the LPE growth of lattice-matched layers on both the InP (100) and (111)B substrates at 650° C. From the data shown in Figure 2.6, $Ga_xIn_{1-x}As_{1-y}P_y$ can be grown on InP over the entire range of lattice-matched compositions, $0 \leq x \leq 0.47, 0 \leq y \leq 1.0$. Figure 2.3 also shows the solution compositions for the growth of lattice-matched $Ga_{0.47}In_{0.53}As$ on the InP (100) and (111)B substrates. The distribution coefficient for Ga is larger on the (100) face than on the (111)B face and the case

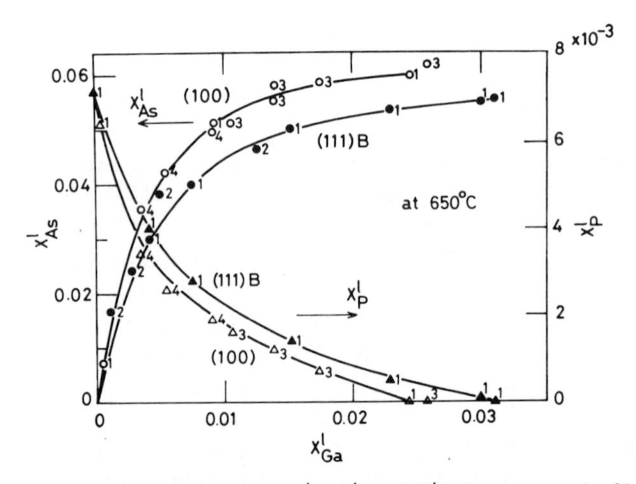

Figure 2.6 Quaternary melt compositions, X^l_{As}, X^l_P, and X^l_{Ga}, for the growth of lattice-matched $Ga_xIn_{1-x}As_{1-y}P_y$ layers on InP (111)B and (100) substrates at 650°C, including points from (1) Nakajima *et al.*,[10,11,34] (2) Sankaran *et al.*,[16] (3) Nagai and Noguchi,[19] and (4) Feng *et al*[31]

for As is the reverse. Over the temperature range between 600 and 660° C, the distribution coefficient for Ga for the growth on the (100) face is larger than that for the growth on the (111)B face.[43] The effect of substrate orientation on the solidus data cannot be ignored, and the free surface and adhesion energies of the epitaxial film onto the substrate[35] must be considered in addition to the chemical free energy of the solid phase when the distribution coefficients are calculated.

2.4 COMPOSITION DEPENDENCE OF BANDGAPS

Nahory *et al.*[36] and Nakajima *et al.*[11] first measured the bandgap versus composition of the GaInAsP system from photoluminescence (PL) spectra. The position of the room-temperature spectral peak is taken as the measure of the bandgap energy. Figure 2.7 shows the measured bandgaps E_g of the lattice-matched $Ga_xIn_{1-x}As_{1-y}P_y$ $(0 \le x \le 0.47, 0 \le y \le 1.0)$ alloy at 300 and 77 K as a function of the alloy composition y. The broken lines are drawn through the data points as linear functions such as[11]

$$E_g = 0.74 + 0.61y \qquad \text{at } 300\,\text{K} \tag{2.1}$$

and

$$E_g = 0.80 + 0.61y \qquad \text{at } 77\,\text{K} \tag{2.2}$$

Both lines have the same slope, and the energy shift between the band gaps at 300 and 77 K is equal to 0.06 eV. The chain curve is drawn as a quadratic equation given by[36]

$$E_g = 0.75 + 0.48y + 0.12y^2 \tag{2.3}$$

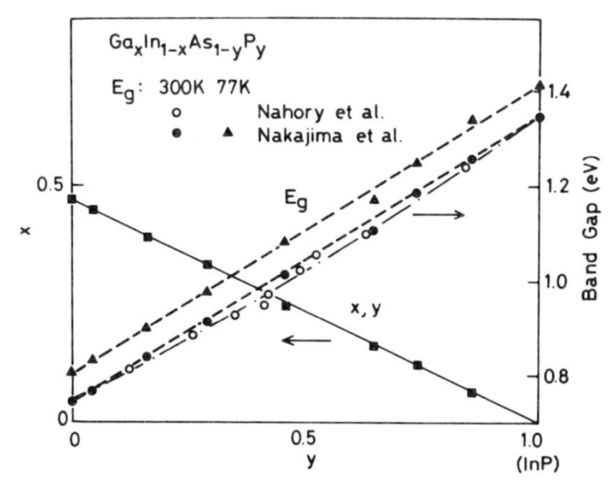

Figure 2.7 Measured bandgaps of lattice-matched $Ga_xIn_{1-x}As_{1-y}P_y$ alloys as a function of alloy composition parameters x and y, including points from Nakajima *et al.*[11] and Nahory *et al.*[36] The lines through the data are given by Pearsall in Chapter 12 of this volume

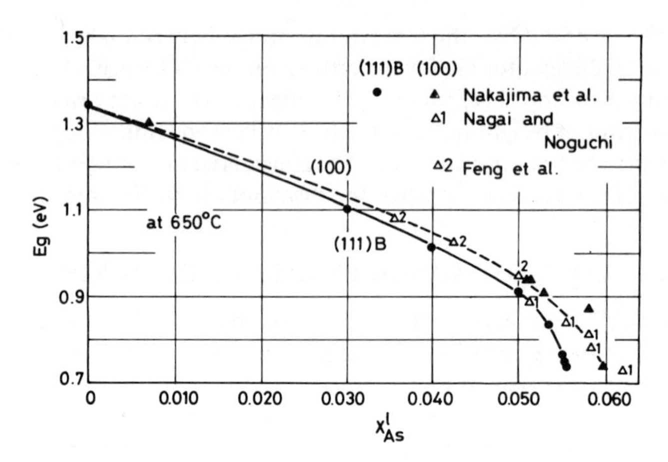

Figure 2.8 Measured bandgaps of lattice-matched $Ga_xIn_{1-x}As_{1-y}P_y$ alloys as a function of X^l_{As}, including points from Nakajima *et al.*, [11,34] Nagai and Noguchi,[19] and Feng *et al*[31]

Thus, the bowing of these lines is not very large, and this feature is also confirmed by electroreflectance measurements of bandgaps.[37] The composition dependence of the quaternary bandgaps can be calculated by geometrical methods[38] using the four ternary boundaries in the quaternary system. However, the bowing of the calculated result is larger than that of the experimental one.[11,36,37] In Chapter 12, Pearsall shows that the composition dependence of the measured bandgaps can be explained accurately by considering only the contribution of alloy disorder to the crystal potential because there is no change in the average bond length of the lattice-matched quaternary alloys. Figure 2.8 shows the bandgap of lattice-matched layers on the (100) and (111) B InP as a function of X^l_{As} at 650°C. The bandgap varies rapidly with increasing X^l_{As} when the bandgap is less than 0.9 eV. The growth conditions for lattice-matched layers with desired bandgaps on both faces can be determined easily from the information provided in Figures 2.6, 2.7, and 2.8.

The solid line shown in Figure 2.7 is the calculated result on the basis of Vegard's law. The experimental data points are in good agreement with the calculated line. Thus, Vegard's law may be considered valid for the calculation of lattice constants of $Ga_xIn_{1-x}As_{1-y}P_y$ over the entire range of composition.[11,36]

2.5 FEATURES OF THE GROWTH

The surface degradation of InP substrates in the growth reactor prior to epitaxy is a principal difficulty in crystal growth since the dissociation pressure of InP is high[39] at typical growth temperatures. High-quality layers cannot be grown directly on InP substrates that are thermally etched by vaporization of P. Growth of GaInAsP on a surface roughened by thermal etching produces local variations in composition, and the resulting variations in lattice constant lead to the formation of special defects[29] because the composition of the layers is sensitive to substrate orientation. A number of techniques have generally been used in an attempt to prevent the surface degradation. One such procedure is, for example, placing a polished InP wafer in contact with the substrate prior to growth and etching off the degraded surface under a pure In or undersaturated In + P solution. Other techniques used to eliminate the surface degradation include PH_3 introduced into the H_2 gas stream in the reactor tube[40] or a relatively high P partial pressure as a result of increased P solubility in In + Sn + P solutions.[41]

The distribution coefficient, growth rate, and surface morphology are strongly dependent on the crystallographic orientation of the substrate. The solution compositions for lattice matching are affected by the substrate orientation because of the orientation dependence of the distribution coefficient (see section 2.3). Layers with large lattice misfits can be more easily grown on the (111)B face than on the (100) face.[42] Even when the lattice misfit is nearly zero, it is very difficult to grow the thick quaternary layers on the (100) face. The surface of the thick layers is very uneven and this unevenness is caused by the irregular growth of extremely lattice-mismatched layers as a result of the depletion of P or Ga in solutions near the growing interface. On the (111)B face, relatively thick layers can be easily grown. However, hillocks with ⟨110⟩ edges often appear on the surface of epitaxial layers grown on the (111)B face irrespective of the degree of lattice matching. On the (100) face, hillocks are rarely observed. Reducing the cooling rate to 0.1°C min^{-1} can eliminate these hillocks on the (111)B face.[3,4] Figure 2.9 shows substrate orientation dependence of growth rate. The solutions

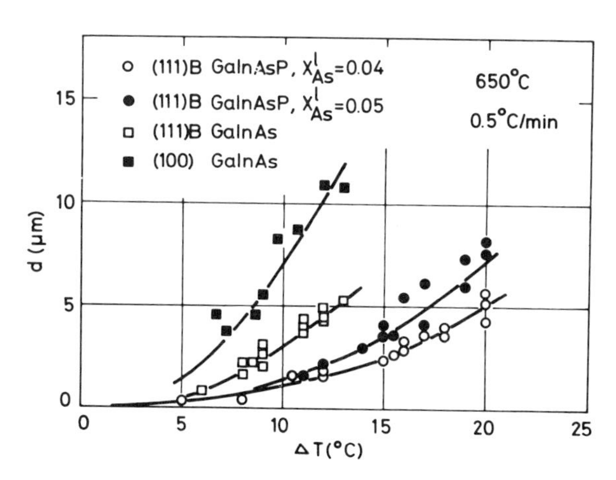

Figure 2.9 Thickness d (μm) of $Ga_xIn_{1-x}As_{1-y}P_y$ and $Ga_xIn_{1-x}As$ layers grown on (11)B and (100) InP substrates as a function of temperature interval ΔT cooled from 650°C

for the growth are just saturated at the starting growth temperature, 650° C, and a constant cooling rate of 0.5° C min^{-1} is used. The thickness d(μm) of the ternary and quaternary layers is displayed as a function of the temperature interval ΔT cooled from 650° C. The growth rate of ternary layers is larger than that of quaternary ones, and the growth rate of ternary layers is about twice as large on the (100) face as on the (111)B face. The growth rate differences may be affected by different nucleation densities on the two orientations.[43] For the quaternary system, the growth rate is also higher on the (100) face than on the (111)B face.

The composition of a thick quaternary layer is easy to grade to lower P concentrations with distance away from the substrate as a result of the depletion of P in the solution near the growing interface. According to Feng's reports,[44, 45] growth at constant temperature from an infinitely large solution will yield a quaternary layer of constant composition. However, solution depletion can lead to compositional grading even using this step-cooling technique. When a constant growth temperature is used, the amount of step cooling does not affect the composition of epitaxial layers at least for values of step cooling smaller than 10° C.[44] However, the maximum layer thickness which can be obtained by the step-cooling technique is limited due to nucleation in the highly supercooled solution. Any growth technique involving growth with changing temperature may result in graded alloy compositions of the epitaxial layers. Hsieh *et al.*[29] investigated the correlation between the composition of epitaxial layers grown using the supercooling technique and the amount of supercooling (ΔT). Figure 2.10 shows the x and y solid compositions as functions of ΔT for $Ga_xIn_{1-x}As_{1-y}P_y$ layers grown on (111)B and (100) substrates from a solution with the composition $X^l_{Ga} = 0.0047$, $X^l_{In} = 0.9599$, $X^l_{As} = 0.0330$, and $X^l_P = 0.0024$, which was determined by Hsieh *et al.*[29] The composition of the (100) layers is relative insensitive to changes in ΔT, although there is an appreciable decrease in y at the highest values of ΔT. The effect of supercooling is much

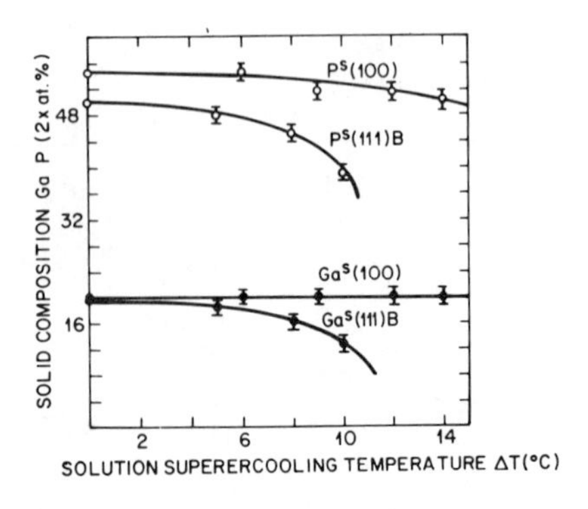

Figure 2.10 Variation of epilayer compositions $Ga^s(x)$ and $P^s(y)$ with degree of solution supercooling ΔT. After Hsieh *et al.*[29] (Reproduced by permission of the Institute of Physics

stronger for the (111)B layers, with both x and y decreasing markedly as ΔT increases. Therefore, the (100) face is perhaps more suitable for the growth of thick layers with constant composition.

For the growth of thin quaternary layers, the two-phase solution method can be used.[33,45,46] This growth method used the same cooling procedure as the equilibrium-cooling method but a piece of InP, in excess of the quantity needed to saturate the solution, is on top of the solution. The floating InP eliminates the need to control exactly the small quantity of P required for saturation, and is in equilibrium with the solution at the time when the growth begins. Therefore, this is one of the successful growth methods in the growth of thin layers. As pointed out by Hsieh,[47] however, it is possible for there to be some supersaturation in the two-phase solution at the time growth begins because of diffusion-limited growth on the precipitates, and this can often produce high growth rates. The near-equilibrium growth technique using a very small amount of supercooling can produce smaller growth rates and is suitable for the growth of thin layers with controlled thickness.[48]

In the growth of $Ga_xIn_{1-x}P$ on GaAs, Stringfellow[49] has observed that over a certain range of solution compositions all the LPE layers had the same solid composition, rather than varying in composition in a manner consistent with the bulk equilibrium phase diagram. This phenomenon was also reported in the LPE growth of $Ga_xIn_{1-x}As$ on InP (111)B substrates,[50] but the contradictory results that no appreciable distortion of the crystal composition can be detected are reported in the Ga–In–As–P[29] and Ga–In–As[4] systems. It is not clear whether or not this phenomenon actually occurs. Stringfellow and Hirth[49,51] suggested that the excess strain energy due to lattice mismatch must be added to the chemical free energy of the solid phase. However, in this quaternary system, the surface and adhesion energy terms[35] have more effect on the total free energy of the nearly lattice-matched epitaxial layer than the strain energy terms because the solidus data are affected much more by substrate orientation than by mismatching.

The InP/GaInAsP/InP double heterostructure must be realized to fabricate such optical devices as lasers. For the growth of this structure in the long-wavelength region (beyond 1.4 μm) by the LPE method, there is a serious problem of dissolution of the GaInAsP layer on bringing the In–P solution into contact with it, because of the larger As solubility in the solution in equilibrium with the longer-wavelength layer. The dissolution occurs more easily on the (111)B face than on the (100) face.[19] The chemical transition width of the heterostructure interface is measured by Auger depth profiles.[52,53] The GaInAsP/GaInAs interface is much broader than the GaInAs/GaInAsP interface. Thus, there is significant dissolution of a layer if it has a high As concentration relative to the next layer to be deposited because interface compositional grading can occur from the dissolution. Compositional gradients in GaInAsP/InP and InP/GaInAsP interfaces with varying amounts of lattice misfit, $\Delta a/a$ (where Δa is equal to the lattice constant of GaInAsP minus the lattice constant of InP, $a = 5.86875\,\text{Å}$[54]), are also studied by Auger depth profiles.[52] A small amount of lattice misfit, $\Delta a/a$ ($-1.1 \times 10^{-3} \leq \Delta a/a \leq 2 \times 10^{-3}$) causes significant grading, particularly at the InP/GaInAsP interface. A close lattice match is essential for the fabrication of GaInAsP/InP heterostructures using LPE. There are several strategies to prevent this dissolution, for example reducing the growth temperature of the InP layer on top of the GaInAsP layer to about 600° C,[53, 55, 56] and using a thin antimeltback layer between the InP and quaternary layers.[57, 58] The incorporated fraction of P in the antimeltback layer is larger than that in the quaternary layer.

Several types of lasers[59–63] have complicated structures which are prepared by two-step LPE growth and/or selective etching techniques. These wafers are exposed to air and etched selectively after the first growth, and the sequential growth is performed on the wafer. The two-step LPE growth is easier to perform for the GaInAsP system than for the AlGa As system because the surface of the quaternary layer contains no Al and is thus not violently oxidized.[63] The fact that proper selective etchants[63–65] are available also facilitates the fabrication of the complicated structures. For the (100) face, the AB (HNO_3:HF = 5:1)[66] and (H_2SO_4:H_2O_2.H_2O = 10:1:1)[63] etchants can be used to etch off selectively GaInAsP on InP, and the (HBr:HF = 1:10)[65, 67] and dilute HCl etchants are useful to remove selectively InP on GaInAsP.

2.6 GROWTH OF MISFIT-DISLOCATION-FREE LAYERS

Misfit-dislocation-free epitaxial layers are required to fabricate optical devices of good quality. The generation of interfacial misfit dislocations depends on both the lattice mismatch $\Delta a/a$ and the layer thickness under a fixed starting growth temperature.[68] The conditions for the growth of thick GaInAs layers without misfit dislocations were determined by Nakajima *et al.*[69] X-ray topographs were used to determine whether misfit dislocations were generated or not in a series of GaInAs/InP heterostructures with different $\Delta a/a$ and layer thicknesses. In Figure 2.11, the open circles show the misfit-dislocation-free layers and the crosses show the layers with misfit dislocations. The samples are divided into two groups depending on whether misfit dislocations are present or not. The lines separating these areas define the threshold for the formation of misfit dislocations. Thus, the area where misfit dislocations are not generated is determined. Thick layers without misfit dislocations cannot be obtained by exact lattice matching to InP at room temperature. Misfit-dislocation-free layers thicker than 10 μm can be grown starting from 650°C only when $\Delta a/a$ is between -6.5×10^{-4} and -9×10^{-4}. The surface morphology of these ternary layers is as good as that of layers with $\Delta a/a \leq 1 \times 10^{-4}$. For thin GaInAsP layers of 0.4 μm, Oe *et al.*[70] reported that misfit dislocations were not found if $\Delta a/a$ was between 5×10^{-3} and -5×10^{-3}. Most misfit dislocations in the GaInAs/InP

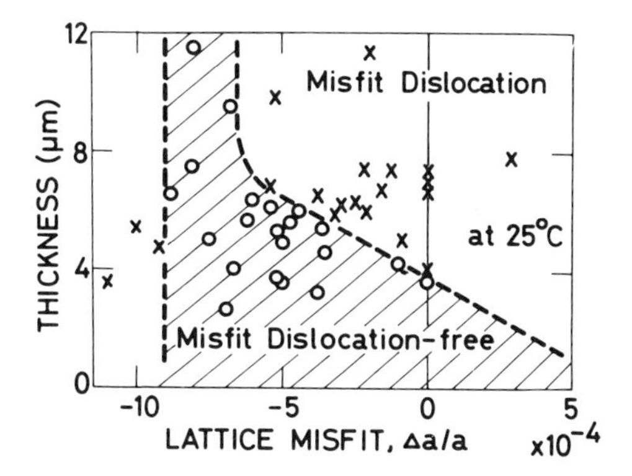

Figure 2.11 Threshold region of both the lattice misfit $\Delta a/a$ and the layer thickness d (μm) for initiation of misfit dislocations into $Ga_xIn_{1-x}As$ layers on InP. O and X represent samples with no misfit dislocations and with misfit dislocations, respectively

heterostructures are introduced from edges of wafers in the early stage of their formation and formed unidirectional arrays.

The reason why the thickest layers without misfit dislocations have the negative lattice misfit at room temperature as shown in Figure 2.11 can be explained by the difference in the thermal expansion coefficients of InP and GaInAs. Thermal expansion coefficients of InP and $Ga_{0.47}In_{0.53}As$ reported by Bisaro *et al.*[71] are $(4.56+0.10)\times 10^{-6}\,°C^{-1}$ and $(5.66+0.10)\times 10^{-6}\,°C^{-1}$ respectively. Assuming the relation between the lattice constant and temperature to be linear, the formula for the thermal expansion is as follows

$$a_T = a_0\,(1+\alpha T) \tag{2.4}$$

where a_T is the lattice constant at any temperature T (° C), a_0 is the lattice constant at 0° C, and α is the thermal expansion coefficient. The temperature dependence of the lattice constants of InP and $Ga_{0.47}In_{0.53}As$ as calculated by equation (2.4) is shown by the lines (a) and (b), respectively, in Figure 2.12. $Ga_{0.47}In_{0.53}As$ has the positive lattice misfit equal to 6×10^{-4} at 650° C. This shows that misfit dislocations are introduced in lattice-matched thick layers by the compressive stress due to the positive lattice misfit at the growth temperature. The thickest layer without misfit dislocations has negative lattice misfit between -6.5×10^{-4} and -9×10^{-4} at room temperature as shown in Figure 2.11, and such a layer almost lattice-matched to InP at the growth temperature, 650° C, is shown by the broken line (c) in Figure 2.12. Therefore, misfit dislocations are more easily introduced in a layer which is lattice-matched at room temperature than in a layer which is lattice-matched at the growth temperature; that is to say, they are more easily formed at high temperature.[71]

The experimentally determined misfit-dislocation-free region has a perpendicular edge on the negative side as shown in Figure 2.11. This shows that, at

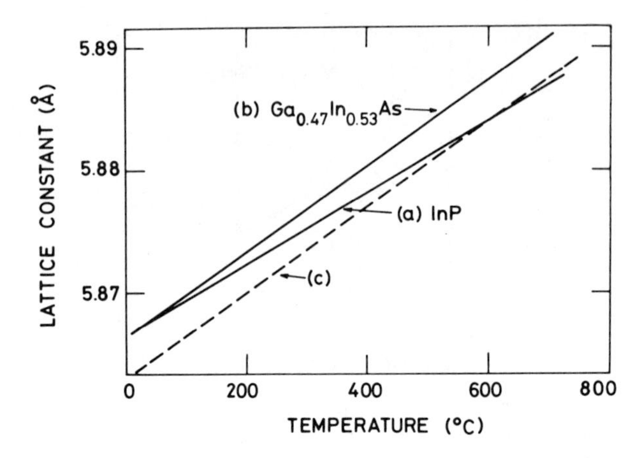

Figure 2.12 Temperature dependence of lattice constant for representative heterojunction combinations. Full lines (a) and (b) are the lattice constants of InP and $Ga_{0.47}In_{0.53}As$. The broken line (c) is a schematic representation of a layer almost lattice-matched to InP at 650°C

the growth temperature, misfit dislocations are much more easily introduced by tension than by compression. This result is consistent with the previously reported result at room temperature for GaInP/GaAs[72] and GaInAs/InP[73] systems that layers in tension cracked at smaller misfit than those in compression. At room temperature, however, misfit dislocations are not more easily formed in tension, and the ternary layers yielded elastically even with a negative lattice misfit of -8×10^{-4}. It is most important for the growth of thick and misfit-dislocation-free epitaxial layers that, at the growth temperature, layers should be lattice-matched to the substrates and that tensile stress should not be incurred at the growth temperature.

2.7 CONCLUDING COMMENTS

Since the GaInAsP/InP alloy system has many advantages, it has received considerable research attention for electro-optical device applications. However, there are still important unsolved problems. One of these is the development of practical LPE techniques to grow reproducibly InP and GaInAsP alloys with a low (10^{15} cm^{-3}) carrier concentration level. This is especially important for fabrication of avalanche photodiodes (APDs). Reliable methods to grow InP on the longer-wavelength layer such as $Ga_{0.47}In_{0.53}As$, without any interface compositional grading layers and antimeltback layers, must be developed for lasers and APDs. The growth conditions to obtain misfit-dislocation-free InP/GaInAsP/InP heterostructures must be determined as soon as possible. The growth of controlled submicrometre layers is also an important problem. When these problems are solved, the GaInAsP/InP alloy system will be of greater value for a wider variety of applications.

ACKNOWLEDGMENTS

I would like to thank the following for permitting reproduction of their experimental results: G. A. Antypas, K. J. Bachmann, M. Feng, J. J. Hsieh, H. Hagai, R. E. Nahory, T. P. Pearsall, R. Sankaran, G. E. Stillman, and Y. Takeda. I am indebted to T. Kusunoki, A. Yamaguchi, S. Komiya, T. Tanahashi, K. Akita, T. Kotani, T. Yamaoka, and O. Ryuzan for their experiments and helpful discussions.

REFERENCES

1. E. K. Muller and J. L. Richards, 'Miscibility of III–V semiconductors studied by flash evaporation', *J. Appl. Phys.*, **35**, 1233–1241, 1964.
2. M. Cardona, K. L. Shaklee, and F. H. Pollack, 'Electroreflectance at a semiconductor–electrolyte interface', *Phys. Rev.*, **154**, 696–720, 1967.
3. R. Sankaran, R. L. Moon, and G. A. Antypas, 'Liquid phase epitaxial growth of InGaAs on InP', *J. Cryst. Growth*, **33**, 271–280, 1976.
4. K. Nakajima, T. Tanahashi, K. Akita, and T. Yamaoka, 'Determination of In–Ga–As phase diagram at 650°C and LPE growth of lattice-matched $In_{0.53}Ga_{0.47}As$ on InP', *J. Appl. Phys.*, **50**, 4975–4981, 1979.
5. R. E. Nahory, M. A. Pollack, and J. C. DeWinter, 'Growth and characterization of liquid-phase epitaxial $In_{1-x}Ga_xAs$', *J. Appl. Phys.*, **46**, 775–782, 1975.
6. L. M. Foster and J. E. Scardefield, 'The solidus boundary in the GaP–InP pseudobinary system', *J. Electrochem. Soc.*, **117**, 534–536, 1970.
7. Y. Nishitani and T. Kotani, 'Chemical etching of InP by H_2O_2–H_2SO_4–H_2O solution', *J. Electrochem. Soc.*, **126**, 2269–2271, 1979.
8. G. A. Antypas and R. L. Moon, 'Growth and characterization of InP–InGaAsP lattice-matched heterojunctions', *J. Electrochem. Soc.* **120**, 1574–1577, 1973.
9. G. A. Antypas and J. Edgecumbe, 'Distribution coefficients of Ga, As, and P during growth of InGaAsP layers by liquid-phase epitaxy', *J. Cryst. Growth*, **34**, 132–138, 1976.
10. K. Nakajima, T. Kusunoki, K. Akita, and T. Kotani, 'Phase diagram of the In–Ga–As–P quaternary system and LPE growth conditions for lattice matching on InP substrates', *J. Electrochem. Soc.*, **125**, 123–127, 1978.
11. K. Nakajima, A. Yamaguchi, K. Akita, and T. Kotani, 'Composition dependence of the band gaps of $In_{1-x}Ga_xAs_{1-y}P_y$ quaternary solids lattice matched on InP substrates', *J. Appl. Phys.*, **49**, 5944–5950, 1978.
12. K. Pak, T. Nishinaga, and S. Uchiyama, 'Thermal-etching effect of InP substrate in LPE saturation process', *Jpn. J. Appl. Phys.*, **14**, 1613–1614, 1975.
13. M. Ilegems and M. B. Panish, 'Phase diagram of the system Al–Ga–P', *J. Cryst. Growth*, **20**, 77–81, 1973.
14. J. J. Hsieh, 'An accurate instrumental version of the visual method for determining liquidus temperatures', *J. Electrochem. Soc.*, **121**, 99c–100c, 1974.
15. J. J. Hsieh, 'Thickness of InP layers grown by LPE from supercooled solutions', *Gallium Arsenide and Related Compounds (St Louis), 1976*, ed. L. F. Eastman, Conf. Ser. no. 33b, Institute of Physics, London, 1977, pp. 74–80.

16. R. Sankaran, G. A. Antypas, R. L. Moon, J. S. Escher, and L. W. James, 'Growth and characterization of InGaAsP–InP lattice matched heterojunctions', *J. Vac. Sci. Technol.*, **13**, 932–937, 1976.
17. Y. Takeda, A. Sasaki, Y. Imamura, and T. Takagi, 'Properties of liquid phase epitaxial $In_{1-x}Ga_xAs$ ($x \simeq 0.5$) on InP', *J. Electrochem. Soc.*, **125**, 130–135, 1978.
18. K. J. Bachmann and J. L. Shay, 'An InGaAs detector for the 1.0–1.7 μm wavelength range', *Appl. Phys. Lett.*, **32**, 446–448, 1978.
19. H. Nagai and Y. Noguchi, 'InP–$Ga_xIn_{1-x}As_yP_{1-y}$ double heterostructure for 1.5 μm wavelength', *Appl. Phys. Lett.*, **32**, 234–236, 1978.
20. T. P. Pearsall and R. W. Hopson, Jr, 'Growth and characterization of lattice-matched epitaxial films of $Ga_xIn_{1-x}As$/InP by liquid-phase epitaxy', *J. Appl. Phys.*, **48**, 4407–4409, 1977.
21. T. P. Pearsall, R. Bisaro, R. Ansel, and P. Merenda, 'The growth of $Ga_xIn_{1-x}As$ on (100) InP by liquid-phase epitaxy', *Appl. Phys. Lett.*, **32**, 497–499, 1978.
22. M. Ilegems and M. B. Panish, 'Phase equilibria in III–V quaternary system—Application to Al–Ga–P–As', *J. Phys. Chem. Solids*, **35**, 409–420, 1974.
23. G. B. Stringfellow, 'Calculation of ternary and quaternary III–V phase diagrams', *J. Cryst. Growth*, **27**, 21–34, 1974.
24. A. S. Jordan and M. Ilegems, 'Solid–liquid equilibria for quaternary solid solutions involving compound semiconductors in the regular solution approximation', *J. Phys. Chem. Solids*, **36**, 329–342, 1975.
25. M. B. Panish and M. Ilegems, 'Phase equilibria in ternary III–V systems', *Progress in Solid State Chemistry*, vol. 7, ed. H. Reiss and J. O. McCaldin, Pergamon Press, New York, 1972, pp. 39–83.
26. K. Osamura and Y. Murakami, 'Phase diagram analysis using a partially associated solution model for III–V binary system', *J. Phys. Chem. Solids*, **36**, 931–937, 1975.
27. K. Osamura, K. Nakajima, and Y. Murakami, 'Experiments and calculation of the Al–Ga–Sb ternary phase diagram', *J. Electrochem. Soc.*, **126**, 1992–1997, 1979.
28. E. H. Perea and C. G. Fonstad, 'Phase diagram calculations for $In_uGa_{1-u}P_vAs_{1-v}$ lattice matched to (111-B) InP, in the temperature range 600–660°C', *J. Appl. Phys.*, **51**, 331–335, 1980.
29. J. J. Hsieh, M. C. Finn, and J. A. Rossi, 'Conditions for lattice matching in the LPE growth of GaInAsP layers on InP substrates', Ref. 15, *ibid.*, pp. 37–44.
30. G. A. Antypas and L. Y. L. Shen, 'Orientation dependence of the incorporation of Ga, As, and Zn during LPE growth of InGaAsP alloys', Ref. 15, *ibid.*, pp. 96–104.
31. M. Feng, T. H. Windhorn, M. M. Tashima, and G. E. Stillman, 'Liquid-phase epitaxial growth of lattice-matched InGaAsP on (100) InP for the 1.15–1.31 μm special region', *Appl. Phys. Lett.*, **32**, 758–761, 1978.
32. M. A. Pollack, R. E. Nahory, J. C. DeWinter, and A. A. Ballman, 'Liquid phase epitaxial $In_{1-x}Ga_xAs_yP_{1-y}$ lattice matched to $\langle 100 \rangle$ InP over the complete wavelength range $0.92 \leq \lambda \leq 1.65$ μm', *Appl. Phys. Lett.*, **33**, 314–316, 1978.
33. T. Yamamoto, K. Sakai, and S. Akiba, '500-h continuous CW operation of double-heterostructure GaInAsP/InP lasers', *Jpn. J. Appl. Phys.*, **16**, 1699–1700, 1977.
34. K. Nakajima, T. Kusunoki, and K. Akita, 'InGaAsP phase diagram and LPE growth conditions for lattice matching on InP', *Fugitsu Sci. Tech. J.*, **16**, 59–83, 1980.
35. J. H. van der Merwe and C. A. B. Ball, 'Energy of interfaces between crystals', *Epitaxial Growth*, part B, ed. J. W. Matthews, Academic Press, New York, 1975, pp. 493–528.
36. R. E. Nahory, M. A. Pollack, W. D. Johnston, Jr, and R. L. Barns, 'Band gap versus composition and demonstration of Vegard's law for $In_{1-x}Ga_xAs_yP_{1-y}$ lattice matched to InP', *Appl. Phys. Lett.*, **33**, 659–661, 1978.
37. Y. Yamazoe, T. Nishino, Y. Hamakawa, and T. Kariya, 'Bandgap energy of InGaAsP quaternary alloy', *Jpn. J. Appl. Phys.*, **19**, 1473–1479, 1980.

38. R. L. Moon, G. A. Antypas, and L. W. James, 'Bandgap and lattice constant of GaInAsP as a function of alloy composition', *J. Electron. Mater.*, **3**, 635–644, 1974.
39. K. Weiser, 'Dissociation pressure and cohesive energy of indium phosphide', *J. Phys. Chem.*, **61**, 513–515, 1957.
40. A. R. Clawson, W. Y. Lum, and G. E. McWilliams, 'Control of substrate degradation in InP LPE growth with PH_3 partial pressure', *J. Cryst. Growth*, **46**, 300–303, 1979.
41. G. A. Antypas, 'Prevention of InP surface decomposition in liquid phase epitaxial growth', *Appl. Phys. Lett.*, **37**, 64–65, 1980.
42. K. Nakajima, S. Yamazaki, S. Komiya, and K. Akita, 'Misfit dislocation-free $In_{1-x}Ga_xAs_{1-y}P_y$/InP heterostructure wafers grown by LPE', *J. Appl. Phys.*, **52**, 4525–4582, 1981.
43. T. P. Pearsall, M. Quillec, and M. A. Pollack, 'The effect of substrate orientation on the liquid–solid distribution coefficients', *Appl. Phys. Lett.*, **35**, 342–344, 1979.
44. M. Feng, L. W. Cook, M. M. Tashima, T. H. Windhorn, and G. E. Stillman, 'The influence of LPE growth techniques on the alloy composition of InGaAsP', *Appl. Phys. Lett.*, **34**, 292–295, 1979.
45. M. Feng, L. W. Cook, M. M. Tashima, and G. E. Stillman, 'Lattice constant, bandgap, thickness, and surface morphology of InGaAsP–InP layers grown by step-cooling equilibrium-cooling, supercooling and two-phase-solution growth techniques', *J. Electron*, **9**, 241–280, 1980.
46. K. Sakai, S. Akiba, and T. Yamamoto, 'Growth of lattice-matched InGaAsP/InP double-heterostructures by the two-phase supercooled solution technique', *Jpn. J. Appl. Phys.*, **16**, 2043–2044, 1977.
47. J. J. Hsieh, 'Thickness and surface morphology of GaAs LPE layers grown by supercooling, step-cooling, equilibrium-cooling, and two-phase solution techniques', *J. Cryst. Growth*, **27**, 49–61, 1974.
48. R. J. Nelson, 'Near-equilibrium LPE growth of low threshold current density $In_{1-x}Ga_xAs_yP_{1-y}$ ($\lambda = 1.35\,\mu m$) DH lasers', *Appl. Phys. Lett.*, **35**, 654–656, 1979.
49. G. B. Stringfellow, 'The importance of lattice mismatch in the growth of $Ga_xIn_{1-x}P$ epitaxial crystals', *J. Appl. Phys.*, **43**, 3455–3460, 1972.
50. Y. Takeda and A. Sasaki, 'Composition latching phenomenon and lattice mismatch effects in LPE-grown $In_{1-x}Ga_xAs$ on InP substrate', *J. Cryst. Growth*, **45**, 257–261, 1978.
51. J. P. Hirth and G. B. Stringfellow, 'A mechanism for liquid-phase epitaxial growth of nonequilibrium compositions producing a coherent interface', *J. Appl. Phys.*, **48**, 1813–1814, 1977.
52. M. Feng, L. W. Cook, M. M. Tashima, G. E. Stillman, and R. J. Blattner, 'Auger profile study of the influence of lattice mismatch on the LPE InGaAsP–InP heterojunction interface', *Appl. Phys. Lett.*, **34**, 697–699, 1979.
53. L. W. Cook, M. Feng, M. M. Tashima, R. J. Blattner, and G. E. Stillman, 'Interface grading in InGaAsP liquid phase epitaxial heterostructures', *Appl. Phys. Lett.*, **37**, 173–175, 1980.
54. G. Giesecke, 'Lattice constants', *Semiconductors and Semimetals*, vol. 2, eds R. K. Willardson and A. C. Beer, academic Press, New York, 1968, pp. 63–73.
55. H. Kawaguchi, T. Takahei, Y. Toyoshima, H. Nagai, and G. Iwane, 'Room-temperature CW operation of InP/InGaAsP/InP double heterostructure diode lasers emitting at 1.55 μm', *Electron. Lett.*, **15**, 669–670, 1979.
56. T. Takahei, H. Nagai, and H. Kawaguchi, 'Low temperature liquid phase epitaxy growth for room-temperature CW operation of 1.55 μm InGaAsP/InP double-heterostructure laser', *Appl. Phys. Lett.*, **36**, 309–310, 1980.
57. S. Akiba, K. Sakai, and T. Yamamoto, '$In_{0.53}Ga_{0.47}As/In_{1-x}Ga_xAs_yP_{1-y}$ double

heterostructure lasers with emission wavelength of 1.67 μm at room temperature', *Jpn. J. Appl. Phys.*, **17**, 1899–1900, 1978.

58. S. Arai, Y. Suematsu, and Y. Itaya, '1.67 μm $Ga_{0.47}In_{0.53}As$/InP DH lasers double cladded with InP by LPE technique', *Jpn. J. Appl. Phys.*, **18**, 709–710, 1979.
59. J. J. Hsieh and C. C. Shen, 'Room-temperature CW operation of buried-stripe double-heterostructure GaInAsP/InP diode lasers', *Appl. Phys. Lett.*, **30**, 429–431, 1977.
60. H. Kano, K. Oe, S. Ando, and K. Sugiyama, 'Buried stripe GaInAsP/InP DH laser prepared by using melt-back method', *Jpn. J. Appl. Phys.*, **17**, 1887–1888, 1978.
61. H. Nishi, M. Yano, Y. Nishitani, K. Akita, and M. Tagusagawa, 'Self-aligned structure InGaAsP/InP DH lasers', *Appl. Phys. Lett.*, **35**, 232–234, 1979.
62. K. Mizuishi, M. Hirao, S. Tsuji, H. Sato, and M. Nakamura, 'Accelerated aging characteristics of InGaAsP/InP buried heterostructure lasers emittings at 1.3 μm', *Jpn. J. Appl. Phys.*, **19**, L429–L432, 1980
63. R. J. Nelson, P. D. Wright, P. A. Barnes, R. L. Brown, T. Cella, and R. G. Sobers, 'High-output power InGaAsP ($\lambda = 1.3$ μm) strip-buried heterostructure lasers', *Appl. Phys. Lett.*, **36**, 358–360, 1980.
64. S. B. Phatak and G. Kelner, 'Material-selective chemical etching in the system InGaAsP/InP', *J. Electrochem. Soc.*, **126**, 287–292, 1979.
65. S. Komiya and K. Nakajima, 'Multiplication of a threading dislocation in the InP/InGaAsP/InP double heterostructure grown on InP (111)B substrate', *J. Cryst. Growth*, **48**, 403–410, 1980.
66. M. S. Abrahams and C. J. Buiocchi, 'Etching of dislocations on the low-index faces of GaAs', *J. Appl. Phys.*, **36**, 2855–2863, 1965.
67. K. Akita, T. Kusunoki, S. Komiya, and T. Kotani, 'Observation of etch pits produced in InP by new etchants', *J. Cryst. Growth*, **46**, 783–787, 1979.
68. W. A. Jesser and D. Kuhlmann-Wilsdorf, 'On the theory of interfacial energy and elastic strain of epitaxial overgrowths in parallel alignment on single crystal substrates', *phys. Stat. Sol.*, **19**, 95–105, 1967.
69. K. Nakajima, S. Komiya, K. Akita, T. Yamaoka, and O. Ryuzan, 'LPE growth of misfit dislocation-free thick $In_{1-x}Ga_xAs$ layers on InP', *J. Electrochem. Soc.*, **127**, 1568–1572, 1980.
70. K. Oe, Y. Shinoda, and K. Sugiyama, 'Lattice deformations and misfit dislocations in GaInAsP/InP double-heterostructure layers', *Appl. Phys. Lett.*, **33**, 962–964, 1978.
71. R. Bisaro, P. Merenda, and T. P. Pearsall, 'The thermal-expansion parameters of some $Ga_xIn_{1-x}As_yP_{1-y}$ alloys', *Appl. Phys. Lett.*, **34**, 100–102, 1979.
72. G. H. Olsen, M. S. Abraham, and T. J. Zamerowski, 'Asymmetric cracking in III–V compounds', *J. Electrochem. Soc.*, **121**, 1650–1656, 1974.
73. H. Nagai and Y. Noguchi, 'Crack formation in InP–Ga_xIn_{1-x}As–InP double-heterostructure fabrication', *Appl. Phys. Lett.*, **29**, 740–741, 1976.

GaInAsP Alloy Semiconductors
Edited by T. P. Pearsall

Chapter 3

$Ga_{0.47}In_{0.53}As$/InP *and* GaInAsP/InP *Double Heterostructures Grown by Low-pressure Metal-organic Vapour-phase Epitaxy*

J. P. HIRTZ, M. RAZEGHI, M. BONNET, AND J. P. DUCHEMIN
Thomson-CSF, LCR, Domaine de Corbeville, BP No. 10, 91401 Orsay, France

3.1 INTRODUCTION

The mixed III–V alloys $Ga_{0.47}In_{0.53}As$ and GaInAsP are potentially useful materials both for heterojunction microwave and optical device applications. Liquid-phase epitaxy (LPE)[1], hydride vapour-phase epitaxy (VPE),[2] and molecular-beam epitaxy (MBE)[3] have been used to grow high-quality ternary and/or quaternary materials. We have developed an original epitaxial technique; the low-pressure metalorganic chemical vapour deposition (LP-MOCVD) technique which has several advantages compared to these other epitaxial processes for the growth of single or multi-heterostructures using these materials. These advantages are outlined here and discussed more fully later in this chapter.

(i) Growth by LP-MOCVD, in common with MBE, takes place far from thermodynamic equilibrium, and growth rates are determined generally by the arrival rate of material at the growing surface rather than by temperature-dependent surface reactions between the gas and solid phases.

(ii) The technique is versatile; numerous starting compounds can be used and the growth is controlled by fully independent parameters. Some of the reported growths by low-pressure and atmospheric-pressure MOCVD are listed in Table 3.1. The range of applications is shown in Table 3.2 which lists microwave and optoelectronic devices made from MOCVD-grown layers.

(iii) Low-pressure MOCVD is well-adapted for the growth of submicrometre layers of uniform thickness and composition. This results first from the ability of the process to produce abrupt composition changes and secondly from the result that the composition and growth rate are generally temperature-independent and can thus be held constant during the growth of a layer.

Table 3.1 Summary of epitaxial growth of III–V compounds by MOCVD

Film	Substrate	Starting materials	References
GaAs	α-Al_2O_3 $MgAl_2O_4$ BeO ThO_2	TMG, AsH_3	6
GaAs	GaAs	TMG, AsH_3 TEG, AsH_3 TEG, TEAs	7 8 9
InP	InP	TEI, PH_3	10
InAs	InAs	TEI, AsH_3	11
GaP	GaP	TMG, AsH_3	12
GaSb	GaSb	TMG, TMSb	13
GaInAs	GaAs	TMG, TEI, AsH_3	14
GaInAs	GaAs	TMG, TEI, TMAs	15
GaInAs	InP	TEG, TEI, AsH_3	16
GaAlAs	GaAs	TMG, TMAl, AsH_3	17
GaAlSb	GaAs	TMG, TMAl, TMSb	18
GaInP	GaAs	TMG, TEI, PH_3	19
GaInAsP	InP	TEI, TEG, AsH_3, PH_3	20
GaAlAsSb	GaAs	TMG, TMAl, TMAs, TMSb	18
InAsSbP	InAs	TEI, TESb, AsH_3, PH_3	21

Table 3.2 List of microwave and optoelectronics devices made from materials grown by MOCVD

Materials	Devices	References
GaAs	MESFETs Miter diodes	22 23
GaAs/GaAlAs	MESFETs Solar cells Phototransistors Photocathodes DH lasers Distributed Bragg confinement Quantum well super lattices Single longitudinal mode	24 25 26 27 28 29 30 31
InP	Gunn diodes Solar cells	37 32
GaInAsP/InP	DH lasers	33
GaInAs	IMPATTs FET active layers	

Table 3.3 Summary of the characteristics of epitaxial methods used for the growth of GaInAs and GaInAsP

Techniques	Advantages	Disadvantages
LPE	Simple, low-cost apparatus Knowledge of the GaInAsP phase diagram Excellent material quality State of the art for 1.3–1.5 μm DH lasers 1.0–1.65 μm photo detectors	Poor thickness uniformity Small scale: small surface per run, 1 wafer per run Difficulty of growing InP on GaInAs and (low E_g) GaInAsP
VPE	Ideal for growth of thick high-purity materials Large-scale production system	Expensive method for the growth of DH (multichumber reactor) Use of toxic gases
MBE	Low growth temperature Excellent thickness control Well-developed technique for GaAlAs/GaAs DH lasers and superlattices First reported GaInAs/AlInAs MESFETS *In situ* characterization	Composition control Small surface area Slow growth rate Expensive technique (UHV) Difficulty of growing III–V compounds containing phosphorous
LP-MOCVD	Well developed technique for GaAlAs/GaAs lasers Ideal for heterostructure growth High versatility Highest GaAs purity $\mu(77\,K) \simeq 140\,000\,cm^2\,V^{-1}\,s^{-1}$ Large-scale production system *In situ* characterization	Use of toxic gases Relatively slowgrowth rate ($2\,\mu m\,h^{-1}$)

In this chapter we briefly introduce the reader to several theoretical concepts concerning the growth mechanism, describe the growth of mixed III–V alloys, then give examples of the results of this technique in terms of material properties and in terms of the characteristics of devices made from LP-MOCVD-grown material. Finally we give a brief assessment of the present state of LP-MOCVD, future trends, and applications.

3.2 MODEL OF GROWTH BY LP-MOCVD

We believe that it will be helpful to the reader to be introduced at this stage to a basic model of growth by LP-MOCVD. It will be seen that this simple model can be applied generally to low-pressure and atmospheric-pressure MOCVD of III–V materials and that it can qualitatively explain the observed experimental results concerning the growth process itself and some of the characteristics of the grown materials.

Generally, growth is achieved by introducing metered amounts of the group III alkyls (e.g. $Ga(CH_3)_3$, $In(C_2H_5)_3$ and the group V hydrides (e.g. AsH_3, PH_3) into a quartz reaction tube which contains a substrate placed on an RF-heated carbon

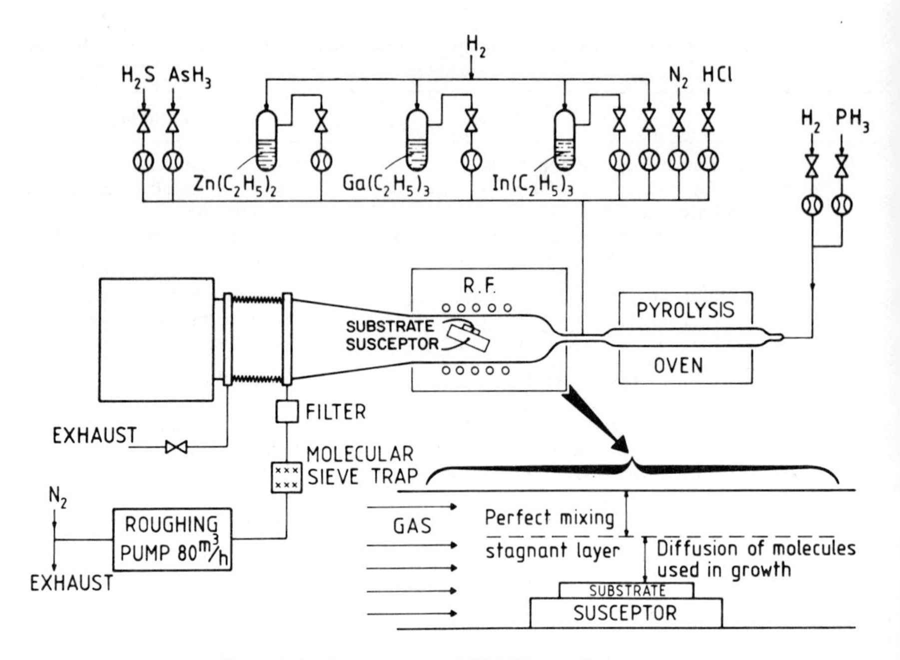

Figure 3.1 Low-pressure MOCVD gas flow system

susceptor. The hot susceptor has a catalytic effect on the decomposition of the gaseous products and the growth therefore primarily takes place at this hot surface. At the working pressure of between 0.1 to 0.5 atm (with the gas flows normally used, i.e. $10 \rightarrow 20\,\mathrm{l\,min^{-1}}$), the gas velocities in the neighbourhood of the susceptor are between 1 and $15\,\mathrm{cm\,s^{-1}}$, with the result that the gas flow is laminar. Under these conditions, the growth process can be schematically represented as shown in Figure 3.1(a).

We introduce the concept of the stagnant boundary layer[4] which is adjacent to the growing surface. Within this layer, concentration changes occur only by diffusion. The existence of this layer has been experimentally verified by Duchemin[5] for the growth of Si from SiH_4 by CVD, and can be deduced for the case of the III–V compounds by an analysis of the growth characteristics (see below). The growth is believed to occur as follows.

(a) The gas molecules diffuse across the stagnant layer to the substrate surface.
(b) At the hot surface the metal alkyls and hydrides are decomposed, thus producing the group III and group V elemental species.
(c) The elemental species move on the hot surface until they find an available lattice site where growth then occurs.

Any of the three processes (a), (b), or (c) can limit the growth rate but in general LP-MOCVD takes place under conditions where (a), diffusion across the stagnant layer, is the rate-limiting step. If (b) or (c) limits the growth, then the growth rate is strongly temperature-dependent because the reaction coefficients are exponentially dependent on temperature. However, where diffusion dominates (i.e. process (a)), the growth rate is virtually independent of temperature and is controlled simply by the rate of arrival of the growing species at the solid–vapour interface. This arrival rate is in turn controlled by the flow rates of the various species injected into the reactor, which can be metered accurately by mass-flow controllers. It will be shown that for LP-MOCVD under normal operating conditions the growth rate is generally temperature-independent and the growth process is therefore in the diffusion-limited regime. The model also predicts that the LP-MOCVD technique should suffer less from the problem of autodoping. Autodoping is an effect which is observed when undoped layers are grown onto a highly doped substrate where the dopant is volatile, e.g. the n-type dopant tellurium in GaAs. Specifically the autodoping effect arises out of the presence of a significant vapour pressure of the dopant in the vicinity of the growing interface, resulting in the incorporation of the dopant in the layer. Growth at low pressure increases the diffusion of this dopant away from the growth surface by increasing the concentration gradient across the stagnant layer. This results in a more abrupt change in doping level which can be critical in devices which utilize very thin layers.

Also as a result of the low pressure in the reactor, the gas speeds are generally higher than for other growth techniques. Thus changes of composition at heterojunctions can be quickly achieved as new gas compositions are rapidly established. The higher gas velocity also results in a more uniform stagnant layer thickness over a large area of substrate, which will ensure a good uniformity of thickness and composition in this diffusion-limited growth process.[34]

Finally the low-pressure technique suffers less from parasitic reactions in the gas phase away from the susceptor as the gas heating and collision probability is lower at lower pressure. This result is experimentally verified by the virtual elimination at low pressure of 'smoke' formation (spontaneous gas-phase nucleation which gives rise to observable fumes), which can be a significant problem in the MOCVD growth of GaAs, InP, and GaAlAs at atmospheric pressure.

3.3 EXPERIMENTAL DETAILS

The metalorganic chemical vapour deposition is based on the thermal cracking of metalorganic compounds and hydrides on a hot substrate. The walls of the silica reactor remain cold and parasitic reactions in the gas phase are reduced as much as possible. In the case of GaAs deposition, there is a co-pyrolysis of a metalorganic compound, usually trimethyl-gallium (TMG) and arsine, following the reaction:

$$Ga(CH_3)_3 + AsH_3 \xrightarrow{\text{heat}} GaAs + 3CH_4$$

The growth of III–V compounds which contain indium (InAs, InP, GaInAs, . . .) is complicated by the fact that In organometallics react very easily at room temperature with hydrides, leading to fumes, formation of addition compounds, and rapid exhaustion of the vapour phase. Care has to be taken with the gas mixture to avoid these problems. Several alternatives have been proposed for the growth of InP to avoid the room-temperature reaction between triethyl-indium (TEI) and phosphine forming an adduct

$$In(C_2H_5)_3 + PH_3 \rightarrow PH_3\text{–}In(C_2H_5)_3$$

as follows:

(i) By mixing PH_3 and TEI just above the substrate the parasitic reaction has no time to take place and is shielded by the fast pyrolysis reaction of $In(C_2H_5)_3$ and PH_3 on the hot surface.[38] However, this method seems to be limited to small-area surfaces. The growth is 4–5 $\mu m\,h^{-1}$, and the reported best mobilities for InP are 36 300 $cm^2\,V^{-1}\,s^{-1}$ at 77 K.

(ii) It may be advantageous to use group V alkyls instead of hydrides as the group V sources. Growth of GaP,[39] GaInAs,[15] and GaAsSb,[18] have been reported by using respectively $P(CH_3)_3$, $As(CH_3)_3$, and $Sb(CH_3)_3$. Using this approach, the copyrolysis of $In(C_2H_5)_3$ and, for example, $P(C_2H_5)_3$ instead of PH_3, could avoid the formation of adduct products.

(iii) Recently the pyrolysis of a new organometallic adduct

$$(CH_3)_3\text{–}In\text{–}P\text{–}(CH_3)_3$$

has been reported[40] as a suitable method for the growth of InP. The advantages of this method are the ease of purification and the lower inherent dangers compared to the metalorganic technique. However, the presence of the pre-existing In–P bond limits the usefulness of this method to binary InP. At 520°C, the growth rate is about 0.6 $\mu m\,h^{-1}$. For non-intentionally doped n–InP, the background carrier concentration is about $5 \times 10^{15}\,cm^{-3}$.

For temperatures in excess of 400°C, incongruent phosphorus evaporation from the substrate leads to a rough surface morphology. If pure phosphine PH_3 is passed through the reaction chamber, we observe indium droplets on the substrate surface because PH_3 is more stable than, for example AsH_3, and does not decompose readily at the surface to provide the required phosphorous overpressure. By cracking PH_3 at 750°C, in a separate furnace before it entered the reactor, Duchemin *et al.*[10] have achieved the growth of device-quality InP in a low-pressure system (~ 100 mbar). At a growth temperature of 510°C, for a hydrogen flow rate of 3 $l\,min^{-1}$ through the $In(C_2H_5)_3$ bubbler and a PH_3 flow of 0.1 $l\,min^{-1}$, the growth rate was 1–2 $\mu m\,h^{-1}$. The background carrier concentration is $(1\text{–}3) \times 10^{15}\,cm^{-3}$ and the liquid-nitrogen Hall mobility is 30 000 $cm^2\,V^{-1}\,s^{-1}$. The low-pressure horizontal reaction chamber (diameter 10 cm in the substrate area) shown in Figure 3.1(b) is pumped by a high-capacity roughing pump (~ 100 $l\,h^{-1}$) to adjust the total pressure to between 10 and

300 mbar. The pressure inside the chamber is controlled by a mechanical gauge and adjusted by varying the flow rate of the pump using a control valve. The substrate is mounted on a pyrolytically coated graphite susceptor which is heated by RF induction at 1 MHz, the temperature being measured by an infrared pyrometer. The gas flow system for the growth of InP, $Ga_{0.47}In_{0.53}As$, and GaInAsP is shown in Figure 3.2. The gas sources used in this study are listed in Table 3.4. The valves used in this first-generation reactor (see figure 3.3) are operated manually. The gas piping is made primarily from $\frac{1}{4}$ inch stainless steel and (Swagelock) fittings are used. Commercially available metalorganics are delivered in stainless steel containers provided with valves and an inlet diptube assembly to allow the bubbling of carring gas, which is high-purity palladium-diffused hydrogen. Flows are monitored by mass-flow controllers. These containers are connected by stainless steel tubes and supported in thermostatically controlled baths in the ranges $+20°C$ to $+40°C$ for triethyl-indium (TEI), $-10°C$ to $+10°C$ for triethyl-gallium (TEG), and $-35°C$ to $-25°C$ for diethyl-zinc (DEZ). For all these alkyl sources, the flows are established first in an exhaust line and then switched to the reactor when required. Hydrogen chloride is used for cleaning the susceptor and the reactor after each run.

Figure 3.2 Photograph of quartz tube used in the LP-MOCVD system

Table 3.4 Starting materials used for the growth of GaInAsP by low-pressure MOCVD

Group III sources	Group V Sources	Dopant sources
$Ga(CH_3)_3$	AsH_3	$Zn(C_2H_5)_3$
or		
$Ca(C_2H_5)_3$	AsH_3	$Zn(C_2H_5)_3$
$In(C_2H_5)_3$	PH_3	H_2S

The growth procedure is as follows:

(i) The mechanochemically polished InP substrates, usually one n^+ Sn-doped slice and one semi-insulating Fe-doped substrate (misoriented 0° to 2° off the ⟨100⟩ towards the ⟨110⟩ direction), are used for each run, cleaned in methanol at room temperature and rinsed in isopropyl alcohol. Just before loading, they are etched in a solution of 1 % bromine in methanol at room temperature and rinsed in warm isopropyl alcohol. After loading the substrates, the reactor is purged with N_2 and the low pressure is established in hydrogen. As for hydride VPE or LPE, etching of the InP substrate is carried out just prior to the MOCVD growth although the optimum *in situ* etching method has not yet been established. We use

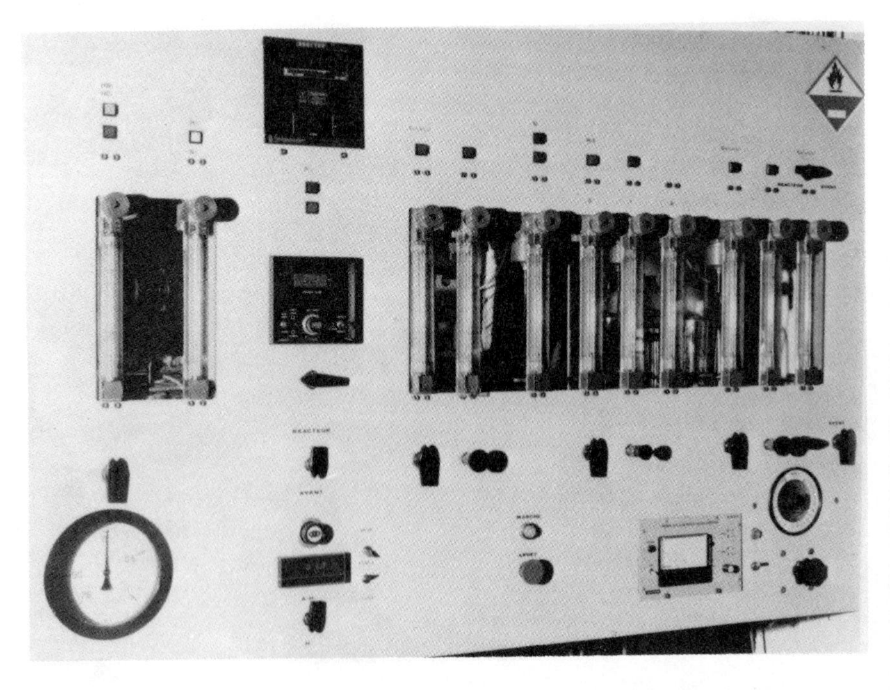

Figure 3.3 Photograph of the gas panel used for the growth of GaInAsP by LP-MOCVD

an HCl etch, but here the surface morphology is strongly influenced by orientation and care has to be taken to avoid parasitic reactions between TEI and HCl.

(ii) When the susceptor temperature is above 300°C, PH_3 is injected into the reactor through an oven heated at 750°C. A flow of between 10 and 100 $cm^3 min^{-1}$ is sufficient to avoid degradation of the InP substrate during thermal treatment.

(iii) The phosphine flow is established during the *in situ* cleaning step. After the heating-up period, the growth starts when the flow of the group III metalorganics is switched into the reactor. For the growth of $Ga_{0.47}In_{0.53}As$, where the growth temperature is in the range 500 to 650°C, an AsH_3 flow is injected and the PH_3 is switched out one minute before the start of growth. Formation of an intermediate $InAs_yP_{1-y}$ has been detected using Auger spectroscopy when the InP substrate is heated for more than 10 min in an arsenic overpressure for temperatures higher than 500°C.

(iv) To stop the growth the group III species are switched off, but the arsine and/or phosphine overpressure is maintained during the cooling of the substrate down to 300°C.

3.4 GROWTH PROPERTIES

The important growth parameters affecting single-layer or heteroepitaxial growth characteristics are substrate orientation, substrate preparation, growth temperature, gas flows, and source material purity. In the following discussion, the effect of various parameters on the properties of InP, $Ga_{0.47}In_{0.53}As$, and GaInAsP layers are examined.

3.4.1 InP

In a previous study,[10] we investigated the possibility of surface-related growth mechanisms of the LP-MOCVD growth of InP. We considered the growth parameters, substrate temperature and reactant partial pressure, and we observed that the growth rate:

(a) varies linearly with the TEI partial pressure (Figure 3.4(a)),
(b) is independent of the phosphorus partial pressure (Figure 3.4(b)), and
(c) is approximately temperature-independent in the range 500–650°C.

These results were reported for a ⟨100⟩ growth surface orientation with various misorientations between 0° and 2° off, towards ⟨110⟩. No significant dependence on the edge PL efficiency was observed until the misorientation exceeded 2°. Thus referring to the growth model we can see that InP growth is limited by the arrival of the group III species with no kinetic effects due to surface reactions,

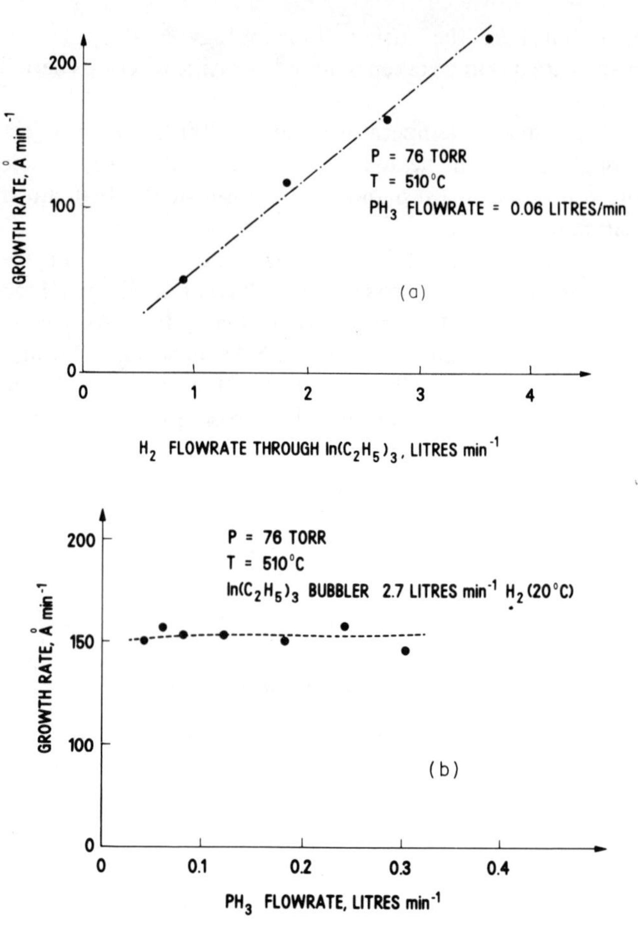

Figure 3.4 (a) The growth rate of InP versus the flow of TEI. (b) The growth rate of InP versus the flow of PH_3

i.e. the decomposition of a TEI alkyl compound on the surface is a much faster process than the TEI diffusion through the boundary layer.

3.4.2 $Ga_{0.47}In_{0.53}As$

The choice of group III alkyl sources used for the growth of $Ga_xIn_{1-x}As$ is critical. We first investigated growth using TEI and TMG and found that for x values close to that required for lattice matching on InP, the surface morphology was poor. No band-edge PL was observed and at low growth temperature (T_S < 600°C) the layers were polycrystalline. We suggested that these problems were related to a surface kinetics limitation such as steric hindrance; the adsorption of

Figure 3.5 Surface quality of typical GaInAs/InP MOCVD material

large ethyl and smaller methyl radicals on the surface may introduce a new rate-limiting step in the growth process which non-uniformly affect the growth rate along the growing surface. By contrast when the growth was performed with TEI and TEG, these problems disappeared and the epilayer surface was featureless for compositions in the range $0.40 < x < 0.60$ (see Figure 3.5). An alternative would be to use TMG and TMI, but we have not as yet tried this. We suggest however that the growth rate could be higher with the methyl compounds because adsorption of ethyl radicals can limit the growth rate.[4] However, trimethyl-indium is a difficult source to use, being a solid at room temperature; also some reports[8] indicate that TEG may be of higher purity than TMG.

The growth rate of $Ga_xIn_{1-x}As$ from TEI, TEG and AsH_3 varies as the sum of the partial pressure of TEG + TEI and does not depend on the arsenic partial pressure (see Figure 3.6). If the TEI flow rate is kept constant, the growth rate is proportional to the hydrogen flow through the triethyl-gallium source. The growth rate of GaInAs is independent of temperature between 525 and 650°C but the distribution coefficient of indium may increase slightly with temperature as shown in Figure 3.6(f). A similar behaviour was reported for the Al distribution coefficient in the MOCVD growth of GaAlAs.[24] The thickness of the epilayer was measured by a bevel-polishing and staining method and was used to determine the growth rate, which is 200–350 Å min^{-1}. We have also measured the relative lattice mismatch for different growth conditions. When the arsine partial pressure is varied between 0.3 and 1.3 Torr, the relative lattice mismatch remains better than 0.1%, i.e. $\Delta a/a \leq 10^{-3}$. Starting from conditions which give a lattice matching of 0.05% at 625°C, the mismatch remains within 0.15% in the temperature range 525–625°C. We measure the lattice mismatch between the thin film (0.5–1.5 μm) and the substrate by single- or double-crystal x-ray diffractometry, using (400), (600), or (711) Cu Kα reflections. To obtain the true lattice mismatch at the growth temperature, it is necessary to obtain the unstrained mismatch at room temperature[42] and then correct for the differential thermal expansion using the data of Pearsall *et al.*[43]

Under optimized growth conditions, a lattice matching of $\Delta a/a < 0.04\%$ is

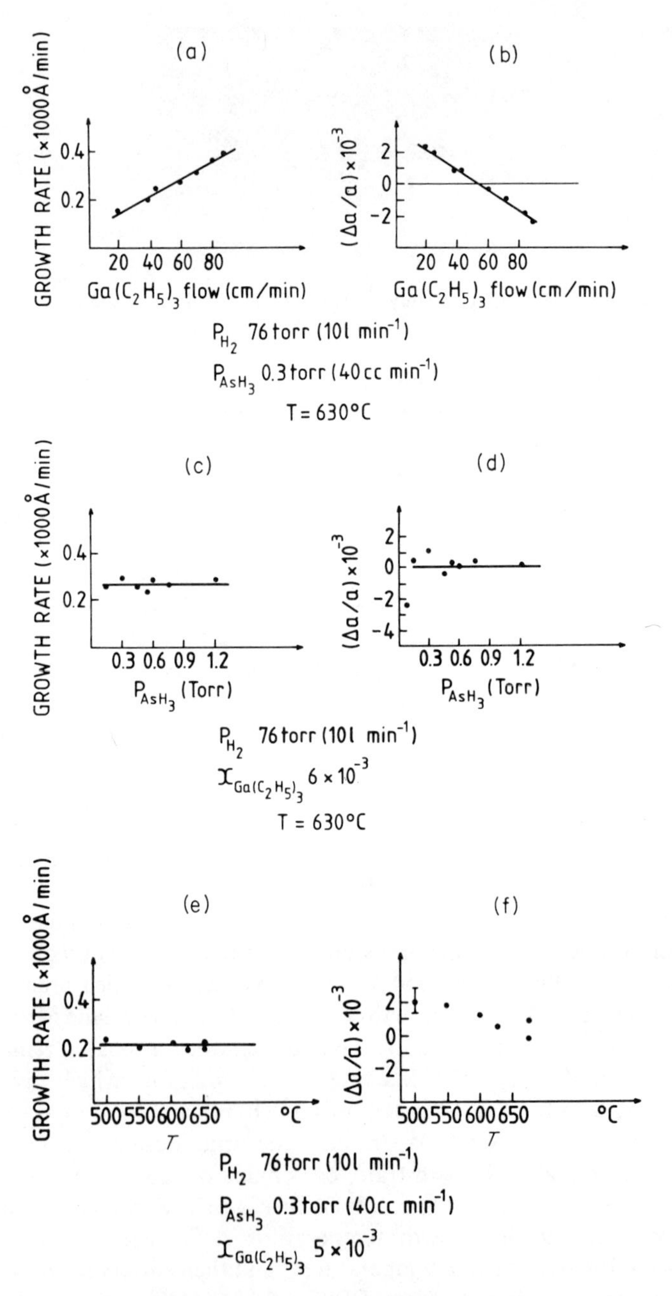

Figure 3.6 Growth rate and lattice mismatch of MOCVD GaInAs as functions of (a, b) $Ga(C_2H_5)_3$ flow rate, (c, d) AsH_3 partial pressure, (e, f) temperature

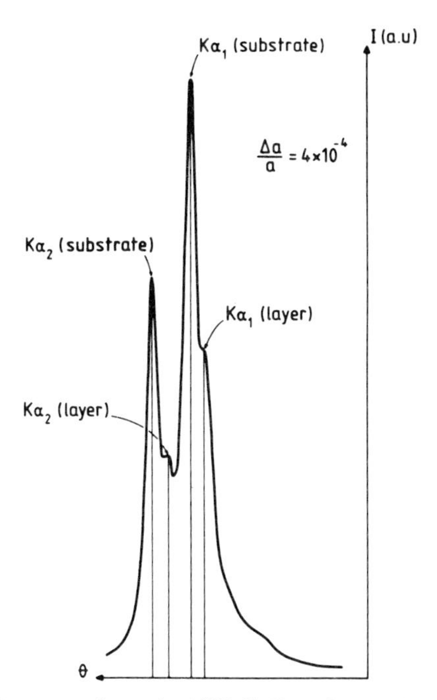

Figure 3.7 X-ray rocking curve about the (400) Cu Kα reflection from $Ga_{0.47}In_{0.53}As$ on InP substrate

achieved as shown in Figure 3.7. A low-temperature photoluminescence spectrum for an MOCVD $Ga_{0.47}In_{0.53}As$ epilayer is shown in Figure 3.8 and is compared with other samples with a similar carrier concentration grown by LPE and hydride VPE.[44] All of these samples show a sharp peak near the band edge (0.81 eV), a broader peak displaced by 18.19 meV, tentatively attributed to residual Zn level, and a deep level 51 meV below the conduction band. No specific evidence of carbon contamination has been observed in MOCVD-grown layers.

To study the Hall mobility, Van der Pauw measurements were made at 300 K and 77 K using clover-leaf-shaped samples, in a magnetic field of 5 kG. Ohmic contacts were formed using pure In or In–Zn alloyed at 450°C for 1 min in flowing H_2. The measured Hall mobilities in n-type and p-type samples 1 μm thick are shown in Figures 3.9 and 3.10 at 300 K and 77 K. A background carrier concentration of $8 \times 10^{15}\,cm^{-3}$ is obtained using standard-purity alkyl and hydride compounds, with a corresponding Hall mobility of approximately $8000\,cm^2\,V^{-1}\,s^{-1}$. For intentionally doped samples the mobility is $\mu_H \simeq 6300\,cm^2\,V^{-1}\,s^{-1}$ for a residual donor doping $N_D - N_A \simeq 1.5 \times 10^{17}\,cm^{-3}$, which represents a 50% improvement over GaAs values. We have measured the Schottky barrier height on GaInAs, using an Al-Au contact, and find a value of 0.3 eV, which is in good agreement with that formed by Kajiyama *et al.*[45] For Zn-

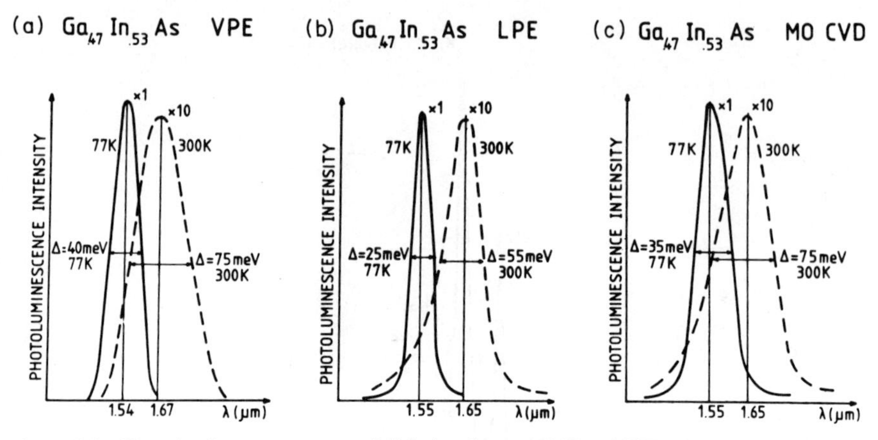

Figure 3.8 Photoluminescence spectra of GaInAs taken at 300 K and 77 K: (a) grown by VPE, $N_D - N_A \simeq 5 \times 10^{17}\,cm^{-3}$; (b) grown by LPE, $N_D - N_A \simeq 10^{17}\,cm^{-3}$; (c) grown by MO-CVD, $N_D - N_A \simeq 5 \times 10^{17}\,cm^{-3}$

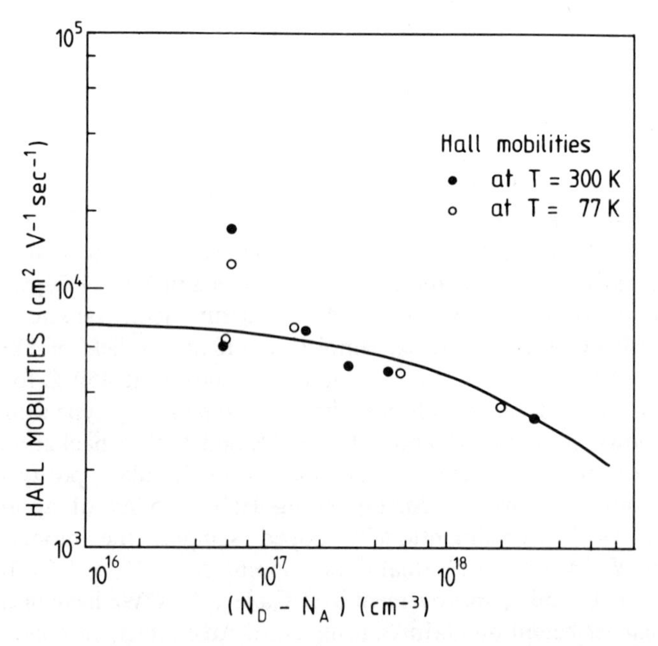

Figure 3.9 300 K and 77 K Hall mobility for n-type $Ga_{0.47}In_{0.53}As$ on InP

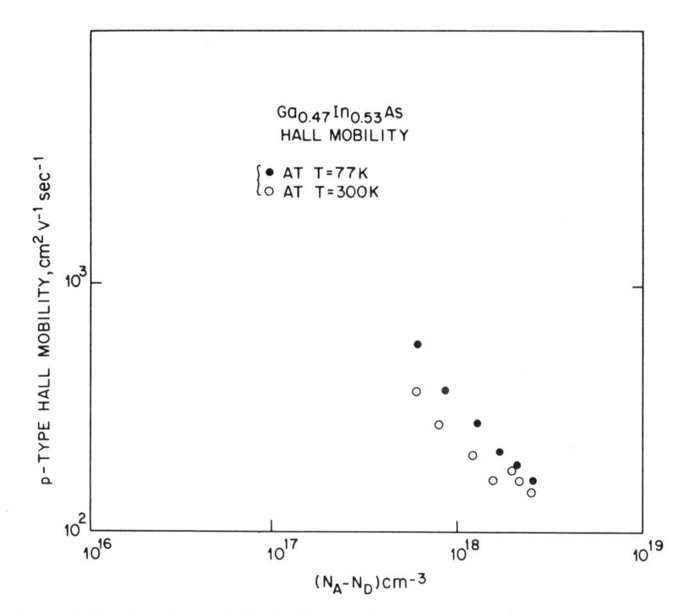

Figure 3.10 300 K and 77 K Hall mobility for p-type $Ga_{0.47}In_{0.53}As$ on InP

doped p-type samples, MOCVD epilayers have the highest mobility reported for $Ga_{0.47}In_{0.53}As$, i.e. $\mu_H \simeq 400\ cm^2\ V^{-1}\ s^{-1}$ for $N_A - N_D \simeq 6 \times 10^{17}\ cm^{-3}$. As previously observed for InP, the carrier concentration for sulphur-doped GaInAs epilayers increases with increasing growth temperature, for a fixed value of the H_2S flow. The opposite behaviour is observed for zinc-doped epilayers, i.e. the acceptor level decreases with increasing temperature. Both isotype and p–n junction InP/GaInAsP/InP double heterostructure (DH) have been grown by MOCVD. Both types of DH show an improvement in PL efficiency by one order of magnitude, as compared with single layers of GaInAs. This indicates good carrier confinement by the heterostructure and a low interfacial non-radiative recombination rate which we believe is related to the abrupt heterointerfaces (characterization presented in next section). Analysis of the *I–V* characteristics of the p–n double heterostructure have shown that the junction is always in the small-bandgap material. Zn diffusion during the growth is controlled by decreasing the growth temperature of the p-InP confining layer.

It has been demonstrated that metalorganic chemical vapour deposition can be used to produce $Ga_{0.47}In_{0.53}As$ lattice-matched on InP, with electronic properties comparable to other epitaxial techniques. In addition the larger area, better uniformity of thickness and composition, good surface flatness, and excellent morphology show that LP-MOCVD is in fact a superior technique for applications where a high yield and high surface area of good material are important.

3.4.3 GaInAsP

For GaInAsP epilayers lattice-matched on InP, the same general growth behaviour was observed as long as the sum of the partial pressures of arsenic and phosphorus does not exceed a critical value. The growth rate is proportional to the sum of the partial pressure of TEG and TEI, and is independent of the arsenic or phosphorus partial pressure. Under these conditions, the growth rate is proportional to the hydrogen flowing through the TEG bubbler when the AsH_3, PH_3, and $In(C_2H_5)_3$ flow rates are kept constant (Figure 3.11).

The behaviour of the incorporation of group V elements is different because both arsenic and phosphorus are volatile and we have found that the probability of incorporation for arsenic is much higher than the phosphorus probability. For example, to achieve the composition $Ga_{0.21}In_{0.79}As_{0.47}P_{0.53}$, the group V flow-rate ratio (PH_3/AsH_3) must exceed 20. When the value of PH_3/ASH_3 was equal to 10 a secondary ion mass spectrometry (SIMS) analysis revealed an atomic phosphorus concentration of only 2%. This result agrees qualitatively with MBE studies where it is well-known that phosphorus has a much lower sticking coefficient than arsenic.[46] For the growth of GaInAsP with a characteristic wavelength of $\lambda = 1.1\ \mu m$, a working pressure of 30 Torr allows us to obtain the high PH_3/AsH_3 ratio.

When the sum of arsenic and phosphorus partial pressures exceeds the critical value, an 'inhibition' effect is observed (see Figure 3.12). Under these conditions the growth composition becomes strongly temperature-dependent, and the growth rate is no longer limited by diffusion of the group III alkyls through the boundary layer. Under these growth conditions (see Figure 3.12), the growth rate

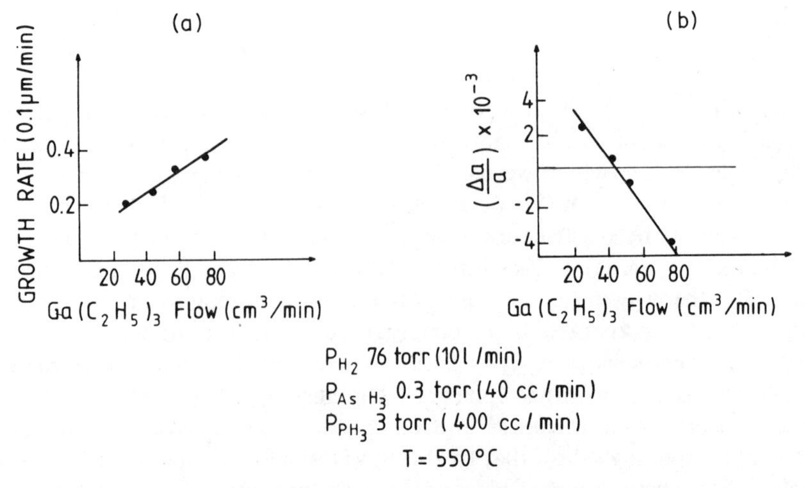

Figure 3.11 (a) Growth rate and (b) lattice mismatch of MOCVD GaInAsP single layer as a function of the flow of $Ga(C_2H_5)_3$

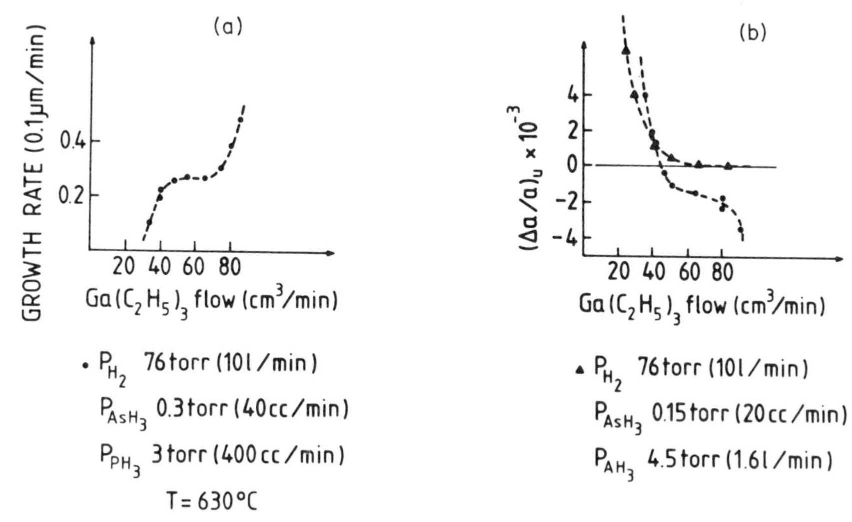

Figure 3.12 (a) Growth rate of GaInAsP versus $Ga(C_2H_5)_3$ flow for PH_3 partial pressure above the critical value. (b) Lattice mismatch behaviour above the critical PH_3 pressure

and the composition are no longer linearly dependent on the $Ga(C_2H_5)_3$ flow rate. When this behaviour is observed, no PL band-edge emission is observed. This critical domain for the growth of GaInAsP on InP corresponds to the quaternary composition range represented by the characteristics wavelength range of 1.35–1.45 μm. It is expected that LP-MOCVD growth in such a composition region would revert to a kinetically limited, rather than mass-transfer-limited, growth. The material would be highly mismatched, dislocated, and would therefore be expected to show poor PL and electrical properties. The room-temperature mobilities of GaInAsP compositions lattice matched to InP are always smaller than those observed for GaInAs. Table 3.5 summarizes these results. Compositions in the range 1.2–1.3 μm are lattice-matched to within 0.04% when the substrate temperature is in the range 640–650°C. Figure 3.13 shows a double-crystal x-ray diffraction spectrum of a 0.6 μm thick, 1.3 μm

Table 3.5 Mobilities of GaInAsP epilayers grown by low-pressure MOCVD

Composition	λ(μm)	$N_A - N_D$(cm^{-3})	μ(cm V^{-1} s^{-1}) at 300 K
InP	0.91	5×10^{15}	4300
GaInAsP	1.15	4×10^{17}	3200
GaInAsP	1.33	2.5×10^{17}	4000
GaInAsP	1.57	8×10^{15}	4100
GaInAs	1.67	8×10^{15}	7750
GaAs	0.88	1×10^{16}	6000

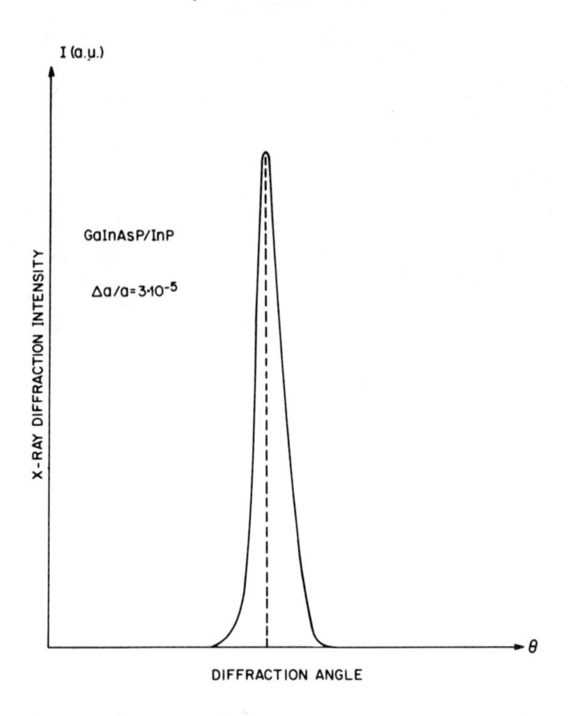

Figure 3.13 Double-crystal x-ray rocking curve about the (600) Cu Kα reflection from a GaInAsP single layer. A low-dislocation Zn-doped InP substrate is used as first crystal

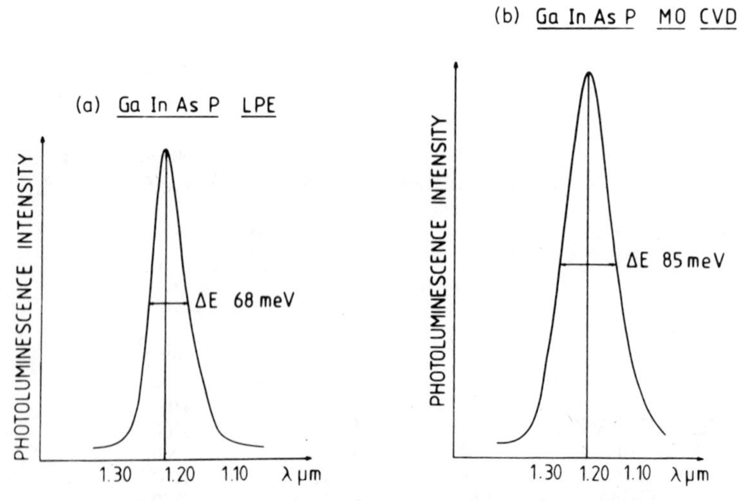

Figure 3.14 Band-edge PL of a 1.3 μm GaInAsP single layer grown (a) by LPE, $N_D - N_A \simeq 10^{17}$ cm^{-3}, and (b) by MOCVD, $N_D - N_A \simeq 5 \times 10^{17}$ cm^{-3}

GaInAsP epilayers, for which the relative lattice mismatch is estimated to be $\Delta a/a \simeq 3 \times 10^{-5}$ from a Cu Kα (600) reflection. A low-dislocation Zn–InP substrate was used as the first diffracting crystal. The 300 K and 77 K band-edge photoluminescence efficiency for 1.2–1.3 μm MOCVD material is comparable to that from an LPE standard (Figure 3.14).

Higher bandgap quaternary materials (i.e. $\lambda = 1.1$ to 1.2 μm) grown as a single layers on InP often exhibited a poor room-temperature PL output. This was attributed to a high surface recombination, in these phosphorus-rich materials, as the PL efficiency significantly increased after a short etch in HCl:H_2O (1:5). Growth of InP on these etched quaternary layers has been attempted with some success.

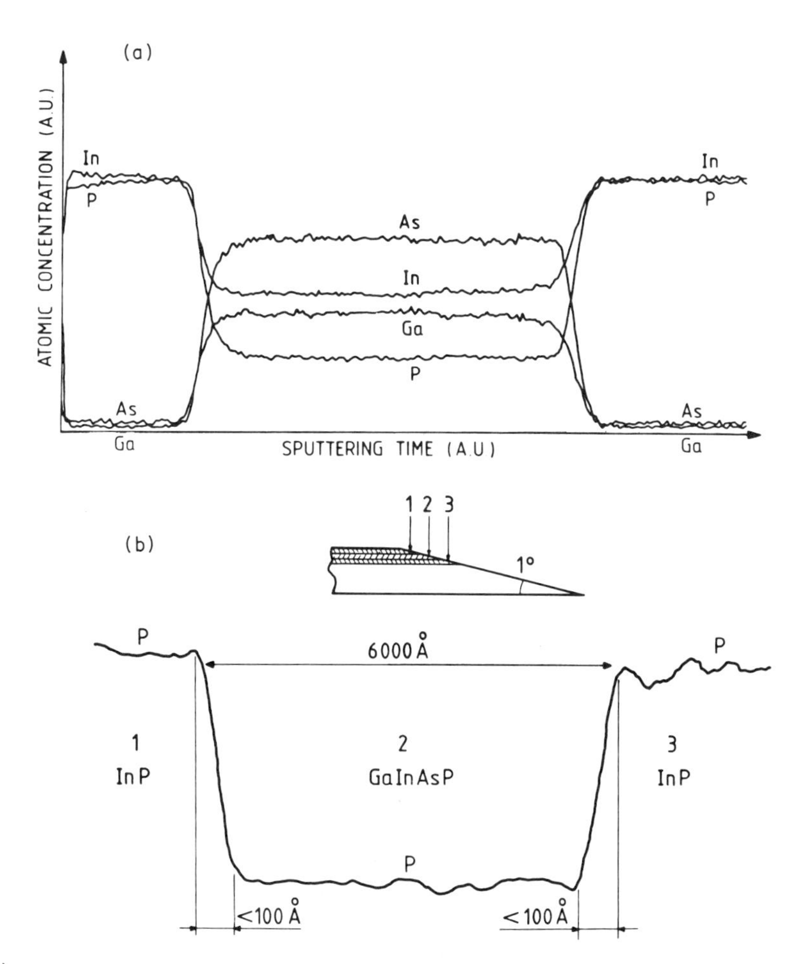

Figure 3.15 Auger profile for GaInAsP/InP double heterostructure: (a) eroded by sputtering, and (b) Auger profile for phosphorus scanned on a 1° angle lapped DH

We have measured the homogeneity of the quaternary layer and interface sharpness for a 1.3 μm double-heterostructure InP/GaInAsP/InP by Auger electron spectroscopy. Two techniques were used and these are shown schematically in Figure 3.15. In the first measurement the composition is measured after successive sputtering steps. In the second measurement a finely focused Auger beam is scanned along a bevelled sample giving a continuous Auger readout. In both cases an interfacial abruptness of better than 100 Å was measured, which is the present limit of resolution for these techniques.

3.5 MOCVD-GROWN GaInAsP/InP LASERS

Broad-area and stripe geometry lasers[36] have been fabricated from MOCVD-grown materials. The following double heterostructure was grown for laser fabrication:

(a) 1.5 μm S-doped InP	$(N_D - N_A \simeq 2 \times 10^{18}\ \text{cm}^{-3})$
(b) 0.2–0.4 μm GaInAsP	(unintentionally doped)
(c) 2 μm Zn-doped InP	$(N_A - N_D \simeq 5 \times 10^{17}\ \text{cm}^{-3})$
(d) 0.2 μm Zn-doped InP	$(N_A - N_D \simeq 2 \times 10^{18}\ \text{cm}^{-3})$

A typical growth cycle for a laser structure is shownFigure 3.16. Some isotype structures were also grown for characterization by photoluminescence. A post-growth Zn diffusion (using a Zn_3P_2 diffusion source) can be used to produce a *p–n* junction in this structure. This doping source permits the location of the junction to a precision of better than 0.2 μm.

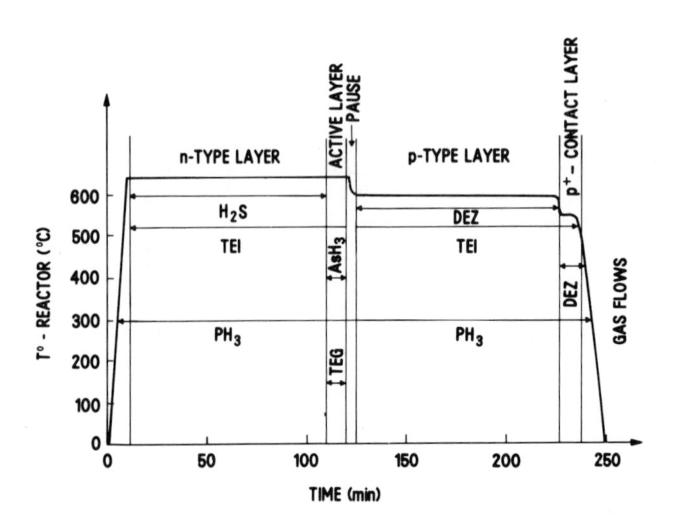

Figure 3.16 Laser growth sequence

Table 3.6 Summary of characteristics of DH GaInAsP/InP lasers grown by low-pressure MOCVD (μm)

Wavelength (μm)	d (active layer)	J_{th}	T_0 (K)
1.15	0.4 pulsed	5.9	
1.22	0.4	1.2 pulsed and CW	65–70
1.27	0.2	1.5 pulsed	65–80
1.29	0.2	1.05 pulsed and CW	65–75

After deposition of the contact metals, 500 Å of Au–Ge (12 at %) on the n-type side and 1000 Å of Au–Zn 8 at %) on the p-type side, the samples are annealed for 10 min at 430°C. This anneal is followed by the deposition of an Au layer (2000 Å) on both sides. Broad-area laser chips are then made by standard sawing and cleaning procedures producing Fabry–Perot cavities 400 μm in length. These lasers are tested with 100 ns driving pulses at a pulse repetition rate of 10 kHz (0.1 % duty cycle) on a probe tester with a heated sample stage. The threshold current values are summarized in Table 3.6. The apparent improvement in J_{th} at longer wavelengths is believed to be related to the superior lattice matching which was achieved rather than to be a true wavelength dependence.

Measurements on a 0.6 μm thick 1.15 μm GaInAsP single test layer showed a lattice match to InP of about 0.1 % where at 1.22–1.29 μm the layers exhibit a

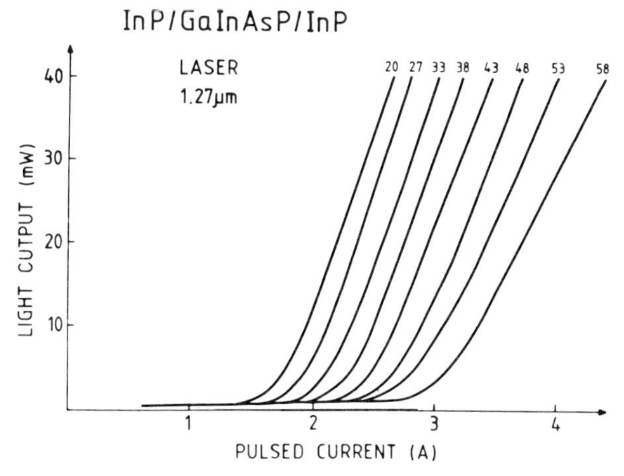

Figure 3.17 Light output versus pulsed current at several heat sink temperatures for a 1.27 μm GaInAsP/InP laser grown by LP-MOCVD

lattice match of better than 0.04% and sometimes as little as 0.0003%. The external quantum efficiency at $1.2 \times J_{th}$ was 13% at room temperature for the lasers emitting at 1.22 μm. As with GaInAsP/InP lasers grown by LPE, the lasing threshold varies exponentially with temperature and the T_0 value for the MOCVD-grown lasers, tested in pulsed operation, typically varies between 65 K and 80 K (Figure 3.17). It has been proposed that this small T_0 value for the temperature dependence of the lasing threshold is related to the intrinsic properties of the material and may be due in particular to non-radiatave Auger recombination.[47]. However, reported T_0 values are in the range 50–70 K between room temperature and 65°C for lasers grown by LPE[1] or hydride VPE[2]. Since the average T_0 value for MOCVD-grown lasers, where the heterointerfaces are more abrupt, is slightly higher than 70 K, it is possible that interface recombination and leakage current may also contribute to the temperature sensitivity of J_{th}. Some stripe lasers have been fabricated using the usual SiO_2-isolated stripe technology.[1] These devices can be operated CW at room temperature and Figure 3.18 shows the emission spectrum of a 12 μm stripe laser operating at between 1.1 and 1.2 times the threshold current.

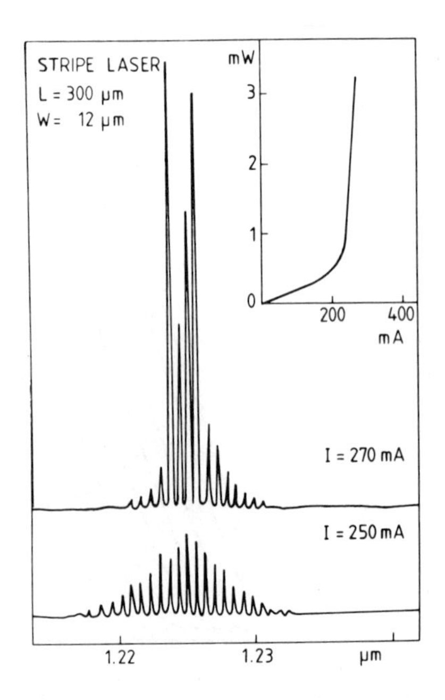

Figure 3.18 Light-current characteristics and spectrum of a 12 μm stripe GaInAsP/InP laser under CW operation

3.6 FUTURE INVESTIGATIONS AND CONCLUSIONS

The low-pressure metalorganic chemical vapour deposition process based on the prolysis of alkyls of group III elements in an atmosphere of the hydrides of group V elements would seem to be a widely applicable growth technique which is well-adapted to the growth of submicrometre layers and heterostructures. By working at low pressure, we have found optimized growth conditions for the preparation of a wide range of III–V materials and heterostructures, GaAs, InAs, GaP, InP, $Ge_{0.47}In_{0.53}As$ on InP, $Ga_{0.48}In_{0.52}P$ on GaAs, GaAlAs on GaAs and GaInAsP on InP. The low-pressure growth offers an improved thickness uniformity and compositional homogeneity, in addition to the supression of autodoping memory effects.[35] This, coupled with the reduction of parasitic decomposition in the gas phase, allows the growth of good-quality material over a large surface area of growth. Abrupt composition changes and low interfacial recombination velocity have allowed the fabrication of broad-area and stripe geometry lasers with low threshold currents and good thermal behaviour. One of the key issues for the usefulness of this method is the possibility of obtaining high-purity and therefore high-mobility $Ga_{0.47}In_{0.53}As$. Recent results obtained for GaAs are very encouraging, where with special effort on the purification of starting materials Nakasihi *et al.*[48] have grown GaAs epilayers with an electron concentration of 4×10^{14} cm^{-3} and an electron mobility of 140 000 cm^2 V^{-1} s^{-1} at 77 K. Early results on selective growth on InP allow us to predict a new generation of single transverse-mode lasers grown by MOCVD, operating at a wavelength of 1.3–1.5 μm. In addition to the GaInAsP system, the MOCVD growth of GaAlInP on GaAs or GaAlInAs on InP respectively present interesting possibilities for optoelectronic devices emitting in the visible range and high-mobility microwave devices with single or multi-heterostructure of submicrometre layers.

ACKNOWLEDGMENTS

The authors would like to acknowledge B. de Cremoux, S. D. Hersee, and P. Hirtz for useful discussions in the preparation of this chapter.

REFERENCES

1. S. Arai, Y. Suematsu, and Y. Itaya, 'Injection lasers prepared by liquid phase epitaxy', *IEEE J. Quantum Electron.*, **QE-16**, 197–205, 1980.
2. G. Olsen, T. Zamerowski., and DiGiuseppe, '1.5–1.7 μm VPE InGaAsP/InP CW lasers', *Electron. Lett.*, **16**, 516–518, 1980.
3. B. Miller, J. McFee, R. Martin, and P. K. Tien, 'Room temperature operation of lattice-matched InP/$Ga_{0.47}In_{0.53}As$ lasers', *Appl. Phys. Lett.*, **33**, 44–47, 1978.
4. F. C. Evershteym, I. W. P. Severin, and C. H. J. Brehel, 'A stagnant layer model for the

epitaxial growth of silicon silane in a horizontal reactor', *J. Electrochem. Soc.*, **117**, 925–931, 1970.

5. J. P. Duchemin, M. Bonnet, and F. Koelsch, 'Kinetics of silicon growth under low hydrogen pressure', *J. Electrochem. Soc.*, **125**, 637–644, 1978.
6. G. W. Cullen, and C. C. Wang (eds), *Heteroepitaxial Semiconductors for Electronic Devices*, Springer-Verlag, Berlin, 1958, pp. 600–649.
7. H. Manasevit, and A. Aurvey, 'A survey of the heteroepitaxial growth of semiconductor films on semi-insulating substrates', *J. Cryst. Growth*, **22**, 125–148, 1974.
8. J. Siki, K. Tanno, and E. Chili, 'Properties of epitaxial GaAs layers from a triethyl gallium and arsine system', *J. Electrochem. Soc.*, **122**, 1108–1113, 1975.
9. L. Skvortsor, *Izv. Akad. Nauk SSSR N Cory*, **12**, 754, 1976.
10. J. P. Duchemin, M. Bonnet, G. Beuchet, and F. Koelsch, 'Organometallic growth of device quality InP by cracking of In $(CrH_5)_3$ and PH_3 at low pressure', *Gallium Arsenide and Related Compunds 1978*, Conf. Ser. No. 45, Institute of Physics, Bristol, 1979, pp. 10–18.
11. H. M. Manasevit and W. I. Simpson, 'The use of metal-organics in the preparation of semiconductor materials, V: the formation of In-group V compounds and alloys, *J. Electrochem. Soc.*, **120**, 135–137, 1973.
12. H. Beneking, and H. Roehle, 'MO-CVD growth of GaP and GaAlP', *J. Cryst. Growth*, **55**, 79–86, 1981.
13. H. Manasevit and K. Hess, 'The use of metalorganics in the preparation of semiconductor materials', *J. Electrochem. Soc.*, **126**, 2031–2033, 1979.
14. B. Baliga, and S. Gandhi, 'Growth and properties of heteroepitaxial GaInAs alloys on GaAs substrates using trimethyl-gallium–triethyl-indium and arsine', *J. Electrochem. Soc.*, **122**, 683–687, 1975.
15. C. Cooper, M. Ludovik, and R. Achi Vand Moon, 'The organometallic VPE growth of $GaAsAs_{1-y}Sb_y$ using trimethyl-antimony and $Ga_{1-x}In_xAs$ using trimethyl-arsenic', *Electron. Lett.*, **16**, 299–309, 1980.
16. J. P. Hirtz, J. P. Larivain, J. P. Duchemin, T. P. Pearsall, and M. Bonnet, 'Growth of $Ga_{0.47}In_{0.53}As$ on InP by low pressure MO-CVD', *Electron. Lett.*, **16**, 415–416, 1980.
17. J. P. Hirtz, Vung Bui Dinh, J. P. Duchemin, P. Hirtz, B. de Cremoux, and P. Merenda, 'A low beam divergence CW (GaAl)As double heterostructure laser grown by low-pressure metallorganic chemical vapor deposition process', *Appl. Phys. Lett.*, **36**, 795–796, 1980.
18. C. Cooper, R. Saxena, and M. Ludovik, 'OM-VPE growth of AlGaSb and AlGaAsSb', *Electron. Lett.*, **16**, 892–893, 1980.
19. J. Yoshino, T. Iwanot, and H. Kukimoto, 'Organometallic VPE growth of $In_{1-x}Ga_xP$', *J. Cryst. Growth*, **55**, 74–78, 1981.
20. J. P. Hirtz, J. P. Larivain, D. Leguen, M. Razeghi, and J. P. Duchemin, 'Low pressure metalorganic growth and characterization on $Ga_xIn_{1-x}As_yP_{1-y}$ on InP substrates', *Gallium Arsenide and Related Compounds 1980*, Conf. Ser. no. 56, Institute of Physics, Bristol, 1981, pp. 29–35.
21. T. Fukui, and Y. Horikoshi, 'InAsSbP–InAs superlattice grown by organometallic VPE method', *Jpn. J. Appl. Phys.*, **19**(9), L551–554, 1980.
22. M. Bonnet, N. Visentin, G. Bessonneau, and J. P. Suchemina, 'Comparison of FET performances versus material growth techniques', *J. Cryst. Growth*, **55**, 246–254, 1981.
23. J. Lacombe, J. P. Duchemin, M. Bonnet, and D. Huyghe, 'Schottky mixer diodes made by a new method', *Electron. Lett.*, **13**, 472–473, 1977.
24. J. Hallais, J. P. André, P. Bandit, and D. Boccan, 'New MESFET devices based on GaAs–(GaAl)As heterostructures grown by metallorganic VPE', *Gallium Arsenide and*

Related Compounds 1978, Conf. Ser. no. 45, Institute of Physics, Bristol, 1979, pp. 361–370.
25. R. Dupuis, P. Dapkus, R. Yurgling, and L. Moundy, 'Solar cells grown by metalorganic chemical vapor deposition', *Appl. Phys. Lett.*, **31**, 201–203, 1977.
26. R. Milano, T. Windhorn, E. Anderson, and G. E. Stillman, '$Al_{0.5}Ga_{0.5}As$–Ga heterojunction phototransistors grown by metalorganic chemical vapor deposition', *Appl. Phys. Lett.*, **34**, 562–565, 1979.
27. C. Riagit, J. Richard, M. Monnier, and P. Quittard, 'Materiaux III–V pour photocathodes fonctionnant en transmission', *Acta Electron.*, **26**, 303, 1977.
28. E. Thursh, and J. Whiteaway, 'Preparation and performance of GaAs (GaAl)As double heterostructures grown by the MO–CVD', *Gallium Arsenide and Related Compounds 1980*, Conf. Ser. no. 56, Institute of Physics, Bristol, 1981, pp. 337–346.
29. R. Dupuis and P. Dapkus, 'Room-temperature operation of distributed Bragg confinement $Ga_xAl_{1-x}As$–GaAs lasers grown by metallorganic chemical vapor deposition', *Appl. Phys. Lett.*, **33**, 68–70, 1978.
30. R. Dupuis, P. Dapkus, R. Kollas, and N. Holonyak, 'Quantum well $Al_xGa_{1-x}As$ heterostructures grown by metalorganic chemical vapor deposition', *IEEE J. Quantum Electron.*, **QE-15**, 756, 1979.
31. J. Coleman, and P. Dapkus, 'Single longitudinal-mode metalorganic chemical-vapor-deposition self-aligned GaAlAs–GaAs double heterostructures lasers', *Appl. Phys. Lett.*, **37**, 262–263, 1980.
32. H. Manasevit, K. Hess, D. Dapkus, R. Ruth, J. Yang, A. Comphall, R. Johanson, and L. Moudy, 'The properties of homoepitaxial InP films prepared by the MO-CVD process for the fabrication of heterojunction solar cells', *Proc. 3rd IEEE photovoltaic Specialists' Conf.*, IEEE, New York, 1978, pp. 165–173.
33. J. P. Hirtz, M. Razeghi, J. P. Larivain, S. Hersee, and J. P. Duchemin, 'Low threshold GaInAsP/InP lasers with good temperature dependence grown by low pressure MO-VPE', *Electron. Lett.*, **17**, 113–115, 1981.
34. J. P. Duchemin, M. Bonnet, and F. Koelsch, 'Kinetics of silicon growth under low hydrogen pressure', *J. Electrochem. Soc.*, **125**, 637–644, 1978.
35. S. Hersee, M. A. Di-Forte Poisson, M. Baldy, and J. P. Duchemin, 'A new approach to the "gettering" of oxygen during the growth of GaAlAs by low pressure MO-CVD', *J. Cryst. Growth*, **55**, 53–57, 1981.
36. J. P. Hirtz, J. P. Duchemin, T. P. Pearsall, P. Hirtz, and B. de Cremoux, 'Room temperature operation of a GaInAsP/InP double-heterostructures laser emitting at 1.15 μm grown by low-pressure metalorganic chemical vapor deposition', *Electron. Lett.*, **16**, 275–277, 1980.
37. P. Moutou, J. Chevrier, and A. Huber, 'Vapor epitaxial growth and characterization of InP multimeter-wave oscillators', *Gallium Arsenide and Related Compounds 1978*, Conf. Ser. no. 45, Institute of Physics, Bristol, 1979, pp. 452–461.
38. T. Fukui, and Y. Horikoshi, 'Properties of InP films grown by organometallic VPE method', *Jpn. J. Appl. Phys.*, **19**, 395–397, 1980.
39. R. Thomas, 'Growth of single crystal GaP from organometallic sources', *J. Electrochem. Soc.*, **116**, 1449–1450, 1969.
40. H: Renz, J. Weidlein, K. Benz, and M. Pilkuhn, 'InP epitaxy with a new metalorganic compound', *Electron. Lett.*, **16**, 228, 1979.
41. G. Stringfellow, and H. Hall, 'VPE growth of $Al_xGa_{1-x}As$', *J. Cryst. Growth*, **43**, 47–59, 1978.
42. E. Estop, A. Izrael, and M. Sauvage, 'Double crystal measurements of lattice parameters and x-ray topography in heterojunctions of GaAs–GaAlAs', *Acta Crystallogr.*, A, **32**, 627–632, 1976.

43. T. Pearsall, R. Bisaro, P. Merenda, G. Laurencin, R. Ansel, J. C. Portal, C. Houlbert, and M. Quillec, 'The characterization of $Ga_{0.47}In_{0.53}As$ grown lattice-matched on InP substrates', *Gallium Arsenide and Related Compounds 1978*, Conf. Ser. no. 45, Institute of Physics, Bristol, 1979, pp. 94–102.
44. T. Pearsall, G. Beuchet, J. P. Hirtz, N. Visentin, and M. Bonnet, 'Electron and hole mobilities in $Ga_{0.47}In_{0.53}As$', *Gallium Arsenide* and Related Compounds 1980, Conf. Ser. no. 56, Institute of Physics, Bristol, 1981, pp. 639–649.
45. K. Kajiyama, Y. Mizushima, and S. Sakata, 'Schottky barrier height of Schottky contacts to n-$In_xGa_{1-x}As$', *Electron. Lett.*, **14**, 738, 1978.
46. C. Foxon, B. Joyce, and M. Norres, 'Composition effects in the growth of Ga(In)AsP alloys by MBE', *J. Cryst. Growth*, **49**, 132–140, 1980.
47. T. Nakanisi, T. Udaqawa, A. Tanaka, and K. Kamei, *J. Cryst. Growth*, **55**, 255–262, 1981.
48. G. Thompson, and G. Henshall, 'Sensitivity of threshold in 1.27 μm (GaIn)(AsP)/InP DH lasers', *Electron. Lett.*, **16**, 42, 1980.

GaInAsP Alloy Semiconductors
Edited by T. P. Pearsall

Chapter 4

III–V Alloy Growth by Molecular-beam Epitaxy

COLIN E. C. WOOD

School of Electrical Engineering,
Cornell University, Ithaca, NY 14853, USA

4.1 INTRODUCTION

Progress demands a search for new, even faster and more versatile devices for high-speed digital, high-frequency microwave and integrated optical applications. In this search, development of device fabrication and systems technology is eventually limited by the fundamental properties of the materials from which a particular semiconductor device is made. Further advances then require alternative materials to be considered. As this book is a testament, the fundamental limits of binary III–V semiconductors are being approached to the extent that research and development are now seriously evaluating III–V ternary and quaternary alloys for many professional and strategic devices. The purpose of this chapter is not to corroborate the doubtless advantages of many ternary and quaternary alloys over binary compounds, but to identify and explain systematic considerations in alloy growth parameters that must be made to obtain the optimum properties by molecular-beam epitaxy (MBE). The extra degree of freedom (e.g. bandgap, lattice parameter, refractive index, etc.) that is allowed in (mostly) complete solid solutions (ternary or quaternary) of one III–V binary in another is not available without strings, however. It is just such advantages in versatility and the associated penalties that this chapter addresses.

4.2 EPITAXIAL GROWTH

4.2.1 Two Types of III–V Alloys

From a MBE point of view there are two distinct types of ternary or quaternary alloy.[1] The (close to) unity sticking and incorporation coefficients for the group III elements at typical growth temperatures,[2] etc., make III,III′–V[3] and

III,III′,III″–V^4 (type I) alloys easier to grow with predetermined composition by simple setting of relevant incident fluxes. On the other hand the non-unity incorporation coefficient of incident group V_4 fluxes[5] and the surface temperature (T_s) and absolute flux J_{V_4} dependences of their residence lifetimes[6] immediately allows one to identify a real problem in controlling the composition of III–V, V′[7–9] ternary III,III′–V,V′[10–12] or III–V,V′,V″[13] (type II) quaternary alloys.

To date, most published works on type I ternaries have centred around $Al_xGa_{1-x}As$ for use in double-heterostructure (DH) lasers[14] with GaAs active layers. Other applications of the matched GaAs/AlGaAs system are in (1) high-output impedance electron confinement FET buffer layers,[15] (2) two-dimensional electron gas-field effects[16–18] and devices[19–21] and (3) quantum-well laser structures.[22] The great interest in this alloy system from the extremely good lattice match of $Ga_xAl_{1-x}As$ alloys of any x to GaAs. Together with the direct–indirect bandgap E_g transition at about 40% Al, the approximately 80% ΔE_g discontinuity of E_c below 40% Al and the large differences in refractive indices, etc., make structures built in this alloy/compound regime both scientifically and technologically interesting.

There are two other ternary systems that form very close lattice-matched alloys across the complete composition range, i.e. $Al_xGa_{1-x}P$ and $Al_xGa_{1-x}Sb$. However, the bandgaps of the former are all indirect, giving rise to high effective masses and a necessity for phonon-assisted optical transitions. The latter have low melting points compared with the minimum substrate temperature (T_s) requirements, so there is little practical interest in either of these systems with the exception of some recent superlattice work on GaSb/AlSb[23] and GaSb/AlGaSb structures.[24]

An alternative cladding system for GaAs active layers that has received attention by MBE is GaInP.[25] However predictably from the temperature window considerations below, very poor-quality GaInP layers resulted using P_4 sources below about 400°C with marginal improvement using a P_2 source at a maximum (non-congruent decomposition-limited) T_s about 450°C.[25] More recently, an increasing interest in integrated optics using low-dispersion-loss fibres at 1.55 μm have led to several attempts to use GaInAs active layer DH lasers clad either by InP[26–28] or AlInAs.[29]

4.2.2 Substrate temperature

It is increasingly clear from Hall, deep-level transient capacitance and photoluminescence spectroscopy measurements that most III–V binary layers grown significantly below their respective congruent sublimation temperature (T_{cs}) are severely degraded.[30,31] For example GaAs grown below about 480°C is highly resistive,[30] has high concentrations of compensating deep levels,[32] and gives little, if any, near-band-edge photoluminescent intensity.[33] What little luminescence

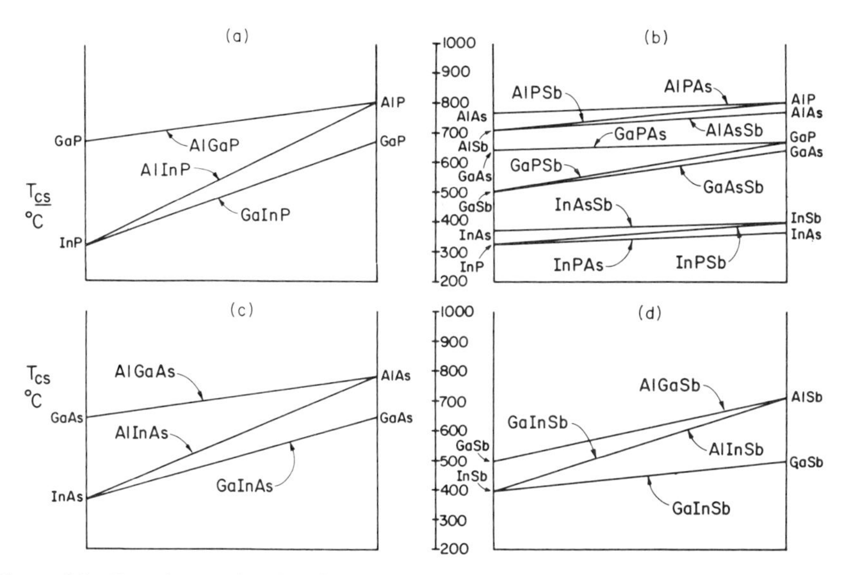

Figure 4.1 Vegard approximations for congruent sublimation temperatures for the type I ternary phosphides (a), arsenides (b), and antimonides (c), and type II ternary alloys (d)

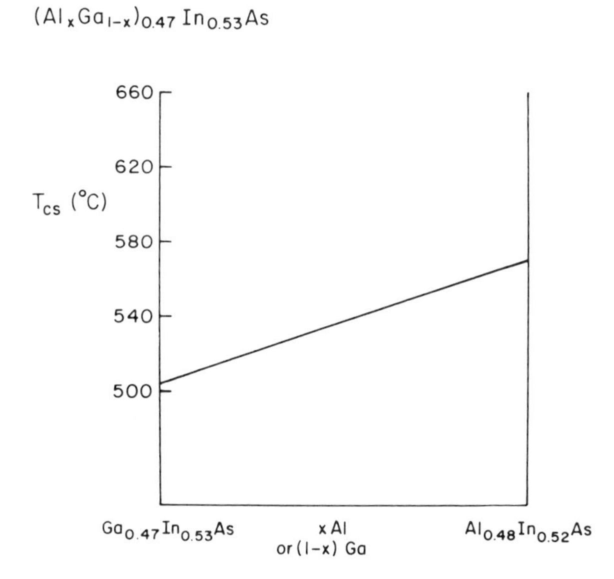

Figure 4.2 Vegard interpolation for congruent sublimation temperatures for $(AlGa)_{0.4}In_{0.53}As$

there is, is usually broad, and at energies significantly less than that of the forbidden energy gap, i.e. high concentrations of non-radiative recombination centres (deep levels).

From considerations above, the substrate temperature must be kept above certain limits which may be above the congruent sublimation temperature of the least stable binary compound even with high excess group V element fluxes. Foxon and Joyce[34] have indicated from modulated-beam experiments that the maximum temperature for alloy growth is set at or close to the congruent sublimation temperature of the least stable binary. Recent results in the author's laboratory have shown this to be essentially correct, and are conveniently described by Vegard's law interpolations between the bounding binary compounds. For example, during the growth of GaInAs–AlInAs multilayer structures, it was found that, for fixed Al,Ga,In and As_4 fluxes and with $J_{Ga} = J_{In}$, the temperature at which the surface reconstruction changed from a 'bulk' (1 × 1) pattern (arsenic-stabilized) to a (4 × 2) metal-stabilized pattern ($\sim T_{cs}$) was 520°C for $Ga_{0.47}In_{0.53}As$ and 565°C for $Al_{0.48}In_{0.52}As$.[35] This value for GaInAs fits very well between the T_{cs} for the end binaries, i.e. about 640°C[2] for GaAs and about 370°C[36] for InAs. The 520°C value is higher than the true T_{cs} as these layers are grown with approximately 50–100% excess As_4 flux. The value for AlInAs can also be used to predict a T_{cs} value for AlAs of 770–780°C (see Figure 4.1(a) to (d)). Similarly T_{cs} for $(GaAl)_{0.47}In_{0.53}As$ alloys is described by a Vegard treatment between the two type I ternaries (see Figure 4.2).

4.2.3 Background oxygen-containing gases and suboxide volatility

Recent mass spectrometric observations and thermodynamic calculations have indicated that one major culprit degrading the electrical properties of MBE GaAs is the volatile oxide Ga_2O, either from the gallium source[37] according to

$$Ga_2O_3 + 4Ga \rightarrow 3Ga_2O\uparrow \tag{4.1}$$

or, as the author believes, formed by reaction of the gallium adatom population on the growing film surface CO, etc., from the background vacuum,[38] thus

$$2Ga^s + CO^s \rightarrow Ga_2O^s + C \tag{4.2}$$

The obvious result of equation (4.2) is the production of free carbon which can then be included in the growing film as simple substitutional acceptors.[39] The result that MBE films almost always have strong C_{As} luminescence peaks and that Hall measurements of nominally undoped GaAs typically indicate residual p-type behaviour[2] with[1] $N_A - N_D \lesssim 10^{15}\ cm^{-3}$ further supports this generalization.

The volatility of Ga_2O is such[40] that for normal values of background CO pressure and arsenic flux, above 500°C ($P_{eq} \sim 10^{-9}$ Torr) most of this suboxide can be evaporated from the growing surface. However at lower values of T_s or higher values of surface arsenic concentrations (τJ_{As_4}) the reaction equilibrium higher values of surface arsenic concentrations (τJ_{As_4}), the reaction equilibrium for (4.3) (below) rapidly moves to the left

$$Ga_2O^s \rightleftharpoons Ga_2O^v \qquad (4.3)$$

where the final form of the oxygen in the grown GaAs is not known but is probably in the form of simple substitution at O atoms in arsenic sites. Neither the kinetic limitations nor the final form of such reactions are known. However, calculations have indicated that for normal MBE conditions, the formation of Ga_2O on the surface will exceed its desorption at temperatures at or below about 480°C for GaAs.[37] A similar series of reactions is possible, indeed probable, with indium and aluminium; however, the free energy for equivalent reactions (1) to (3) are very different.[41] For example, In_2O_3 is thermodynamically unstable above about 500°C whereas Al_2O_3 is well-known to be an exceptionally involatile and stable refractory oxide. Thus the position of the equilibrium constant K for equation (4.3) depends on ΔG according to

$$\Delta G = \Delta G_0 + RT \ln K \qquad (4.4)$$

where K is defined by

$$K_3 = \frac{P_{Ga_2O^v}}{[Ga_2O]^s} \qquad (4.5)$$

For low background CO pressures, Ga_2O^s is a direct function of the background CO pressure and an inverse function of its surface residence lifetime τ_{CO} (which is itself an exponentially decreasing function of surface temperature). Thus the dependence of incorporated oxygen concentration can be qualitatively described as

$$[O] \propto P_{CO} \tau_{CO} K_3^{-1} \qquad (4.6)$$

Table 4.1 Optimum conditions empirically found* or predicted† for some III–V binary compounds

Material	T_{cs} (°C)	$T_{max}(V_4)$ (°C)	$T_{min(V_4)}$ (°C)	$T_{max(V_2)}$ (°C)	T_{opt} (°C)	J_{V_4}/J_{III}
AlP	>700†	>700†	640†	⇄700†	~800†	~1.0/1†
GaP	670*	>700†	490†	~750†	>650†	~1.0/1†
InP	363*	~400*	>450†	>500†	>450*	~3.0/1*
AlAs	>750†	>800*	630†	⇄750†	>700*	~1.5/1*
GaAs	~620*	~680*	480*	>760*	~630*	~1.0/1*
InAs	~370*	~450*	~450†	~550†	~500†	~5.0/1†

T_{cs} = congruent sublimation temperature.
$T_{max(V_4)}$ = maximum practical substrate temperature using group V elemental sources (tetramers).
$T_{min(V_4)}$ = minimum practical substrate temperature set by competitive desorption of group III suboxide
$T_{max(V_2)}$ = maximum practical substrate temperature using dimer sources (typically more than 100°C higher than $T_{max(V_4)}$.
T_{opt} = optimum substrate temperature considering J_V, P_{m_2O}, interface diffusion, surface segregation, and preferential desorption.
J_{V_4}/J_m = group III flux ratio for optimum properties.

It is obvious that the incorporation of oxygen is more energetically favourable for AlAs and much less for InAs than GaAs at comparable values of T_s and J_{As_4}/J_{III} flux ratio.

To avoid the compensating effects of 'oxygen' incorporation and a high concentration of either simple or complexed group III vacancies, ternary and (even more so) quaternary alloys must therefore be grown in windows of T_s values set, at the upper limit, by the rate of non-congruent decomposition (see Table 4.1) and, at the lower limit, by the incorporation of group III vacancies and/or oxygen (via the most stable group III suboxide).

4.2.4 III/V flux ratio

In practice a $T_{max} \sim 50°C$ greater than T_{cs} can be employed if sufficiently high excess group V fluxes are used to balance the group V non-congruent sublimation overpressures.[35] Recent results of Panish[42] shows that the use of As_2 and P_2 beams stabilize GaAs and InP surfaces respectively up to approximately 150°C (Figure 4.3), higher than equivalent tetrameric arsenic or phosphorous fluxes, thus indicating lower excess group V_2 fluxes than V_4 fluxes above T_{cs}. Dimer sources have also been found helpful in reducing the group III vacancy-related trap densities and increasing the PL intensities of MBE $Ga_{0.47}In_{0.53}P$ grown on GaAs.[26] In cases where the group III elements are adjacent in the periodic table, e.g. Al,Ga or Ga,In, then the arsenides are relatively easily grown, e.g. two-dimensional, stoichiometric growth of device-quality $Al_{0.5}Ga_{0.5}As$ can be achieved up to and above 700°C[32] by using $J_{As_4}/(J_{Ga}+J_{Al}) \sim 2$. However, growth much below about 600°C leads to inferior electrical and optical properties

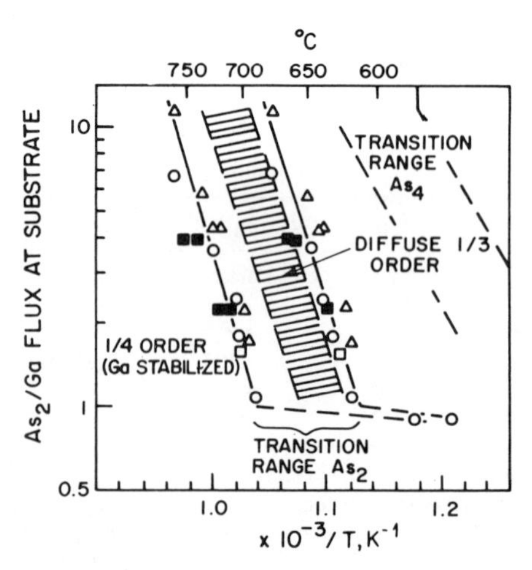

Figure 4.3 Arsenic dimer to gallium flux ratio dependence of As → Ga stabilized surface structures. Similar dependence is shown for elemental As_4 tetramer species. (After Panish[42]; reprinted by permission of the publisher, The Electrochemical Society, Inc.)

because of the Al(O) and/or V_{Al} problems at higher arsenic pressures, i.e. GaAs $T_{max} \simeq 600°C$, AlAs $T_{min} \simeq 630°C$.

Vegard's law interpolations between $T_{cs}^{GaAs}(620°C)$ and $T_{cs}^{InAs}(390°C)$ indicate that for $Ga_{0.47}In_{0.53}As$ $T_{cs} \sim 505°C$. Arsenic-stabilized growth up to 560°C has been achieved in the author's laboratory with high arsenic fluxes. However 520°C is normally used with $J_{As_4}/(J_{Ga}+J_{In}) \sim 5$ (see Figure 4.4). It appears then that above T_{cs} of the lowest stability binary components, the growing layer increasingly moves towards the group-III-rich side of the stoichiometric limit (i.e. toward LPE conditions). Increasing J_{As_4} is therefore needed to reduce arsenic vacancy-related complexes and deep (non-radiative) centres. On the other hand, an empirical value of $T_{cs} \sim 560°C$ was determined for $Al_{0.47}In_{0.53}As$ which is below T_{min} (630°C) for significant concentration of V_{Al} or O_{As} species in Al-containing III–V compounds by about 70°C. Indeed growth of AlInAs matched to InP produces electron trap densities in excess of 10^{17} cm^{-3} associated with very poor PL (see Figure 4.5(a)) unless very high J_{As_4} values are employed to allow increased T_s (Figure 4.5(b)).[43,44] Studies are on going at the author's laboratory to test the Panish[42] result and to improve the quality of this important ternary and the quaternary AlGaInAs alloy using an As_2 source.

To the author's knowledge, growth of AlGaP/GaP or AlInP/GaAs has not been attempted by MBE. From considerations discussed above and reference to Table 4.1, it can be confidently predicted that the AlGaP alloys will be very easy to prepare at temperatures in excess of 650°C with either dimer or tetramer sources,

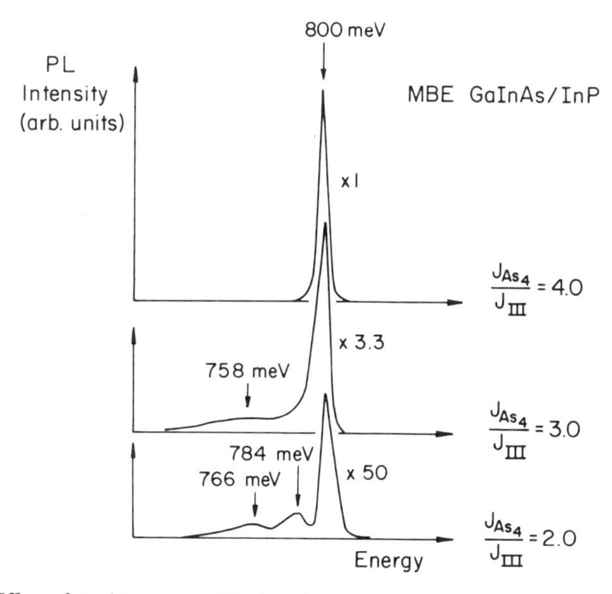

Figure 4.4 Effect of J_{As}/J_{Ga+In} on 4 K photoluminescence of $Ga_{0.47}In_{0.53}As$ grown above the congruent sublimation temperature

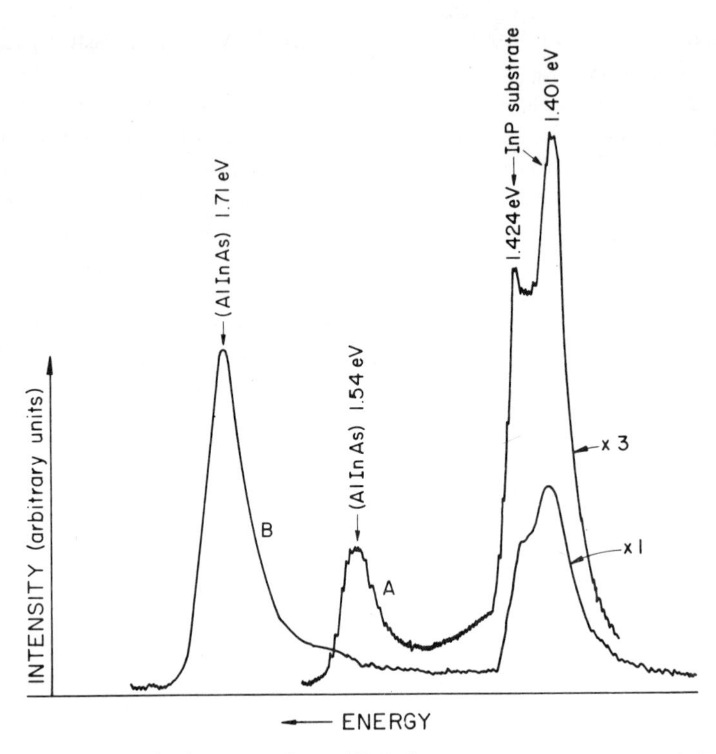

Figure 4.5 5 K photoluminescence of two AlInAs layers grown at (A) 520°C, and (B) 565°C

whereas it will probably be possible to grow device-quality AlInP only under extremely high P_2 fluxes at temperatures in excess of 580°C. The relatively low vapour pressure of free antimony makes successful growth of III, III′–Sb alloys conceivable without high Sb_2 or Sb_4 fluxes. More information on conditions for successful growth of the individual binary compounds AlSb and GaSb are needed before real predictions can be made for ternary or quaternary type I antimonides. The very low volatility of antimony can set practical lower limits to the acceptable T_s windows unless great care is taken to balance J_{Sb} to J_{III}. This new restriction is caused by the condensation of excess antimony flux as a second phase[45–47] which severely degrades or interrupts epitaxial growth.[47] There is some doubt about the completeness of solid solubility across the compositional range of AlInSb as well as AlPSb, GaPSb, and InPSb[48] which may cloud the issue (as well as layer surfaces) and could benefit from further study.

4.2.5 GaInAs

The aperiodic nature of atomic distributions in alloy lattices reduces the low-field carrier mobility significantly below that predicted from Vegard interpolations.[49]

This alloy (or disorder) scattering becomes more apparent at low carrier concentrations and at lower temperatures where ionized impurity and phonon scattering become less important respectively. Indeed, the highest mobilities obtained in the GaInAsP system have been achieved in one of the bounding ternaries GaInAs.[50] Unfortunately the ternary alloy will be lattice-matched to InP substrates at only one fixed composition, $Ga_{0.47}In_{0.53}As$. Despite the problematical InP substrate,[51] GaInAs is sufficiently attractive for optical detectors and microwave and high-speed digital device applications[50] that it has (after AlGaAs alloys) received most attention by the MBE community.[23–29, 51, 53] It is this, and the equivalent AlInAs alloy, that is discussed briefly below and at length in Chapter 17.

The growth of high-quality GaInAs on InP first requires that the InP be cleaned of its native oxide which can only be achieved thermally above 500°C. Using a flux of arsenic (first demonstrated by Calawa) it is possible to stabilize the InP[51] against non-congruent decomposition ($T_{cs} \sim 360$°C) by forming a surface

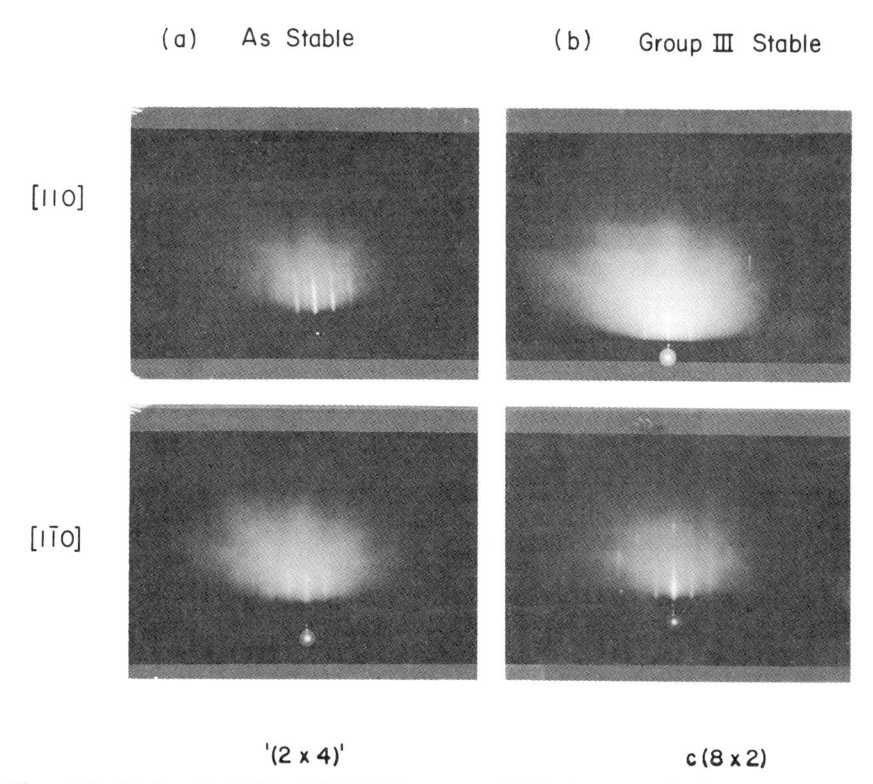

Figure 4.6 (a) Arsenic-stabilized 10 kV RED patterns of (AlGa)InAs along [110] (upper) and [110] (lower) directions. (b) Al(Ga)/In-stabilized 10 kV RED patterns of Al(Ga)InAs along [110] (upper) and [1$\bar{1}$0] (lower) directions

InAs layer. Initial nucleation of GaInAs growth then usually proceeds without degradation of (2 × 4) (Figure 4.6(a)) reflection electron diffraction (RED) patterns if carried out at about 480°C. T_s can subsequently be raised as high as 520°C under normal arsenic fluxes or up to about 560°C with the highest practical As_4 flux ($\sim 10^{16}$ cm^{-2} s^{-1}) before arsenic-deficient surface roughness (cloudiness) becomes apparent in association with a change of the RED pattern to C(8 × 2) (Figure 4.6(b)). J_{As}/J_{In+Ga} flux ratios of approximately 5/1 were found necessary[52] for the highest luminescent efficiency GaInAs (Figure 4.4) films (in direct contrast to GaAlAs alloys),[33] presumably because As vacancy-related defects (V_{As}) behave as non-radiative recombination centres more effectively in this alloy than simple or complex group III vacancies. To date, most nominally undoped GaInAs grown by MBE has had high (10^{15} cm^{-3}) to low (10^{16} cm^{-3}) residual donors[27–28, 44] with significantly lower μ_{77} values than best LPE or VPE layers. However, recently, μ_{77} values as high as the better LPE or VPE values have been achieved by MBE.[54] The source of residual donors is not known but is probably associated with residual tin from the In source or oxygen (from volatile In_2O) which may be a well-behaved shallow donor in GaInAs.

4.2.6 Type I quaternaries

As little as 1% Al in GaInAs alloys reduces PL efficiency by several orders of magnitude[4] when grown at or below 550°C. There is also an associated reduction in the background free-carrier concentration such that layers demonstrate semi-insulating properties. Intentionally added dopant above 10^{17} cm^{-3} does provide electrically active layers with improved PL intensity (see Figure 4.7). Despite the real restriction in lattice-matching considerations, there is currently great interest

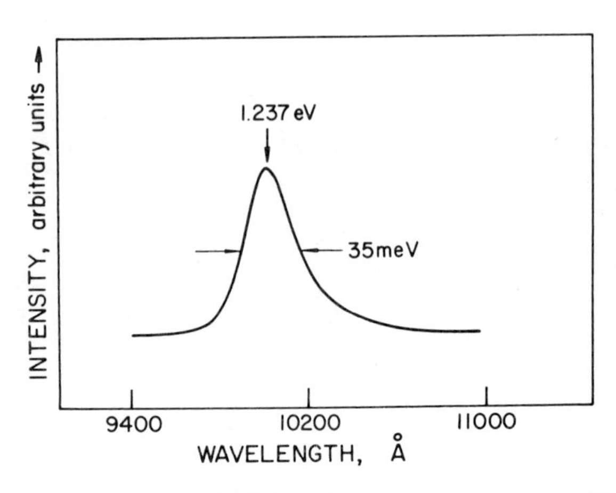

Figure 4.7 5 K photoluminescence of $(AlGa)_{0.47}In_{0.53}As$ layer grown on (100) InP substrate at 543°C

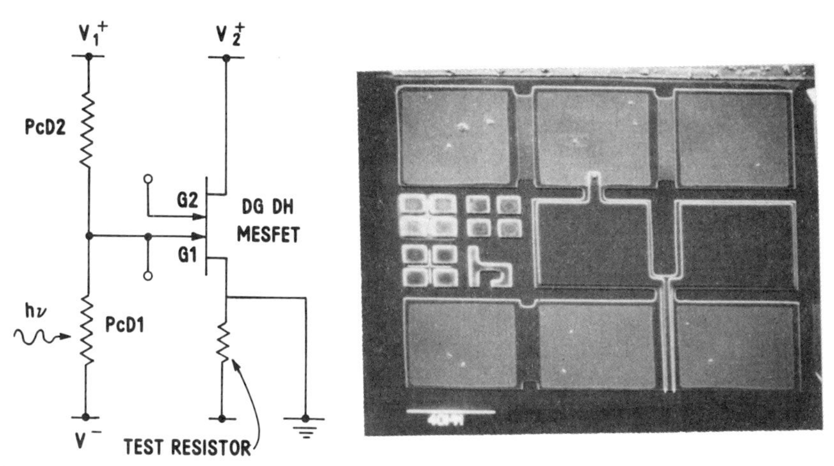

Figure 4.8 Double heterostructure (GaInAs/AlInAs) balanced OPFET–dual gate MESFET integrated photoreceiver for photoreception above 0.75 eV with external gain control

in growth of the $(Al,Ga)_{0.47}In_{0.53}As$ quaternary alloys as laser materials that can be used with GaInAs detectors and receivers in integrated optical applications (see Figure 4.8). However, until dimer sources are used in this quaternary $(Al,Ga)_{0.47}In_{0.53}$ As alloy and its end member $Al_{0.47}In_{0.53}As$, the practical excess fluxes of arsenic that are required for stabilization against surface decomposition above T_{min} ($\sim T_s > 580°C$) will in practice be very high.

Other type I quaternaries, i.e. Al, Ga, In phosphides and antimonides, are interesting by virtue of the extra degree of freedom in bandgap, by partial substitution of Al for Ga in GaIn–V ternary alloys, and the freedom from competitive group V accommodation problems. These quaternaries, however, do present the most difficult problem of type I alloy growth as they contain both Al and In which, from considerations above and Table 4.1, put demanding limits on the available growth parameter windows.

4.2.7 Type II ternaries

Of the possible type II ternary alloys only three have been successfully grown by MBE at the time of writing: GaAsP,[2,7] GaAsSb[8,9] and InAsSb.[53] All evidence suggests[34] no radical difficulty in epitaxial crystal growth. However, the incorporation coefficients of group V elements on growing III–V surfaces are less than unity under normal MBE conditions, usually requiring little more than twice[6] as many group V atoms than group III (if this were not the case then stoichiometric growth could not occur). There is also a wide dispersion between the individual equilibrium vapour pressures[55] and hence the surface residence lifetimes of the three important group V tetrameric species, P_4, As_4, and Sb_4. In addition, for

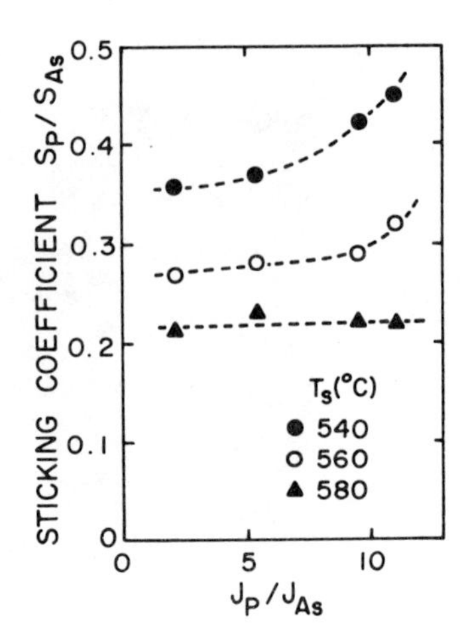

Figure 4.9 Arsenic/phosphorus flux ratio versus film composition for MBE growth of GaPAs. (After Neave and Joyce[7]; reproduced by permission of North-Holland Publishing Company)

each binary and especially for the ternary compounds, there is a temperature-dependent minimumJ_{V_4}/J_{III} ratio for stoichiometric crystal growth that is further complicated by an absolute J_{III} flux dependence[6] (see Figure 4.9). It can be seen then that successful growth of III–V,V′ ternaries with controlled compositions is further complicated (over type I ternaries) by competition for the group V sites (see Figure 4.9 and 4.10). In general it appears that the heavier group V_4 molecule will dominate owing to lower vapour pressure and hence its longer residence lifetime. Thus for type II alloys of known composition, it is necessary to find the exact J_V/J_{III} flux ratio (R) for stoichiometric growth of the heavier III–V binary at the required T_s and set the practical ratio to be X times

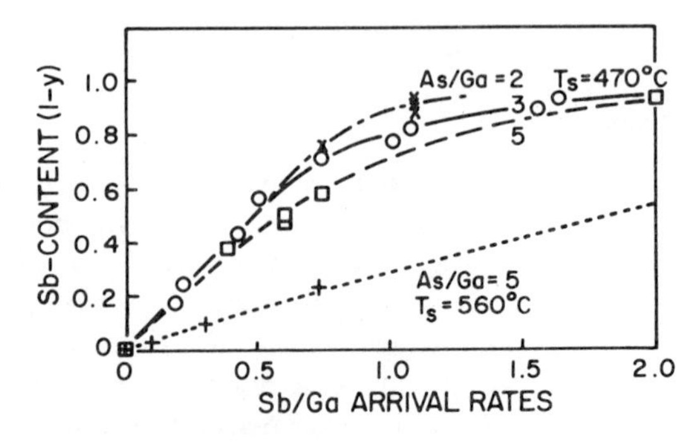

Figure 4.10 Antimony/arsenic flux ratio and temperature dependence of film composition

that (XR); and then to use a slight excess ($\rightarrow (1-X)R$) of the lighter group V_4 flux in order to maintain overall stoichiometry. It can now be appreciated that both group V_4 flux uniformities across the substrate, as well as their absolute stability with time, are vitally important, as are their spatial uniformity and absolute value of T_s.

4.2.8 Dimer sources

The greatly increased surface lifetimes (lower equilibrium pressures) of excess dimer species on III–V surfaces compared to tetramer species makes their use more attractive for both type I and type II alloys, by simplifying the surface competition reaction for the available group V sites. The use of P_2 and As_2 beams generated from compound sources,[2] e.g. InP or GaAs, typically have associated contributions of the group III element and cannot easily be used to grow type I alloys with accurately controlled composition. However, significantly improved optical and electrical properties have been demonstrated in certain cases.[26] In practice, quaternary alloys of the GaInAsP system can be lattice matched to GaAs and InP substrates and have been successfully grown by liquid-phase epitaxy for optical applications. There have been three reports[10–12] of growth of alloys in this system by MBE. However, poor compositional control was reported except where thermally cracked gas sources ($AsH_3 + PH_3$) were used for the group V elements.[12]

4.2.9 Type II quaternaries

To the author's knowledge no work has been reported or is even being considered to grow Al or Ga or In–PAsSb quaternary compounds by MBE (perhaps as well).

4.2.10 Spatial uniformity

The off-axis intensity distributions from practical effusion cells are complex functions of many geometric and thermal characteristics as well as cell charge state (liquid, solid) and quantity. Control and increase of spatial uniformity of effusion cell fluxes has improved markedly in recent years but is still not much better than about 10% over 1 inch diameter for most practical and commercial MBE systems. This has been amply demonstrated by measuring thickness and doping level variations for MBE-grown binary compounds.[56] When ternary and (even more so) quaternary alloys with good uniformity ($< 1\%$) are required, concentric cells such as used for GaInAs[57–59] and continuous azimuthal substrate rotation[60] during growth must be considered. Mixing of components (Ga + In) in one cell has been used[1,28] but is a temporal solution for group III elements only, because the source composition changes as the more volatile element is depleted.

4.2.11 Depth uniformity

Even with continuous substrate rotation, a further practical problem has to be considered if control of depth uniformity is to be obtained. As most cell temperatures are controlled from a sensing thermocouple at the back of the furnaces, opening a shutter in close proximity to the cell causes an increase in radiational heat loss from the front of the cell (as the shutter acts as a heat shield or reflector). There are short thermal transients depending roughly on T^4 and on the thermal mass, etc., of the cell, and on permanent changes in the effective temperature of the effusing material within the cell as the power input, and the thermal profile of the cell, change to accommodate the new heat loss. Attempts to control abrupt changes in dopant or composition accurately by opening and closing cells should consider this, as transient effects will occur. In addition, calibration of flux monitors[61] should be carried out only after allowing sufficient time to restabilize thermal equilibrium, giving rise to depth variations of concentration close to interfaces. In addition to individual cell shutters, a plea is made here for a main shutter between, and remote from, cells and substrate to be incuded in future commercial machines.

A word of caution is necessary on the subject of alloy composition determination. Despite the many indirect techniques for determining the composition of a particular alloy, such as photoluminescence, X-ray fluorescence, X-ray rocking curves, and Auger electron spectroscopy, the author has found only one relatively error-free calibration technique, and then only for type I alloys. This involves the growth of thick ($\gtrsim 2\,\mu m$) layers of the individual binary compounds, having separately measured the individual group III fluxes using an ion gauge in the beam paths (making allowance for background pressure contributions), measuring the thickness by optical or scanning microscopy of a cleaved edge, and, if necessary, staining the interfaces. A knowledge of the respective lattice parameters then allows calculation of alloy composition. The percentage of Al in $Al_xGa_yIn_{1-x-y}As$ that would be grown can be calculated from

$$x = \frac{d_{AlAs}}{a_{AlAs}} \left(\frac{d_{AlAs}}{a_{AlAs}} + \frac{d_{GaAs}}{a_{GaAs}} + \frac{d_{InAs}}{a_{InAs}} \right)^{-1} \tag{4.7}$$

where d_{AlAs}, d_{GaAs}, and d_{InAs} are layer growth rates of AlAs, GaAs, and InAs respectively, and a_{AlAs}, a_{GaAs}, and a_{InAs} are the lattice parameters of AlAs, GaAs, and InAs respectively.

There are three further complications that need consideration, especially when the high substrate temperatures advocated in this chapter are used in type I alloy growth.

4.2.12 Heterojunction interdiffusion

Exchange of Al for Ga and/or In or exchange of Ga for In has been observed when the elements Al or Ga are deposited on III–V compound surfaces.[62] Such

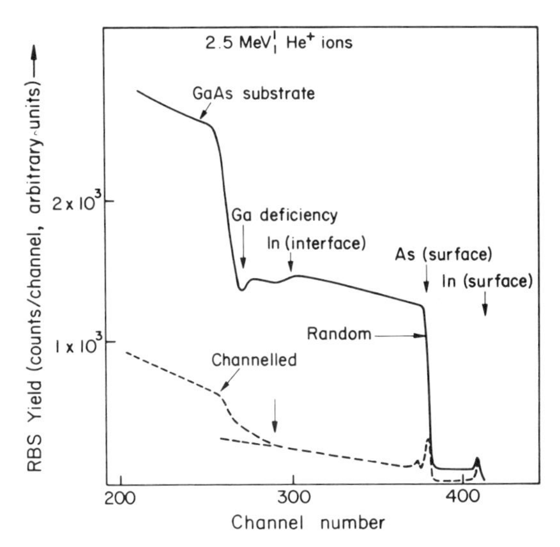

Figure 4.11 2.5 MeV He^+ ion dechannelling yield spectrum of epitaxial GaInAs/InP substrate interface showing the dip in the indium composition caused by exchange with gallium upon nucleation of the film

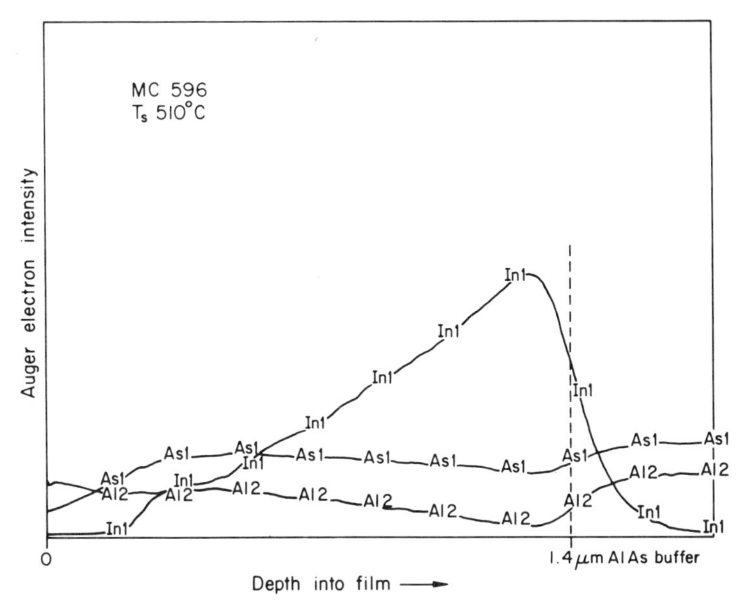

Figure 4.12 Sputter/Auger electron spectral intensity profile for stepped composition (AlGa)InAs grown above 500°C. (Note the virtually complete interdiffusion.)

energetically favourable exchange reactions can also occur upon nucleation of, for example, AlInAs or GaInAs (see Figure 4.11). This process is believed to be partially responsible for the differences between two-dimensional electron gas mobilities using modulation-doped interfaces of GaAs/AlGaAs compared with AlGaAs/GaAs.[63] This interdiffusion can be reduced under higher arsenic fluxes; however, it is doubtful that it can ever be completely suppressed. There is also significant evidence for enhanced interdiffusion when large mismatch exists between layer and substrate[44] (Figure 4.12).

4.2.13 Surface accumulation

Surface accumulation of the group III element of the least stable binary has been observed by varying substrate temperature during growth of GaInAs. Again this can be suppressed by increasing the arsenic flux, as it is a manifestation of the decompostion of InAs and 'floating' of the In on the growing surface.

4.2.14 Congruent desorption

The congruent desorption pressure of In from InAs is higher than Ga from GaAs which is in turn higher than Al from AlAs. The desorption of In from GaIn–V or AlIn–V or AlGaIn–V alloys and the desorption of Ga from AlGa–V alloys can therefore be sufficiently higher than the other group III elements that the resultant proportion of the more volatile element can be significantly lower than that predicted from the incident $J_{III}/J_{III'}$ flux ratio (see Figure 4.13). This deviation increases with T_s and with decreasing alloy growth rate.

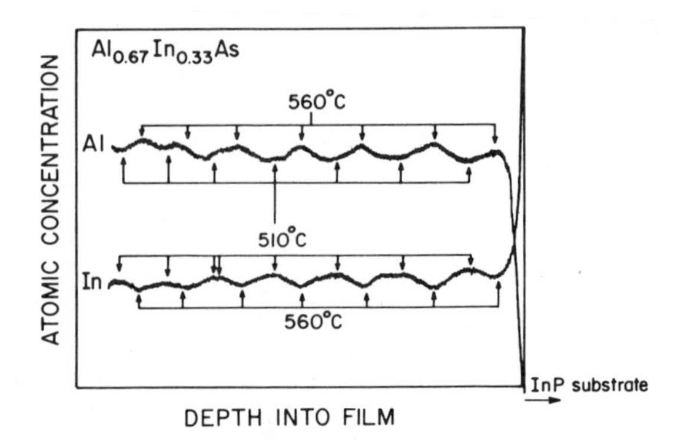

Figure 4.13 Sputter/Auger electron intensity profile of an AlInAs layer growth with 7 × 2 min periods at $T_s \sim 565°C$ and 7 × 2 min periods at 510°C. Note the reduction of In signal at the high T_s

4.3 SUMMARY

In summary, an attempt is here made to define conditions for successful epitaxial growth of ternary and quaternary III–V alloys. General problems associated with background-gas incorporation, surface accumulation, competitive incorporation, preferential and non-congruent sublimation, interface exchange, and strain-induced interdiffusion are identified by example. The conclusions are encouraging, and there is every reason to believe that most if not all III–V alloys can be successfully grown with device quality by MBE providing that the various problems mentioned herein are accounted for.

Note added in proof

Since the completion of this manuscript, evidence for significant bowing of T_{cs} with composition has been obtained for the AlInAs and GaInAs alloy systems. The full extent of these findings is not yet quantified however the reader is cautioned to treat the relationships of Fig. 4.1 as only indications. Further work is continuing in this materials system and will be published elsewhere.

The absoulte values of T_{cs} for the alloys are not exactly known, however the discussion of the temperature windows for successful optical and electronic quality III–V alloys still holds true.

REFERENCES

1. Colin E. C. Wood, 'Progress, problems and applications of MBE', *Phys. Thin Films*, **11**, 35–103, 1980.
2. A. Y. Cho and J. R. Arthur, *Prog. Solid State Chem.*, **10**, 157, 1975.
3. A. Y. Cho, M. B. Panish, and I. D. Hayashi, 'Molecular beam epitaxy of GaAs, AlGaAs and GaP', *Gallium Arsenide and Related Compounds, 1970*, Conf. Ser. No. 9, Institute of Physics, Bristol, 1971, pp. 18–29.
4. C. E. C. Wood, G. W. Wicks, H. Ohno, and L. F. Eastman, 'MBE AlGaInAs for optical applications', *Appl. Phys. Lett.*, submitted.
5. J. R. Arthur, 'Surface stoichiometry and structure of (100) GaAs', *Surf. Sci.*, **43**, 449–461, 1974.
6. J. H. Neave and B. A. Joyce, 'Structure and stoichiometry of (100) GaAs surfaces during molecular beam epitaxy', *J. Cryst. Growth*, **44**, 387–397, 1978.
7. Y. Matsushima and S. Gonda, 'Molecular beam epitaxy of GaP and GaAsP', *Jpn. J. Appl. Phys.*, **15**, 2093–2097, 1976.
8. C. A. Chang, R. Ludeke, L. L. Chang, and L. Esaki, 'Molecular beam epitaxy (MBE) of InGaAs and GaSbAs', *Appl. Phys. Lett.*, **31**, 759–761, 1977.
9. A. Y. Cho, H. C. Casey, and P. W. Foy, 'Back surface emitting GaAsSb LEDs ($\lambda = 1.0\,\mu m$) prepared by molecular beam epitaxy', *Appl. Phys. Lett.*, **30**, 397–399, 1977.
10. A. Y. Cho, 'Recent developments in molecular beam epitaxy', *J. Vac. Sci. technol.*, **16**, 275–284, 1979.
11. D. W. Covington and W. Cox, *Workshop on Compound Semiconductors for Active Microwave Devices*, San Francisco, CA, January 1980; and private communication, May 1981.
12. A. R. Calawa, private communication, 1981.

13. It is unlikely that alloys in these systems will ever be of sufficient interest to be studied seriously.
14. W. T. Tsang, 'High-through-put, high-yield and highly-reproducible (AlGa)As double heterostructure laser wafer grown by molecular beam epitaxy', *Appl. Phys. Lett.*, **38**, 587–589, 1981.
15. W. I. Wang, S. Judaprawira, C. E. C. Wood, and L. F. Eastman, 'Molecular beam epitaxial GaAs–$Al_xGa_{1-x}As$ heterostructures for metal semiconductor field effect transistor applications', *Appl. Phys. Lett.*, **38**, 708–710, 1981.
16. R. Dingle, H. L. Stormer, A. C. Gossard, and W. Wiegman, 'Electron mobilities in modulation doped semiconductor heterojunction superlattices', *Appl. Phys. Lett.*, **33**, 665–667, 1978.
17. S. Hyamizu, T. Mimura, T. fujii, and K. Nanbu, 'High mobility of two-dimensional electrons at the GaAs/n-AlGaAs heterojunction interface', *Appl. Phys. Lett.*, **37**, 805–807, 1980.
18. W. I. Wang, C. E. C. Wood, and L. F. Eastman, 'Extremely high electron mobilities in modulation doped GaAs–AlGaAs heterojunction superlattices', *Electron. Lett.*, **17**, 36–37, 1981.
19. S. Judaprawira, W. I. Wang, P. C. Chao, C. E. C. Wood, D. W. Woodard, and L. F. Eastman, 'Modulation doped MBE GaAs–GaAlAs MESFETs', *IEEE Electron Devices Lett.*, **EDL-2**, 14–15, 1981.
20. P. Deslescluses and N. T. Linh, 'Modulation doped MBE FET structures', *Workshop on Compound Semiconductors for Active Microwave Devices*, New Orleans, 1980.
21. T. Mimura, S. Hyamizu, T. Fujii, and K. Nanbu, *Jpn. J. Appl. Phys.*, **19**, L225–L227, 1980.
22. R. C. Miller, R. Dingle, A. C. Gossard, R. A. Logan, W. A. Nordland, Jr, and W. Wiegman, 'Laser oscillation with optically-pumped, very thin GaAs/AlGaAs multilayer structures and conventional double heterostructures', *J. Appl. Phys.*, **47**, 4509–4517, 1976.
23. Y. Suzuki, M. Naganuma, and H. Okamoto, *Proc. Annu. Fall Meet. Jpn. Appl. Phys. Soc.*, **29p**(M15) 537, 1980.
24. M. Naganuma, Y. Suzuki, and H. Okamoto, *Proc. Annu. Fall Meet. Jpn. Appl. Phys. Soc.*, **29p**(M14) 537, 1980.
25. G. B. Scott and J. S. Roberts, *Gallium arsenide and Related Compounds 1978*, Conf. Ser. no. 45, Institute of Physics, Bristol, 1979, pp. 181–189.
J. S. Roberts and C. E. C. Wood, 'Growth of GaInPAs/GaAs structures by MBE', unpublished.
26. H. Asahi, Y. Kawamura, M. Ikeda, and H. Okamoto, 'Molecular beam epitaxial growth of InP homo epitaxial layers and their electrical and optical properties', *Jpn. J. Appl. Phys.*, **20m**, L181–L183, 1981.
27. B. I. Miller, J. H. McFee, R. J. Martin, and P. K. Tien, 'Room temperature operation of lattice matched InP/GaInAs/InP double heterostructure lasers grown by MBE', *Appl. Phys. Lett.*, **33**, 44–46, 1978.
28. B. I. Miller and J. H. McFee, 'Growth of GaInAs/InP heterostructures by MBE', *J. Electrochem. Soc.*, **125**, 1310–1314, 1978.
29. G. W. Wicks, J. Barnard, and C. E. C. Wood, 'AlInAs clad GaInAs laser structures by MBE', to be published.
30. C. E. C. Wood, J. Woodcock, and J. J. Harris, 'Low compensation n-type and flat surface p-type Ge-doped GaAs by MBE', *Gallium Arsenide and Related Compounds* 1978, Conf. Ser. no. 45, Institute of Physics, Bristol, 1979, pp. 28–37.
31. T. Murotani, T. J. Shimanoe, and S. Mitsui, 'Growth temperature dependence in molecular beam epitaxy of GaAs', *J. Cryst. Growth*, **45**, 302–308, 1978.
32. R. A. Stall, C. E. C. Wood, P. D. Kirchner, and L. F. Eastman, 'Growth parameter

dependence of deep levels in MBE GaAs', *Electron. Lett.*, **16**, 171–172, 1980.
33. G. W. Wicks, W. I. Wang, C. E. C. Wood, and L. F. Eastman, 'Photoluminescence of AlGaAs grown by MBE', *J. Appl. Phys.*, **52**, 5792–5796, 1981.
34. C. T. Foxon and B. A. Joyce, 'Composition effects in the growth of Ga(In)AsP alloys by MBE', *J. cryst. Growth*, **49**, 132–140, 1980. **Erratum**: **50**, 774–775, 1980.
35. C. E. C. Wood, and G. W. Wicks, 'A unified pseudobinary approach to III–V alloy growth by MBE', Gallium Arsenide Conf. Ser. no. 63, *J. Appl. Phys.*, to be published.
36. B. T. Meggit, E. H. C. Parker, and R. M. King, 'Thin InAs epitaxial layers grown on (100) GaAs substrates by molecular beam deposition', *Appl. Phys. Lett.*, **33**, 528–530, 1978.
37. P. D. Kirchner, J. M. Woodal, J. L. Freeouf, and G. D. Pettit, 'Volatile metal oxide incorporation in layers of GaAs, GaAlAs and related compounds grown by molecular beam epitaxy', *Appl. Phys. Lett.*, **38**, 427–429, 1981.
38. G. B. Stringfellow, R. A. Stall, and W. Koschel, 'Carbon contamination in MBE GaAs', *Appl. Phys. Lett.*, **38**, 156–157, 1981.
39. H. Kunzel, A. Fisher, and K. Ploog, 'Quantitative evaluation of substrate temperature dependence of Ge in corporation in GaAs during molecular beam epitaxy', *Appl. Phys.*, **22**, 23–30, 1980.
40. C. N. Cochran and L. M. Foster, 'Vapour pressure of gallium, stabilities of gallium suboxide vapour and equilibrium of some reactions producing Ga_2O vapour', *J. Electrochem. Soc.*, **109**, 145, 1962.
41. G. B. Stringfellow, 'Calculation of ternary phase diagrams of III–V systems', *J. Phys. Chem. Solids*, **33**, 665–677, 1972.
42. M. B. Panish, 'Molecular beam epitaxy of GaAs and InP with gas sources for As and P', *J. Electrochem. Soc.*, **127**, 2729–2733, 1980.
43. C. E. C. Wood, H. Ohno, G. W. Wicks, and L. F. Eastman, 'MBE ternary arsenides for μ-wave applications', *2nd Int. MBE Workshop*, Cornell University, October 1980.
44. H. Ohno, C. E. C. Wood, L. Rathbun, D. V. Morgan, G. W. Wicks, and L. F. Eastman, 'GaInAs/AlInAs structures grown by MBE', *J. Appl. Phys.*, **52**, 4033–4037, 1981.
45. S. Baba, H. Horita, and A. Kinbara, 'Synthesis of stoichiometric InSb thin films by a simple molecular beam technique', *J. Appl. Phys.*, **49**, 3632–3635, 1978.
46. K. Oe, S. Ando, and Sugiyama, 'RHEED study of InSb films grown by molecular beam epitaxy', *Jpn. J. Appl. Phys.*, **19**, L417–L420, 1980.
47. A. Noreika, M. Francombe, and C. E. C. Wood, 'Growth of Sb and InSb by molecular beam epitaxy',*J. Appl. Phys.*, **52**, 7416–7420, 1981.
48. L. M. Foster, 'A lattice parameter criterion for miscibility gaps in the III–V and II–VI pseudo-binary solid solutions', *J. Electrochem. Soc.*, **121**, 1665, 1974.
49. M. A. Littlejohn, L. A. Arledge, T. H. Glisson, and J. R. Hauser, *Electron. Lett.*, **15**, 586–587, 1979.
J. H. Marsh, P. A. Houston, and P. N. Robson, 'Compositional dependence of the mobility, peak velocity and threshold field in InGaAsP', *Gallium Arsenide and Related Compounds* 1980, Conf. Ser. no. 56, Institute of Physics, Bristol, 1981, pp. 621–630.
50. T. P. Pearsall, R. Bisaro, P. Merenda, G. Laurencin, R. Ansel, J. C. Portal, C. Houlbert, and M. Quillec, 'The characterization of $Ga_{0.47}In_{0.53}As$ grown lattice-matched on InP substrates', *Gallium Arsenide and Related Compounds* 1978, Conf. Ser. no. 45, Institute of Physics, Bristol, 1979, pp. 94–102.
J. D. Oliver and L. F. Eastman, 'Liquid-phase epitaxial growth and characterization of high purity lattice-matched GaInAs on (iii) InP', *J. Electron.*, **9**, 693–712, 1980.
51. G. J. Davies, R. Heckingbottom, H. Ohno, C. E. C. Wood, and A. R. Calawa, 'Arsenic stabilization of InP substrates for growth of GaInAs by MBE', *Appl. Phys. Lett.*, **37**, 290–292, 1980.
52. G. W. Wicks, C. E. C. Wood, and H. Ohno, 'Optical device quality GaInAs grown by

molecular beam epitaxy', *J. Electron. Mat.*, 1982.
53. A. Noreika, M. Francombe, and C. E. C. Wood, 'Bi incorporation in MBE for long wavelength (12.14 μm) detectors', *Appl. Phys. Lett.*, submitted, 1981.
54. K. Y. Cheng, A. Y. Cho, and W. R. Wagner, 'Molecular beam epitaxial growth of uniform $Ga_{0.47}In_{0.53}As$ with a rotating sample holder', *Appl. Phys. Lett.*, **39**, 607–609, 1981.
55. R. E. Honig and D. A. Kramer, 'Vapour pressure data for the solid and liquid elements', *RCA Rev.*, **30**, 285–305, 1969.
56. Y. C. Chai, R. Chow, and C. E. C. Wood, 'The effect of growth conditions of Si incorporation in MBE GaAs', *Appl. Phys. Lett.*, **39**, 800–803, 1981.
57. H. Ohno, J. Barnard, C. E. C. Wood, and L. F. Eastman, 'Double heterostructure GaInAs MESFETs by MBE', *IEEE Electron Devices Lett.*, **EDL-1**, 154–155, 1980. J. Barnard, H. Ohno, C. E. C. Wood, and L. F. Eastman, 'Double heterostructure GaInAs MESFETs with submicron gates', *IEEE Electron Devices Lett.*, **EDL-1**, 174–176, 1980.
58. J. Barnard, H. Ohno, C. E. C. Wood, and L. F. Eastman, 'Integrated double heterostructure GaInAs photoreceiver with automatic gain control', *IEEE Electron Devices Lett.*, **EDL-2**, 7–9, 1981.
58. K. Y. Cheng, A. Y. Cho, W. R. Wagner, and W. A. Bonner, 'Molecular beam epitaxial growth of uniform InGaAs on InP with a coaxial In–Ga oven', *J. appl. Phys.*, **52**, 1015–1021, 1981.
60. A. Y. Cho and K. Y. Cheng, 'Growth of extremely uniform layers by rotating substrate holder with MBE for electro-optical and microwave devices', *Appl. Phys. Lett.*, **38**, 360–362, 1981.
61. C. E. C. Wood, 'Surface exchange doping of MBE GaAs from S and Se captive sources', *Appl. Phys. Lett.*, **33**, 770–772, 1978.
62. J. R. Waldrop and R. W. Grant, 'Chemistry of Schottky barrier formation on GaAs', Paper presented at *21st Electronic Materials Conf.*, Boulder, CO, 27–29 June, 1979.
63. C. E. C. Wood, W. I. Wang, G. W. Wicks, and L. F. Eastman, 'MBE GaAs/AlGaAs heterostructures for MESFET's application', *Workshop on Compound Semiconductors for Active Microwave Devices*, New Orleans, 1981.

GaInAsP Alloy Semiconductors
Edited by T. P. Pearsall

Chapter 5

Ion Implantation

F. H. EISEN AND L. R. TOMASETTA

Rockwell International Microelectronics Research and Development Center, Thousand Oaks, CA 91360, USA

5.1 INTRODUCTION

Ion implantation has been used in the fabrication of a variety of devices in III–V compounds and their alloys. It has been most widely studied and applied in GaAs, although limited studies of ion implantation in several other compounds have been carried out. The use of implantation for both n- and p-type doping in GaInAsP has been reported.[1, 2] The only published implantation studies in GaInAsP, involving characteristic such as dopant activation or carrier concentration profiles and/or the effects of varying annealing temperatures, are for the use of Be as a p-type dopant. In view of the small amount of work directly in GaInAsP, in addition to discussing the work which has been carried out in that material, some general features of implantation in III–V compounds will be reviewed in this chapter and attempts will be made to extrapolate these to application in GaInAsP. More complete reviews of implantation in III–V compounds may be found in references 3 to 5. Proton bombardment isolation and a recent application of Be implantation in GaInAsP to the fabrication of avalanche photodiodes will also be discussed.

5.2 GENERAL FEATURES OF IMPLANTATION IN III–V COMPOUNDS

Before beginning the review of specific results of ion implantation doping of III–V compounds, it is of interest to discuss some distinct features of implantation in these materials such as the volatility of the group V constituents during annealing and the necessity of performing some implantations at elevated temperature. In III–V compounds such as GaAs or InP, substantial vaporization of the As or P can occur at the temperatures required for post-implantation annealing. This dissociation may be prevented or minimized by encapsulating the semiconductor wafer with a suitable material, or by maintaining an overpressure of the volatile constituent at a level which prevents appreciable loss of that constituent from the wafer during annealing. Si_3N_4,[6] SiO_2,[7] AlN,[8] and Al have all been used as encapsulants on GaAs. The activation of high doses of the group V n-type dopants (S, Se, and Te) in GaAs has been found to be strongly affected by the choice and method of deposition of the encapsulating material.[4] For InP the use of SiO_2,[10] phosphosilicate glasses,[11] and Si_3N_4[12] as encapsulants has been reported and SiO_2, Si_3N_4, and Ga-saturated SiO_2 have been used for n-type implants in InAsP.[13] The n-type doping of InAsP seems less dependent on the nature of the encapsulant than is the case for GaAs.[13] It may be that a similar insensitivity of n-type implantation results to encapsulating material will also be observed for GaInAsP.

Several methods of maintaining a suitable overpressure of the volatile constituent of compound semiconductors during annealing have been employed. These include the use of systems similar to vapour epitaxial reactors which provide the required overpressure,[14] inclusion of solid materials within the heated volume to provide the overpressure,[15] and placing two or more wafers of the same material in close proximity,[15, 16] sometimes with an overpressure of the volatile constituent provided by one of the above-mentioned methods. These techniques seem to give generally acceptable results, so that it is often possible to avoid the problem of depositing a material which will remain on the semiconductor wafers during annealing and have suitable encapsulant properties. GaInAsP implanted with Be has been successfully annealed by sandwiching the implanted wafer between two wafers of InP.[2]

The temperature required for post-implantation annealing varies somewhat with the particular III–V material. Ion-implanted GaAs is usually annealed at temperatures between 800 and 900°C, although both higher and lower temperatures have been used in some special cases.[3, 4] Annealing of InP and InAsP has been successfully carried out at 700 to 750°C.[4, 13] The Be-implanted GaInAsP mentioned in the preceding paragraph was annealed at 560°C.

In order to obtain good activation of high doses of group V dopants implanted into GaAs, it is necessary to carry out the implantation at a temperature of 150°C or higher.[4] This requirement for an elevated implantation temperature is probably associated with the fact that, when implantations are performed at temperatures below 150°C, an amorphous layer is often produced, whereas no amorphous layer is formed when the implantation temperature is above about 150°C.[17] Implantation at high temperatures is not required, however, for the successful activation of high doses of Si or p-type (group III) dopants[4] in GaAs. It is not clear at present whether the same behaviour would be observed for group V dopants in other III–V compounds or alloys such as GaInAsP. However, there is some indication that the regrowth of amorphous layers on GaAs is similar to that of amorphous layers on InP,[18] suggesting that high-temperature implants may well be necessary for high doses of group V dopants in InP also. It should be possible, however, to employ implantation doping in GaInAsP for a wide variety of purposes using Si as an n-type dopant and Be as a p-type dopant without the requirement for high-temperature implantations.

5.3 N-TYPE DOPING

Before beginning the discussion of specific n-type dopant effects, it is useful briefly to introduce some information concerning ion ranges. The range distribution of implanted ions can be approximated by a Gaussian distribution:[19]

$$n(x) = \frac{N_d}{\Delta R_p (2\pi)^{1/2}} \exp\left(-\frac{(x - R_p)^2}{2\Delta R_p^{\,2}} \right) \tag{5.1}$$

where $n(x)$ is the dopant concentration as a function of the distance x measured along the incident beam direction, N_d is the dose in ions cm^{-2}, R_p is the projected range of the implanted ions, and ΔR_p is the range straggling. Values of R_p and ΔR_p can be estimated theoretically using the LSS range theory.[19] Tables of R_p, ΔR_p, and higher moments of the range distribution have been published by several authors.[20, 21] The projected range and range straggling of n-type dopants implanted in GaAs at 400 keV, which is a common maximum energy for implanters in use for compound semiconductor work, are listed in Table 5.1. The corresponding values for InAs, GaP, and InP lie within about $\pm 15\%$ of those for GaAs. The numbers in this table can be used to estimate the approximate maximum n-type layer thickness that can be achieved with various n-type dopants. This thickness may be taken as approximately the projected range plus twice the range straggling. The maximum concentration achieved in a given implantation is related to the implanted dose by:

$$n(R_p) = \frac{0.4\, N_d}{\Delta R_p} \tag{5.2}$$

The above-mentioned maximum layer thickness and the maximum dopant concentration values are only valid if there is little or no diffusion of the

Table 5.1 Range parameters for n-type dopants implanted in GaAs at 400 keV

	$R_p(\mu m)$	$\Delta R_p(\mu m)$	$R_p + 2\Delta R_p(\mu m)$
Si	0.350	0.121	0.593
S	0.306	0.110	0.526
Se	0.137	0.0557	0.248
Te	0.0917	0.0364	0.165

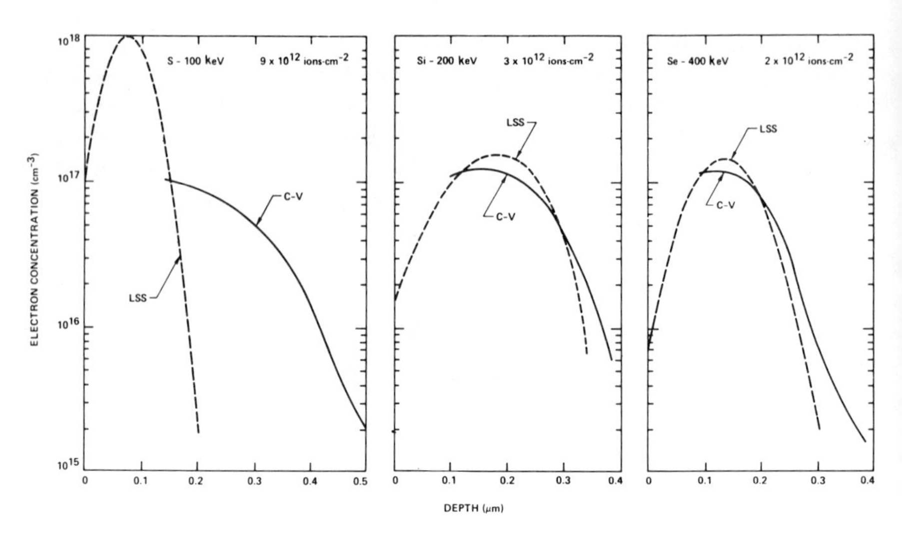

Figure 5.1 Comparison of electron concentration profiles derived from *C–V* data for S, Si, and Se[22] implanted in GaAs at the indicated dose and energy with estimates of the profiles of the implanted ions derived from LSS range parameters[19–21]

implanted species either during the implantation or during post-implantation annealing.

In GaAs very high activation of n-type implants with peak dopant concentrations less than about 10^{17} cm^{-3} can be achieved.[4] Typical electron concentration profiles for such low-dose implants are shown in Figure 5.1 for Si, S, and Se.[22] It can be seen that the electron concentration profiles obtained for implanted Si or Se are close to the theoretical distributions of the respective dopants. However, the electron concentration profile measured for S implantation is appreciably deeper than the theoretical S range distribution and the maximum doping level is about an order of magnitude lower than the predicted maximum S concentration. This is believed to be due to extensive diffusion of the S during post-implantation annealing, including outdiffusion of the S to the surface of the GaAs.[23]

As the implanted n-type dopant concentration is raised above about 10^{17} cm^{-3} in GaAs, the maximum electron concentration attained begins to fall significantly below the dopant concentration.[4] Si and Se have yielded the highest electron concentrations for implanted n-type dopants in GaAs. For implant doses of these dopant ions of 10^{14} to 10^{15} cm^{-2}, the electron concentrations obtained are generally no higher than about $(3–4) \times 10^{18}$ cm^{-3}, even though the maximum concentration of the implanted dopant may be well above 10^{19} cm^{-3}.[24, 25] Typical implantation profiles for high doses of Si or Se compared with theoretical calculations of the implanted dopant distributions[24, 25] are shown in Figure 5.2.

The behaviour of implanted n-type dopants in InP is somewhat similar to that described above for GaAs. Good activation of low doses of implanted dopants

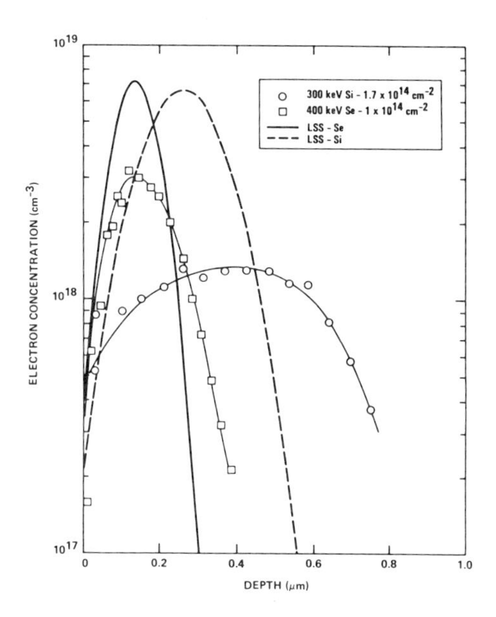

Figure 5.2 Electron concentration profiles obtained from differential Hall effect measurements on GaAs implanted with Se[24] or Si,[25] as indicated, and annealed for 30 min at 850°C (Si) or 900°C (Se). Estimates of the profiles calculated from LSS range statistics[20] are also shown

has been observed but the activation decreases somewhat as the implanted ion dose is raised above about 10^{14} cm^{-2}.[26,27] The maximum electron concentrations which are readily achieved in InP by implantation are about 2–3 times larger (i.e. about 10^{19} cm^{-3}) than those obtained in GaAs.[26,27] It has been observed that the electron mobility in samples implanted at elevated temperatures (200°C) is about a factor of 2 higher than those measured in samples implanted at room temperature.[27] Implantation at 200°C also seems to result in some broadening of the electron concentration profile, as compared to that observed for samples implanted at room temperature for Si doses of 10^{15} cm^{-2}. Limited high-dose implantation experiments on InAsP also indicate that good activation of high-dose n-type implants are more readily obtained in this material than in GaAs.[13]

The use of Si as an implanted dopant in GaInAsP to increase the carrier concentration in FET source and drain regions has been reported.[1] However, no information was given in this work on the carrier concentration obtained nor on the activation of the implanted dopants. Based on the work discussed above for GaAs, InP, and InAsP, however, it may be anticipated that electron concentrations between 10^{18} and 10^{19} cm^{-3} could be achieved in GaInAsP by implanting high doses of Si or perhaps Se. It is probable that low doses of ions (10^{12} to 10^{13} cm^{-2}) would be almost fully activated following a suitable post-implantation anneal. The depths of the n-type layers which could be produced by implantation of n-type dopants would be similar to those listed in Table 5.1 for GaAs, assuming little or no diffusion of the implanted species.

5.4 P-TYPE DOPANTS

The use of the group III elements, Be, Mg, Zn, and Cd, as implanted p-type dopants in GaAs has been fairly extensively investigated.[3–5] Implantation doping with Be in GaAs has resulted in formation of p–n junctions with excellent properties.[28] Be implantation has also been applied to a large number of other III–V materials including InP,[29], GaInAsP,[2, 29] GaAsP,[30] InAsSb[31] GaInSb,[31] and GaAlSb.[32] An important factor in the successful application of Be implantation for p-type doping in these materials is the fact that a very small fraction of the energy of the implanted Be ions is dissipated in the processes which can cause displacement of the atoms in the implanted substrate. This is a consequence of the low mass of the Be and results in the introduction of substantially less damage or disorder in the implanted layer than is the case for any of the other p-type dopants. It should, therefore, be easier to restore the implanted layer to its original crystalline quality by post-implantation annealing.

Values for the projected range and projected range straggling[21] for Be implanted in GaAs are listed for several energies in Table 5.2. It is more difficult to calculate range parameters accurately for Be than for heavier ions because of the importance of electronic stopping in determining the range of low-mass ions. The range distributions are expected to deviate significantly from a Gaussain distribution at higher energies. However, the values given for $R_p + 2\Delta R_p$ should serve to indicate the approximate thickness of Be-implanted layers in GaAs and other III–V materials such as InP, InAs, and GaInAsP.

Because of the wide applicability of Be as a p-type dopant in III–V materials, the discussion here will be limited to the use of Be for this purpose. It is easiest to illustrate some features of Be implantation doping by considering the data available for GaAs. Figure 5.3 shows electron concentration profiles for several different implant doses following annealing at 900°C. Be implants in GaAs can be activated at lower annealing temperatures; however, there is evidence that a temperature of about 900°C is required to remove some of the defects from the implanted region.[34] From the data in Figure 5.3 it can be seen that, when the

Table 5.2 Range parameters for Be implanted in GaAs

Energy (kev)	R_p (μm)	ΔR_p (μm)	$R_p + 2\Delta R_p$ (μm)
20	0.0668	0.0477	0.162
100	0.323	0.136	0.597
250	0.765	0.216	1.198
400	1.119	0.249	1.618

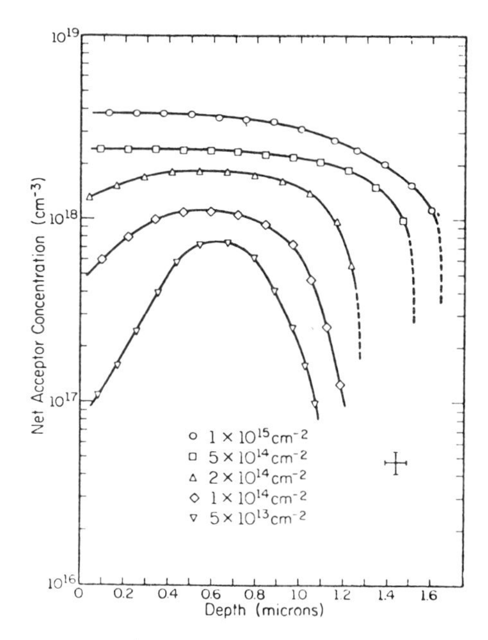

Figure 5.3 Net acceptor concentration profiles obtained from differential Hall effect measurements on GaAs implanted with Be at 250 keV to the doses shown and annealed for 30 min at 900°C [33]

maximum hole concentration is lower than about 10^{18} cm^{-3}, the hole concentration profile is similar to a Gaussian distribution. However, when the maximum hole concentration rises above this value, extensive broadening of the profile is observed. This effect, which results from Be diffusion during post-implantation annealing, is not understood at present but is a feature of Be implantation doping of GaAs which must be taken into account in device fabication.

High activation (close to 100%) of implanted Be as an acceptor in GaAs has been observed. In InP and GaInAsP the doping efficiencies are not as high as in GaAs (often only about 50%); nevertheless it is possible to dope these materials effectively with implanted Be.[29] Measurements on Be-implanted InP suggest that the diffusion behaviour of implanted Be in that material is qualitatively similar to that observed in GaAs, in that extensive diffusion seems to be observed when the Be concentration exceeds about 3×10^{18} cm^{-3}.[29] There is limited evidence that similar behaviour may occur in Be-implanted GaInAsP annealed at 700°C.[29] Evidence of Be diffusion has also been observed in GaInSb and InAsSb.[31]

The data discussed above suggest that it may be possible to achieve hole concentrations in GaInAsP of the order of $(2\text{–}3) \times 10^{18}$ cm^{-3} but that, when one utilizes implantation doses which should result in substantially higher hole concentration, extensive diffusion of the Be may take place. Such an effect would limit the maximum hole concentration and result in junctions deeper than predicted from LSS range statistics. It should be mentioned that implantation at a single energy may result in a buried p-type region. However, use of several different implantation energies can result in a p-type layer with relatively uniform hole concentration extending to the surface of the implanted material.[35]

5.5 PROTON BOMBARDMENT ISOLATION

Proton bombardment is a commonly employed isolation technique for GaAs and GaAlAs devices. The technique has been widely applied to lasers to confine the current to a very narrow stripe and therefore increase the current density and lower total device current. More recently, the technique has been successfully used to reduce edge breakdowns in GaAlAs avalanche photodiodes[36] and has also become a powerful technique in isolating the rows of GaAs charge-coupled device (CCD) area imagers.[37] Without the passive bombardment isolation, large-area depletion capacitors would be required, significantly reducing the yield and performance of these devices. The use of proton bombardment as an isolation technique for InP has only been successful in p-type InP;[38] however, the implantation of Fe to produce high-resistivity layers in n-type InP has been reported.[39]

Attempts to apply proton bombardment as an isolation technique in InGaAsP or GaSb and its alloys have not proved satisfactory. While some increase in resistivity results from proton bombardment of these materials, it has not been possible to achieve the very high resistivities obtained in GaAs and GaAlAs. It may be that the energy levels associated with proton bombardment defects in these materials do not lie near the centre of the bandgap, as they do in GaAs and GaAlAs, so that a high resistivity cannot be achieved using proton bombardment.

5.6 APPLICATION OF Be IMPLANTATION TO DEVICE FABRICATION

The use of Be as an implanted p-type dopant in GaInAsP photodetectors provides a good example of the improvements in device performance and advantages in fabrication which can be achieved by the application of ion implantation doping techniques. Recent interest in 1.2–1.6 μm fibre communication systems has encouraged development of high-performance lasers and detectors, ternary (GaInAs)[40] and quaternary (GaInAsP[41] and GaAlAsSb[32]) alloys. The performance of GaInAsP photodetectors has not been suitable for multi-gigabit data rate ($t_r < 100$ ps) communication systems because of a problem resulting from the diffusion of Zn which is used as a p-type dopant in LPE growth systems. One solution to this problem is to incorporate a suitable p-type dopant by a technique other than LPE. Be, which has a limited solubility in indium, cannot be used as a p-type dopant in the liquid-phase epitaxial growth of GaInAsP but is an excellent p-type dopant if ion-implanted into InGaAsP as indicated above. The implanted Be can be activated at a low temperature (560°C), thus avoiding the problem of severe dissociation of InP which occurs above 580°C.

Undoped layers of $Ga_{0.16}\ In_{0.84}\ As_{0.34}\ P_{0.66}$ for implantation were grown by the liquid-phase epitaxial technique. To provide the top p^+ layer, a dose of $5 \times 10^{14}\ cm^{-2}$ of 100 KeV Be ws implanted into the GaInAsP Layer. The implanted wafer was then sandwiched between two InP wafers and was annealed in an argon atmosphere at 560°C for 2 h. No evidence of phosphorus dissociation could be observed. Junctions were defined by mesa etching as shown in Figure 5.4(a).

An alternative planar process was also used. A 1 μm thick SiO_2 layer was used for ion implantation mask definition. Holes of a desired size were etched in an SiO_2 mask by standard photolithography techniques. The photoresist of approximately 3 μm was intentionally left on to provide additional masking for the implantation. (This should provide enough masking for 100 KeV Be ions.) After the Be implantation, the photoresist and 5000 Å of SiO_2 were stripped off. The sample was then annealed in a similar fashion to that described above. The planar structure is shown in Figure 5.4(b). In this structure, the surface leakage of the diode should be minimized. However, the mesa structure has so far exhibited the best performance. An additional guardring formed by proton bombardment might be needed to improve the planar diodes.

The residual donor concentration of the undoped $Ga_{0.16}\ In_{0.84}\ As_{0.34}\ P_{0.66}$

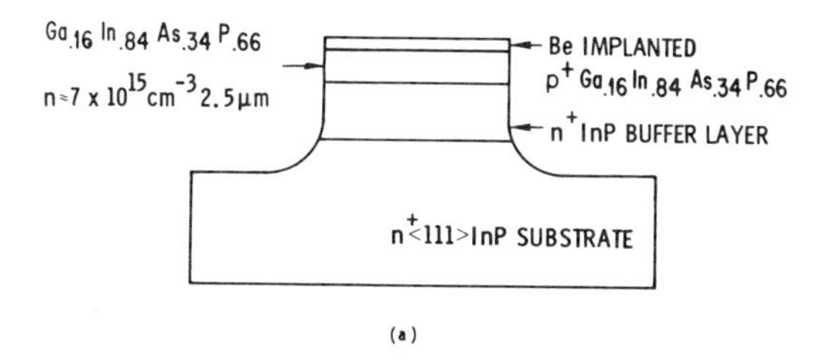

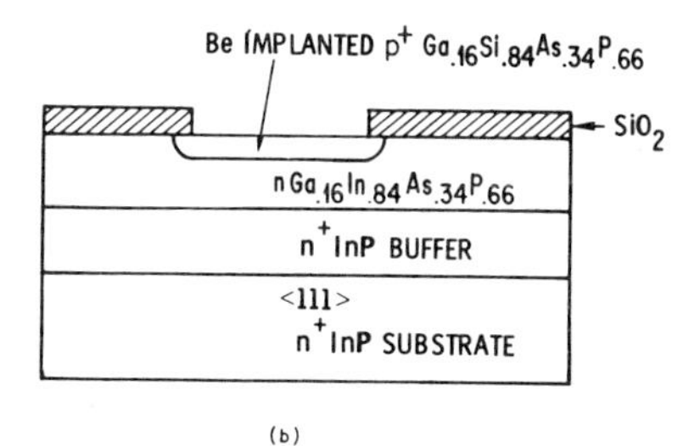

Figure 5.4 (a) Mesa structure of the Be-implanted InGaAsP APD. (b) Planar structure of the Be-implanted InGaAsP APD[2]

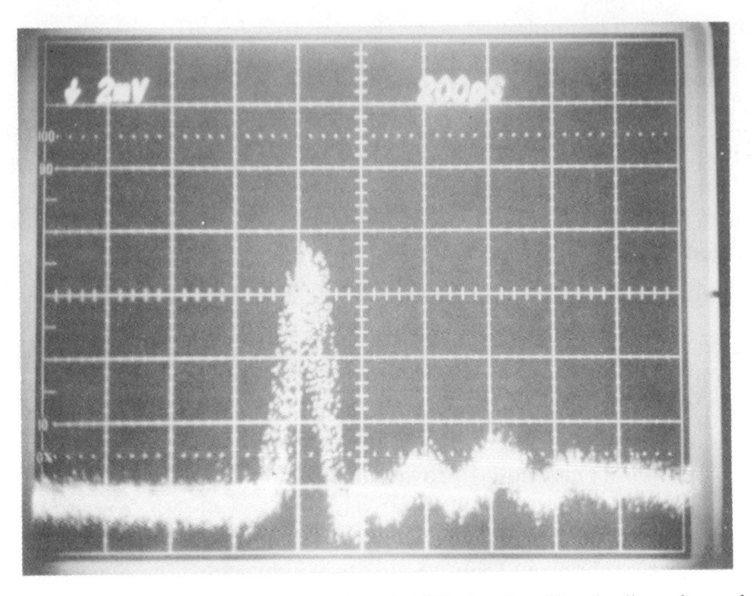

Figure 5.5 The pulse shape of the ion-implanted APD showing 60 ps leading edge and 120 ps FWHM[2]

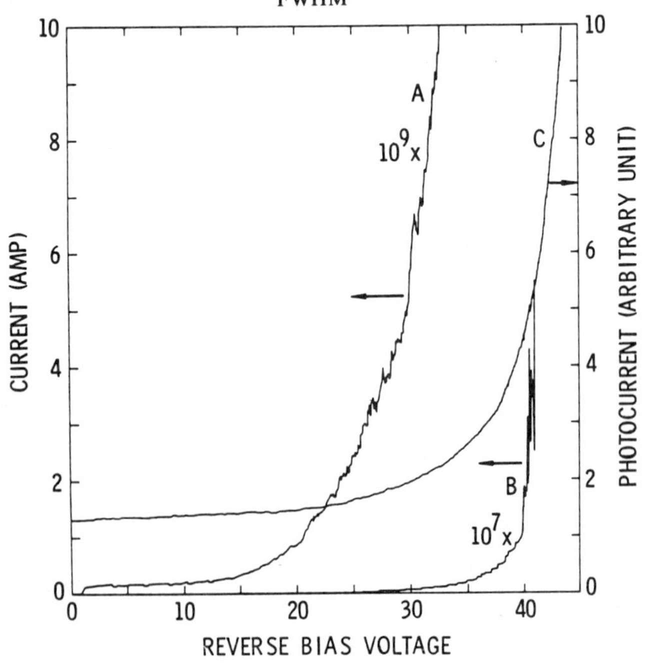

Figure 5.6 Trace A is the dark current of a mesa-implanted InGaAsP APD; trace B is the less-sensitive scale dark current of the same diode extending to breakdown. Trace C is the corresponding electron multiplication curve, take by illuminating the p^+ layer of the diode by 0.63 μm light [2]

active layer was 7×10^{15} cm^{-3}. For high-speed and high-quantum-efficiency operation of an avalanche photodiode, photons should be absorbed in the depletion region in order to eliminate minority-carrier diffusion. Therefore, the thickness of active layers must be between 2 and 3 μm. The incoming 1.06 μm light is absorbed in the active layer which, for the punch-through structure, is also the space-charge region. In addition to high quantum efficiency, the response of the diode when driven by a mode-locked Nd:YAG laser, operating at a 275 MHz repetition rate, shows no evidence of a 'diffusion current tail' or 'back porch'. Figure 5.5 shows the 60 ps leading edge of the photodiode response to the laser pulse. The full width of the response is approximately 120 ps.

The dark current is a good indication of the quality of the junction as well as the quality of the surface. Trace A in Figure 5.6 is the dark current of a mesa diode at room temperature. The dark current density at 10 V is 4.0×10^{-6} A cm^{-2}. It increases to 1.9×10^{-5} A cm^{-2} at 20 V. Trace B of Figure 5.6 shows the dark current close to breakdown. Trace C shows the corresponding optical gain of the electron current. This was done by directing the 0.63 μm light on top of the p^+ region of the diode. The fact that the dark current multiplies slightly faster than the optical current is not fully understood at present, although it has been suggested that this effect is related to tunnelling.[42] External quantum efficiences of 60–65% (non AR-coated) have been obtained by this technique.

Be-implanted GaAlAsSb APDs of designs similar to the InGaAsP-implanted have also been fabricated.[32] Gain, response time, and dark current, are all similar to LPE-grown junctions (Ge used as a p-type dopant). The major advantage is the ability to fabricate planar devices or arrays.

5.7 SUMMARY

There is only limited direct information available on ion implantation doping of GaInAsP. However, extrapolation of results on implantation of n-type dopants in other III–V materials suggests that Si may be an effective implanted n-type dopant in GaInAsP and that it should be possible to achieve electron concentrations greater than 10^{18} cm^{-3} and perhaps concentrations as high as 10^{19} cm^{-3}. Be has been used successfully as an implanted p-type dopant in GaInAsP and it may be possible to achieve hole concentrations as high as (2–3) $\times 10^{18}$ cm^{-3}. Phosphosilicate glass, SiO_2, or Si_3N_4 may be useful as encapsulants for both n- and p-type implants. Successful annealing of Be implants at 560 and 700°C has been reported.

REFERENCES

1. Y. Shinoda, M. Okamura, E. Yamaguchi, and T. Kobayashi, 'InGaAsP *n*-channel inversion-mode MISFET', *Appl. Phys. Lett.*, **19**, 2301–2302, 1980.
2. H. D. Law, L. R. Tomasetta, and K. Nakano, 'Ion-implanted InGaAsP avalanche photodiode', *Appl. Phys. Lett.*, **33**, 920–922, 1978.

3. J. P. Donnelly, 'Ion-implantation in GaAs,' *Gallium Arsenide and Related Compounds (St Louis) 1976*, Conf. Ser. no. 33b, Institute of Physics, Bristol, 1977, pp. 166–190.
4. F. H. Eisen, 'Ion-implantation in III–V compounds', *Radiat. Effects*, **47**, 99–116, 1980.
5. D. V. Morgan, F. H. Eisen, and A. Ezis, 'Prospects of ion bombardment and ion-implantation for GaAs and InP device fabrication', *Proc. IEE*, Part I, 107–130, 1981.
6. J. S. Harris, F. H. Eisen, B. Welch, J. D. Haskell, R. D. Pashley, and J. W. Mayer, 'Influence of implantation temperature and surface protection on tellurium implantation in GaAs', *Appl. Phys. Lett.*, **21**, 601–603, 1972.
7. A. G. Foyt, J. P. Donnelly, and W. T. Lindley, 'Efficient doping of GaAs by Se^+ ion implantation', *Appl. Phys. Lett.*, **14**, 372–374, 1969.
8. F. H. Eisen, B. M. Welch, H. Muller, K. Gamo, T. Inada, and J. W. Mayer, 'Tellurium implantation in GaAs', *Solid-State Electron.*, **20**, 219–223, 1977.
9. B. J. Sealy and R. K. Surridge, 'A new thin film encapsulant for ion-implanted GaAs', *Thin Solid Films*, **26**, L19–L22, 1975.
10. T. Nishioka and Y. Ohmachi, 'Silicon-ion implantation in InP and annealing with CVD SiO_2 encapsulation', *Appl. Phys. Lett.*, **51**, 5789–5791, 1980.
11. J. P. Donnelly and C. E. Hurwitz, 'Ion-implanted n- and p-type layers in InP', *Appl. Phys. Lett.*, **31**, 418–420, 1977.
12. K. R. Gleason, H. B. Dietrich, R. C. Henry, E. D. Cohen, and M. L. Bark, 'Ion-implanted *n*-channel InP metal semiconductor field effect transistor', *Appl. Phys. Lett.*, **32**, 578–581, 1978.
13. D. E. Davies, J. K. Kennedy, and L. F. Lowe, 'Low-resistivity n-type layers in InAsP by ion implantation', *Electron. Lett.*, **11**, 462–463, 1975.
14. J. Kasahara, M. Arai, and N. Watanabe, 'Capless anneal of ion-implanted GaAs in controlled arsenic vapor', *J. Appl. Phys.*, **50**, 541–543, 1979.
15. J. M. Woodall, H. Rupprecht, and R. J. Chicotka, 'Proximate capless annealing of GaAs using a controlled-excess As vapor pressure source', *Appl. Phys. Lett.*, **38**, 639–641, 1981.
16. F. H. Doerbeck, H. M. Macksey, G. E. Brehm, and W. R. Frensley, 'Ion-implanted GaAs *x*-band power FETs', *Electron. Lett.*, **15**, 576–577, 1979.
17. F. H. Eisen, J. S. Harris, B. Welch, R. D. Pashley, D. Sigurd, and J. W. Mayer, *properties of Tellurium Implanted Gallium Arsenide*, Plenum, New York, 1973, pp. 631–640.
18. E. F. Kennedy, 'Damage and reordering of ion-implanted layers of InP', *Appl. Phys. Lett.*, **38**, 375–377, 1981.
19. J. Lindhard, M. Sharff, and M. E. Schiott, 'Range concepts and heavy ion ranges', *Mat.-Fys. Medd. K. Dan. Vidensk Selsk*, **33**, no. 14, 1963.
20. J. F. Gibbons, W. S. Johnson, and S. W. Mylroie, *Projected Range Statistics: Semiconductors and Related Materials*, Dowden, Hutchinson and Ross, Stroudsburg, 1975.
21. K. B. Winterbon, *Implantation Range and Energy Deposition Distributions*, vol. 2, *Low Incident Ion Energies*, IFI/Plenum, New York, 1975.
22. J. A. Higgins, R. L. Kuvas, F. H. Eisen, and D. R. Ch'en, 'Low-noise GaAs FETs prepared by ion implantation', *IEEE Trans. Electron. Devices*, **ED-25**, 587–596, 1978. A. A Immorlica, private communication (Rockwell International, Thousand Oaks, CA 91360).
23. F. H. Eisen, and B. M. Welch, 'Radiotracer profiles in sulfur implanted GaAs', *Ion Implantation in Semiconductors and Other Materials*, Plenum, New York, 1977, pp. 97–106.
24. K. Gamo, T. Inada, S. Krekeler, J. W. Mayer, F. H. Eisen, and B. M. Welch, 'Selenium implantation in GaAs', *Solid-State Electron.*, **20**, 213–217, 1977.

25. J. L. Tandon, M-A. Nicolet, and F. H. Eisen, 'Silicon implantation in GaAs', *Appl. Phys. Lett.*, **34**, 165–167, 1979.
26. J. P. Donnelly and G. A. Ferrante, 'The electrical characteristics of InP implanted with the column IV elements', *Solid-State Electron.*, **23**, 1151–1154, 1980.
27. D. E. Davies, J. J. Comer, J. P. Lorenzo, and T. G. Ryan, 'Improved electrical mobilities from implanting InP at elevated temperatures', *Appl. Phys. Lett.*, **35**, 192–194, 1979.
28. A. A. Immorlica, and F. H. Eisen, 'Planar passivated GaAs hyperabrupt varactor diodes', *Proc. Sixth Biennial Cornell Electrical Engineering Conf.*, Cornell School of Electrical Engineering, Ithaca, NY. 1977, pp. 151–159.
29. J. P. Donnelly and C. A. Armiento, 'Beryllium-ion implantation in InP and InGaAsP', *Appl. Phys. Lett.*, **34**, 96–99, 1979.
30. P. K. Chatterjee, W. V. McLevige, and B. G. Streetman, 'Electrical properties of Be-implanted GaAsP', *Solid-State Electron.*, **19**, 961–964, 1976.
31. L. O. Bubulac, E. E. Barrowcliff, W. E. Tennant, J. G. Pasko, G. Williams, A. M. Andrews, D. T. Cheung, and E. R. Gertner, 'Be ion implantation in InAsSb and GaInSb', *Gallium Arsenide and Related Compounds 1978*, Conf. Ser. no. 45, Institute of Physics, Bristol, 1979, pp. 519–529.
32. H. D. Law, R. Chin, K. Nakano, and R. A. Milano, 'The GaAlAsSb quaternary and GaAlSb ternary alloys and their application to infrared detectors', *IEEE. Quantum Electron.*, **QE-17**, 275–283, 1981.
33. W. V. McLevige, M. J. Helix, K. V. Vaidyanathan, and B. G. Streetman, 'Electrical profiling and optical activation studies of Be-implanted GaAs', *J. Appl. Phys.*, **48**, 33423346, 1977.
34. P. K. Chatterjee, K. V. Vaidyanathan, W. V. McLevige, and B. G. Streetman, 'Photoluminescence from Be-implanted GaAs', *Appl. Phys. Lett.*, **27**, 567–569, 1979.
35. C. A. Armiento, J. P. Donnelly, and S. H. Groves, 'P–n junction diodes in InP and InGaAs fabricated by beryllium-ion implantation', *Appl. Phys. Lett.*, **34**, 229–231, 1979.
36. R. A. Milano, M. J. Helix, T. H. Winelhorn, B. G. Streetman, K. V. Vaidyanathan, and G. E. Stillman, 'Planar ion-implanted avalanche photodiodes in GaAs', *Gallium Arsenide and Related Compounds* 1978, Conf. Ser. no. 45, Institute of Physics, Bristol, 1979, pp. 411–419.
37. Y. Z. Liu, I. Deyhimy, R. J. Anderson, R. A. Milano, M. J. Cohen, J. S. Harris, Jr, and L. R. Tomasetta, 'A backside illuminated imaging AlGaAs/GaAs charge-coupled device', *Appl. Phys. Lett.*, **37**, 803–805, 1980.
38. J. P. Donnelly and C. E. Hurwitz, 'Proton bombardment of InP', *Solid-State Electron.*, **20**, 727–730, 1977.
39. J. P. Donnelly and C. E. Hurwitz, 'High-resistivity layers in n-InP produced by Fe ion implantation', *Solid-State Electron.*, **21** 475–478, 1978.
40. N. Susa, H. Nakagome, H. Ando, and H. Kanbe, 'Characteristics in InGaAs/InP avalanche photodiodes with separated absorption and multiplication regions', *IEEE. Quantum Electron.*, **QE-17**, 243–250, 1981.
41. T. P. Lee, C. A. Burrus, Jr, and A. G. Dentai, 'InGaAsP/InP photodiodes: microplasma-limited avalanche multiplication at 1–1.3 μm wavelength', *IEEE. Quantum Electron.* **QE-15**, 30–35, 1979.
42. S. R. Forrest, M. DiDomenico, Jr, R. G. Smith, and H. J. Stocker, 'Evidence for tunneling in reverse-biased III–V photodetector diodes', *Appl. Phys. Lett.*, **36**, 580–582, 1980.

GaInAsP Alloy Semiconductors
Edited by T. P. Pearsall

Chapter 6

High-purity Material

G. E. STILLMAN, L. W. COOK, T. J. ROTH, T. S. LOW, AND B. J. SKROMME

Electrical Engineering Research Laboratory,
University of Illinois at Urbana-Champaign,
Urbana, IL 61801, USA

6.1 INTRODUCTION

The growth of high-purity material is important for several reasons. First, high-purity material is essential for particular device applications, such as Gunn effect or transferred electron oscillators, buffer layers in FET structures, and wide depletion widths with uniform electric fields in p–i–n photodetectors. Secondly, the capability of growing high-purity material makes it easier to grow material with uniform doping concentrations at the desired level. Of nearly equal importance to device applications requiring high-purity material is the requirement of high-purity material for the characterization of these new semiconductor alloys. Studies of band structure, low- and high-field transport, impurities, defects, and other material properties are made easier and the results more reliable if they are done on high-purity material.

In order to determine the possible sources of impurities in high-purity crystal growth so that these sources can be eliminated or at least so that their contribution can be minimized, it is important to be able to characterize the purity of the crystals that are grown. This is not a trivial matter. Most of the direct chemical identification techniques are not sensitive enough for the detection of electrically active impurities in state-of-the-art high-purity III–V compounds, although the sensitivity of secondary ion mass spectrometry (SIMS) measurements (about 1 part in 10^8 or better for the detection of Si and S) is however great enough to be of some use in detecting particular impurities. The most useful techniques for determining the impurities in these materials are less direct, and there are still many uncertainties in interpretation even for GaAs, which is probably the highest-purity, best characterized, compound semiconductor material available.

In this chapter we describe briefly the three characterization techniques most useful for evaluating the purity and identifying the residual donor and acceptor levels in high-purity semiconductors. The information that can be obtained from

Hall effect, photoluminescence, and photothermal impurity ionization photoconductivity measurements is discussed, and the limitations of each of these techniques are presented. The techniques used for growth of high-purity binary compound semiconductors that are applicable to the growth of high-purity GaInAsP alloys are then summarized. The results for the two most highly developed binary compounds in this quaternary alloy system, GaAs and InP, are presented, and the present state of development of high-purity GaInAsP is described. The purpose of this chapter is to give a summary of our present understanding of high-purity GaAs and InP with the hope that the techniques described will provide a basis for future improvements in these materials and for the growth of high-purity GaInAsP.

6.2 CHARACTERIZATION TECHNIQUES

The characterization techniques that are of most value in estimating the purity of semiconductor materials each contribute different types of information. Hall effect measurements yield information about the total electrically active impurity concentrations and the total concentrations of donors and acceptors. For reasons that will be discussed in more detail below, Hall effect measurements on the compound semiconductor materials of interest here cannot help in the identification of specific chemical donor species nor provide reliable information about the identification of acceptor species. The photoluminescence technique is best suited to the detection of different chemical acceptor species, and when used with careful doping experiments the transitions due to particular chemical acceptor species can be identified. Photoluminescence is not very useful for the identification or detection of different shallow donor chemical species, although shallow donors have been observed by this method. The best method for the identification of donor impurities is photothermal ionization photoconductivity measurements in a high magnetic field. Although much work remains to be done in applying these techniques, particularly to the characterization of InP and GaInAsP, they have already yielded information of significant value in pointing towards improved growth techniques for the preparation of higher-purity crystals of these materials.

6.2.1 Hall effect measurements

The experimental techniques used for Hall effect measurements are well-known,[1] and the most common technique in use now is the method developed by van der Pauw.[2, 3] To obtain the most information from the Hall coefficient and resistivity measurements, they should be made over a wide temperature range—for GaAs and InP, over the temperature range $4.2\ \mathrm{K} \lesssim T \lesssim 400\ \mathrm{K}$. Interpretation of measurements at higher temperatures must include the influence of the higher conduction band minima, and such measurements contribute little additional information about the purity of the sample. Measurements at temperatures less than 4.2 K also contribute very little additional information about the purity of the epitaxial layer. For measurements in the lower part of the temperature range, especially from 4.2 to 10 K, precautions must be taken to ensure, as far as possible, that the sample is at a uniform temperature, that the background is at the same temperature as the sample, and that the actual temperature of the sample is accurately measured. The design and analysis of the measurement must take into account unwanted thermo-electric effects,[4] non-ohmic contacts,[5,6] and non-ideal sample geometries.[7,8]

Determination of carrier concentration and mobility

The parameters determined directly from the experimental measurements are the Hall coefficient and resistivity, while the quantities of interest for the determination of the electrically active impurity concentrations are the free-carrier concentration and the free-carrier mobility. Using n-type material as an example, the free-carrier concentration n is determined from the measured Hall coefficient R_{H} using the expression

$$n = r_{\mathrm{H}}/eR_{\mathrm{H}} \tag{6.1}$$

in which e is the electronic charge and r_{H}, the Hall coefficient scattering factor, is a numerical factor close to unity which relates the carrier concentration to the measured Hall coefficient. Hall coefficient factors for various scattering mechanisms have been discussed in detail by Beer.[8] When the scattering process is such that the energy change of the scattered carriers is relatively large compared with their initial energy, a universal relaxation time cannot be defined. Scattering by optical phonons through the polar interaction is one such process which is important for GaAs and other III–V compound semiconductors. The Hall coefficient factor for polar-mode scattering is temperature dependent,[9–11] and this can produce significant errors in the calculated donor and acceptor concentrations, N_{D} and N_{A}.

The measured Hall coefficient and resistivity can be combined to calculate the Hall mobility

$$\mu_{\mathrm{H}} = R_{\mathrm{H}}/\rho$$

The conductivity mobility is then given by

$$\mu = \mu_{\mathrm{H}}/r_{\mathrm{H}}$$

where $r_{\mathrm{H}} \geq 1$. Experimentally, the Hall coefficient factor at a given magnetic field B can be determined from

$$r_{\mathrm{H}}(B) = R_{\mathrm{H}}(B)/R_{\mathrm{H}}(\infty)$$

in which $R_{\mathrm{H}}(\infty)$ is the Hall constant in the high magnetic field limit.

Determination of donor and acceptor concentrations

To determine the donor and acceptor concentrations from the temperature variation of the carrier concentration, an equation for the theoretical carrier concentration $n_0(T)$ must be fitted to the experimental carrier concentration $n(T)$ determined from equation (6.1). For a non-degenerate n-type semiconductor with a shallow donor concentration N_D and acceptor concentration N_A, the acceptors are fully ionized at all temperatures of interest. The theoretical concentration $n_0(T)$ of free electrons in the conduction band at temperature T, including the influence of excited states and neglecting terms in n_i^2, is given by the solution of

$$\frac{n_0(T)[n_0(T)+N_A]}{N_D-N_A-n_0(T)}=\frac{N_c\exp(-E_D/kT)}{g_1(1+F)} \tag{6.2}$$

where

$$F=\sum_{r=2}\frac{g_r}{g_1}\exp\left(-\frac{E_r}{kT}\right). \tag{6.3}$$

In this equation, $N_c = 2(2\pi m_D^* kT/h^2)^{3/2}$ and m_D^* is the conduction band density-of-states effective mass, g_r is the degeneracy factor for the rth state of the impurity ($r = 1$ is the ground state, $r = 2$ is the first excited state, etc.), and E_r is the energy of the rth state measured above the energy of the ground state. The sum over r in equation (6.3) is over all discrete excited states (i.e. omitting those excited states which are banded) of the impurity centre.

For semiconductors in which the donor levels are relatively deep and where different donor species have significantly different ionization energies, such as Si, a least-squares fit of the appropriate theoretical equation to the experimental data for $n(T)$ will yield reliable values for N_D, N_A, and E_D. The value of E_D can then be used to identify the donor species present. For GaAs, InP, and other materials of interest here however, the donor ionization energies are small (about 5–6 meV), and even for the highest-purity material available there is a significant amount of interaction among the different donor atoms. Thus, the thermal activation energy E_D determined from the analysis of Hall effect data can be much different than the ionization energy E_I of an isolated donor atom and the variation in E_D with donor concentration N_D is much greater than the maximum variation in donor ionization energy between donor species. As the donor concentration increases, the higher excited states overlap and form a quasicontinuum. This quasicontinuum results in an effective lowering of the conduction band edge and a decrease in the thermal activation energy. This picture of the behaviour of shallow donor levels has been verified for GaAs, and Figure 6.1 shows the energy level model of hydrogenic donors in GaAs. The experimental variation of the width ΔE of the quasicontinuum with donor concentration is shown in the right side of this diagram. Thus, for donor concentrations greater than about 7×10^{14} cm^{-3} there are no discrete excited states. For $2\times10^{14} \lesssim N_D \lesssim 7\times10^{14}$ cm^{-3} there is only one discrete excited state, and for $5\times10^{13} \lesssim N_D \lesssim 2\times10^{14}$ cm^{-3} there are two,

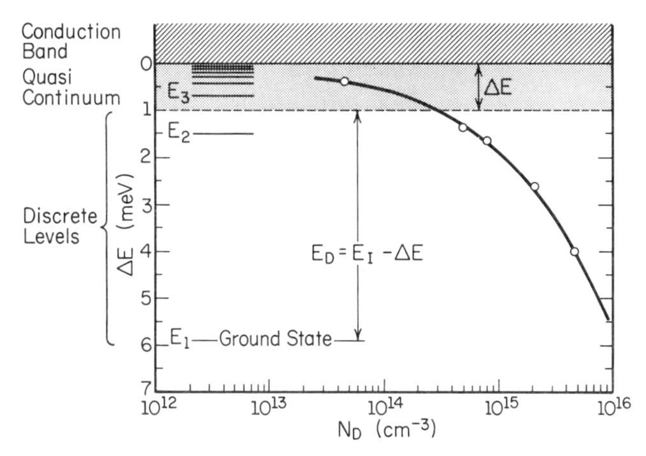

Figure 6.1 Diagram showing the energy levels of shallow donors in GaAs. The quasicontinuum results from overlap of the donor excited states and causes a decrease in the thermal activation energy E_D with increasing donor concentration

etc. These measurements also showed that the energy separation between the donor ground state and the excited states did not change with donor concentration. Thus, E_r in equation (6.3) is given by

$$E_r = (1 - r^{-2})E_I \tag{6.4}$$

and

$$E_D = E_I - \Delta E \tag{6.5}$$

There are many computer curve-fitting routines available and it is a simple procedure to determine the values of N_D, N_A, and E_D (or ΔE) which give the best fit to the data throughout the entire temperature range. These curve-fitting procedures minimize the error between the calculated free-carrier concentration $n_0(T)$ and the experimental value $n(T)$ at the same temperature. To give the data points throughout the entire temperature range equal weight, the function which is minimized by the curve-fitting procedure is usually some root mean square such as

$$\left[\sum_i \left(\frac{n_0(T_j) - n(T_j)}{n(T_j)}\right)^2\right]^{1/2}$$

or

$$\left\{\sum_j \left[\log\left(\frac{n_0(T_j)}{n(T_j)}\right)\right]^2\right\}^{1/2}$$

in which $n_0(T_j)$ is the calculated carrier concentration at temperature T_j, $n(T_j)$ is the experimental value, and the sum is over the data points which are to be included in the fit.

The model used in the analysis to determine N_D and N_A is very important. For example the use of a model with none, one, or two excited states can yield calculated total ionized impurity concentrations which differ by as much as 17%. The choice of the values used for the density-of-states effective mass and use of corrections for non-parabolicity can also result in significant differences in the values determined for N_D and N_A. The calculated thermal ionization energy is much less dependent on the model, and the ionization energies for the above models differ by less than 0.5%. Thus, for accurate estimates of N_D and N_A, a realistic model must be used, and the influence of any change in the Hall coefficient factor with temperature must also be minimized. With careful model selection and analysis of $n(T)$ data, it is possible to obtain reasonably reliable estimates of N_D and N_A.

Mobility analyses

For GaAs, analyses of mobility and Hall coefficient measurements vs temperature have been correlated so that over a significant impurity concentration range the donor and acceptor concentrations can be reliably determined from a single measurement of n and μ at 77 K. Figure 6.2 shows the empirical and calculated variation of the 77 K mobility for GaAs with 'screened' total impurity concentration. Using this figure and n_{77} and μ_{77}, both N_D and N_A can be determined. Unfortunately, similar experimental data do not exist for InP or the GaInAsP alloy systems. Nevertheless, the mobility at 77 K is widely used as a measure of the purity of these materials. The 77 K mobility for InP has been calculated by

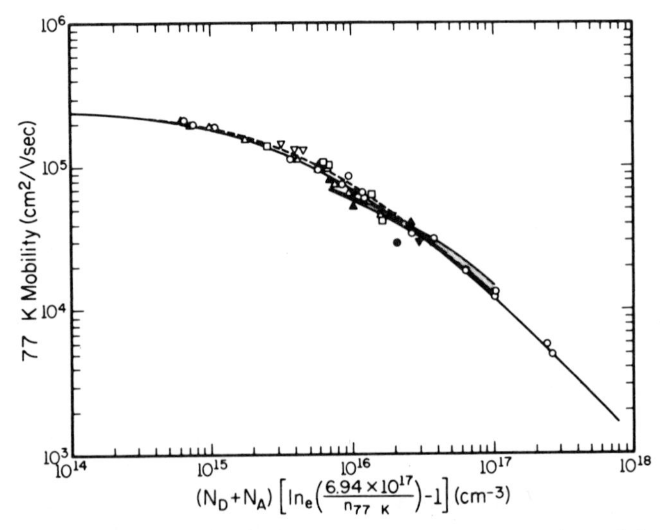

Figure 6.2 The empirical (----) and theoretical (———) curves relating the total electrically active impurity concentration in GaAs to the Hall mobility at 77 K. The shaded band extending from 8 $\times 10^{15}$ to 10^{17} on the abscissa is from the theoretical calculations of Rode[12] for InP

Rode[12] as a function of carrier concentration for different compensation ratios, and these curves permit relative comparisons of sample purity, at least. Recent experimental work has indicated that the curve for GaAs in Figure 6.2 may be similar to that for determining N_D and N_A in InP.[13] The shaded band in Figure 6.2 shows the variation of mobility with screened total impurity concentration for InP from the calculations of Rode.[12]

A disadvantage of simply using the liquid-nitrogen-temperature mobility as an indicator of sample purity is that inhomogeneities and other effects which influence the mobility can often go undetected. Most types of inhomogeneities lead to an overestimate of N_D and N_A determined from the carrier mobility and concentration and in this way at least give an indication of lower sample quality. However, in GaAs in particular, samples have been measured which have room-temperature mobilities over a factor of 2 greater than the generally accepted theoretical limit of about $8000 \pm 1000\,\mathrm{cm^2\,V^{-1}\,s^{-1}}$ and experimental mobilities at liquid-nitrogen temperature that are too high for the true purity of the sample. Wolfe and Stillman[14] have shown that conducting inhomogeneities can result in apparent mobility enhancement. These inhomogeneities may be too small to be detected directly, but they must be considered whenever (1) the experimental mobilities exceed theoretical expectations, (2) measurements from different portions of the same wafer or with different sample geometries give significantly different results, (3) the Hall coefficient shows an anomalous magnetic field dependence, or (4) the degree of compensation appears to be unusually low. Thus, although mobility measurement for routine assessment of sample purity remains very useful, it is clear that the quality or purity of a GaAs sample cannot be established conclusively solely by a high mobility at 77 K (or at room temperature), and it is likely that these same considerations apply to InP and GaInAsP as well.

6.2.2 Photoluminescence measurements

From the above discussion it is clear that although values of the total electrically active donor and acceptor concentrations N_D and N_A can be determined from the analysis of Hall coefficient data, it is impossible to use this technique to identify the donor and acceptor impurity species through their characteristic ionization energies. The technique most useful for the identification of donor impurities will be described in section 6.2.3, and this section will describe the technique most useful for the identification of acceptor species in GaAs and similar compounds–photoluminescence.

Near-band-edge photoluminescence processes

Photoluminescence studies of GaAs are very extensive. The determination in the earlier work of the various processes involved has formed a good basis for the identification of the acceptor species related to the various photoluminescence bands through measurements on both n- and p-type specially doped samples. Figure 6.3 shows the principal near-band-edge photoluminescence bands in GaAs for the identification of acceptor impurities. The higher energy exciton transitions are designated as follows: free exciton recombination (FE); recombination of an exciton bound to a neutral donor leaving the neutral donor in the $1S_{1/2}$ ground state $(D^0, X)_{n=1}$; recombination of an exciton bound to an ionized donor (D^+, X); recombination of an exciton bound to a neutral acceptor, leaving the acceptor in its ground state (A^0, X); and recombination of an exciton bound to a neutral donor, leaving the donor in the first excited $2S_{1/2}$ state $(D^0, X)_{n=2}$. The doublet in the (A^0, X) recombination results from the two initial exciton states, $J = \frac{3}{2}$ and $J = \frac{5}{2}$. The donor two-electron transitions in GaAs were first identified by Rossi *et al.*[15] However, the energy differences between the 1s and the 2s and 2p states for shallow donor states are all very nearly the same, regardless of the chemical identity of the donor species (i.e. the chemical shift or central cell correction for shallow donors is small), so these transitions cannot be used easily for the identification of shallow donor impurities in GaAs.

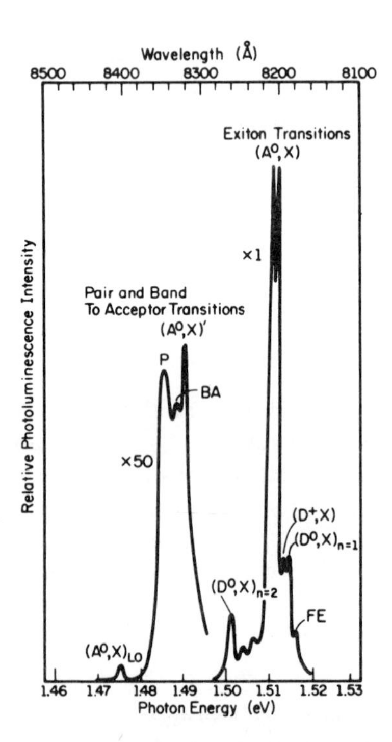

Figure 6.3 Near-band-edge low-temperature photoluminescence peaks and processes of importance for the identification of acceptor impurities in GaAs. The same processes are observed in photoluminescence measurements on high-purity InP. (See text for details.)

The lower energy near-band-edge photoluminescence bands shown in Figure 6.3 are due to the following processes: recombination of an exciton bound to a neutral acceptor which leaves the acceptor in its first excited state $(A^0, X)'$. This mechanism also produces a doublet because of the two ($J = \frac{3}{2}$ and $J = \frac{5}{2}$) initial exciton states. These lines are particularly useful for the identification of acceptor impurities because of the significantly larger and characteristic central cell corrections for acceptors. The next two bands result from free-to-bound or band–acceptor (BA) transitions and donor–acceptor pair (P) transitions, where an electron bound to a donor recombines with a hole bound to a nearby acceptor. The last small peak, $(A^0, X)_{LO}$, is a phonon replica of the neutral acceptor bound exciton recombination peak.

The donor–acceptor (P) and (BA) bands in GaAs were first studied and identified in high-purity material by Rossi *et al.*[16] Figure 6.4 shows the photoluminescence results of Ashen *et al.*[17] for these bands at two different temperatures for an LPE GaAs sample with three different acceptor impurities, Ge, Si, and C. As the temperature is increased from 6 K to 15 K the intensity of the donor–acceptor pair bands (P) decreases relative to the intensity of the band–acceptor (BA) transitions because of the ionization of the shallow donor levels. The more rapid decrease of the (BA) transition with increasing temperature for C than for Si and Ge has been attributed to differential ionization of the shallower C acceptor as the temperature is increased.

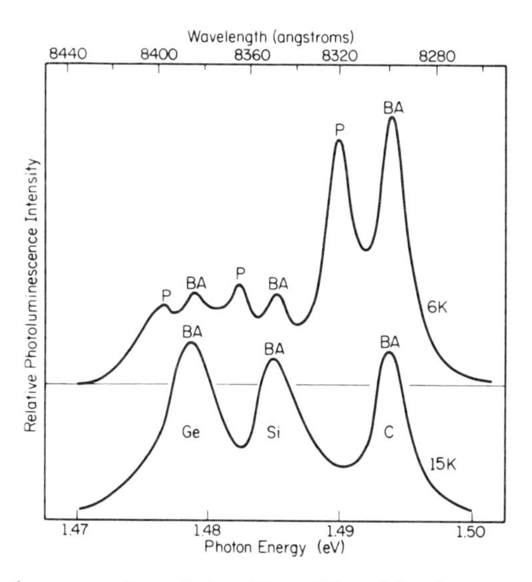

Figure 6.4 Bound donor–acceptor pair transitions (P) and band-to-acceptor (BA) or free-to-bound transitions in high-purity GaAs with Ge, Si, and C acceptors at two different temperatures. (After Ashen *et al.*[17] Reproduced by permission of Pergamon Press and the Controller of Her Majesty's Stationery Office)

Identification of acceptor impurities in GaAs

Using the photoluminescence processes described above which involve acceptor impurities, the bands or peaks and ionization energies of most of the important acceptors in GaAs have been identified through careful doping experiments with high-purity epitaxial material. Figure 6.5 shows the (BA) and (P) transitions measured by Ozeki *et al.*[18] for high-purity n-type GaAs containing Ge, Si, Cd, Zn, and C acceptors. The subscripts 1 and 2 in the figure indicate the (BA) and (P) transitions respectively. These transitions are relatively broad compared with the bound exciton transitions, and although measurements at different temperatures can help in separating overlapping peaks due to different impurities, the identification of residual acceptor impurities when more than one species is present may be difficult using only these transitions.

Figure 6.6 shows low-temperature, high-excitation-intensity photoluminescence measurements on C-, Zn-, and Cd-doped high-purity n-type GaAs samples, also measured by Ozeki *et al.*,[18] for the entire near-band-edge spectral region. The mechanisms responsible for the various peaks have been described above and are designated in the figure. The bound exciton doublet peaks (A^0, X) for each of the acceptor species (X_A, X_B, and X_C) lie in a very narrow energy range. These peaks are subject to broadening and shifting by sample strain, electric fields due to

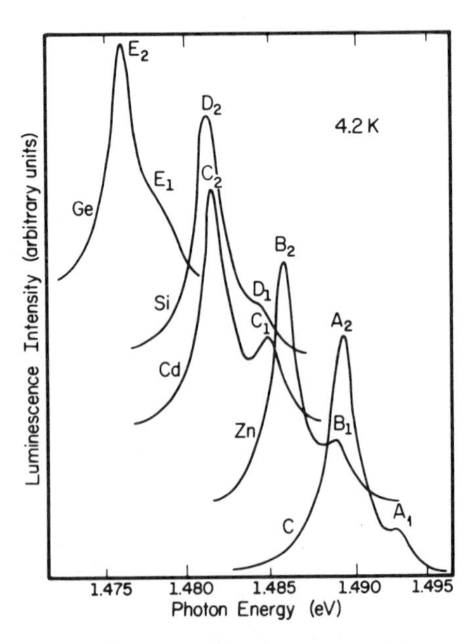

Figure 6.5 Photoluminescence (BA) and (P) transitions for Ge, Si, Cd, Zn, and C acceptors in high-purity GaAs. The subscripts 1 and 2 indicate the (BA) and (P) transitions respectively. (After Ozeki *et al.*[18] Reproduced by permission of *Japanese Journal of Applied Physics*)

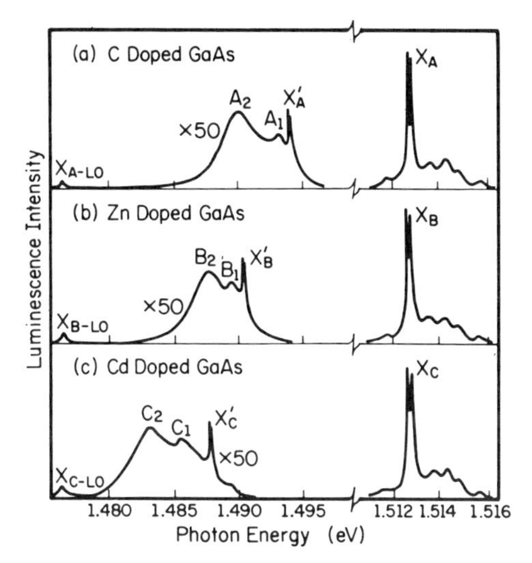

Figure 6.6 Near-band-edge photoluminescence measurements for three different high-purity GaAs samples doped with C, Zn, and Cd at $T = 1.8$ K under high excitation intensity ($\sim 5\ \mathrm{W\,cm^{-2}}$). The acceptors can be easily identified by the bound exciton–two-hole transitions X'_A, X'_B, and X'_C. (After Ozeki *et al.*[18] Reproduced by permission of *Japanese Journal of Applied Physics*)

ionized impurities, etc., so it is difficult to make positive impurity identifications using only these transitions.

The transitions labelled $(A^0, X)'$, the two-hole transitions (X'_A, X'_B, and X'_C), are separated from the (A^0, X) transitions by the energy difference between the $1S_{3/2}$ and $2S_{3/2}$ acceptor states. Because of the significant difference in 1S energies for different acceptor species due to central cell effects, these peaks are very useful for identification of different acceptors.

The most complete tabulation of acceptor photoluminescence bands and ionization energies has been given by Ashen *et al.*[17] Their results were obtained by careful measurements of the near-band-edge photoluminescence from a large number of different samples, some intentionally doped and some undoped, grown by several different techniques in many different laboratories. A summary of the photoluminescence transitions observed in GaAs is shown in Figure 6.7.[17–21] In general, Ashen *et al.*[17] observe that when the photoluminescence peaks are sharp and the (P) and (BA) peaks are mutually well-resolved, the peak positions are reproducible within about 1 meV from sample to sample. For other samples, the shift of the photoluminescence peaks can be greater than the chemical shifts for the different acceptor impurities, making identification unreliable unless other information concerning the acceptors, such as intentional doping, growth conditions, etc., is available.

Not surprisingly, there are distinct differences in the residual acceptor impurities in samples prepared by the different growth techniques, and these

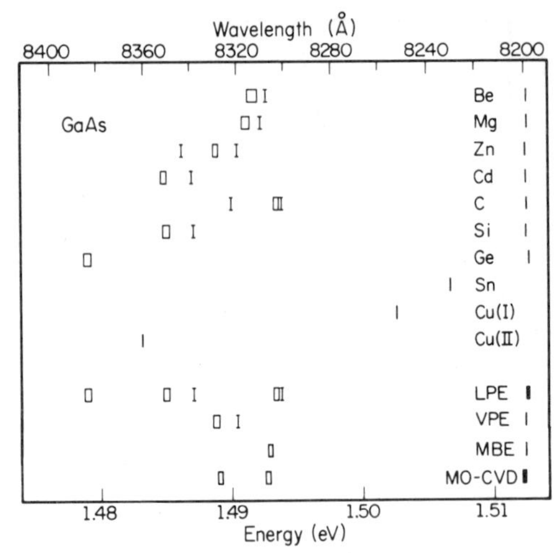

Figure 6.7 Energies of the bound exciton and bound exciton–two-hole transitions at 1.5 K and the band-to-acceptor (BA) transitions ($T \simeq 5$ K) for the common acceptor impurities in GaAs. The photoluminescence lines observed in undoped high-purity GaAs grown by LPE, VPE, MBE, and MOCVD are also shown in the lower part of this figure. (After Ashen *et al.*[17] Reproduced by permission of Pergamon Press and the Controller of Her Majesty's Stationery Office). Additional data added for excitons bound to copper complexes (Willman *et al.*[19]), and recent results on MBE (Künzel and Ploog[20]) and MOCVD (Bishop and Klein[21]) material. The exciton lines are all doublets, corresponding to $J = \frac{3}{2}$ and $J = \frac{5}{2}$ initial states, except for the Sn and Cu lines, which are singlets corresponding to the $J = \frac{1}{2}$ initial state

differences seem to be consistent for samples prepared in various laboratories. Each high-purity growth technique commonly produces epitaxial layers which contain trace residual acceptor impurities that are characteristic of the particular technique.

Ashen *et al.*[17] note that for unintentionally doped vapour-phase epitaxially grown (VPE) samples (generally Ga–$AsCl_3$–H_2), Zn acceptors are almost universally present, Si acceptors are never present, and there is usually a photoluminescence line associated with oxygen. In two VPE layers grown under low As partial pressures, faint traces of C were found. Ozeki *et al.*[18] reported the presence of photoluminescence bands due to residual Zn, Si, and Ge acceptor impurities in undoped VPE layers grown using the Ga–$AsCl_3$–H_2 technique, but in undoped VPE layers grown using the Ga–$AsCl_3$–N_2 technique, the only residual acceptor impurity detected was C.

For high-purity, undoped liquid-phase epitaxially grown (LPE) samples, C acceptors were always present and were generally the dominant shallow acceptor, even when the LPE layers were grown in non-C-containing apparatus. It is thought[22] that a possible source of C in these layers could be relatively large amounts of CO that might be present in Pd-diffused H_2.[23] Si acceptors were also regularly found, and Ge acceptors were also detected in many of the LPE layers examined. Zn was never identified as a residual acceptor in any of the LPE samples studied.

The source of many of these residual acceptor impurities is not known. In particular, the source of Zn in Ga–$AsCl_3$–H_2 VPE material, the source of C in LPE material grown in non-C-containing apparatus and in Ga–$AsCl_3$–N_2 VPE material, and the source of Ge in LPE material have not been positively identified. Although GaAs is the best characterized and most highly developed compound semiconductor, it is clear that much work remains to be done to obtain a complete understanding of the residual acceptor impurities.

Identification of acceptor impurities in InP

The photoluminescence characteristics of high-purity InP are very similar to those just described for GaAs.[24] Using the same techniques as for GaAs, with intentionally doped and undoped samples, many acceptor impurities in InP have been tentatively identified.[25–30] The results of these measurements are summarized in Figure 6.8. It is interesting to note again that LPE and VPE samples

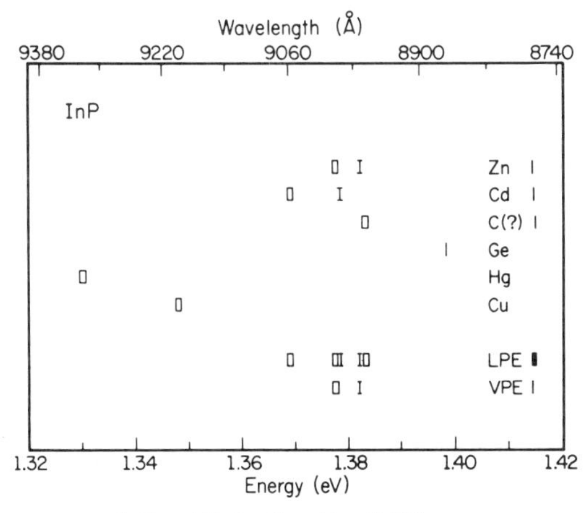

Figure 6.8 Energies of the bound exciton and bound exciton–two-hole transitions at 1.5 K and the band-to-acceptor (BA) transitions ($T \simeq 5$ K) for the common acceptor impurities in InP. Data for Zn and Cd are taken from White *et al.*[25]; for the impurity tentatively identified as C from Hess *et al.*[26]; for Ge from White *et al.*[29]; for Hg and Cu from Williams *et al.*[30] The shallow exciton lines and their replicas are doublets, except for the deeper Ge line, which is a $J = \frac{1}{2}$ singlet. The photoluminescence lines observed for the residual acceptors in undoped high-purity InP grown by VPE (from reference 25) and LPE (from reference 26) are also shown in the lower part of this figure

are generally characterized by different residual acceptor species although, in contrast to GaAs, one acceptor species, Zn, is common to the two growth methods. Also, Zn is the only residual acceptor species common to both VPE GaAs and VPE InP. The common acceptor species in LPE InP are Cd, Zn, and possibly C, and in LPE GaAs the common residual acceptor species are Ge, Si, and C. It should be pointed out that Si acceptors have not been observed in InP.[27]

6.2.3 Photothermal ionization donor spectroscopy

A simple hydrogenic model can be used to estimate the impurity energy levels for an isolated shallow donor in GaAs or InP. Using the dielectric constant and effective mass appropriate for the semiconductor (instead of for the vacuum), a model Hamiltonian for the donor electron can be written as

$$H = \frac{-\hbar^2}{2m^*}\nabla^2 - \frac{e^2}{\varepsilon_0} \tag{6.6}$$

Except for the replacement of the free electron mass m by m^* and e^2 by e^2/ε_0, this is the familiar Hamiltonian of the hydrogen atom. The replacements simply rescale the energy and distance scales of the hydrogenic solutions giving the energy spectrum and Bohr radius as

$$E_n = m^*e^4/(2\varepsilon_0^2\hbar^2 n^2) = [(m^*/m)/\varepsilon_0^2 n^2] \times 13.6\ \text{eV} \tag{6.7}$$

and

$$a_B = (\hbar^2/m^*e^2)\varepsilon_0 = [\varepsilon_0/(m^*/m)] \times 0.529\ \text{Å} \tag{6.8}$$

where (m^*/m) is the effective mass ratio, and ε_0 is the relative static dielectric constant of the semiconductor.

For a direct-bandgap semiconductor with a nearly isotropic and parabolic conduction band minimum in which the donor ionization energy is small compared to the bandgap, such as GaAs or InP, the simple hydrogenic model should predict the observed donor transition energies quite well. This is the case, as can be seen from the agreement between the transition energies predicted from the hydrogenic theory with those observed in the photoconductivity spectrum of a high-purity GaAs sample shown in Figure 6.9.

The simple hydrogenic model above predicts identical ionization energies for different donor species. However, when the electron is within a few lattice constants of the ion core, the potential felt by the donor electron is no longer the simple Coulombic potential in equation (6.6) but rather is dependent on the details of the electronic structure of the donor ion core and the local distortion of the lattice around it which gives rise to the central cell correction or chemical shift of the particular donor species. The small shoulder on the high-energy side of the 1s–2p transition in Figure 6.9 is due to a second donor species with a slightly larger ground-state energy. The impurity-species dependence of the central cell

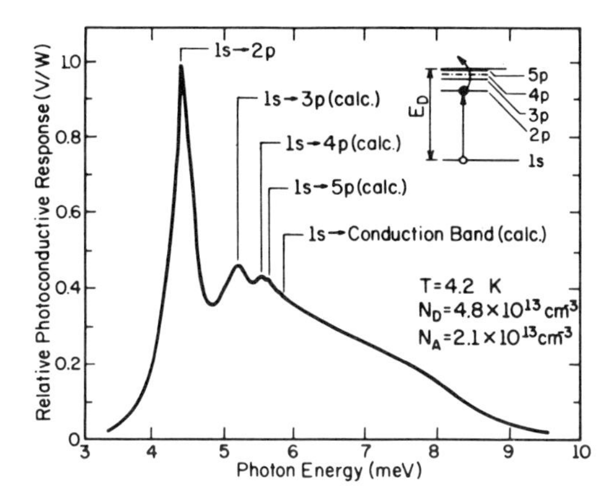

Figure 6.9 Typical photothermal-ionization far-infrared photoconductivity spectrum of a high-purity GaAs sample with zero magnetic field. The inset is an energy level diagram of a shallow donor showing the transition responsible for the dominant photoconductivity peaks. For this sample, $\mu_{77} = 210{,}000\ cm^2V^{-1}s^{-1}$

corrections is of great interest even though it is small, however, because it provides a method for the identification of donors in GaAs and InP through photothermal ionization spectroscopy at high magnetic fields.[31]

Photothermal ionization mechanism

Photothermal ionization spectroscopy is a technique by which the energy level spectra and the associated central cell corrections may be measured through far-infrared photoconductivity in high-purity semiconductors. This is an extremely sensitive method for detection of shallow donors in high-purity semiconductors. Other methods, such as optical absorption, secondary ion mass spectrometry (SIMS), and spark source mass spectrometry (SSMS) measurements, are inadequate to identify impurities at the extremely small concentrations present in high-purity epitaxial material. The extreme sensitivity of this technique is due to the fact that the total photoresponse is relatively independent of the impurity concentration down to very low concentrations. This is because the voltage sensitivity of the extrinsic photoconductivity is proportional to $\Delta n/n$, where n is the total carrier concentration without illumination and Δn is the change in carrier concentration under illumination. Both n and Δn are proportional to the concentration of uncompensated majority impurities, and thus this dependence cancels out of the voltage sensitivity. This feature of the method enables the detection and identification of impurity species even in the highest-purity semiconductor material which can be grown.

The mechanism of photothermal ionization can be understood by considering a donor electron in its ground state as shown in the inset of Figure 6.9. Such an electron can make a transition to one of the donor's excited states by absorption of a far-infrared photon of appropriate energy. Although this electron is still bound, it can be excited into the conduction band (by absorption of phonons) where it contributes to the sample's conductivity as indicated. For this process to occur, the sample temperature must be chosen so that there are ample phonons energetic enough to excite an electron from the donor excited state of interest into the conduction band, and yet few phonons sufficiently energetic to excite electrons directly from the impurity ground state into the conduction band. For GaAs and InP this limits the temperature to a range of about 1.5 to 6 K.

The first excited state of a hydrogenic donor is four-fold degenerate. When a magnetic field is applied, the 2p states split into the 2p, $m = -1$, 2p, $m = 0$, and 2p, $m = +1$ states, as shown in Figure 6.10, in the same way as for the corresponding states in the hydrogen atom. The splitting of these states is reflected in the photothermal ionization spectra, as can be seen in Figure 6.11. As the magnetic field is increased from zero, the dominant 1s→2p peak shown in the zero-field spectrum of Figure 6.9 splits into three major groups of peaks corresponding to 1s→2p, $m = -1, 0$, and $+1$ transitions, as shown in the spectra for the same sample at two different magnetic fields in Figure 6.11. Each group of 1s → 2p transitions in this spectrum consists of three well-resolved peaks, each of which results from the excitation of an electron from the ground state of a different chemical donor species with a slightly different ground-state energy to the nearly degenerate 2p excited states. Some of the other transitions are also

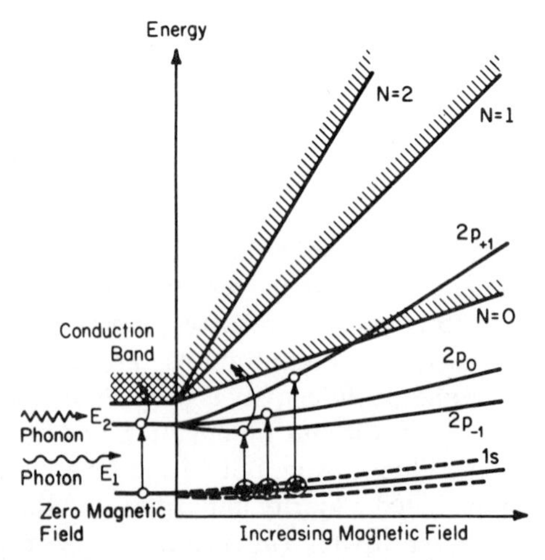

Figure 6.10 Diagram showing the donor energy levels at zero magnetic field and the splitting of the 2p and conduction band Landau levels with magnetic field. The 1s ground-state energy levels indicated by full and broken curves represent different donor species with slightly different ground-state energies due to different central cell corrections. The excited states are essentially identical for all of the impurity species

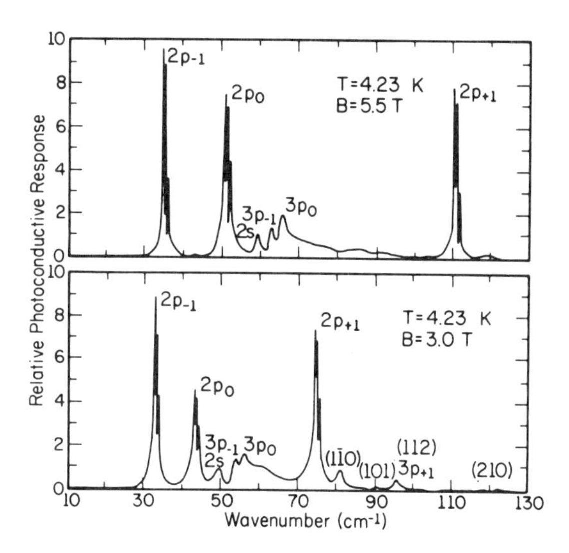

Figure 6.11 High-resolution far-infrared photoconductivity spectra of the same sample as in Figure 6.9 at two different magnetic fields. Each peak in the three 2p groups, 2p, $m = -1$, 2p, $m = 0$, 2p, $m = +1$, corresponds to a different donor impurity with a different ground-state energy as indicated schematically in Figure 6.10

labelled. Figure 6.10 shows the additional ground-state energy levels responsible for the extra 1s→2p transitions in these spectra as broken curves. Each of the curves for the 1s states in Figure 6.10 results from a different chemical donor species. The width and shape of the peaks in the spectra are determined by the random distribution of electric fields at the various donor sites in the crystal. These fields originate from the spatial distribution of neighbouring charged acceptors and ionized donors. Larsen has predicted lineshapes for 1s→2p, $m = -1$ transitions which agree well with lineshapes observed in high-purity GaAs samples on the basis of the resulting Stark shifts of the donor energy levels.[32]

The 1s → 2p, $m = -1$ transitions are of particular interest since they have the largest amplitude and narrowest linewidths for magnetic fields of about 5.0 T. At high magnetic fields, three conditions facilitate impurity identification using the 1s → 2p, $m = -1$ transitions: (1) the 1s → 2p, $m = -1$ lines have been Zeeman-split away in energy from other broader 1s → 2p lines; (2) the electric quadrupole moment of the 2p, $m = -1$ state has been reduced nearly to zero and so the main contribution to Stark broadening of the 1s → 2p, $m = -1$ transition has been removed;[32] and (3) the compression of the 1s wavefunctions by the magnetic field has increased the overlap with the central cell and thus has enhanced the central cell shifts of each of the 1s states. As a result, it is possible to resolve peaks in this transition corresponding to different donor species through the use of high magnetic fields.

Identification of donor impurities in GaAs

The relative position of a peak within the $1s \rightarrow 2p$, $m = -1$ envelope is characteristic of a given donor species and this information can be used to identify different chemical donors in high-purity GaAs. Figure 6.12 shows high-resolution spectra of two high-purity GaAs samples in a magnetic field of 5.35 T. The spectrum in the lower part of the figure is for a high-purity GaAs sample grown by Wolfe[33] using the chloride ($AsCl_3$) process, and the spectrum in the upper part of the figure is for a sample grown by Dapkus *et al.*[34] using the metalorganic chemical vapour deposition (MOCVD) technique. The $1s \rightarrow 2p$, $m = -1$ transition in each sample has a well-resolved fine structure which is the result of the species-dependent central cell shifts of the donor ground-state energies.

The three peaks labelled X_1, X_2, and X_3 in these spectra are the commonly observed residual donor impurities in VPE GaAs, and the small peak between X_1 and X_2 in the spectrum for the MOCVD sample has been previously identified with Sn donors.[33] Back-doping, transmutation doping experiments, and correlation with various growth parameters have been used to attempt to identify the

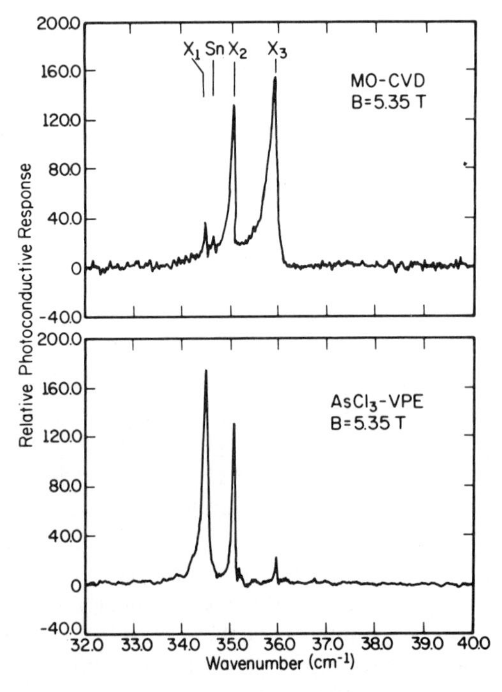

Figure 6.12 High-resolution spectra of just the $1s \rightarrow 2p, m = -1$ transitions for two different high-purity GaAs samples prepared by different growth techniques, at a magnetic field of 5.35 T. The MOCVD sample had $N_D - N_A = 7.0 \times 10^{13}$ and $\mu_{77} = 99{,}100\ cm^2V^{-1}s^{-1}$ and the $AsCl_3$ sample had $N_D - N_A = 4.5 \times 10^{13}$ and $\mu_{77} = 201{,}000\ cm^2\ V^{-1}\ s^{-1}$

chemical donor species responsible for the peaks X_1, X_2, and X_3 as well as other peaks observed in high-purity samples grown by LPE and MBE techniques.[33–39]

The $1s \rightarrow 2p, m = -1$ transition energies observed for GaAs in a magnetic field of about 5.5 T are summarized in Figure 6.13. The donor peaks observed in unintentionally doped high-purity GaAs are for the most part independent of the laboratory in which the samples were prepared. The residual donors characteristic of each particular growth technique are shown in the lower part of the figure by vertical lines at the wavenumber appropriate for a magnetic field of 5.5 T. The lengths of the lines indicate the relative concentrations of the different residual impurities generally observed. The broad Sn line present in n-type molecular-beam epitaxial (MBE) samples results from intentional Sn doping since for this growth technique undoped samples are typically p-type.

Considerable work obviously remains to obtain absolute identification of the donor impurities in GaAs and in particular to identify the source of the residual impurities in the various growth techniques. The present results and identifications pose many interesting and new problems concerning impurity incorporation in epitaxial GaAs. For example, if X_1 is due to Si, then Si is not a significant residual donor impurity in LPE GaAs. If X_3 is due to Ge, then neglecting possible degeneracies of different chemical donor species (e.g. C and Ge), Ge is the

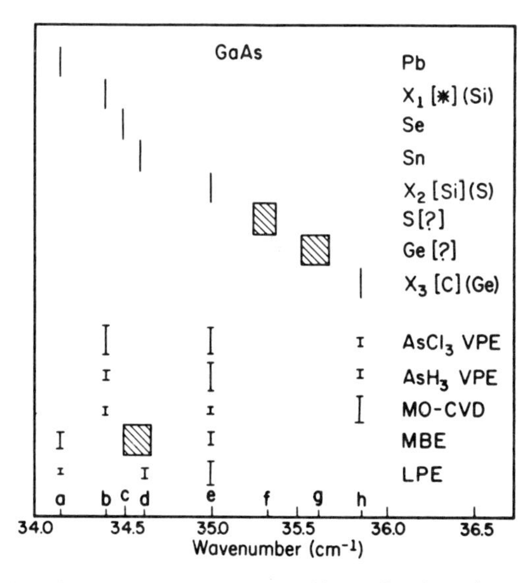

Figure 6.13 Energies of the $1s \rightarrow 2p, m = -1$ transitions of various donor species in GaAs at a magnetic field of 5.5 T. The donor identifications enclosed in parentheses (□) are due to Ozeki *et al.*[37] Those in brackets [□] are due to Wolfe *et al.*[33] and Stillman *et al.*,[36] and are in general agreement with the identification of Cooke *et al.*[35] The common residual donor transitions in samples grown by various techniques are shown in the lower part of this figure

dominant residual donor in MOCVD GaAs. If X_2 is due to S as identified by Ozeki *et al.*,[37] then the two other peaks attributed to S by Stillman *et al.*[36] and Afsar *et al.*[39] are unexplained. In spite of these complications and uncertainties, photothermal ionization spectroscopy is the only characterization method that can reliably distinguish and identify the electrically active chemical donor species present in high-purity epitaxial GaAs. This characterization technique will certainly be more widely used in the future.

Identification of donor impurities in InP

The photothermal ionization technique can also be applied to the identification of shallow donor impurities in InP. Early measurements were made on material that was not pure enough to be able to resolve the central cell corrections,[39] but measurements are presented in section 6.3.3 where two different residual impurities have been resolved in both LPE- and hydride (PH_3) VPE-grown epitaxial samples. Stradling has made laser magnetospectroscopy measurements on intentionally doped InP samples and has identified the same two peaks observed in our work with Si and S. He also observed several peaks which he associated with stoichiometic defects.[35]

As higher-purity InP becomes available, the photothermal ionization technique will permit the detection and identification of the donor impurities associated with the various growth techniques.

Effective-mass measurements

Other information which can be extracted from the photothermal ionization spectra includes the band edge effective mass m_0^*. It is easy to show, for a hydrogenic donor in an arbitrary magnetic field, that the splitting between the 2p, $m = +1$ and the 2p, $m = -1$ states is given by

$$\Delta E_{2p\pm} = \frac{e\hbar B}{m_0^* c} \tag{6.9}$$

To the extent that donors behave according to the hydrogenic model, and to the accuracy with which the magnetic field can be measured, the bottom of the band effective mass m_0^* can be determined spectroscopically from the difference between the $1s \rightarrow 2p$, $m = +1$ and $1s \rightarrow 2p$, $m = -1$ transition energies.

Table 6.1 Conduction band effective masses determined from photothermal ionization measurements

GaAs	0.0665	± 0.0005
InP	0.0796	± 0.0006
$Ga_{0.47}In_{0.53}As$	0.0437	± 0.001
$Ga_{0.17}In_{0.83}As_{0.40}P_{0.60}$	0.0635	± 0.001

This technique has been utilized in GaAs, InP, $Ga_{0.47}In_{0.53}As$, and $Ga_{0.17}In_{0.83}As_{0.40}P_{0.60}$ lattice matched to InP to determine m_0^* for these materials subject to the conditions mentioned above. Table 6.1 shows the values of effective masses determined in this way.

6.3 HIGH-PURITY GROWTH TECHNIQUES

As indicated in the previous section, essentially all of the common epitaxial growth techniques have been used to grow reasonably high-purity epitaxial GaAs. These techniques include tipping, dipping, and slider boat LPE, Ga–$AsCl_3$–H_2, Ga–$AsCl_3$–N_2, and Ga–HCl–AsH_3–H_2 VPE growth, metal-organic chemical vapour deposition (MOCVD), and molecular-beam epitaxy (MBE). The capability of achieving high-purity growth is of course dependent on the purity of the starting materials, but the impurity incorporation can be influenced by substrate orientation and other effects, as discussed by Wolfe and Stillman.[40] However, one of the most important factors in achieving high-purity VPE GaAs material appears to be the cleanliness and leak integrity of the growth systems. Once these requirements are satisfied, it is usually possible to adjust the growth conditions (gas composition) to obtain high-purity, high-mobility material. For LPE growth, however, the growth conditions cannot be changed significantly, as is possible in VPE, so the purity in this growth technique may be influenced more strongly by impurities in the boat, substrate, starting materials, and growth reactor in general. However, careful baking techniques have resulted in LPE GaAs semiconductor layers with purity comparable to the best VPE material, and molecular-beam epitaxy and metalorganic chemical vapour deposition have also produced high-purity epitaxial GaAs. Although the growth techniques for high-purity InP are not as highly developed as those for GaAs, many of the same considerations apply to both materials.

6.3.1 General considerations

Silicon contamination

Si is generally recognized as a ubiquitous residual impurity in GaAs because of the reaction of H_2 and HCl with the silica reactor tubes generally used for Ga–$AsCl_3$–H_2 and Ga–HCl–AsH_3–H_2 VPE and LPE growth techniques. The thermochemistry of the SiO_2–Ga–H_2 and SiO_2–Ga–HCl systems has been studied in detail by many authors.[41–43] The various Si contamination reactions with H_2, HCl, and Ga are given in Table 6.2. Similar reactions also occur for In. These reactions apply to both VPE and LPE growth in SiO_2 reactors, although reaction (6.16) only applies to the case where liquid Ga (or In) is in direct contact with SiO_2. Reaction (6.13) shows that even for a system that is perfectly dry and oxygen-free at room temperature, there can be a significant concentration of

Table 6.2 Silicon contamination reactions[42]

Reaction	
From $H_2(g)$:	
$SiO_2(s) + H_2(g) \rightleftarrows SiO(g) + H_2O(g)$	(6.13)
$2Ga(l) + H_2O(g) \rightleftarrows Ga_2O(g) + H_2(g)$	(6.14)
$2Ga(l) + SiO(g) \rightleftarrows Ga_2O(g) + Si(l)$	(6.15)
From silica boat:	
$4Ga(l) + SiO_2(s) \rightleftarrows 2Ga_2O(g) + Si(l)$	(6.16)
From HCl(g) and $H_2(g)$:	
$nHCl(g) + SiO_2(s) + (4-n)H_2(g) \rightleftarrows SiCl_nH_{4-n}(g) + 2H_2O(g)$ $n = 1, 2, 3, 4$	(6.17)
$(4-n)Ga(l) + SiH_nCl_{4-n}(g) \rightleftarrows (4-n)GaCl(g) + Si(l) + (n/2)H_2(g)$	(6.18)
$(n-2)H_2(g) + SiCl_nH_{4-n}(g) \rightleftarrows Si(\text{in GaAs}) + nHCl(g)$	(6.19)

water vapour in the system when it is hot, simply due to dissociation of the SiO_2 in the presence of H_2 to form gaseous SiO and H_2O. Wrick *et al.*[44] followed the work of Hicks and Greene[41] to calculate the water vapour pressure that will be present in a perfectly leak-tight silica system at a particular temperature simply from reduction of the SiO_2 by the hydrogen gas. Subsequent work has shown that their calculated values are at least two orders of magnitude too high (they calculate that at 750 °C there will be more than 1 ppm H_2O vapour in a leak-tight silica reactor),[22] but this work emphasized the importance of this process.

It has been observed by many workers that the purity of VPE GaAs grown using the Ga–$AsCl_3$–H_2 technique varies inversely with the partial pressure of $AsCl_3$ in the reactor. DiLorenzo and Moore[45] explained this 'mole fraction' effect using the reactions (6.13)–(6.19) to calculate the activity of Si in the vapour and hence in the solid GaAs. In this model, the activity of Si is controlled by the partial pressure of HCl in the growth zone through the formation of the chlorosilanes, $SiCl_nH_{4-n}$. Studies of the mole fraction effect show that Si can be a dominant residual donor in GaAs grown by the Ga–$AsCl_3$–H_2 technique. Recent work by Ashen *et al.*[46] using a BN furnace liner with the $AsCl_3$ system indicates that it is the attack of the SiO_2 by the HCl which is the primary source of Si using this growth technique.

The problem of Si contamination of high-purity LPE GaAs grown in Spectrosil quartz boats was studied by Hicks and Greene.[41] They considered the net result of reactions (6.13) and (6.15) and showed that Si contamination is dependent on the H_2O content of the gas stream and the temperature. The work of Hicks and Greene demonstrated that Si acts as both a donor and an acceptor in LPE GaAs in a ratio determined by the growth temperature. Predominantly donor behaviour occurs at growth temperatures greater than about 860° C, and predominantly acceptor behaviour occurs for growth temperatures less than about 860° C. The work of Hicks and Greene indicated that contamination of a Ga melt by Si through reduction of the silica boat can be limited to an 'acceptable' level through increasing the water content of the H_2 gas and working at a sufficiently low temperature. Using these techniques they could reproducibly grow n-type layers with n in the low $10^{13}\ \text{cm}^{-3}$ range and $\mu_{77} > 100\,000\ \text{cm}^2\ \text{V}^{-1}\ \text{s}^{-1}$. This behaviour is not consistent, however, with Si being the only residual donor or even the dominant residual donor in high-purity LPE GaAs.

Impurity incorporation

It is well-known that many different factors can influence the incorporation of impurities in compound semiconductors. One factor that is often considered, using GaAs as an example, is the relative concentration of Ga and As vacancies at the growth temperature. By considering only Ga and As vacancies (V_{Ga} and V_{As}) Ashen *et al.*[17] calculated the equilibrium ratio of the concentration of Ga to As vacancies ($[V_{Ga}]/[V_{As}]$) as a function of temperature and As pressure (P_{As_2}). Using this result, they calculated that for LPE growth at 750° C the ratio of Si donors to acceptors is $[Si_{Ga}]/[Si_{As}] = 0.25$, while for typical $AsCl_3$ VPE growth conditions ($P_{AsCl_3} = 6 \times 10^{-3}$ atm, $T = 750°$ C) $[Si_{Ga}]/[Si_{As}] = 1.6 \times 10^3$. These results imply that for LPE crystals the ratio of Si donors to Si acceptors should be 0.25 while for VPE crystals the ratio of Si donors to Si acceptors should be 1.6×10^3. Thus, if this were the correct mechanism and Si was the only impurity present, VPE GaAs would be n-type and relatively uncompensated while LPE GaAs would be p-type with $N_A/N_D \simeq 4$.

These calculations were based on equilibrium growth conditions and it is well-known that these conditions are rarely achieved in practice. Different crystal orientations exhibit vastly different growth rates, net carrier concentrations, and compensation ratios,[47] and such a dependence on substrate orientation can only be explained by non-equilibrium growth processes. Two models of non-equilibrium impurity incorporation have been considered:

(1) Adsorption—this model assumes that impurity atoms which are adsorbed at the growth interface are incorporated in the epitaxial layer to a greater or lesser extent depending on the growth rate or interface velocity.[48,49] In this case orientation dependence in impurity incorporation would come from orientation dependence in surface or interface adsorption.

(2) Band-bending or Fermi level effects—this model assumes that the Fermi level is fixed at the growth interface or surface by a Schottky barrier-like effect in LPE[50,51] or by surface states in VPE.[52] In this case, orientation dependence in impurity incorporation would come from orientation dependence of the Schottky barrier height or of the surface state density, respectively.

The band-bending model has been used to obtain quantitative agreement between the calculated and experimental distribution coefficients for LPE growth,[51] and between the calculated and experimental incorporation of donors and acceptors, including the compensation ratio, for both group IV and group VI impurities in GaAs.[53] No quantitative treatment of the adsorption model is available so it cannot be compared directly with experimental results. More work remains to understand the important impurity incorporation mechanisms in compound semiconductors and in semiconductor alloys in particular.

Contamination from substrates

In addition to impurities from the growth reactor and starting materials, another source of contamination that can degrade the purity of the epitaxial layer is the substrate itself. The properties of high-purity layers grown on semi-insulating substrates can be dominated by substrate effects, and it is not uncommon for a large fraction of a high-purity layer grown on a Cr-doped semi-insulating substrate actually to be high-resistivity or semi-insulating also. This behaviour is due to out-diffusion of acceptors from the substrate during epitaxial growth, and it is generally agreed that Cr is responsible for this behaviour although there are still some puzzling questions concerning the behaviour of the fast-diffusing acceptor.[54] Autodoping from both semi-insulating and conducting substrates can also influence the purity of epitaxial layers.

6.3.2 High-Purity VPE

Vapour-phase epitaxy of GaInAsP quaternary compounds is now being studied extensively in several laboratories throughout the world. So far, most of this work has been directed towards device fabrication techniques, and no high-purity results have been reported. However, because vapour-phase techniques will undoubtedly be important in future applications of GaInAsP, some of the high-purity VPE results for GaAs and InP will be briefly summarized even though space does not permit a complete review of this important work.

Gallium arsenide

Chloride ($AsCl_3$) growth techniques Results on high-purity VPE growth of GaAs prior to 1970 have been reviewed previously.[40] At that time the only VPE technique that had produced high-purity GaAs was the $AsCl_3$–Ga–H_2 vapour-phase system. Epitaxial material with carrier concentrations in the mid 10^{13} cm^{-3} range and liquid-nitrogen-temperature mobilities of $\mu_{77} \simeq 200{,}000$ cm^2 V^{-1} s^{-1} could be routinely produced. The highest-purity results were initially obtained for growth on ⟨211⟩Ga oriented substates, but subsequent work on the $AsCl_3$–Ga–H_2 technique has yielded samples with equally high liquid-nitrogen-temperature mobilities for ⟨100⟩ and ⟨111⟩ Ga oriented material. Growth of high-purity material using this technique has been subsequently accomplished in many laboratories.

Since 1970 high-purity vapour-phase GaAs has been prepared by many other techniques, and much research is still being done to refine these methods. One variation of the $AsCl_3$–Ga–H_2 technique is the use of N_2 as the carrier gas instead of H_2.[55] This method appears to have the following advantages: (a) the Si contamination caused through reduction of the silica reactor tubes by H_2 can be decreased (equations (6.13)–(6.15) in Table 6.2); (b) the growth temperature can be lower for this system, also reducing impurities; and (c) there is more complete gas mixing and uniformity with N_2 as the carrier gas. Photoluminescence studies by Ozeki and coworkers[18] of samples grown in the same $AsCl_3$ growth system with either H_2 or N_2 as the carrier gas indicated that the residual acceptors for the samples grown with H_2 as the carrier gas were C, Zn, Si and Ge, while samples grown with N_2 as the carrier gas contained only C as a residual acceptor. The samples grown with H_2 appeared to be considerably more compensated than the samples grown with N_2, and these workers attributed this to a reduction of sample inhomogeneities, which result in space-charge scattering, due to more thorough gas mixing with N_2. Later studies of residual donors in samples grown by these two techniques were interpreted to indicate that Si donors were also eliminated from the samples grown with the N_2 system, with S and Ge the remaining residual donors.[37] This technique has produced samples with μ_{77} in the range from 180,000 to over 200,000 $cm^2\,V^{-1}\,s^{-1}$.

Another recent variation of the chloride VPE technique that has produced high-purity material is the 'single flat temperature zone' technique where the source and substrate are maintained at the same temperature.[56] This system permits simple control of the furnace temperature since only one flat temperature zone is required. However, a more important advantage is that Si contamination of the Ga in the source boat may be reduced because the source temperature can be lower than for the standard $AsCl_3$ growth techniques. Samples with liquid-nitrogen-temperature mobilities as high as $\mu_{77} = 198{,}300\,cm^2\,V^{-1}s^{-1}$ have been obtained with this technique.

Hydride (Ga–AsH_3–HCl–H_2) growth technique While the chloride vapour-phase growth technique has produced the highest-purity GaAs, other VPE growth techniques are of more interest for the growth of GaInAsP alloys. The hydride (Ga–AsH_3–HCl–H_2) growth technique is particularly interesting because of the ability to vary the III/V ratio independently and because of the ability to change alloy composition abruptly, at least compared to the chloride systems where there is a considerable transient during any change in $AsCl_3$ (or PCl_3) flow over the Ga (or In) boat due to the requirement of source saturation. The purity obtained using this technique has generally not been as high in early work as with the chloride technique, but layers with $\mu_{77} \simeq 100{,}000\,cm^2V^{-1}s^{-1}$ were grown.[57] Subsequently, layers with $\mu_{77} = 167{,}000\,cm^2V^{-1}s^{-1}$ were grown by Kennedy[58] using this technique, and samples with $\mu_{77} \simeq 200{,}000\,cm^2V^{-1}s^{-1}$ have recently been grown.[59] These results make the hydride growth technique very promising for the growth of high-purity GaInAsP.

Metalorganic chemical vapour deposition This vapour-phase epitaxial growth technique has been extensively studied for GaAs and GaAlAs device applications, but there have only been two reports of high-purity growth using this technique. The first report by Seki *et al.*[60] used triethyl-gallium as the metalorganic source, and samples with $\mu_{77} \simeq 120{,}000\,\mathrm{cm^2V^{-1}s^{-1}}$ were grown. In more recent work, Dapkus *et al.*[34] used trimethyl-gallium as the metalorganic source to grow samples with $\mu_{77} = 125{,}000\,\mathrm{cm^2V^{-1}s^{-1}}$. Far-infrared photoconductivity measurements indicated that the dominant residual donor impurities in undoped MOCVD material were X_2 and X_3, and photoluminescence measurements revealed that C and Zn were the dominant residual acceptors. These results indicate that this technique for the growth of high-purity GaInAsP is very promising.

Molecular-beam epitaxy High-purity, unintentionally doped GaAs grown by MBE with a metallic As source is p-type, with net acceptor concentrations in the low $10^{14}\,\mathrm{cm^{-3}}$ range. For applications requiring high-purity *n*-type material, low levels of Sn doping are used. Morkoç and Cho[61] achieved an electron concentration of $5 \times 10^{14}\,\mathrm{cm^{-3}}$ and $\mu_{77} = 60{,}000\,\mathrm{cm^2V^{-1}s^{-1}}$ for a $3\,\mu$m thick layer. A 27 μm thick layer grown under similar conditions for a period of 30 h had a carrier concentration of $4 \times 10^{14}\,\mathrm{cm^{-3}}$ but with $\mu_{77} = 105{,}000\,\mathrm{cm^2V^{-1}s^{-1}}$. Subsequent photoluminescence measurements[62] have shown that the dominant residual acceptor in this sample is C, and far-infrared photoconductivity measurements on this same sample have been used to identify tentatively Pb, X_1, and X_2 as residual donors in addition to the intentionally added Sn.[63] The increase in the mobility from the 3 μm to the 27 μm thick samples may have been influenced by several factors in addition to the thickness. The long growth time (27 h) may have served to outgas the effusion cells more effectively and may also have 'passivated' the walls of the growth chamber with a thin GaAs coating, so that the thicker layer may actually be of higher purity than the thinner one.

The growth of high-purity MBE GaAs using AsH_3 as the As source has recently been reported by Calawa.[64] The AsH_3 is decomposed into As_1, As_2, and As_4 when it flows through a cracking furnace as it is introduced into the growth chamber. A GaAs sample 5.4 μm thick grown in a 5 h period using this technique has a liquid-nitrogen-temperature mobility of $110{,}000\,\mathrm{cm^2\,V^{-1}\,s^{-1}}$ and an electron concentration of $2.4 \times 10^{14}\,\mathrm{cm^{-3}}$ when allowance was made for surface and interface depletion in this thin sample. The undoped epitaxial layers grown using this technique were n-type in contrast to samples grown using metallic As as a source. Far-infrared photoconductivity measurements on these samples indicate that the residual impurities are also Pb, X_1, Sn, and X_2.[63] The Sn in the sample grown by Calawa probably results from previous use of Sn doping in the growth chamber. These results show that thin, high-purity epitaxial layers can be grown by MBE, and thus this technique is promising for future use in the growth of high-purity GaInAs and possibly GaInAsP alloys.

Indium phosphide

Chloride growth technique The VPE growth technique for InP that has been most thoroughly studied is the chloride (PCl_3) technique. This technique has produced the highest-purity InP that has been prepared. The first report of high-purity results using this method was by Clarke *et al.*[65] and they achieved a mobility of 44,000 $cm^2 V^{-1} s^{-1}$ at 77 K. Joyce and Williams[66] subsequently achieved values of μ_{77} as high as 60,100 $cm^2 V^{-1} s^{-1}$. Fairman *et al.*[67] studied the characteristics of this growth technique and found that the mole fraction effect previously reported[68] was due to the same mechanism as in GaAs, and postulated that Si is also the donor impurity responsible for the background doping dependence observed in the PCl_3 mole fraction effect. They obtained one sample with μ_{77} = 140,220 $cm^2 V^{-1} s^{-1}$ the highest value ever reported for InP, but the next highest mobility they obtained was μ_{77} = 73,950 $cm^2 V^{-1} s^{-1}$. A summary[67] of published data up to that time (1976) indicated only four samples with μ_{77} > 60,000 $cm^2 V^{-1} s^{-1}$. A recent summary [69] of PCl_3 growth of InP, however, indicates that there is much conflicting information in the literature and that 'high values of electrical characteristics are rare'. The influence of halide concentration on the electrical characteristics as previously described could not be reproduced in this later work, although low carrier concentrations and high purity were often obtained. (There are more high-purity InP samples reported in reference 69 than in all previous papers combined.) The highest mobility reported was 121,050 $cm^2 V^{-1} s^{-1}$. The authors attributed the unreproducibility to variations in the oxyhalides present in the system. The water produced in the reactor by the oxyhalides is considered to be the main source of variations in the sample electrical characteristics. Other variations in growth rate with the PCl_3 growth technique are attributed to unpredictable reactions in the dissociation of the PH_3 formed in the reactor.

As in the case of the $AsCl_3$ growth of GaAs, the III/V ratio is nearly fixed in the PCl_3 growth technique, although it can be varied somewhat by changing the PCl_3 flux.[70] This technique leads to source desaturation/saturation problems, but Clarke has described a system based on a PCl_3 source which does permit independent variation of the growth parameters by using the PCl_3 as separate sources of HCl and P through reactions to form yellow P and HCl.[71] Although the major motivation for developing this system rather than simply utilizing the hydride (PH_3) technique, which also provides independent variation of P and HCl concentrations, was the higher purity of the source material in the chloride technique, the highest mobility achieved at 77K was only about 33,000 $cm^2 V^{-1} s^{-1}$ for a sample with $n = 1 \times 10^{15}$ cm^{-3}.

Hydride growth technique Another disadvantage of the ordinary PCl_3 technique for applications to alloy growth is the inability to make rapid changes in gas composition because of the required saturation/desaturation of the metallic source when the PCl_3 (or $AsCl_3$) fluxes are changed. The hydride growth technique, as previously discussed for GaAs, is particularly well-suited for this application. However, there have only been a few reports of high-purity InP prepared in this way. Olsen and Zamerowski[72] have reported samples with μ_{77} as high as 20,000 $cm^2 V^{-1} s^{-1}$ and Zinkiewicz *et al.*[73] have reported a sample with $\mu_{77} = 56{,}000\ cm^2 V^{-1} s^{-1}$. Subsequently, Roth *et al.*[74] have obtained samples with μ_{77} as high as 71,000 $cm^2 V^{-1} s^{-1}$. All of these results were obtained with hydride VPE systems designed for the growth of GaInAsP. These results for InP, and the previously described results for GaAs using the hydride technique, make this VPE growth method very promising for the growth of high-purity quaternary alloys.

The other two VPE growth techniques, MOCVD and MBE, have been used for the growth of InP, but the results are not extensive enough to permit the evaluation of these techniques for the growth of high-purity InP. There is considerable research on these growth methods, however, and they will undoubtedly be widely used for high-purity growth in the future.

GaInAsP alloys

The hydride vapour-phase growth techniques used for epitaxial growth of GaInAsP are described by Olsen in Chapter 1, and the MOCVD and MBE growth techniques are discussed in Chapters 3 (Hirtz *et al.*) and 4 (Wood) respectively. None of the present results would qualify as high-purity. However, this lack of results probably reflects an orientation of current research towards materials that are relatively heavily doped for laser and LED applications. Zinkiewicz *et al.*[73] reported preliminary results of $\mu_{77} = 18{,}170\ cm^2 V^{-1} s^{-1}$ for $Ga_{0.47}In_{0.53}As$ and Olsen and Zamerowski[72] and Susa *et al.*[75] have reported samples with $\mu_{77} = 33{,}500$ and 35,400 $cm^2 V^{-1} s^{-1}$ respectively. Future work on high-purity growth techniques should lead to significant improvements in these results.

6.3.3 High-purity LPE

Gallium arsenide

There has been considerable effort and some success in the LPE growth of high-purity GaAs since this material first became of interest for Gunn effect or transferred electron devices. Kang and Greene[76] reported the growth of material with $\mu_{77} = 95{,}000\ cm^2 V^{-1} s^{-1}$ using a graphite boat (vacuum-baked at 1400°C for several hours prior to the experiment) and a modified tipping method. They initially speculated that the shallow donors and acceptors were silicon, since this element was always present in the spectrographic analysis of the gallium used for the epitaxial growth, but subsequently they suggested that sulphur from the graphite crucible could be the main residual donor. Mobilities as high as $\mu_{77} = 106{,}000\,cm^2\,V^{-1}\,s^{-1}$ were reported by Andre and LeDuc[77] and Solomon[78] using similar techniques.

In 1969, Hicks and Manley[79] reported the growth of 'ultra-pure' GaAs using the tipping method, but with a Spectrosil quartz boat instead of carbon. The highest mobility they obtained, $\mu_{77} = 175{,}000\,cm^2\,V^{-1}s^{-1}$, was for a sample which had been intentionally doped with Sn. Subsequently, Hicks and Greene[41] studied the conditions required to achieve high purity using a silica boat. As discussed previously, their work indicated that for growth at temperatures higher than 860° C Si was primarily a donor, while for growth below 860° C Si was primarily an acceptor. The relative concentration ratio of Si donors to acceptors varied with growth temperature (and increased with increasing temperature), and the Si concentration in the solution could be reduced by baking at lower temperatures. Also, the Si concentration could be decreased by the addition of water vapour during baking to inhibit reduction of the silica by hydrogen. It should perhaps be noted that Mattes *et al.*[80] observed apparently considerably different behaviour in their work on a fused quartz–graphite–hydrogen LPE system. They noted a change from n–type to p–type conductivity as the *baking* temperature was increased from below to above about 775° C. The transition temperature in this case was not shifted significantly by different *growth* temperatures. They did not identify the impurities involved, but attributed them to chemical reactions between the growth system components. This behaviour is not inconsistent with that reported by Hicks and Greene[41] if there are other donors present. In 1971, Miki and Otsubo[81] reported growth of high-purity GaAs using the tipping method and glassy carbon boat. They obtained values for μ_{77} as high as 214,000 and $244{,}000\,cm^2\,V^{-1}\,s^{-1}$ although these extreme values may have been influenced by sample inhomogeneities. Nevertheless, high-purity layers were grown and subsequently, they have reported reproducible growth of samples with carrier concentrations of the order of $10^{12}\,cm^{-3}$ and mobilities of $170{,}000\,cm^2\,V^{-1}\,s^{-1}$ at 77 K using a glassy carbon boat, but with an additional boat with a saturated melt to provide excess arsenic vapour over the growth boat.[82] They attributed this behaviour to a decrease in oxygen donors on arsenic sites because of increased arsenic pressure, and therefore fewer arsenic vacancies for group VI donors with the mixed gas flow. They also have shown that at 800° C the apparent distribution coefficient of Si acceptors is decreased by almost a factor of 2 when the arsenic mixed gas flow system is used, and this result has been attributed to a reduction in the incorporation of Si acceptors because of fewer available arsenic vacancy sites.

All of the LPE results which we have described so far utilized the horizontal tilt or 'tipper' reactor configuration, and this is probably best-suited for the growth of high-purity, thick epitaxial layers. For many device applications, good surface morphology and layer thickness uniformity are of equal importance to high purity, however. Rosztoczy *et al.*[83] investigated the use of a multiple-well graphite boat for the growth of high-purity layers and multilayer structures. They also used methods to increase the arsenic pressure over the substrate, as well as elevated-temperature baking of the Ga–GaAs solution in H_2 for long times to reduce the Si contamination, to obtain samples with μ_{77} as high as 75,000 $cm^2 V^{-1} s^{-1}$. Subsequently, Morkoç and Eastman[84] also studied the growth of high-purity GaAs in the widely used horizontal slider graphite boat. The boat in their system was machined from Ultra Carbon UT6-ST high-density, high-purity graphite, and before use it was vacuum-baked at a pressure of about 10^{-7} Torr and a temperature of 1400°C. Samples were grown at temperatures of 750°C, 735°C, and 700°C with various baking times. Their results indicate that additional impurities are introduced each time the reactor is opened for loading, and that a baking time between 10 h at 750°C and 29 h at 700°C with the substrate in the reactor is necessary to outgas most of these impurities. They obtained 77 K mobilities of 105,000 $cm^2 V^{-1} s^{-1}$ consistently after baking 10 h or more at 750°C, and 163,000 $cm^2 V^{-1} s^{-1}$ after baking 24 h or more at 700°C.

In this early work they hypothesized that carbon donors were the dominant residual impurity due to formation of CO at growth temperatures after the boat is exposed to oxygen (air) during loading. However, in a subsequent paper, Morkoç *et al.*[85] re-examined these samples, using photoluminescence measurements and far-infrared measurements by Stradling, and concluded that the dominant donor impurity was Si, with Sn and perhaps Pb also being present but in lower concentrations. The photoluminescence measurements indicated that by growing at 700°C after baking for 24 h carbon acceptors were nearly eliminated, leaving Si as the main acceptor. When baking was carried out for shorter times at 700°C, more carbon acceptors were observed and carbon acceptors were also seen when the baking was carried out at 780°C for 12–16 h. Epitaxial layers grown in a Spectrosil (high-purity silica) slider boat were also examined, and Si and Zn acceptors, but apparently not carbon, were detected in these samples.

Other reports also indicate that high-purity material can be grown in graphite slider boats. Abrokwah *et al.*[86] reported samples with $\mu_{77} > 100{,}000$ $cm^2 V^{-1} s^{-1}$, obtained by baking both source and substrate for about 24 h at 775°C, with growth at 700°C. Contrary to the behaviour expected from the Si contamination model, baking longer times (38.5 h) at 700°C followed by growth at 700°C only produced layers with $\mu_{77} \lesssim 82{,}000$ $cm^2 V^{-1} s^{-1}$. Higher purity was achieved in this work with higher H_2 flow rates, indicating that a residual impurity reacts readily with H_2 at these temperatures to yield a species which is carried away by the gas stream. Far-infrared photoconductivity measurements on these samples indicated that X_2 was the only remaining residual donor species.[63]

Indium phosphide

High-purity InP has also been grown by LPE with techniques similar to those used for GaAs. However, there are two distinct differences in Si incorporation in LPE InP.[27] The first is that Si apparently is not incorporated as an acceptor in InP. The second is that the distribution coefficient for Si in In melts is much greater than 1, $k_{Si} \simeq 30$. For these reasons, and because of direct chemical detection of Si at high concentrations in In, Si has been considered to be the dominant residual impurity in LPE InP.

The first report of reasonably high-purity LPE InP was by Ip *et al.*[87] They obtained mobility values as high as $\mu_{77} = 67{,}000\,cm^2V^{-1}s^{-1}$ after very long baking of the In to lower the Si background level below the ~ 0.1–0.04 ppm level present in high-purity commercially available In. (With a Si concentration of 0.1 ppm in the In and $k_{Si} \simeq 30$, the resulting donor concentration is about $10^{17}\,cm^{-3}$.) Wrick *et al.*[44] developed a baking schedule for the In which enabled them to reduce [Si] in In to 1 ppb and the Si donor concentration to about $10^{15}\,cm^{-3}$, utilizing only residual H_20 vapour in the reactor. Groves and Plonko[88] studied the controlled addition of O_2, which is converted to H_2O in the reactor at the 600–700°C temperatures used for baking and growth, and they were able to obtain samples with μ_{77} as high as $61\,000\,cm^2\,V^{-1}\,s^{-1}$. In subsequent work, Groves and Plonko[89] reported routine achievement of 77 K mobility in the 50,000–70,000 $cm^2\,V^{-1}\,s^{-1}$ range by baking for 50 h or more with about 0.7 ppm of added H_2O, independent of the source of In. (Mass spectrographic analysis of the In sources indicated impurities no higher than the 1 ppm level. The graphite boat used for the growth of high-purity samples contained about 1 ppm S.) They also noted that the simple Si contamination model apparently does not apply for carrier concentrations below the $1 \times 10^{15}\,cm^{-3}$ level.

Eastman's group at Cornell University has continued the study of some of the InP crystals grown previously and further developed the techniques for LPE growth of high-purity InP.[90] Studies of long-term baking and melt saturation with different InP material indicated that the material used to resaturate the melt after the long baking times was the source of further Si contamination. By using thick epitaxial layers as the source for this resaturation, they were able to obtain samples with μ_{77} as high as 94,000 $cm^2\,V^{-1}\,s^{-1}$. A peculiarity of the high-purity InP observed in Eastman's work[90] and also in our work at the University of Illinois[91] is that the mobility tends to increase slowly over a period of several months when the samples are stored at room temperature. Eastman reported[90] that the increase in mobility could be accelerated to a 3–4 day period by elevating the temperature to 150°C. The increase in mobility was accompanied by a decrease in the carrier concentration. For one sample with an initial mobility of 94,000 $cm^2\,V^{-1}\,s^{-1}$ the highest value reached was 144,000 $cm^2\,V^{-1}\,s^{-1}$ and for another with an initial mobility of 84,000 $cm^2\,V^{-1}\,s^{-1}$ the mobility increased to

$124{,}000\,cm^2\,V^{-1}\,s^{-1}$. Any high-temperature processing, such as 460°C for alloying contacts, reduced the mobility to close to its original value, and then the mobility slowly increased with time again. The acceptor concentration, C at about $1 \times 10^{14}\,cm^{-3}$, did not change. Eastman speculated that the change may have been due to Si donors changing condition, such as from shallow (ionized) to deep (neutral). No such increase in mobility with time has been observed in high-purity GaInAsP epitaxial layers.

GaInAsP alloys

The highest-purity alloy material available so far has been grown by LPE, utilizing solution-baking techniques similar to those described above for high-purity LPE GaAs and InP. The same behaviour is not always observed for all systems within the same laboratory and certainly not for different types of systems in different laboratories and by different crystal growers. Because of these differences, it is difficult to make valid comparisons between different techniques. In this section we will briefly describe the techniques used at the University of Illinois to grow high-purity LPE InP ($n \simeq 9 \times 10^{14}\,cm^{-3}$, $\mu_{77} \simeq 67{,}000\,cm^2\,V^{-1}\,s^{-1}$) and GaInAs ($n \simeq 2 \times 10^{14}\,cm^{-3}$, $\mu_{77} \simeq 70{,}000\,cm^2\,V^{-1}\,s^{-1}$). These techniques are similar to those used in other laboratories, but where there are differences they will be mentioned. The electrical and far-infrared photoconductivity measurements on these samples will be discussed.

Growth apparatus and procedure Because of the ultimate interest in multiple-layer device structures, a conventional multiple-well horizontal slider-boat, machined from high-purity Poco graphite, is used for high-purity InP and GaInAs growth described here. The gas lines consist of stainless steel tubing coupled with Swagelok fittings except at the input and output ports of the quartz reactor tube where Ultratorr connectors are used. A palladium hydrogen purifier was positioned as close to the input of the quartz reactor tube as possible to minimize the length of the stainless steel tubing and the number of fittings to reduce the possibility of the hydrogen picking up impurities from the tubing and to help ensure a leak-tight system. To monitor the water vapour content of the hydrogen flowing through the quartz reactor tube, a Panametrics hygrometer sensor was used. It was installed in the output line as close to the output port of the reactor tube as possible to reduce the effect of water vapour condensation on the walls of the tubing. However, due to effects such as this, it is really only a qualitative measure of the actual water vapour content. Typical measured values were 0.1 to 1.0 ppm. A quartz push rod fits through an Ultratorr connector on the end-cap of the reactor tube and is used to control the slider motion. The temperature of the growth solution is measured using a thermocouple in a small-diameter sealed quartz tube which extends from the end-cap of the reactor down through a hole in the body of the graphite boat. In this way the thermocouple can be positioned under any of the eight wells in the boat. A transparent furnace (Transtemp Corp., Chelsea, MA) is used to permit direct visual liquidus measurements of the growth solutions. This is an important part of the growth procedure to be described later. This lightweight furnace was mounted on rails to provide rapid heating or cooling at the beginning and end of the growth cycle by sliding the furnace, which is kept hot at all times, on or off the reactor tube.

The high-purity growth procedure requires baking the growth solution with a continuous flow of dry hydrogen over the boat for 24 to 48 h before loading the substrate and starting the growth cycle. It is important that all the components which are to be included in the growth solution are included in the baking cycle. Baking the indium alone has not been effective in obtaining high-purity layers, especially for InP. For example, baking the indium alone for this prescribed time and adding undoped InP ($N_D - N_A = 5 \times 10^{15}\ \mathrm{cm}^{-3}$) at the beginning of the growth cycle results in InP epitaxial layers with $N_D - N_A = 1 \times 10^{17}\ \mathrm{cm}^{-3}$! This high doping level probably results from the high silicon content of the undoped InP and the high distribution coefficient of silicon. It has been found that baking both the indium and InP source materials results in layers with much lower doping levels. The source materials consist of Johnson Matthey A1A indium and undoped InP, GaAs, and InAs. Details of the baking technique will be described below.

After baking the growth solution, an Fe-doped semi-insulating InP substrate is loaded into the boat and the system is purged with a continuous flow of hydrogen for several hours. The growth cycle is started by sliding the transparent furnace over the reactor tube and the temperature is initially raised 10–15°C above the liquidus temperature of the growth solution and held constant for 45 min to ensure a homogeneous solution. To reduce the thermal decomposition of the substrate, it is slid beneath the cover of the graphite boat as soon as it can be visually determined that the growth solution is completely molten. After this initial 45 min heating period, the temperature is lowered over another 45 min period to the beginning growth temperature. This short 'baking' time after the substrate is loaded is different from the procedures developed for GaAs.[84] Before growth of the epitaxial layer begins, the substrate is etched for 15–20 s using a pure indium melt to remove the thermally damaged material and to provide a smooth surface on which the epitaxial layer can be grown. This procedure removes approximately 15 μm of the substrate. The step-cooling growth technique has been employed for the growth of the GaInAs and GaInAsP layers in this work, since growth at a constantly changing temperature may result in a compositionally graded layer.[92] The GaInAs layers were grown at either 627.5 or 687.5°C from solutions with liquidus temperatures which were 2.5°C above the corresponding growth temperatures. The GaInAsP layers were grown at 640°C from solutions with a liquidus temperature of 650°C. The growth times for these layers were equal to or less than 30 min, and the layer thicknesses were small enough that the effects of the finite-size growth solution did not disturb the composition of the epitaxial layer.[93] The supercooling growth technique was employed for the growth of the InP epitaxial layers with an initial supersaturation of 3°C and cooling rates of 0.1 to 0.25°C min^{-1}, because this technique resulted in much better surface morphology for InP growth. Typical growth times of the InP epitaxial layers were from 1 to 2 h. The epitaxial layers characterized in this work were typically 5–10 μm thick.

Baking of InP growth solutions presents a particular problem because of the high vapour pressure of phosphorus over the indium-rich solution. This means the solution liquidus temperature will decrease significantly during baking. As mentioned earlier, we find it essential to bake the InP source material along with the indium melt in order to achieve low doping levels. Thus, any attempt to resaturate the growth solution with more InP source material after baking will defeat the purpose. Therefore, it is necessary to determine the rate at which the solution liquidus temperature is decreasing for the particular baking temperature in order to estimate the amount of InP source material required to obtain a particular final liquidus temperature after baking.

In many of the actual baking cycles, rather than hold the temperature constant and have the liquidus temperature continue to drop well below this value, the baking temperature was either decreased stepwise or linearly ramped to maintain a nearly saturated solution, reducing the overall loss of saturation during the bake cycle since the dissociation rate increases rapidly with temperature. This is also a natural way to achieve the lowest equilibrium silicon concentration in the growth solution in the shortest time, as discussed by Wrick *et al.*[44] and Oliver and Eastman.[94] Baking at higher temperatures increases the elimination rate of silicon from the solution, but the equilibrium silicon concentration is higher. On the other hand, lower temperatures result in lower equilibrium concentrations, but require a much longer baking time. Thus, by baking in stages, the silicon concentration can initially be reduced at a faster rate at high temperatures and finally be reduced to a lower equilibrium value at lower temperatures, thereby reducing the overall required baking time.

A typical baking cycle for InP growth solutions in this work begins with an initial baking temperature of 700°C and a liquidus temperature of 690°C, and after 30 h ends with a final baking temperature of 670°C and a liquidus temperature of 660°C. At the end of the baking cycle the liquidus temperature is visually measured by rapidly cooling the solution until InP crystallization appears on the surface of the solution and then slowly reheating the solution until all the crystallized material has completely dissolved. The temperature at which this occurs is taken to be the liquidus temperature. The initial growth temperature is 3°C below the measured liquidus temperature and the solution is cooled 20–25°C during a 2 h growth time.

The composition of the growth solution used to grow lattice-matched $Ga_{0.17}In_{0.83}As_{0.40}P_{0.60}$ ($\lambda_g = 1.15$ μm) epitaxial layers on a (100) InP substrate at 640°C with a step-cooling temperature of $\Delta T = 10$°C consisted of X^l_{Ga}

$= 0.00398$, $X^l_{As} = 0.03527$, and $X^l_P = 0.00381$. Again, a significant amount of phosphorus is lost from the solution during the baking cycle and the initial composition of the solution must be adjusted to compensate for this loss. Since the liquidus temperature is strongly influenced by the amount of phosphorus in the GaInAsP solution, the rate of phosphorus loss can be calibrated by measuring the decrease of the liquidus temperature in the same way as above for InP solutions. It has been found that by increasing X^l_P to 0.00453, which increases the initial liquidus temperature to 660°C, a 36 h bake at 665°C, reduces the liquidus temperature to the desired value of 650°C.

The $Ga_{0.47}In_{0.53}As$ epitaxial layers were grown from two different solutions with very different liquidus temperatures. The highest-purity GaInAs layers were grown at the highest temperature. One solution consisting of $X^l_{Ga} = 0.02203$ and $X^l_{As} = 0.04995$ has a liquidus temperature of 630°C and the other one consisted of $X^l_{Ga} = 0.02902$ and $X^l_{As} = 0.08477$ and has a liquidus temperature of 690°C. Each solution produced lattice-matched growth on a (100) InP substrate at the appropriate growth temperature. For either solution, the epitaxial layers were grown at constant temperature with $\Delta T = 2.5$°C. These solutions were baked for 30 h at 650°C and 700°C respectively and showed no indication of loss of saturation.

In the above baking experiments, it has been found that the quality of the source material is not of extreme importance. Equally good layer characteristics have been observed in samples grown using any of four different batches of Johnson Matthey AlA indium as well as undoped InP source material with $N_D - N_A$ either 5×10^{15} or 3×10^{14} cm^{-3}. In one experiment, performed in an attempt to obtain lightly doped p-type InP, InP doped with 1×10^{18} cm^{-3} Zn was used for source material. Utilizing the same baking technique as described above, an epitaxial layer was grown which was determined to be n-type with as low a doping level and as high a mobility as those grown using undoped source material! Thus, the use of this baking technique seems to eliminate effectively the influence of the variations in purity of the starting source materials with a high degree of reproducibility.

Other workers have used similar baking procedures to obtain material of relatively low carrier concentration. Oliver and Eastman[94] have used very long baking times and a baking schedule similar to that of Wrick *et al.*[44] for growth of high-purity GaInAs on (111)B InP. In this work the only H_2O vapour present was the residual water vapour in the system, either because of leaks or reduction of silica by H_2. Groves has pointed out that the baking schedule that works with one particular system may not work with another similar system, apparently because of differences in the residual water vapour in the two systems.[89] For some cases it may be necessary to add O_2 or H_2O to obtain the desired results.[22] Lee *et al.*[95] have also used similar baking techniques to obtain low doping in GaInAs material. They have found that the carrier concentration can be lowered to 2×10^{16} cm^{-3} by (1) selecting high-purity In, and (2) baking the In at 700°C for 16 h in flowing H_2 prior to adding the solutes. A further order-of-magnitude

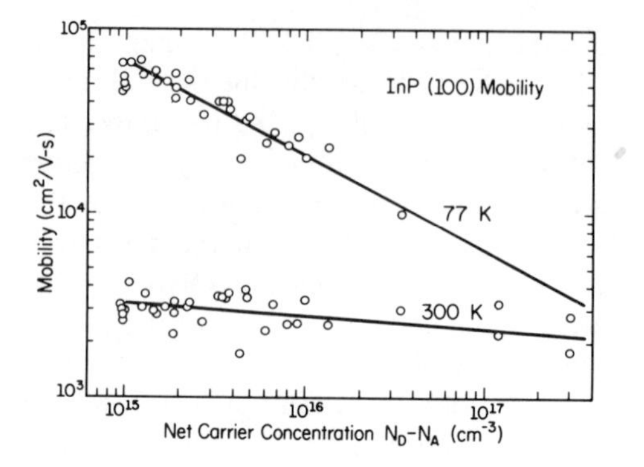

Figure 6.14 Variation of room-temperature and liquid-nitrogen-temperature Hall mobility with net carrier concentration for (100) InP samples grown at the University of Illinois using the technique described in the text

reduction in net carrier concentration was obtained by prebaking the growth solution and the substrate at about 675°C for 16 h just before use. A preliminary high-temperature (~ 1600°C) bake-out of the graphite boats further reduced the carrier concentration to the 2–3 × 10^{14} cm^{-3} range. These high-temperature bake-outs are repeated every six or seven growth runs.[96]

Electrical measurements The electrical characteristics of the epitaxial layers were measured by the van der Pauw technique in a 0.67 T magnetic field. Pure tin spheres (0.05 cm in diameter) are placed near opposite edges of the sample and alloyed in a hydrogen atmosphere at 400°C for one minute. The sample dimensions are approximately 0.5 cm on a side. The expected error in the measurement of the mobility is less than 5 %.

A summary of the InP and GaInAs van der Pauw measurements are shown in Figures 6.14 and 6.15. All of the undoped layers are n-type, and the InP liquid-

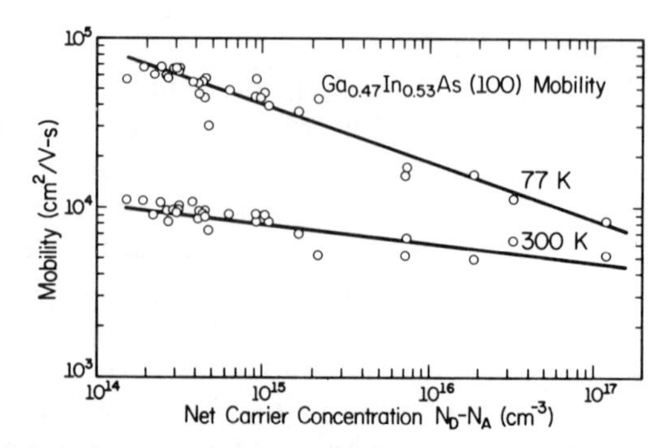

Figure 6.15 Variation of room-temperature and liquid-nitrogen-temperature Hall mobility with net carrier concentration for $Ga_{0.47}In_{0.53}As$ samples lattice-matched to (100) InP grown at the University of Illinois using the technique described in the text

nitrogen-temperature mobilities, when compared with the theoretical curves by Rode, indicate relatively little compensation. The freeze-out ratios, $n(300\text{ K})/n(77\text{ K})$, of the InP and GaInAs layers are generally between 1.0 and 1.2. The GaInAs layers in Figure 6.15 with $N_D - N_A$ from 2×10^{14} to $4 \times 10^{14}\text{ cm}^{-3}$ and liquid nitrogen mobilities above $60{,}000\text{ cm}^2\text{ V}^{-1}\text{ s}^{-1}$ were grown at 687.5°C, and the rest were grown at 627.5°C. The growth of InP and GaInAs above 650°C is of higher purity, perhaps because of a lower distribution coefficient for silicon at higher growth temperatures. The GaInAs layers with $N_D - N_A \simeq 1 \times 10^{16}\text{ cm}^{-3}$ were grown from unbaked solutions and those with $N_D - N_A > 1 \times 10^{16}\text{ cm}^{-3}$ were doped with tin. Typically, InP layers grown from unbaked solutions yield doping levels closer to $1 \times 10^{17}\text{ cm}^{-3}$, presumably because of the use of lower-quality InP source material.

The growth of high-purity material has been very reproducible. In recent attempts to grow high-purity InP layers using the technique described above, over 85% of the samples have had liquid-nitrogen-temperature mobilities between 50,000 and $70{,}000\text{ cm}^2\text{ V}^{-1}\text{ s}^{-1}$ and values of $N_D - N_A$ between 9.0×10^{14} and $2.5 \times 10^{15}\text{ cm}^{-3}$. Also, over 86% of the GaInAs layers grown at 687.5°C have had liquid-nitrogen-temperature mobilities between 55,000 and $70{,}000\text{ cm}^2\text{ V}^{-1}\text{ s}^{-1}$ with net carrier concentrations between 1.5×10^{14} and $4.5 \times 10^{14}\text{ cm}^{-3}$. These results have been obtained with two different systems. Fewer $Ga_{0.17}In_{0.83}As_{0.40}P_{0.60}$ ($\lambda = 1.15\ \mu\text{m}$) layers have been prepared for van der Pauw measurements, but of those which have been measured the highest-purity layer was determined to have $N_D - N_A = 5.6 \times 10^{14}\text{ cm}^{-3}$ with room-temperature and liquid-nitrogen-temperature mobilities of 4600 and $24{,}800\text{ cm}^2\text{ V}^{-1}\text{ s}^{-1}$ respectively. Most of these layers were grown between runs where device structures, which included layers heavily doped with either zinc or tin, were grown on InP substrates also doped with either zinc or tin. Thus, the use of dopants in the boats does not produce a memory effect reducing the purity of layers grown in subsequent runs or the reproducibility of these results.

The mobilities of these GaInAs layers grown on (100) InP are in general agreement with the mobilities of GaInAs layers grown on (111)B InP by Oliver and Eastman,[94] with the exception that for equally high liquid-nitrogen-temperature mobilities the room-temperature mobilities of our layers are usually lower. Oliver's highest room-temperature mobility is $13{,}800\text{ cm}^2\text{ V}^{-1}\text{ s}^{-1}$ for a sample with $N_D - N_A = 1.9 \times 10^{15}\text{ cm}^{-3}$, and in this work the highest is $11{,}000\text{ cm}^2\text{ V}^{-1}\text{ s}^{-1}$. There have been other reports of high room-temperature mobility values. Pearsall *et al.*[97] reported a room-temperature mobility of $15{,}000\text{ cm}^2\text{ V}^{-1}\text{ s}^{-1}$ for a sample with $N_D - N_A = 10^{16}\text{ cm}^{-3}$. The samples of Oliver and Eastman[94] and Pearsall *et al.*[97] were grown using LPE techniques similar to those described above. High room-temperature mobility values have also been reported for GaInAs layers grown using the hydride technique. Room-temperature mobilities of $15{,}800\text{ cm}^2\text{ V}^{-1}\text{ s}^{-1}$ and $15{,}200\text{ cm}^2\text{ V}^{-1}\text{ s}^{-1}$ were reported[98] for samples with $N_D - N_A = 1.2 \times 10^{17}\text{ cm}^{-3}$. Although Oliver[99] reported that techniques had been developed to achieve high-purity GaInAs reproducibly, he was only able to obtain two samples which had a liquid-nitrogen-temperature mobility greater than $50{,}000\,cm^2\,V^{-1}\,s^{-1}$. The other reports of high room-temperature mobilities pertain to samples that were grown in systems that did not produce really high-purity (low carrier concentration, high μ_{77}) material. These factors make it unlikely that the reported high room-temperature mobilities are real. Conducting inhomogeneities can produce anomalous room-temperature mobilities (as well as anomalous liquid-nitrogen-temperature mobilities, but these are not as easily detected),[14] and it may be that such effects are responsible for the high values of room-temperature mobility discussed above, especially since the theoretical upper limit on the room-temperature mobility for GaInAs is about $13{,}500\,cm^2\,V^{-1}\,s^{-1}$ (see Chapter 9).

Far-infrared photoconductivity measurements The high-purity InP grown in this work is sufficiently pure to be able to measure the spectra of residual donors using far-infrared photothermal-ionization donor spectroscopy. Figure 6.16 shows the spectrum of an LPE InP sample with $N_D - N_A = 1.2 \times 10^{15}\,cm^{-3}$ and $\mu_{77} = 67{,}500\,cm^2\,V^{-1}\,s^{-1}$ taken at a magnetic field of 3.0 T and at a temperature of 4.2 K. This magnetic field and temperature are the same as those for the GaAs samples in the lower part of Figure 6.11, and many of the transitions identified in Figure 6.11 can also be observed in the InP spectrum in Figure 6.16. The three widely spaced peaks labelled in this figure correspond to the ls → 2p transitions

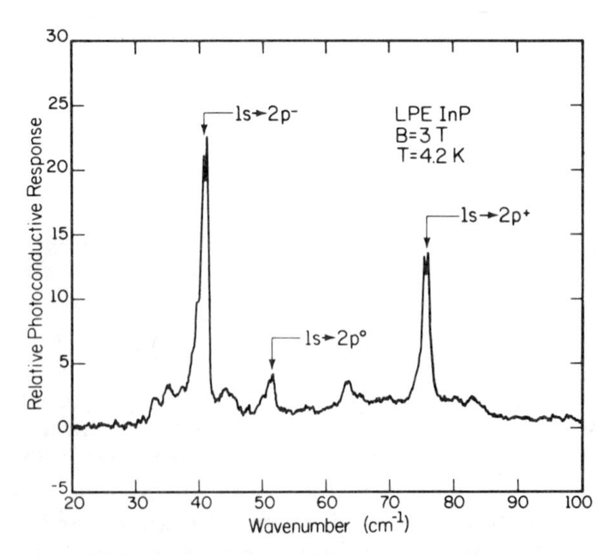

Figure 6.16 Relative far-infrared photoconductive response for a high-purity (100) InP sample at a magnetic field of 3.0 T and a temperature of 4.2 K. This sample has a net carrier concentration of $1.2 \times 10^{15}\,cm^{-3}$ and a liquid-nitrogen-temperature mobility of $67{,}500\,cm^2\,V^{-1}s^{-1}$

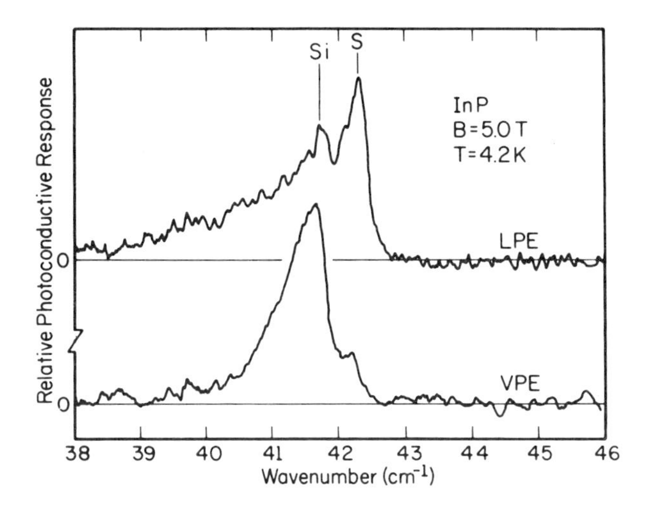

Figure 6.17 High-resolution far-infrared photoconductivity spectra of the 1s→2p, $m = -1$ transition for the high-purity LPE sample of Figure 6.16 and a hydride VPE sample. The peaks identified as S and Si by Stradling[35] are indicated. For the LPE sample, $N_D - N_A = 1.2 \times 10^{15}\ \text{cm}^{-3}$ and $\mu_{77} = 67{,}500\ \text{cm}^2\ \text{V}^{-1}\ \text{s}^{-1}$; and for the VPE sample, $N_D - N_A = 1.3 \times 10^{15}\ \text{cm}^{-3}$ and $\mu_{77} = 57{,}500\ \text{cm}^2\ \text{V}^{-1}\ \text{s}^{-1}$

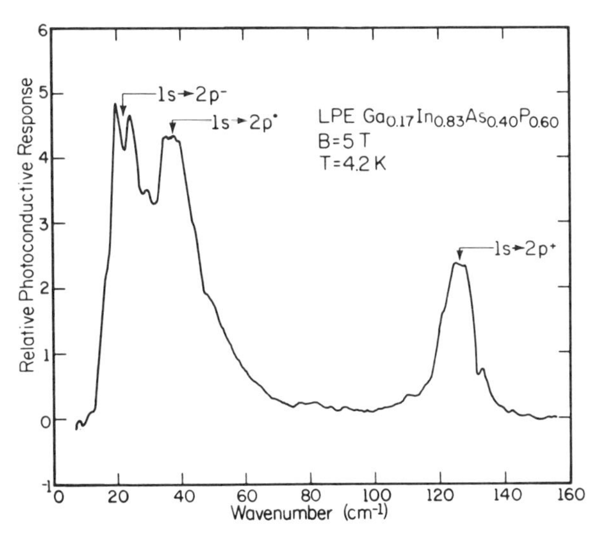

Figure 6.18 Relative photoconductive response for a $Ga_{0.17}In_{0.83}As_{0.40}P_{0.60}$ sample lattice-matched to (100) InP at a magnetic field of 5.0 T. This sample has a net carrier concentration of $5.6 \times 10^{14}\ \text{cm}^{-3}$ and $\mu_{77} = 24{,}800\ \text{cm}^2\ \text{V}^{-1}\ \text{s}^{-1}$

whose degeneracy is split by the magnetic field. In all of the high-purity InP samples that have been measured, two closely spaced peaks can be resolved in each of the 2p transitions. As described for GaAs, each of these peaks results from a different chemical donor species. The ground-state energies of the different chemical donors are slightly different because the central cell correction depends on the particular donor species.

High-resolution photoconductivity measurements of just the ls → $2p^{-1}$ transition for two different InP samples at a magnetic field of 5.0 T are shown in Figure 6.17. The upper curve in this figure is for the same LPE sample as in Figure 6.16, and the lower curve is for a high-purity VPE InP sample grown by the hydride technique in our laboratory. From this figure, it is clear that the same two donor species are residual impurities in both LPE and hydride VPE epitaxial InP samples. This is in contrast to high-purity GaAs where there is only one residual donor species common to LPE and VPE material, identified as Si in the work of Wolfe *et al.*[33] and as S in the work of Ozeki *et al.*[37] The identification of the residual donor impurities in InP is also shown in Figure 6.16, based on the work of Stradling *et al.*[35] With this identification the S concentration is higher in the LPE material, while the Si concentration is higher in the VPE material. Markunas *et al.*[100] have detected both S and Si in quantitative SIMS measurements on high-purity InP at the 5×10^{14} cm^{-3} level, supporting the identification of S and Si as the dominant residual impurities in LPE and VPE InP.

Similar spectra are observed for high-purity GaInAsP and GaInAs as shown in Figures 6.18 and 6.19. However, besides the differences expected because of the

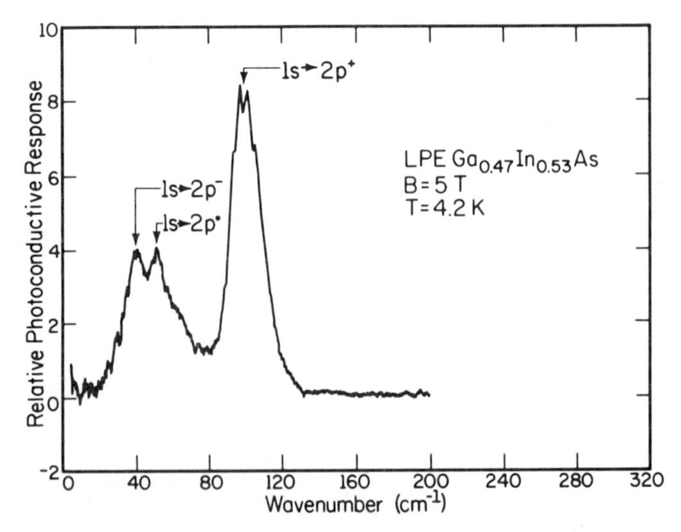

Figure 6.19 Relative photoconductive response for a $Ga_{0.47}In_{0.50}As$ sample lattice-matched to (100) InP at a magnetic field of 5.0 T. This sample has a net carrier concentration of 1.9×10^{14} cm^{-3} and $\mu_{77} = 67\,900$ cm^2 V^{-1} s^{-1}

different effective masses for these materials compared to InP or GaAs, there are other significant differences in the alloy spectra, the major one being that the linewidths observed are much greater than for comparable purity GaAs or InP. This may be the influence of alloy disorder, but remains to be investigated more fully. Correlation of measurements of this type with mobility measurements and photoluminescence should increase our understanding of these materials significantly.

ACKNOWLEDGMENTS

The authors would like to thank Dr M. Feng and Dr L. M. Zinkiewicz for their contributions to the early part of this work, and especially to thank Ms B. L. Payne for her assistance in the preparation of this manuscript. Many helpful and continuing discussions with Professor N. Holonyak, Jr, are greatly appreciated.

This work was supported by the National Science Foundation, the Office of Naval Research, and the Naval Research Laboratory, and we would particularly like to acknowledge the interest and support of Howard Lessoff of NRL.

REFERENCES

1. H. H. Wieder, *Laboratory Notes on Electrical and Galvanomagnetic Measurements*, Elsevier, New York, 1979.
2. L. J. van der Pauw, 'A method of measuring specific resistivity and Hall effect of discs of arbitrary shape', *Phillips Res. Rep.*, **13**, 1–9, 1958.
3. L. J. van der Pauw, 'A method of measuring the resistivity and Hall coefficient on lamellae of arbitrary shape', *Phillips Tech. Rev.*, **20**, 220–224, 1958.
4. O. Lindberg, 'Hall effect', *Proc. IRE*, **40**, 1414–1419, 1952.
5. D. J. Oliver, 'Electric properties of n-type gallium arsenide', *Phys. Rev.*, **127**, 1045–1052, 1962.
6. R. A. Reynolds, 'Impact ionization breakdown of n-type epitaxial GaAs at liquid helium temperatures', *Solid-State Electron.*, **11**, 385–390, 1968.
7. H. H. Wieder, 'Electrical and galvanomagnetic measurements on thin films and epilayers', *Thin Solid Films*, **31**, 123–138, 1978.
8. A. C. Beer, *Galvanomagnetic Effects in Semiconductors*, Academic Press, New York, 1963.
9. G. E. Stillman, C. M. Wolfe, and J. O. Dimmock, 'Hall coefficient factor for polar mode scattering in n-type GaAs', *J. Phys. Chem. Solids*, **31**, 1199–1204, 1970.
10. B. F. Lewis and E. H. Sondheimer, 'The theory of the magneto-resistance effects in polar semiconductors', *Proc. R. Soc. London*, A, **227**, 241–251, 1954.
11. S. S. Devlin, 'Transport properties', in *Physics and Chemistry of II–VI compounds*, eds M. Aven and J. S. Prener, North-Holland, Amsterdam, 1967, pp. 551–609.
12. D. L. Rode, 'Electron transport in InSb, InAs, and InP', *Phys. Rev.*, B, **3**, 3287–3299, 1971.
13. G. Lindemann, E. Gornik, R. Schawarz, and D. C. Tsui, 'Cyclotron emission from GaAs and InP', *Gallium Arsenide and Related Compounds 1980*, Conf. Ser. no. 56, Institute of Physics, Bristol, 1981, pp. 631–638.
14. C. M. Wolfe and G. E. Stillman, 'Apparent mobility enhancement in inhomogeneous

crystals', in *Semiconductors and semimetals*, vol. 10, eds R. K. Willardson and A. C. Beer, Academic Press, New York, 1975, pp. 175–220.
15. J. A. Rossi, C. M. Wolfe, and J. O. Dimmock, 'Acceptor luminescence in high-purity n-type GaAs', *Phys. Rev. Lett.*, **25**, 1614–1617, 1970.
16. J. A. Rossi, C. M. Wolfe, G. E. Stillman, and J. O. Dimmock, 'Identification of exciton-neutral donor complexes in the photoluminescence of high purity GaAs', *Solid State Commun.*, **8**, 2021–2024, 1970.
17. D. J. Ashen, P. J. Dean, D. T. J. Hurle, J. B. Mullin, A. M. White, and P. D. Greene. 'The incorporation and characterization of acceptors in epitaxial GaAs', *J. Phys. Chem. Solids*, **36**, 1041–1053, 1975.
18. M. Ozeki, K. Nakai, K. Dazai, and O. Ryuzan, 'Photoluminescence study of carbon doped gallium arsenide', *Jpn. J. Appl. Phys.*, **13**, 1121–1126, 1974.
19. F. Willmann, D. Bimberg, and M. Blatte, 'Optical properties of excitons bound to copper-complex centers in gallium arsenide', *Phys. Rev.*, **137**, 2473–2480, 1973.
20. H. Kunzel and K. Ploog, 'Sharp-line luminescence transitions due to growth induced point defects in MBE GaAs', *Gallium Arsenide and Related Compounds 1980*, Conf. Ser. no. 56, Institute of Physics, Bristol, 1981, pp. 519–528.
21. S. G. Bishop and P. Klein, private communication.
22. S. H. Groves, private communication, 1981.
23. D. A. Stevenson, B. L. Mattes, and H. Dun, 'Control of impurites in the epitaxial growth of high quality GaAs', *Tech. Rep.* CMR-77-4, Center for Materials Research, Stanford University, 1977 (unpublished).
24. A. M. White, P. J. Dean, L. L. Taylor, R. C. Clarke, D. J. Ashen, and J. B. Mullin, 'The photoluminescence spectrum of bound excitons in indium phosphide and gallium arsenide', *J. Phys. C: Solid State Phys.*, **5**, 1727–1738, 1972.
25. A. M. White, P. J. Dean, B. D. Joyce, R. C. Clarke, and L. L. Taylor, 'The photoluminescence of bound excitons in indium phosphide', *Proc. Int. Conf. on the Physics of Semiconductors*, Warsaw, 1972, Warsaw, Polish Scientific Publishers, Warsaw, 1973, pp. 190–195.
26. K. Hess, N. Stath, and K. W. Benz, 'Liquid phase epitaxy of InP', *J. Electrochem Soc.: Solid-State Sci. Technol.*, **121**, 1208–1212, 1974.
27. G. G. Baumann, K. W. Benz, and M. H. Pilkuhn, 'Incorporation of Si in liquid phase epitaxial InP layers'. *J. Electrochem. Soc.: Solid-State Sci. Technol.*, **123**, 1232–1235, 1976.
28. P. J. Dean, D. J. Robbins, and S. G. Bishop, 'Dye laser selective spectroscopy in bulk-grown indium phosphide', *J. Phys. C: Solid State Phys*, **12**, 5567–5575, 1979.
29. A. M. White, P. J. Dean, and B. Day, 'On the binding of excitons to neutral acceptors in narrow direct gap semiconductors', F. G. Fumi, ed., *Proc. Int. Conf. on the Physics of Semiconductors*, Rome, 1976, North-Holland, Amsterdam, 1976, pp. 1057–1060.
30. E. W. Williams *et al.*,'Indium phosphide I. A photoluminescence materials study', *J. Electrochem. Soc.: Solid-State Sci. Technol.*, **120**, 1741–1749, 1973.
31. G. E. Stillman, C. M. Wolfe, and J. O. Dimmock, 'Far infrared photoconductivity in high purity GaAs', *Semiconductors and Semimetals*, vol. 12, eds. R. K. Willardson and A. C. Beer, Academic Press, New York, 1976, pp. 169–290.
32. D. M. Larsen, 'Inhomogeneous line broadening in donor magneto-optical spectra', *Phys. Rev.*, B, **8**, 535–552, 1973.
33. C. M. Wolfe, G. E. Stillman, and D. M. Korn, 'Residual donors in high purity GaAs', *Gallium Arsenide and Related Compounds* (*St Louis*) *1976*, Conf. Ser. no. 33b, Institute of Physics, Bristol, 1977, pp. 120–128.
34. P. D. Dapkus, H. M. Manasevit, K. L. Hess, T. S. Low, and G. E. Stillman, 'High purity GaAs prepared from trimethylgallium and arsine', *J. Cryst. Growth*, **55**, 10–23, 1981.

35. R. A. Stradling, L. Eaves, R. A. Hoult, N. Miura, P. E. Simmonds, and C. C. Bradley, 'Zeeman spectroscopy of the shallow donor states in GaAs and InP', *Gallium Arsenide and related Compounds 1972*, Conf. Ser. no. 17, Institute of Physics, Bristol, 1973, pp. 65–74.
R. A. Cooke, R. A. Hoult, R. F. Kirkman, and R. A. Stradling. 'The characterization of the donors in GaAs epitaxial films by far-infrared photoconductive techniques', *J. Phys. D: Appl. Phys.*, **11**, 945–953, 1978.
R. A. Stradling, private communication, 1981.
36. G. E. Stillman, C. M. Wolfe, and D. M. Korn, 'Donor impurities in GaAs: evidence for a stoichiometric hydrogenic shallow donor level', F. G. Fumi, ed., *Proc. Int. Conf. on the Physics of Semiconductors*, North-Holland, Amsterdam, 1976, Rome, 1976, pp. 623–626.
37. M. Ozeki, K. Kitahara. N. Nakai, A. Shibatomi, K. Dazai, S. Okawa, and O. Ryuzan, 'Residual donors in high purity gallium arsenide epitaxially grown from vapor phase', *Jpn. J. Appl. Phys.*, **16**, 1617–1622, 1977.
38. J. H. M. Stoelenga, D. M. Larsen, W. Walukiewicz, and R. L. Aggarwal, 'Magneto-optical study of shallow donors in transmutation-doped GaAs', *J. Phys. Chem. Solids*, **39**, 873–877, 1978.
39. M. N. Afsar, K. J. Button, and G. L. McCoy, 'Use of far infrared photoconductivity and high magnetic fields to study donors in epitaxial gallium arsenide', *Gallium Arsenide and Related Compounds 1980*, Conf. Ser. no. 56, Institute of Physics, Bristol, 1981, pp. 547–555.
40. C. M. Wolfe and G. E. Stillman, 'High purity GaAs', *Gallium Arsenide and Related Compounds 1970*, Conf. Ser. no. 9, Institute of Physics, Bristol, 1971, pp. 3–17.
41. H. G. B. Hicks and P. D. Greene, 'Control of silicon contamination in solution growth of gallium arsenide in silica', *Gallium Arsenide and Related Compounds 1970*, Conf. Ser. no. 9, Institute of Physics, Bristol, 1971, pp. 92–99.
42. M. E. Weiner, 'Si contamination of open flow quartz systems for the growth of GaAs and GaP', *J. Electrochem. Soc.: Solid-State Sci. Technol.*, **119**, 496–504, 1972.
43. H. Seki, A. Koukitu, H. Seki, and M. Fujimoto, 'Thermodynamic analysis for silicon contamination of VPE GaAs', *J. Cryst. Growth*, **45**, 159–163, 1978.
44. V. L. Wrick, K. T. Ip, and L. F. Eastman, 'High purity in LPE InP', *J. Electron. Mater.*, **7**, 253–261, 1978.
45. J. V. DiLorenzo and G. E. Moore, Jr, 'Effects of the $AsCl_3$ mole fraction on the incorporation of germanium, silicon, selenium, and sulfur into vapor grown epitaxial layers of GaAs', *J. Electrochem. Soc.: Solid-State Sci. Technol.*, **118**, 1823–1830, 1971.
46. D. J. Ashen, P. J. Dean, D. T. J. Hurle, J. B. Mullin, A. Royle, and A. M. White, 'The incorporation of residual impurities in vapour grown GaAs', *Gallium Arsenide and Related Compounds 1974*, Conf. Ser. no. 24, Institute of Physics, Bristol, 1975, pp. 229–244.
47. J. V. DiLorenzo and A. E. Machala, 'Orientation effects of the electrical properties of high purity epitaxial GaAs', *J. Electrochem. Soc.: Solid-State Sci.*, **118**, 1516–1517, 1971.
48. A. Trainor and B. E. Bartlett, 'A possible mechanism of crystal growth from the melt and its application to the problem of anomalous segregation of crystal facets', *Solid-State Electron.*, **2**, 106–114, 1961.
49. P. J. Holms, 'A competitive absorption model of steady-state growth of a crystal from a lightly doped melt', *J. Phys. Chem. Solids*, **24**, 1239–1250, 1963.
50. H. C. Casey, Jr, M. B. Panish, and K. B. Wolfstirn, 'Influence of surface band bending on the incorporation of impurities in semiconductors: Te in GaAs', *J. Phys. Chem. Solids*, **32**, 571–580, 1971.

51. K. -H Zschauer and A. Vogel, 'Dependence of impurity incorporation on growth rate and crystal orientation during GaAs liquid-phase epitaxy', *Gallium Arsenide and Related Compounds 1970*, Conf. Ser. no. 9, Institute of Physics, Bristol, 1971, pp. 100–107.
52. C. M. Wolfe, and G. E. Stillman, 'Effect of surface states on the amphoteric behavior of Sn in vapor epitaxial GaAs', *Solid State Commun.*, **12**, 283–285, 1973.
53. C. M. Wolfe and G. E. Stillman, 'Self-compensation of donors in high-purity GaAs', *Appl. Phys. Lett.*, **27**, 564–567, 1975.
54. H. M. Cox and J. V. Di Lorenzo, 'Review of techniques of epitaxial growth of high-resistivity GaAs—Growth systems, problems and substrate effects', in *Semi-Insulating III–V Materials*, Nottingham, 1980, ed. G. J. Rees, Shiva Pub. Ltd, Orpington, 1980, pp. 41–56.
55. K. Dazai, N. Ihara, and M. Ozeki, 'Vapor phase epitaxial growth of GaAs in an nitrogen atmosphere', *Fujitsu Sci. and Tech. J.* **10**, 125–143, 1974.
56. H. Seki, A. Kookitu, K. Ohta, and M. Fujimoto, 'New methods of vapor phase epitaxial growth of GaAs', *Jpn. J. Appl. Phys.*, **15**, 11–17, 1976.
57. J. K. Kennedy, W. D. Potter, and D. E. Davies, 'The effect of the hydrogen carrier gas flow rate on the electrical properties on epitaxial GaAs prepared in a hydride system', *J. Cryst. Growth*, **24/25**, 233–238, 1974.
58. J. K. Kennedy, private communication, 1979.
59. J. K. Abrokwah, private communication (unpublished), 1981.
60. Y. Seki, K. Tanno, K. Iida, and E. Ichiki, 'Properties of epitaxial GaAs layers from a triethylgallium and arsine system', *J. Electrochem. Soc.: Solid State Sci. Technol.*, **122**, 1108–1112, 1975.
61. H. Morkoc and A. Y. Cho, 'High purity GaAs and Cr-doped GaAs epitaxial layers by MBE', *J. Appl. Phys.*, **50**, 6413–6416, 1979.
62. R. Dingle, C. Weisbuch, H. Morkoç, and A. Y. Cho, "Characterization of high purity GaAs grown by molecular-beam epitaxy", *Appl. Phys. Lett.*, **40**, 507–510, 1982.
63. T. S. Low and G. E. Stillman, unpublished, 1981.
64. A. R. Calawa, 'On the use of AsH_3 in the molecular beam epitaxial growth of GaAs', *Appl. Phys. Lett.*, **38**, 701–703, 1981.
65. R. C. Clarke, B. D. Joyce, and W. H. E. Wilgoss, 'The preparation of high purity epitaxial InP', *Solid State Commun.*, **8**, 1125–1128, 1970.
66. B. D. Joyce and E. W. Williams, 'The preparation and photoluminescent properties of high purity vapor grown indium phosphide layers', *Galium Arsenide and Related* Compounds 1970, Conf. Ser. no. 9, Institute of Physics, Bristol, 1971, pp. 57–63.
67. R. D. Fairman, M. Omori, and F. B. Frank, 'Recent progress in the control of high-purity VPE InP by the $PCl_3/In/H_2$ technique', *Gallium Arsenide and Related Compounds* (*St Louis*) 1976, Conf. Ser. no. 33b, Institute of Physics, Bristol, 1977, pp. 45–54.
68. R. C. Clarke, 'A study of the molar fraction effect in the PCL_3-In-H_2 system', *J. Cryst. Growth*, **23**, 166–168, 1974.
69. K. Fairhurst, D. Lee, D. S. Robertson, H. T. Parfitt, and W. H. E. Wilgoss, 'A study of the vapor phase epitaxy of indium phosphide', *Proc. 1980 NATO Sponsored InP Workshop*, June, 1980, unpublished.
K. Fairhurst, D. Lee, D. S. Robertson, H. T. Parfitt, and W. H. E. Wilgoss, 'A study of the vapor phase epitaxy of indium phosphide', *J. Mater. Sci.*, **16**, 1013–1022, 1981.
70. R. C. Clarke and L. L. Taylor, 'Pure and doped indium phosphide by vapor phase epitaxy', *J. Cryst. Growth*, **43**, 473–479, 1978.
71. R. C. Clarke, 'Chemistry of the In-H_2-PCl_3 process', *Gallium Arsenide and Related Compounds* (*St Louis*) 1978, Confr. Ser. 45, Institute of Physics, Bristol, 1979, pp. 19–27.

72. G. H. Olsen and T. J. Zamerowski, 'Crystal growth and properties of binary, ternary and quaternary (In,Ga)(As,P) alloys grown by the hydride vapor phase epitaxy technique', *Progress in Crystal Growth and Characterization*, vol. II, ed. B. R. Pamplin, Pergamon, London, 1979, pp. 309–375.
73. L. M. Zinkiewicz, T. J. Roth, B. J. Skromme, and G. E. Stillman, 'The vapor phase growth of InP and $In_xGa_{1-x}As$ by the Hydride ($In-Ga-AsH_3-PH_3-HCl-H_2$) technique', *Gallium Arsenide and Related Compounds 1980*, Conf. Ser. No. 56, Institute of Physics, Bristol, 1981, pp. 19–28.
74. T. J. Roth, B. J. Skromme, L. M. Zinkiewicz, T. S. Low, and M. M. Tashima, 'Vapor phase growth of high purity InP using the hydride technique', presented at the *5th Int. Conf. on Vapor Growth and Epitaxy*, San Diego, 1981.
75. N. Susa, Y. Yamauchi, H. Ando, and H. Kanbe, 'Vapor-phase epitaxial growth of InGaAs on (100) InP substrate', *Jpn. J. Appl. Phys.*, **19**, L17–L22, 1980.
76. C. S. Kang and P. D. Greene, 'Preparation and properties of high-purity epitaxial GaAs grown from Ga solution', *Appl. Phys. Lett.*, **11**, 171–173, 1967.
C. S. Kang and P. D. Greene, 'Tin and tellurium doping characteristics in gallium arsenide layers', *Gallium Arsenide 1968*, Conf. Ser. no. 7, Institute of Physics, Bristol, 1969, pp. 18–21.
77. E. Andre and J. M. LeDuc, 'Arseniure de gallium de haute mobilite obtenue par epitaxie en phase liquide', *Mater. Res. Bull.*, **3**, 1–6, 1968.
78. R. Solomon, 'Factors influencing the electrical and physical properties of high quality solution grown GaAs', *Gallium Arsenide* 1968, Conf. Ser. no. 7, Institute of Physics, 1969, pp. 11–17.
79. H. G. B. Hicks and D. F. Manley, 'High purity GaAs by liquid phase epitaxy', *Solid State Commun.*, **7**, 1463–1465, 1969.
80. B. L. Mattes, Y. M. Houng, and G. L. Pearson, 'Growth and properties of semi-insulating epitaxial GaAs', *J. Vac. Sci. Technol.* **12**, 869–875, 1975.
81. H. Miki and M. Otsubo, 'High purity GaAs crystal growth by liquid phase epitaxy', *Jpn. J. Appl. Phys.*, **10**, 509, 1971.
82. M. Otsubo and H. Miki, 'Liquid phase epitaxial growth of GaAs crystals under a mixed gas atmosphere', *Jpn. J. Appl. Phys.*, **14**, 621–627, 1975.
83. F. E. Rosztoczy, S. I. Long, and J. Kinoshita, 'LPE GaAs for microwave applications', *J. Cryst. Growth*, **27**, 205–214, 1974.
84. H. Morkoç and L. F. Eastman, 'Purity of GaAs grown by LPE in a graphite boat', *J. Cryst. Growth*, **36**, 109–114, 1976.
85. H. Morkoç, L. F. Eastman, and D. Woodard, 'Residual impurities in high purity GaAs epitaxial layers grown by liquid phase epitaxy', *Thin Solid Films*, **71**, 245–248, 1980.
86. J. K. Abrokwah, M. L. Hitchell, J. E. Borell, and D. R. Schulze, 'High purity and chrome-doped GaAs buffer layers grown by liquid phase epitaxy for MESFET applications', *J. Electron. Mater.* **10**, 723–746, 1981.
87. K. T. Ip, L. F. Eastman, and V. L. Wrick, 'Very high-purity InP LPE layers', *Electron. Lett.*, **13**, 682–683, 1971.
88. S. H. Groves and M. C. Plonko, 'LPE growth of nominally undoped InP and $In_{0.8}Ga_{0.2}As_{0.5}P_{0.5}$ alloys', *Gallium Arsenide and Related Compounds* (*St Louis*) 1978, Conf. Ser. no. 45, Institute of Physics, Bristol, 1979, pp. 71–77.
89. S. H. Groves and M. C. Plonko, 'Liquid phase epitaxial growth of InP and InGaAsP alloys', *J. Cryst. Growth*, **54**, 81–87, 1981.
90. L. F. Eastman, 'High purity InP grown by liquid phase epitaxy', *Proc.* 1980 *NATO Sponsored InP Workshop*, June 1980, unpublished.
91. L. W. Cook, M. M. Tashima, N. Tabatabaie, T. S. Low, and G. E. Stillman, 'High purity InP and InGaAsP grown by liquid phase epitaxy', *J. Cryst. Growth*, **56**, 475–484, 1982.
92. M. Feng, L. W. Cook, M. M. Tashima, and G. E. Stillman, 'Lattice constant, bandgap,

thickness, and surface morphology of InGaAsP–InP layers grown by step-cooling, equilibrium-cooling, supercooling and two-phase-solution growth techniques', *J. Electron. Mater.*, **9**, 241–280, 1980.
93. L. W. Cook, M. M. Tashima, and G. E. Stillman, 'Variation of the thickness and composition of LPE InGaAsP, InGaAs, and InP layers grown from a finite melt by the step-cooling technique', *J. Electron. Mater.*, **10**, 119–140, 1981.
94. J. D. Oliver, Jr. and L. F. Eastman, 'Liquid phase epitaxial growth and characterization of high purity lattice matched $Ga_xIn_{1-x}As$ on ⟨111⟩B InP', *J. Electron. Mater.*, **9**, 693–712, 1980.
95. T. P. Lee, C. A. Burrus, and A. G. Dentai, 'InGaAs/InP p-i-n photodiodes for light wave communications at the 0.95–1.65μm wavelength', *IEEE J. Quantum Electron.*, E-17, 232–238, 1981.
96. A. G. Dentai, private communication, 1981.
97. T. P. Pearsall, R. Bisaro, P. Merenda, G. Laurencin, R. Ansel, J. C. Portal, C. Houlbert, and M. Quillec, 'The characterization of $Ga_{0.47}In_{0.53}As$ grown lattice matched on InP substrates', *Gallium Arsenide and Related Compounds (St Louis) 1978*, Conf. Ser. no. 45, Institute of Physics, Bristol, 1979, pp. 94–102.
98. G. Beuchet, M. Bonnet, P. Thébault, and J. P. Duchemin, 'GaInAs/InP heterostructures grown in a four-barrel reactor by the hydride method', *Gallium Arsenide and Related Compounds* 1980, Conf. Ser. no. 56, Institute of Physics, Bristol, 1981, pp. 37–44.
99. J. D. Oliver, Jr, L. F. Eastman, P. D. Kirchner, and Wm. J. Schaff, 'Electrical characterization and alloy scattering measurements of LPE $Ga_xIn_{1-x}As$ for high frequency device applications', *J. Cryst. Growth*, **54**, 64–68, 1981.
100. R. J. Markunas, K. Tabatabaie-Alavi, and C. G. Fonstad, 'Quantitative SIMS analysis of epitaxial InP', at *J. Electron. Mater.*, 1982.

III GaInAsP MATERIALS PROPERTIES

GaInAsP Alloy Semiconductors
Edited by T. P. Pearsall

Chapter 7

Defect Motion and Growth of Extended Non-radiative Defect Structures in GaInAsP

W. D. JOHNSTON, JR

Bell Laboratories,
Murray Hill, NJ 07974, USA

The primary motivation for the study of GaInAsP alloys is their practical application in components for light-wave communication systems—light-emitting diodes (LEDs), diode lasers, and photodetectors. These are minority-carrier devices, operated at high levels of injection current, with high local energy densities which create the potential for steep thermal gradients and significant strain fields. Radiative recombination of injected minority carriers inevitably occurs at less than unit efficiency, and hence a high volumetric rate of non-radiative carrier recombination is characteristic of the 'active' region of light-wave source devices.

These three factors—strain, thermal gradients, and a high density of recombining carriers—promote the motion, multiplication, and growth of defects into network clusters. This has been shown to give rise to the dark line defect (DLD) and dark spot defect (DSD) structures in AlGaAs/GaAs heterostructure LEDs and lasers. Observation[1] and identification[2] of the DLD problem in AlGaAs laser diodes was undoubtedly the single most important step in taking these devices from the status of laboratory curiosity to the mainstay of first-generation light-wave systems. DLD and DSD structures also occur in the GaInAsP material.[3] It is reasonable to believe that understanding of this degradation mode will also be important for the realization of practical, long-lived quaternary devices.

In this chapter we will discuss the observations reported to date of defect motion and the development of extended defect structures in GaInAsP. Ordinary diffusive motion of point defects and impurity atoms is important (and in many aspects not clearly understood) for GaInAsP and InP device material. This is not an area we will consider in detail here, however, except insofar as it contributes to the growth of extended defect structures or causes such structures to become non-radiative recombination sites.

7.1 THE 'DARK LINE DEFECT'

The 'dark line defect' is, as the name suggests, a region of greatly reduced radiative efficiency with a roughly linear aspect. In the initial observations in AlGaAs stripe-geometry laser diodes,[1] striking linear dark features were seen to develop, crossing the luminscent stripe at a characteristic 45° angle. The active stripes in these devices are oriented along ⟨110⟩ directions so as to be bounded by the (011) cleavage planes which form the laser cavity mirrors. Thus the DLDs are oriented along ⟨100⟩ crystallographic directions. In addition to this striking geometrical aspect, DLDs were seen to form, after an apparently random nucleation or induction time, at sites which initially did not appear to differ in radiative efficiency from surrounding regions. The rate of post-nucleation development was also seen to vary markedly from site to site, even on the same epitaxial wafer. Identical phenomena were independently observed in optically excited AlGaAs semiconductor lasers at about the same time.[4] In that case there was no need for electrical contacts, a p–n junction, or particular doping levels. It could readily be shown that the DLD phenomena in the optically excited material were intrinsic in origin, and were not caused by contact migration or metallurgy, electric fields associated with the p–n junction, or other details of diode device processing. Improper processing or incautious handling could, of course, exacerbate the problem, giving rise to DLDs in otherwise degradation-resistant material or drastically accelerating the defect development and growth. Subsequent transmission electron microscopy (TEM) studies confirmed that the DLD structures resulting from normal electrical device operation and those from optical excitation were indeed identical in microscopic detail,[5] and thus almost certainly resulted from the same generation process(es).

In addition to the classic ⟨100⟩ DLD, occasional ⟨110⟩ oriented DLDs were also observed in AlGaAs heterostructures. These typically propagated much more rapidly (at rates up to 10^3 μm s^{-1} with scanned optical excitation) and were seen to originate at physical defects such as scratches, tool marks, or regions of mirror facet damage resulting from catastrophic degradation. 'Dark spot defects' were also observed. These lacked the linear character of the DLDs and were generally associated with regions of identifiably poor epitaxial growth, inclusions, etc.

Extensive studies have ensued on these degradation modes of AlGaAs material, and a general consensus has emerged on the features and mechanism of DSDs and the less common ⟨110⟩ oriented form of DLD. The ⟨110⟩ DLDs are believed to result from glide of fresh dislocations in the strain field associated with post-growth plastic deformation, while the DSDs appear to arise from glide of dislocations related to precipitate inclusions. In contrast, there is still some controversy as to the details of ⟨100⟩ DLD formation. Typically, a ⟨100⟩ DLD consists of a vacancy dipole network which originates from a grown-in, inclined (or 'threading') dislocation by a process combining glide and climb. The controversy hinges on the details of point defect involvement and on whether climb is involved in an essential[6] or minor[7] way in the network development. Threading dislocations will in general be propagated from defects in the substrate through the epitaxial layers during growth. While it is possible to avoid an increase in the density of these dislocations by taking care to control the epitaxial growth process appropriately, one cannot expect to reduce the density below that in the starting substrate simply by being careful. The successful extension of AlGaAs laser lifetime into the 10^6 h range resulted from elimination of precipitates by careful substrate cleaning and prevention of air leaks into the growth apparatus (to control DSDs), from careful post-growth handling and development of strain-free mounting procedures (to control $\langle 100 \rangle$ DLDs), and from selection of substrate material having as low a density of dislocations as possible (to reduce the probability that a threading dislocation would intersect the active region of fabricated devices).[8]

In the latter 1970s, as interest turned to the 1.1–1.6 μm wavelength region for second-generation optical systems, there was considerable speculation as to whether devices fabricated in the GaInAsP/InP system would present major lifetime problems, or virtually none at all. On the one hand, the lower energy gap in the quaternary compounds provides reduced energy relative to that available in the AlGaAs system from a non-radiative recombination event. This was expected to be insufficient to activate vacancy motion. On the other hand, InP substrate material suffered from many problems in comparison to the relatively high-quality GaAs material commercially available. Dislocation densities were two to three orders of magnitude higher, precipitation and segregation of dopant impurities were commonly observed, cutting and polishing operations were problematic because of the relative ease with which work damage occurs and propagates in InP, and interfacial problems were known to occur during epitaxial growth as a result of the low temperature at which phosphorus loss occurs. Insufficiently precise compositional control of the quaternary alloy leads to the introduction of strain and/or generation of misfit dislocation arrays, known to be associated with DLD generation in AlGaAs material.

Questions were also raised about the long-term stability of the quaternary alloy against precipitation into energetically favoured binary or ternary phases. Remarkably, 1000 h lifetimes for continuously operated GaInAsP lasers were reported almost as soon as continuous room-temperature operation was obtained.[9] Unfortunately the further leap to the 10^6 h projected lifetime range needed for reliable telecommunications applications has not been made with equal ease, and it is now clear that there are indeed significant degradation modes of the quaternary lasers and LEDs. As was the case with AlGaAs devices, optical and electron-beam excitation studies coupled with optical metallography and transmission electron microscopy are proving particularly enlightening. The study of the degradation modes of quaternary devices and characterization of the salient defect structures and mechanisms is still in an early stage. Results so far are largely of a descriptive, and only tentatively of an interpretative, nature.

7.2 EXPERIMENTAL TECHNIQUES

As discussed in the preceding section, dislocations and defect networks may be propagated by thermal effects, strain, minority-carrier injection across a p–n junction, or electron–hole pair creation obtained with electron- or photon-beam excitation. The 'dark line' and 'dark spot' defect features are most easily observed by their darkness in electro-, photo- or cathodoluminescence; but the defects may be seen and studied in detail by traditional metallographic techniques or transmission electron microscopy as well. Their presence may also be detected by scanning deep level transient spectroscopy (DLTS)[10] with a scanning electron microscope (SEM) or scanning transmission electron microscope (STEM) and in the electron beam induced current (EBIC) mode of an SEM.

For the techniques using luminescence microscopy a photodetector or TV camera sensitive to near-infrared light in the 1.1–1.6 μm range is required. Ge photodiodes or vidicons with PbS or PbS/PbO targets are commercially available and fulfil this requirement. GaInAs or GaInAsP photodiodes are not commercially available but offer reduced dark current and faster response than is possible with Ge detectors. A number of arrangements which are capable of providing satisfactory results are shown schematically in Figure 7.1. The choice of a particular arrangement is apt to be dictated by the ease of access to a suitable laser (Kr II or Nd:YAG with approximately 1 W output) or SEM, availability of scanning mirrors and drivers vs an infrared TV camera, etc.

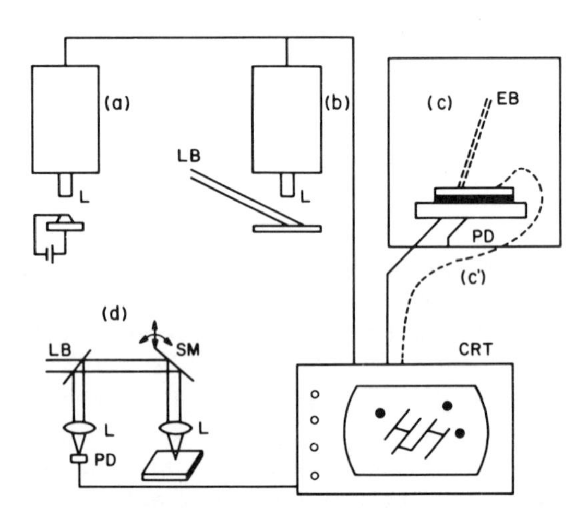

Figure 7.1 Configurations for DLD microscopy. (a) Observation of window laser sidelight or LED with IR microscope. (b) Photoexcitation of wafer with non-scanning laser beam and observation of photoemission through long-wavelength-passing optical filter, (c) Transmission cathodoluminescence excited by the electron beam (EB) in an SEM observed with a large-area photodetector (PD) on which the sample is mounted. (c') EBIC display from top contact on wafer containing p–n junction. (d) Photoluminescence excited with a raster-scanned laser

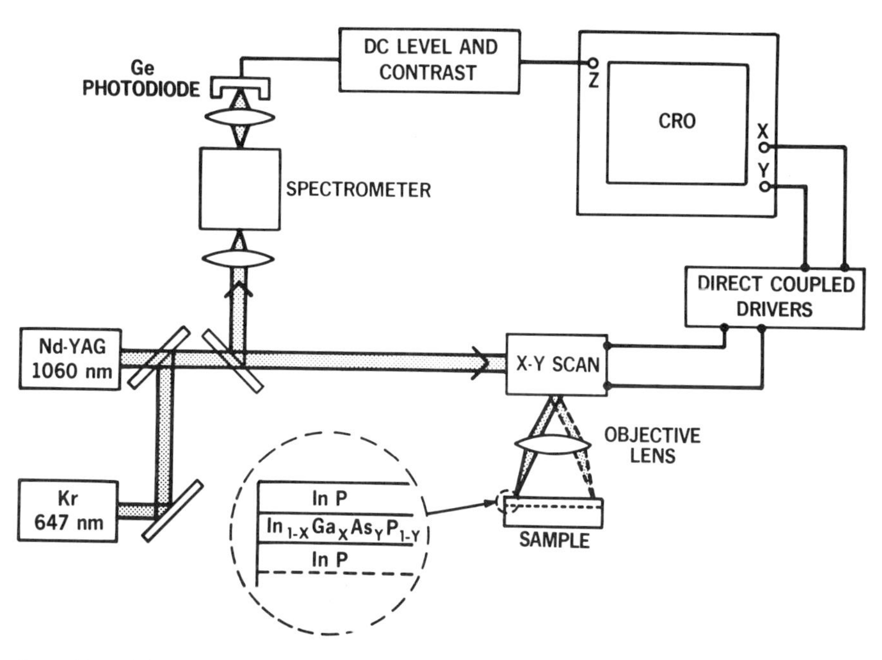

Figure 7.2 Schematic diagram of apparatus for scanned photoluminescence studies and DLD generation in GaInAsP. (From Johnston *et al.*[3] Reproduced by permission of the American Institute of Physics)

The scanning photoluminescence mode (Figure 7.2) is probably the most versatile and convenient and was used to make the first observations of DLD phenomena in GaInAsP.[3] By suitable choice of optical wavelength, the position or depth at which electron–hole pairs are created can be controlled. Laser power, intensity, and scan velocity are all easily and continuously variable. Since no vacuum chamber is required, hot or cold stages are easily implemented and non-destructive observation is possible. Epitaxial wafers can be examined immediately after growth as there is no need for processing or special mounting (Figure 7.3). The development of a film of polymerized pump oil on the excited region of the sample which is commonly encountered in diffusion-pumped SEM systems is totally avoided. The resolution in a scanning photoluminescence system can be increased by appropriate choice of focusing lens and scan amplitude to the limit imposed by the minority-carrier diffusion length. Under conditions of high-intensity excitation, the resolution can actually be seen to increase as the minority-carrier lifetime is reduced by the onset of stimulated emission. Features in the 1–2 μm size range can be observed directly with this technique.

The scanning transmission cathodoluminescence (TCL) technique[11] is similar in concept, differing only in using a scanned electron beam rather than a laser for

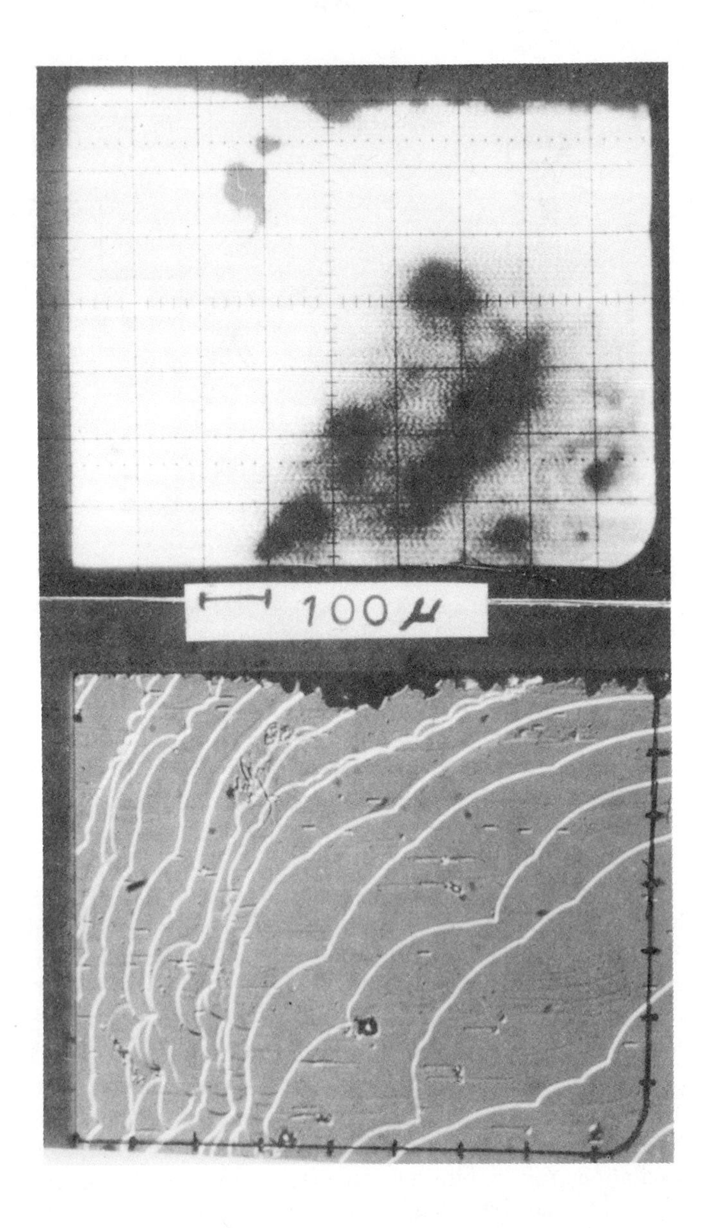

Figure 7.3 Visible light (Nomarski interference contrast) low-power micrograph of the top surface of an LPE GaInAsP/InP double-heterostructure wafer and scanning photoluminescence display of the same area. The DSD features in the active layer are only partially correlated with visible defects

excitation. As the name implies, the emitted light is collected after transmission through the sample. This permits observation of certain kinds of opaque defects within the substrate which are not otherwise accessible, but the primary reason for back-surface light collection is the requirement that the electron beam have unimpeded access to the sample front surface. Use of a broad-area photodetector below the sample is possible, permitting scans of entire epitaxial wafers. The large detector capacitance imposes a restriction on the useful video bandwidth and hence on scan speed and frame rate, however, so smaller detectors, about 3–6 mm in diameter, are generally employed. These are quite satisfactory for high-resolution, limited-field use. The disadvantage of having to work in the SEM vacuum chamber is offset by the opportunity to combine TCL imaging of defects with the standard very-high-resolution secondary electron imaging mode and with X-ray fluorescence analysis of the chemical composition of precipitates associated with defect networks. Required sample preparation is limited to polishing the back surface, so that this technique may also be used to screen epitaxial wafers prior to device processing. Precautions must be taken to avoid pump-oil contamination, however. In practice this requires a dry (sorption and ion-pumped) vacuum system.

With either of the above approaches, DLDs may be created as well as observed by halting the beam scan and allowing the beam to dwell at a chosen place for a suitable length of time. There is no clear threshold for DLD generation. In practice one can define 'threshold' exposure as that energy deposition by which the photo- or cathodoluminescence signal has fallen to some standard fraction, say one-tenth, of its initial value. This is conveniently monitored via the DC level of the video signal as degradation proceeds.

In the case of EBIC or scanning DLTS imaging modes, either a p–n junction or a Schottky barrier metallization must be formed and appropriate contacts to the sample surface must be made. A scanned photocurrent display equivalent to an EBIC image could easily be obtained, but the EBIC mode is more convenient in this case because it is easy to relate the EBIC details to 'visible' sample features by switching back and forth to the secondary electron image. The scanning DLTS mode requires a temperature-controlled stage and sophisticated signal processing electronics. It provides the capability of relating deep-level electron (or hole) traps to the presence or growth of a defect network. This has yielded information coupling a specific donor–vacancy complex (the Dx centre) to dislocation climb in AlGaAs material,[2] but has not as yet seen application to the GaInAsP alloys.

The observation of DLDs in AlGaAs and GaAs has for the most part been restricted to double-heterostructure epitaxial wafers. Single layers of GaInAsP readily exhibit DLD generation,[3] as do InP epilayers and even polished substrate material. Single epilayers of AlGaAs or GaAs do not exhibit the ⟨100⟩ type of DLD. Initiation of ⟨110⟩ DLDs in GaAs is difficult in the absence of heterointerfaces even when fresh dislocations are introduced by room-temperature deformation. (There is a photo-oxidative effect which reduces the level of photoluminescence from freshly cleaved or carefully cleaned GaAs surfaces. This results from a change in surface recombination velocity and is unrelated to defect motion, however.)

When DLDs are generated in a single GaInAsP epilayer, the dislocations can propagate by slip to the surface and are then visible through a light microscope set up for top-illuminated Nomarski interference contrast. Selective defect-sensitive etching can be used to heighten the contrast and/or to reveal the dislocation tangles. In combination with non-selective etching the sample may thus be sectioned and stained for metallographic examination. Dilute solutions of Br-methanol are relatively non-selective and with care can be used for sectioning, although precise control of depth is difficult. A mixture of $KFe(CN)_3$ and KOH, or A-B etchant,[13] appropriately (i.e. by trial and error) diluted for the particular quaternary composition and conductivity type, are suitable for staining purposes. DLDs may be localized in this way to a particular layer of AlGaAs or quaternary heterostructures.[14]

Sample preparation for TEM study is notoriously time-consuming and the level of care exercised is directly reflected in the quality of the results that can be obtained. The semiconductor wafer or chip must be thinned down to a few hundred nanometres or less, with the region of interest surrounded by thicker material which can serve as a mechanical support. It becomes very difficult to locate a particular DLD and correlate the TEM field of view, which at high magnification is of course very limited, with specific regions on the wafer which have been given particular treatment. One solution is to work with pre-thinned samples mounted on indexed copper grids. Alternatively, clear reference features on a visible micrograph of the wafer surface may be used as a mutual reference for photoluminescence and TEM study. Since a great deal of time must be taken to get TEM results, and since they represent a highly detailed view of an extremely small region, great care must be exercised in the judicious selection of samples for TEM characterization. It is usually worthwhile to get as much information as possible with luminescence studies before turning to the TEM for resolution of the precise nature of particular DLD and DSD networks.

7.3 OBSERVATIONS TO DATE

There are relatively few published reports on the properties of DLDs and DSDs in GaInAsP or InP material. This author is aware, however, of several unpublished observations made in other laboratories which confirm the salient features. As yet no reports have appeared on detailed studies of DLDs resulting from operation of quaternary diode lasers. The published reports describe DLD development in normal and accelerated LED life tests, or under conditions of optical excitation. The DLDs were observed in electroluminescence or photoluminescence modes respectively. Transmission cathodoluminescence has been used to study defects in InP material[11] and in stress-aged GaInAsP LEDs,[15] and

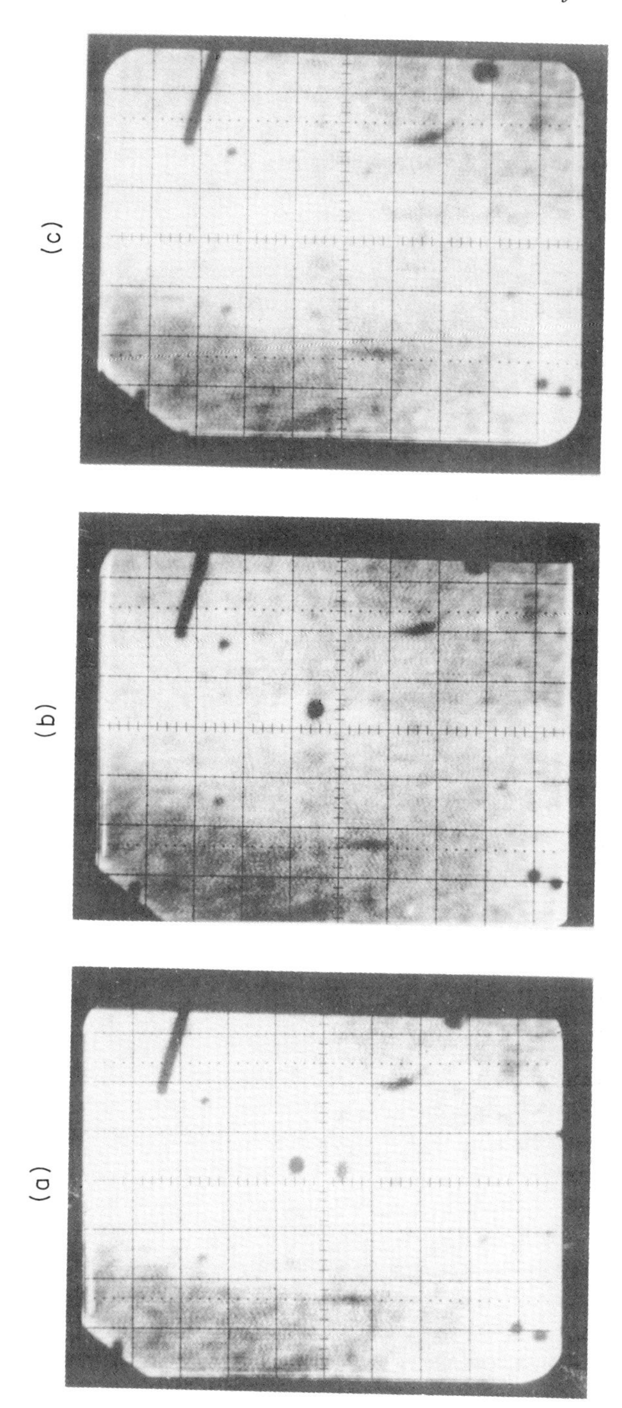

Figure 7.4 Scanning photoluminescence display of DSD generation: (a) non-degraded central region, (b) 1 min exposure to 10^3 J cm^{-2} at left centre (c) 1 min exposure to 10^2 J cm^{-2} at right centre. Scale: 100 μm per grid division

transmission electron microscopy has been used to define the crystallography of these affected regions.[16,17]

7.3.1 Results with optical excitation

The initial observations that GaInAsP material suffered degradation in luminescent efficiency during optical excitation were made during the course of studies evaluating scanning photoluminescence characterization as a screening tool for rapid evaluation of liquid-phase epitaxial (LPE) growth quality.[3] Double-heterostructure laser wafers as well as single-layer samples were exposed to the focused scanning beam from a Kr II (674 nm) or Nd: YAG (1064 nm) laser. Certain wafers were found to show reduced output from the raster-scanned region and initially a surface effect was assumed, since no preferential relation to crystallographic geometry was apparent. The damage could not be removed by surface etching or polishing, however. Although the beam intensity was moderately high in these experiments (about 10^4 W cm^{-2}), the dwell time was very short, due to the scanning pattern, and the damage had to have occurred within a fraction of a second of exposure time.

Subsequent exposure of these easily degraded samples to a stationary, defocused beam verified that exposure doses of a few hundred joules per square centimetre caused an irreversible, order-of-magnitude drop in luminescence intensity. The appearance of these degraded regions is shown in Figure 7.4. These were LPE samples in which the quaternary layer was known to be mismatched in lattice constant by a few tenths of one per cent relative to the InP substrate, or

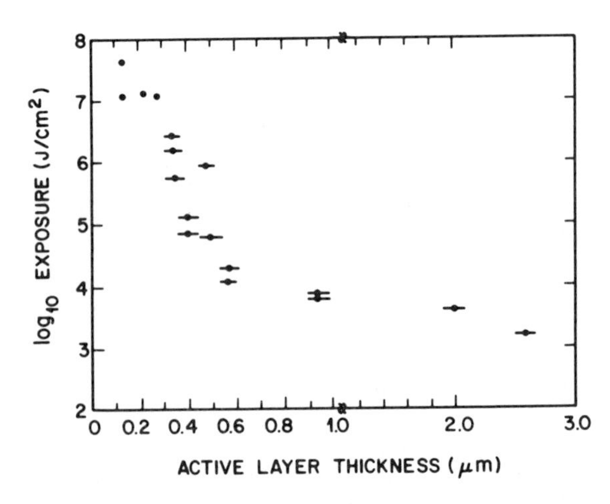

Figure 7.5 Threshold exposure for optically induced degradation in several samples of double-heterostructure and single-layer GaInAsP epitaxial material as a function of the quaternary layer thickness. (From Johnston *et al.*[3] Reproduced by permission of the American Institute of Physics)

samples with rather thick (~ 2 μm) quaternary layers. The particular LPE step-cooling growth procedure used[18] was such that the thicker layers tended to vary from the desired composition due to melt depletion and kinetic effects, so that a graded mismatch and strain was inferred to be present.

Other double-heterostructure samples were investigated and the degree of resistance to degradation was found to vary over nearly five orders of magnitude, with some material being non-degradable within the practical experimental limitations (100 h exposure at 5×10^5 W cm^{-2}, just below the physical 'burn' limit at about 10^6 W cm^{-2}). The dependence of degradation threshold exposure on layer thickness is shown in Figure 7.5. This dependence was interpreted in terms of increasing strain with layer thickness as discussed above. Dislocations will of

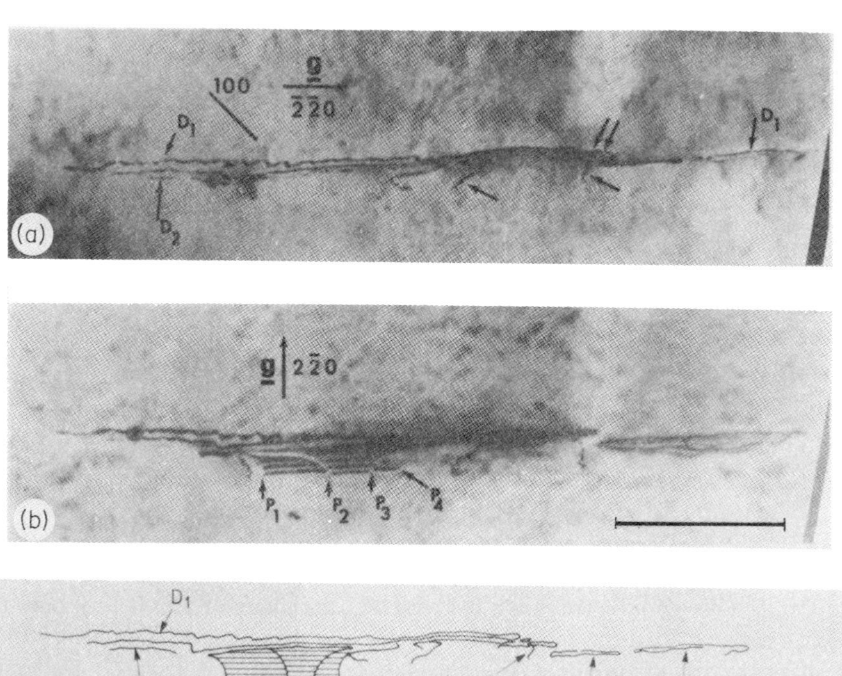

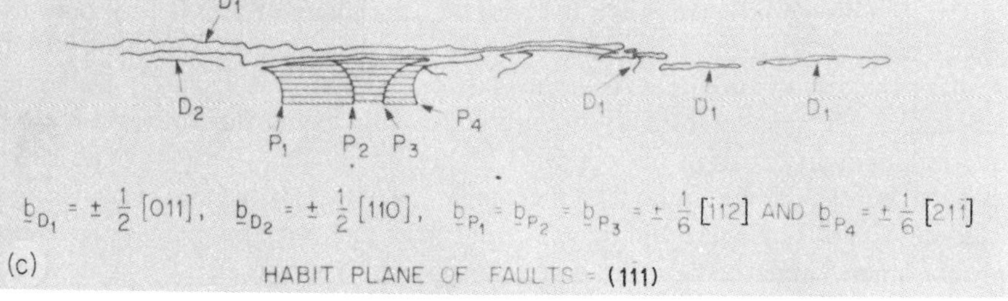

Figure 7.6 Transmission electron micrographs showing an optically degraded region in a GaInAsP single layer. The scale in (a) and (b) is the same; the bar in (b) represents 1 μm. Dislocations D_1, D_2 and partials P_1–P_4 are labelled and their Burgers vectors identified in (c). (From Mahajan *et al.*[16] Reproduced by permission of the American Institute of Physics.)

course glide in a strain field, and glide becomes more and more likely as the strain field increases.

No preferential direction of propagation within the (001) plane was observed, and the optically produced dark regions were referred to as a 'broad area' degradation in the initial report. Subsequently several of these samples were examined by transmission electron microscopy, and the small-scale $\langle 110 \rangle$ linear orientation became clear.

Principal characteristics of the degraded region (see Figure 7.6) were dislocations extending towards the substrate, faulted regions, sausage-link structures, jagged dislocation segments, and dipoles aligned along the $\langle 110 \rangle$ direction. These features were seen only in the degraded regions and were common to all the degraded regions. Threading dislocations and dislocations arising from inclusions were common to both degraded and non-degraded regions. An explanation for the degradation structure based on the recombination enhanced glide of threading and inclusion generated dislocations on $\{111\}$ planes was presented. This rapid glide gives rise to dislocation clusters aligned to $\langle 110 \rangle$, assuming that most of the residual debris is in the form of 60° dislocations and dipoles. This seems plausible given the low temperature (relative to the melting point) and the apparently high strain rate. Essentially identical structure was found in a degraded 'thin' layer ($t = 0.3\ \mu$m) and a 2 μm thick layer, even though these two samples differed by more than 10^3 in degradation threshold.

7.3.2 LED results

The initial device aging results on quaternary LEDs and lasers were most encouraging. There can be little doubt that the LEDs in particular will provide useful life well in excess of communication system requirements. When degradation does occur, however, it is variable in development and the situation appears at present to be somewhat complex and not altogether well understood. Yamakoshi and coworkers have reported that they found no degradation whatsoever over 10 000 h for LEDs operating at temperatures between 20 and 170° C.[19] This would seem consistent with the attainment of 10 000 h continuous laser operation at room temperature[20] and would suggest to the unwary that perhaps the discussion of DLDs in the GaInAsP system was of purely academic interest. These results certainly do provide a convincing demonstration that quaternary light-emitting devices with highly practicable lifetimes have been realized, but they do not imply that there are no problems with yield of such long-lived devices or that there will be no serious problems with reliability assurance programmes. The latter problem becomes apparent when accelerated aging tests are conducted. The former problem, that of yields of long-lived devices, may be a 'simple' matter of careful control of all growth and processing steps. Indeed, it is all too easy to make short-lived quaternary diode lasers (even easier to grow wafers which yield only non-lasers!), and there is naturally little incentive to elaborate publicly on the cause of such failures after striking success has already been reported. DLD and DSD problems do indeed occur in quaternary lasers and LEDs; when they occur they are indeed life-limiting; and we do not presently know enough about them to be confident that we understand how to assure that they will not be a recurrent practical problem.

At temperatures greater than 170° C Yamakoshi *et al.* observed some DSD and DLD generation of the $\langle 110 \rangle$ type.[19,21] The apparent activation energy for slow degradation derived from their results above 170° C was 0.9–1.0 eV, with no defects found at lower temperatures. This would suggest an extrapolated median lifetime at room temperature exceeding 5×10^9 h (about 5×10^5 yr for 50% failure), more than adequate for light-wave systems applications foreseen today. A similar failure mode in quaternary LEDs has been reported by Temkin *et al.*,[15] but with the difference that the LEDs were degraded during storage at 200° C without current injection. This suggests that the degradation is the result of 'annealing' of some instability. Temkin *et al.* pointed out that this effect had an apparent activation energy also of 0.9–1.0 eV, but that in their opinion it did not seem appropriate to describe this process in such terms. It appeared more likely in their case that a thermally induced critical stress gave rise to dislocation slip and $\langle 110 \rangle$ DLDs. The situation in sum appears to be that there is at least one mode of thermal DLD formation which has a very steep temperature dependence but very low probability of occurrence at room temperature. There may be other modes with less pronounced temperature dependence which in fact dominate the room-temperature lifetime.

Temkin *et al.* traced the rapid high-temperature $\langle 110 \rangle$ DLD degradation mode in their LED samples to the development of dislocation clusters originating at inclusion-like defects (see Figure 7.7). They were able to localize the latter to the first-grown n-InP layer and/or the interfaces between that layer and substrate or the quaternary active layer. No $\langle 100 \rangle$ type DLDs were observed, consistent with the optical degradation results. They also found that there was no apparent spatial correlation of the DLDs with the location of the p-side contact or the wirebond thereto. In the case of thermal aging, no change in the I–V characteristics were seen, while devices aged with forward current injection became quite leaky. As a practical matter this permitted EBIC imaging of the thermally degraded structures, which showed clearly the presence of DSDs associated with inclusions and related $\langle 110 \rangle$ DLDs.

An extended defect of a different sort which did show a $< 100 >$ orientation has been seen by Ueda *et al.*[22] This consists of bar- or cross-shaped structures up to 20 μm in length which occur above 200° C in devices operated with forward bias. These defects were found in the quaternary layer rather than in the InP buffer layer. Examination of the X-ray fluorescence spectra from the defect region together with contrast analysis in the TEM showed that these defects were precipitates of a secondary phase with greater In and P content than the background matrix material. No indication of elements other than In, Ga, As, or

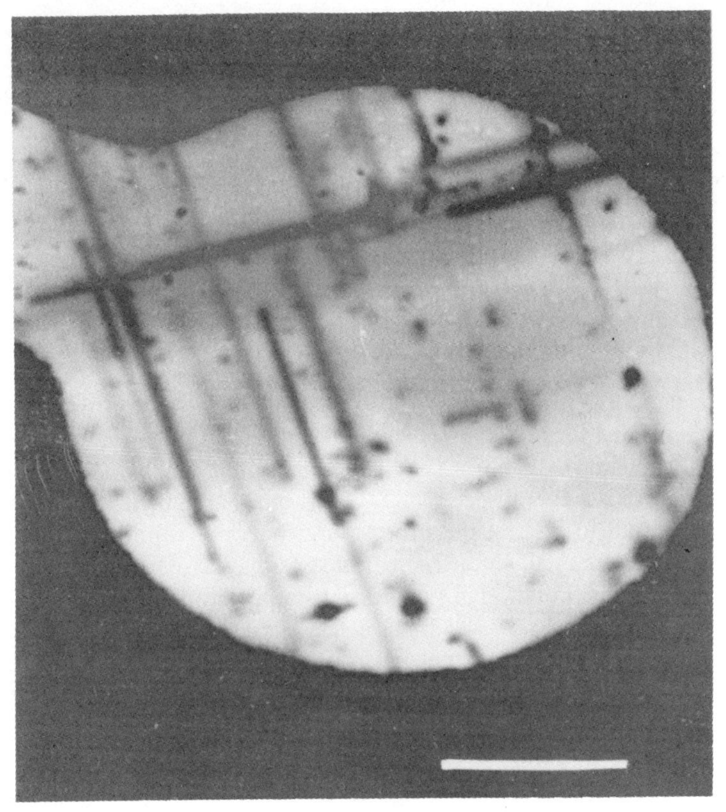

Figure 7.7 ⟨110⟩ oriented DLDs as observed in EBIC image of a thermally aged quaternary LED. Scale bar equals 100 μm. (With thanks to H. Temkin)

P was seen, demonstrating, for example, that electromigration of contact metals into the quaternary layer with subsequent precipitation of some 'extrinsic' compound was not occurring. Associated with these precipitate inclusions was a tangled, fibrous dislocation structure which tended towards simple edge dislocations far from the inclusion. Ueda *et al.* suggested that the onset of this problem is a result of local melting of In- or P-rich regions, which perhaps arise in the first place from thermal damage to the substrate incurred just prior to growth or from damage to the InP buffer layer during the brief time between its growth and the growth of the quaternary active layer in which the precipitates later form.

Quaternary material of device quality has been grown by vapour-phase epitaxy (VPE) as well as the more commonplace LPE process.[23] VPE growth affords a potentially greater range of stoichiometry (i.e. III/V ratio) and thus a possibility to control stoichiometry-related defects. The presence and possible incorporation of excess In or the separation of In-rich phases cannot easily be circumvented in growth from In solution. No published reports of DLD generation specifically in VPE material appears in the current literature, but the reports of the properties of laser diodes fabricated from VPE material suggest that this material can be at least as good as the LPE material.[24] In optical degradation studies we have found that VPE and LPE material behave similarly. Rapid broad-area degradation has been observed for quaternary layers when there exists a significant negative lattice mismatch $(a - a_0)/a_0 \approx -0.1\,\%$ or when a more sizable positive mismatch is present. Inclusions resulting from improper subtrate preparation or thermal damage before growth commences also serve to nucleate $\langle 110 \rangle$ defect structures.

Misfit dislocation arrays in the VPE material are not as a rule seen in photoluminescence. Whether quaternary layers, InP layers grown on quaternary

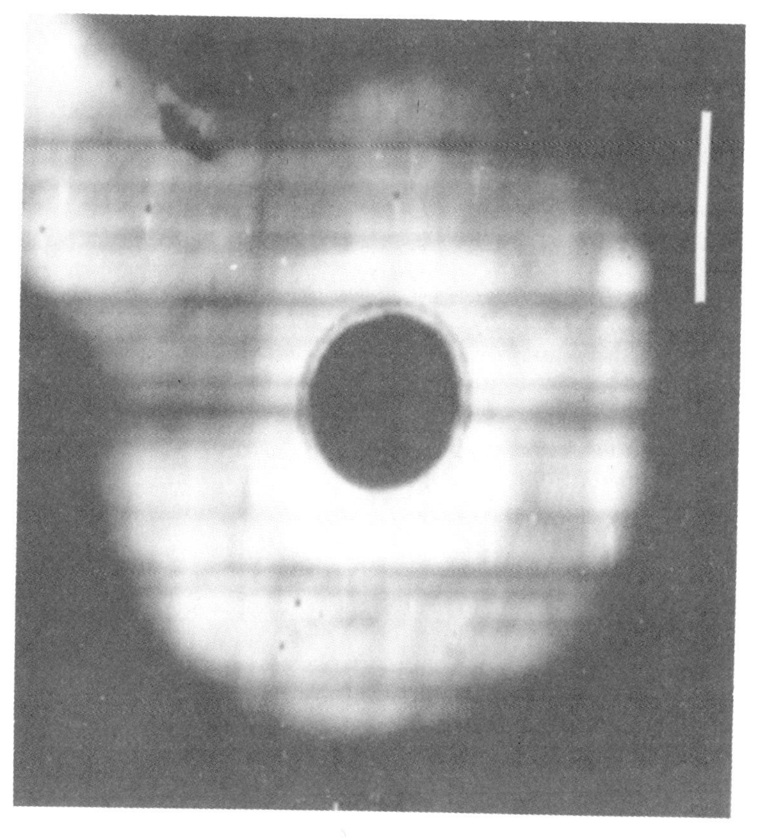

Figure 7.8 Misfit dislocation array evidencing DLD behaviour in transmission cathodoluminescence. The quaternary layer in this LED structure is doped with zinc to the 10^{18} level. The large central dark region is the p-contact metallization and the dislocation lines are oriented along $\langle 100 \rangle$ directions in the (001) plane. (With thanks to A. K. Chin)

layers, or InP homoepitaxial layers are the subject of investigation, misfit arrays are generally ineffective as non-radiative recombination centres. When impurity dopant levels are high ($\sim 10^{18}$ for Te or Zn) this is no longer the case, as shown in Figure 7.8, probably because of dopant segregation at and decoration of the edge dislocations which make up the misfit array.[25] The fact that misfit dislocation arrays (and a lattice mismatch, as confirmed by X-ray diffraction measurements) are seen for some homoepitaxial InP VPE layers but have never been observed in LPE homoepitaxy confirms that different stoichiometric regimes are accessed by the two growth techniques.[26]

Some VPE material which is highly resistant to optically induced degradation has also been examined in the author's laboratory. These samples could not be degraded at the 100 h, 10^6 W cm^{-2} exposure level, and included quaternary layers as thick as 7 μm. The VPE growth technique obviously does not suffer from the effects of melt depletion or from kinetic effects within the melt associated with LPE, but precisely reproducible compositional control is still a formidable challenge. Threading dislocations from the substrate and inclusions at interfaces from thermal damage are also problems common to the two growth techniques. At this juncture there does not appear to be any unique difference between DLD or DSD formation in VPE or LPE material.

7.4 DISCUSSION AND SUMMARY

Degradation of light-emissive devices resulting from defect motion has been extensively studied in the case of AlGaAs material. In spite of the fact that the literature is much less extensive regarding defect motion and degradation in GaInAsP, several conclusions can safely be drawn at this point.

Of perhaps the greatest practical significance is the fact that the 'classic' $\langle 100 \rangle$ DLD has not been seen in quaternary devices—indeed it has not been reported in the binary InP layers either. Short $\langle 100 \rangle$ lines which have been seen correspond to projections of $\langle 011 \rangle$ or $\langle 101 \rangle$ DLDs onto the (001) plane. It is probably safe to say that the climb or climb/glide process responsible for the usual demise of AlGaAs light emitters by DLD formation is not operative in InP and GaInAsP. Perhaps this is a result of the reduced bandgap energy, although there is not a great deal of difference between GaAs and InP in that regard (1.43 vs 1.34 eV), or perhaps it is a result of another difference in bond energy, lattice phonon energy or density of states, etc. Since the phenomenology of $\langle 100 \rangle$ DLD formation in AlGaAs is not clearly understood at this time, it appears more fruitful to ask why it occurs in GaAs than to ask why it does not occur in GaInAsP. In this context it should be noted that a first report of AlGaInAs lasers emitting at 1.65 μm, prepared by molecular-beam epitaxy (MBE), has appeared.[27] Information on DLD development or degradation in this material should shed light on the importance of P vs Al, for instance in moderating dislocation climb, or might confirm or deny the importance of the bandgap energy. The other heterostructure laser system for this longer wavelength range is AlGaAsSb.[28] No DLD or degradation results have been reported in this case either, although continuous operation near room temperature has been obtained.

The ⟨ 110 ⟩ DLDs that are observed in GaInAsP may all be associated with strain or growth pathology, and hence they may properly be considered to fall into the class of problems that can be solved by careful attention to growth technique. It certainly seems desirable to have InP substrates with lower dislocation densities than 10^4–$10^5\,cm^{-2}$, but there is no clear indication yet that this is really necessary for long-lived lasers or LEDs. In contrast, thermal damage to the substrate prior to growth is clearly detrimental, as it leads to In-rich phases and local strains which have been shown to give rise to slip and ⟨ 110 ⟩ DLD and SDS behaviour at elevated temperatures or under minority-carrier injection or optical excitation.

Residual lattice mismatch is as a rule simply accepted in the fabrication of AlGaAs devices, but it must be avoided by proper compositional matching for GaInAsP. Whether this match should be intended for the growth temperature or for room temperature or somewhere in between is not, however, apparent. What is apparent is that the presence of mismatch strain of the order of 0.1 % or less makes a drastic difference in the susceptibility of GaInAsP to DLD and DSD generation under optical excitation. No systematic studies have been reported of quaternary laser lifetime even as a qualitative function of mismatch strain in the active layer, but the known early failure pattern of otherwise apparently good devices from certain wafers makes this cause-and-effect relation plausible.

The behaviour of GaInAsP devices in 'hot storage' is puzzling. In terms of practical device operation it is true that storage and operating temperatures need not exceed about 70–100° C. Accelerated life tests based on thermal stress are rendered difficult to implement and may be less valuable as a life assurance mechanism, given the possibility of local melting of In- or P-rich microphases.

It is disconcerting to face the possibility that the GaInAsP alloy may not be stable against segregation or precipitation of binary or ternary phases, after all. It is probably a poor first approximation to treat the quaternary as a random distribution of In and Ga on the A sublattice, uncorrelated to another random array of P and As on the B sublattice. Bond length arguments would favour In–P and Ga–As clustering, while bond strength arguments would suggest a tendency for Ga–P clustering. The fact that these two effects tend to offset is, of course, what permits a continuous solid solution, but there is no reason to assume the cancellation is exact. Point defect clustering should also be affected by this consideration. All in all there seems to be a significant probability that dopant segregation or precipitation and vacancy or interstitial clustering as well as compositional phase microsegregation should take place. All of these could be expected to act as sites for the initiation of lattice slip, given the relatively low temperatures (400° C) and strain rates required for plastic deformation of InP.[29]

In conclusion, it appears that no new modes of defect motion lead to degradation of GaInAsP light emitters through DLD or DSD generation and indeed that a major mode observed in AlGaAs material does not occur. The principal limitation on the utility of quaternary LEDs and lasers now appears to be the reproducibility with which 'correct' epilayer growth conditions (proper compositional match and uniformity, prevention of thermal damage, etc.) can be met. Reproducibility of 'correct' AlGaAs growth by MBE, LPE, or metalorganic chemical vapour deposition (MOCVD) has been a recurring problem, and the quaternary system is at least as complex. It remains to be seen whether absence of the highly reactive Al compensates in a practical sense for the additional degree of control required for quaternary growth. In the absence of growth problems or processing-induced strain, the GaInAsP/InP devices show promise of lifetime substantially in excess of presently envisioned requirements for optical communications systems.

ACKNOWLEDGMENT

This paper is the result in part of continuing collaborations the author has enjoyed with S. Mahajan and M. A. Pollack, as well as stimulating discussions with C. L. Zipfel and H. Temkin. The support and encouragement provided by C. V. Shank and B. C. DeLoach during various phases of this work are also very much appreciated.

REFERENCES

1. B. C. DeLoach, Jr, B. W. Hakki, R. L. Hartman, and L. A. D'Asari, 'Degradation of CW GaAs double heterojunction lasers at 300K', *Proc. IEEE*, **61**, 1042–1044, 1973.
2. P. M. Petroff and R. L. Hartman, 'Defect structure introduced during operation of heterojunction GaAs lasers', *Appl. Phys. Lett.*, **23**, 469–471, 1973.
3. W. D. Johnston, Jr, G. Y. Epps, R. E. Nahory, and M. A. Pollack, 'Spatially resolved photoluminescence characterization and optically induced degradation of InGaAsP DH laser material', *Appl. Phys. Lett.*, **33**, 992–994, 1978.
4. W. D. Johnston, Jr, and B. I. Miller, 'Degradation characteristics of CW optically pumped AlGaAs heterostructures', *Appl. Phys. Lett.*, **23**, 192–194, 1973.
5. P. M. Petroff, W. D. Johnston, Jr, and R. L. Hartman, 'Nature of optically induced defects in GaAlAs/GaAs double-heterostructure laser structures', *Appl. Phys. Lett.* **25**, 226–228, 1974.
6. P. M. Petroff and L. C. Kimerling, 'Dislocation climb model in compound semiconductors with zinc-blende structure', *Appl. Phys. Lett.*, **29**, 461–463, 1976.
7. J. Matsui, K. Ishida, and Y. Nannichi, 'Rapid degradation in double-heterostructure lasers. (1) Proposal of a new model for the directional growth of dislocation networks', *Jpn. J. Appl. Phys.*, **14**, 1555–1560, 1975.
   Y. Nannichi, J. Matsui, and K. Ishida, 'Rapid degradation in double-heterostructure lasers. (2) Semiquantitative analyses of the propagation of dark-line defects', *Jpn. J. Appl. Phys.*, **14**, 1561–1566, 1975.
8. H. C. Casey and M. B. Panish, *Heterostructure Lasers*, Academic Press, New York, 1978, Chap. 8.

9. C. C. Shen, J. J. Hsieh, and T. A. Lind, '1500-h continuous CW operation of double-heterostructure GaInAsP/InP lasers', *Appl. Phys. Lett.*, **30**, 353–354, 1977.
10. P. M. Petroff and D. V. Lang, 'A new spectroscopic technique for imaging the spatial distribution of non-radiative defects in a scanning transmission electron microscope', *Appl. Phys. Lett.*, **31**, 60–62, 1977.
11. A. K. Chin, H. Temkin, and R. J. Roedel, 'Transmission cathodoluminescence: a new SEM technique to study defects in bulk semiconductor samples', *Appl. Phys. Lett.*, **34**, 476–478, 1979.
12. D. V. Lang, P. M. Petroff, R. A. Logan, and W. D. Johnston, Jr, 'Recombination enhanced interactions between point defects and dislocation climb in semiconductors', *Phy. Rev. Lett.*, **42**, 1353–1356, 1979.
13. M. S. Abrahams and C. J. Buiocchi, 'Etching of dislocations on the low-index faces of GaAs', *J. Appl. Phys.*, **36**, 2855–2863, 1965.
14. W. D. Johnston, Jr, W. M. Callahan, and B. I. Miller, 'Observation of dark-line degradation sites in AlGaAs/GaAs DH laser material by etching and phase-contrast microscopy', *J. Appl. Phys.* **45**, 505–507, 1974.
15. H. Temkin, C. L. Zipfel, and V. G. Keramidas, 'High temperature degradation of InGaAsP/InP LED's', *J. Appl. Phys.*, **52**, 5377–5380, 1981.
16. S. Mahajan, W. D. Johnston, Jr, M. A. Pollack, and R. E. Nahory, 'The mechanism of optically induced degradation in InP/InGaAsP heterostructures', *Appl. Phys. Lett.*, **34**, 717–719, 1979.
17. O. Ueda, S. Yamakoshi, and T. Yamaoka, 'Transmission electron microscope observation of mechanically damaged InGaAsP/InP double-heterostructure light emitting diodes', *Jpn. J. Appl. Phys.*, **19**, L251–254, 1980.
18. M. A. Pollack, R. E. Nahory, J. C. DeWinter, and A. A. Ballman, 'Liquid phase epitaxial InGaAsP lattice matched to (100) InP over the complete wavelength range $0.92 \leq \lambda \leq 1.65\,\mu m$', *Appl. Phys. Lett.*, **33**, 314–316, 1978.
19. S. Yamakoshi, M. Abe, S. Komiya, and Y. Toyama, 'Degradation of high radiance InGaAsP/InP LED's at 1.2–1.3 μm wavelength', *Proc.* 1979 *IEDM*, December 1979, Washington, DC, IEEE Press, New York, 1979, pp. 122–125.
20. T. Yamamoto, K. Sakai, and S. Akiba, '10,000-h continuous CW operation of InGaAsP/InP DH lasers at room temperature', *IEEE J. Quantum Electron.*, **QE-15**, 684–687, 1979.
21. S. Yamakoshi, M. Abe, O. Wada, S. Komiya, and T. Sakurai, 'Reliability of high radiance InGaAsP/InP LED's operating in the 1.2–1.3 μm wavelength range', *IEEE J. Quantum Electron.*, **QE-17**, 167–173, 1981.
22. O. Ueda, S. Yamakoshi, S. Komiya, and T, Kotani, 'Nature of dark defects revealed in the InGaAsP/InP double-heterostructure light emitting diodes aged at high temperature', *Defects and Radiation Effects in Semiconductors 1980*, Conf. Ser. no. 59, Institute of Physics, Bristol, 1981.
23. G. H. Olsen, C. J. Nuese, and M. Ettenberg, 'Reliability of vapor-grown InGaAs and InGaAsP heterojunction laser structures', *IEEE J. Quantum Electron.*, **QE-15**, 688–693, 1979.
24. W. D. Johnston, Jr, K. E. Strege, and A. A. Ballman, 'Uniformly low-threshold diode lasers at 1.5–1.55 μm from VPE InGaAsP material', *Device Research Conf.*, Ithaca, New York, 1980, *IEEE Trans. Electron Devices*, **ED-27**, 2191, 1980.
25. H. Temkin, private communication, 1980.
26. S. N. G. Chu, S. Mahajan, K. E. Strege, W. D. Johnston, Jr, and A. A. Ballman, 'Reduction of threading dislocations in iso-epitaxial layers grown on (001) InP substrates by misfit stresses', *Appl. Phys. Lett.* **38**, 766–768, 1981.
27. W. T. Tsang, 'AlInAs/GaInAs/AlInAs double heterostructure lasers grown by

molecular beam epitaxy with lasing wavelength at 1.65 μm', *J. Appl. Phys.* **52**, 3861–3864, 1981.
28. R. E. Nahory, M. A. Pollack, E. D. Beebe, J. C. DeWinter, and R. W. Dixon, 'Continuous operation of 1.0 μm-wavelength GaAsSb/AlGaAsSb double-heterostructure injection lasers at room temperature', *Appl. Phys. Lett.*, **28**, 19–21, 1976.
29. D. Brasen and F. A. Thiel, private communication, 1977.

GaInAsP Alloy Semiconductors
Edited by T. P. Pearsall

Chapter 8

Low-field Carrier Mobility

J. R. HAYES*, A. R. ADAMS*, AND P. D. GREENE†

* *Department of Physics, University of Surrey, Guildford, Surrey, UK*
† *Standard Telecommunication Laboratories Ltd, London Road, Harlow, Essex, UK*

8.1 INTRODUCTION

Virtually all experimental work on the low-field mobilities in the quaternary alloy $Ga_xIn_{1-x}As_yP_{1-y}$ has been devoted to those alloy compositions that can be grown lattice-matched to InP substrates. In such alloys the composition parameters x and y obey the relationship $y \simeq 2.1x$, and the composition is conveniently identified by the parameter y which can take values from 0 to 1. Layers grown by liquid-phase epitaxy (LPE) have been the basis of most investigations of electron and hole mobilities.[1–8] Some measurements of electron mobility have also been made on material grown by vapour-phase epitaxy (VPE), with hydrides as sources of As and P and either organometallic compounds or the reaction of the elements themselves with HCl as the sources of In and Ga. No significant differences in carrier mobility between LPE and VPE material have been found.[9–11]

The Hall mobilities of both electrons and holes have been determined over the full range of alloy compositions from $y = 0$ to $y = 1$, although certain regions, particularly y around 0.7, have been relatively neglected because of lack of technological applications. Measurements of the temperature dependence of mobility have mainly covered the range 77 to 300 K but some investigations[11] have extended from 30 to 600 K. In addition to temperature variations, a few samples have been studied over a range of hydrostatic pressures up to 15 kbar in order to determine the variation of mobility with effective mass.[5,6]

Most of the effort has been directed towards the study of electron mobilities in n-type material. This is primarily because the behaviour of electrons is of greater technological significance, particularly with respect to microwave devices such as field-effect transistors (FETs) and transferred electron oscillators (TEOs), although the mobility of holes as well as of electrons plays a significant role in optoelectronic devices. The most striking feature of both electron and hole mobilities in the quaternary alloys is that they do not vary monotonically across the alloy range, but pass through a minimum.

In principle, at low magnetic fields, Hall mobilities μ_H are not identical to drift mobilities μ_D, but the ratio μ_H/μ_D has been determined for electrons by extending the measurements to large magnetic fields. Three independent investigations[1,3,11] have yielded results very close to unity, indicating that for practical purposes the Hall mobility provides a close approximation to the drift mobility in n-type material. Equivalent measurements of the ratio μ_H/μ_D have not yet been made for p-type material. However, in this chapter experimental Hall mobilities for both electrons and holes are directly compared to calculated drift mobilities.

Carrier mobilities are dependent upon both the effective mass of the carrier and the scattering processes which occur. Nicholas *et al.*[12] have measured the effective mass of electrons and concluded that there was an almost linear variation across the alloy range. This result has been confirmed by Restorff *et al.*[13] The effective mass of heavy holes does not vary greatly between the binary constituents of the alloy, and it therefore appears that the downward bowing in the mobility of both electrons and holes must be explained in terms of the scattering mechanisms involved. A more detailed acccount of the effective masses as a function of alloy composition is given in Chapter 12.

Pickering[14] has studied the relative intensities of the phonon peaks observed in infrared reflectivity spectra, and his results, which are discussed in some detail later, suggest that electron–phonon scattering is likely to vary smoothly across the alloy range and cannot explain the observed downward bowing of the electron and hole mobilities. It would therefore seem necessary to include one or more further scattering mechanisms whose strength varies with alloy composition, being stronger towards the centre of the alloy range. The three most likely mechanisms are alloy scattering, space-charge scattering, and ionized-impurity scattering. Measurements of the temperature dependence of the mobility show that ionized-impurity scattering is not very significant at room temperature in the lightly doped samples and cannot explain the effect.

Measurements[6] of the change in mobility produced by increased pressure, which modifies the effective mass, provide convincing evidence in favour of alloy scattering rather than space-charge scattering. The inclusion of alloy scattering of the functional form given by Littlejohn *et al.*[15] has been found by Adams *et al.*[7] adequate to explain almost all of the data on the temperature and pressure dependence of mobility in these quaternary alloys.[16]

8.2 MATERIAL PREPARATION AND CHARACTERIZATION

Except at the boundaries of the alloy $y = 0$ and $y = 1$, nearly all the results presented here were obtained by the authors and their coworkers, and it therefore seems useful to describe in some detail the sample growth, characterization, and preparation techniques.

Single layers of the quaternary alloy were grown by LPE on semi-insulating Fe-doped (100) substrates in horizontal furnaces supplied with Pd-diffused hydrogen. The special features of the graphite slider boats were the incorporation of a large InP cover slice to reduce thermal erosion of the substrate during the heating-up period, and a hopper system to deliver InP into the quaternary melt after the other two source materials (InAs and GaAs) had dissolved in the In metal. The quantity of InP was in excess of that required for saturation (usually at 679° C), so that growth was from a two-phase melt. After the melt had been step-cooled to 659° C, it was brought into contact with the substrate for long enough to produce layers 5 to 10 μm thick. The composition and its uniformity were established by microprobe analysis in the scanning electron microscope (SEM) (see Figure 8.1). The growth technique selected can be seen to provide layers of constant composition, in agreement with the work of Feng *et al.*[17] Infrared transmission measurements to determine the wavelength λ_g of the direct absorption edge provided a routine confirmation of the alloy composition. The quantities of As and Ga in the melt were adjusted empirically to achieve a lattice match of 1 in 10^3 or better.

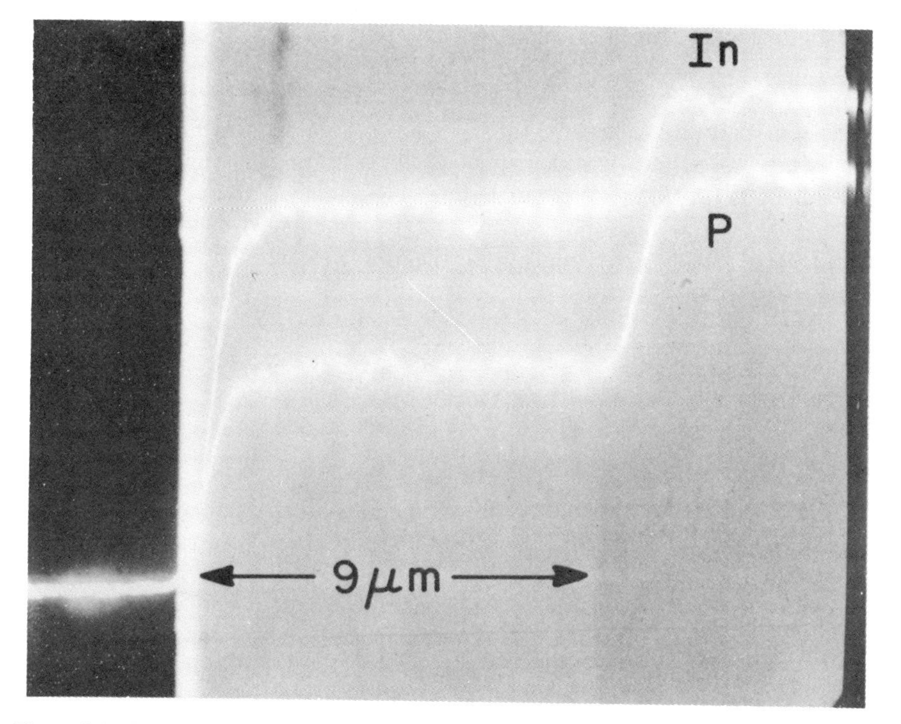

Figure 8.1 Epilayer composition profile measured in the scanning electron microscope by the microprobe method. The step-cooled LPE growth technique provides uniform composition

The growth temperature of 659°C was somewhat higher than that usually adopted for LPE of these materials, but LPE growth at lower temperatures[2] has yielded material with similar electrical properties. It may also be noteworthy that VPE growth at 660°C by Enda[9] produced samples with the composition parameter $y = 0.72$ and mobility consistent with the authors' LPE results.

The n-type samples were nominally undoped. The background free-electron concentration occurs normally in the low 10^{16} cm^{-3} range. However, baking of the melt at 750°C for 10 h in the usual furnace atmosphere (purified H_2 with a few ppm H_2O) regularly produced background carrier concentrations below 1×10^{16} cm^{-3}. This same technique was used in the growth of p-type layers except that Zn or Mn was added as a p-type dopant. The acceptor level[18] associated with Mn appears to have a smaller activation energy in the quaternary than in InP, e.g. just below 0.05 eV for y in the range 0.8 to 0.9. A shallower acceptor is provided by Zn. Two hole concentration ranges were selected for particular investigation, viz. around 4×10^{16} cm^{-3} and 2×10^{18} cm^{-3}.

For Hall effect measurements, 4 mm van der Pauw four-lobed clover-leaf samples were cut. Ohmic contacts were made to the lobes of n-type samples by alloying indium dots while for the p-type material an indium–zinc alloy was used. The Hall effect and resistivity measurements were made in a cryostat with helium exchange gas and the magnetic field routinely used was 0.2 T. The samples showed a high degree of symmetry even down to 77 K. Measurements on some n-type samples were made at high fields (up to 9 T) which cause the Hall mobility μ_H and the drift mobility μ_D to become equal. At both room temperature and 77 K the Hall voltage varied linearly with magnetic field, indicating that the ratio μ_H/μ_D at 0.2T was unity to within the 2% experimental accuracy. Values of μ_H/μ_D in the range 1.00 to 1.03 were found in analogous experiments by Marsh *et al.*[3] and by Bhattacharya *et al.*,[11] so it would appear that in such material μ_H may be taken as identical to μ_D in this temperature range.

8.3 ELECTRON MOBILITY

8.3.1 Composition dependence of μ_e

Figure 8.2 shows the measured, room-temperature variation of the electron mobility μ_e as a function of the composition parameter y for samples with an electron concentration in the low 10^{16} cm^{-3} range. InP samples ($y = 0$) grown by the method described in section 8.2 had a mobility of about 4000 cm^2 V^{-1} s^{-1} in agreement with the accepted value for material of this carrier concentration. The mobility does not vary monotonically between $y = 0$ and $y = 1$ but shows a downward bowing, with a minimum near $y = 0.3$. The dependence of mobility on composition shown in Figure 8.2 is similar to that found by Leheny *et al.*[2] but significantly different from the results of Bhattacharya *et al.*[11] which show a

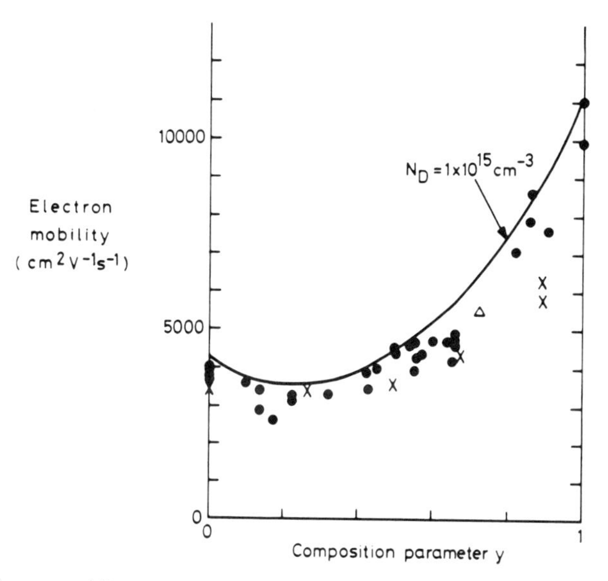

Figure 8.2 Electron mobility at room temperature in the quaternary alloy $Ga_xIn_{1-x}As_yP_{1-y}$ as a function of the alloy composition parameter y. The experimental points correspond to samples with free carrier concentrations in the range 1×10^{16} to 5×10^{16} cm^{-3}: ●. authors' own work; ×, Leheny *et al.*[2]; Δ, Enda[9]. The full curve shows the predicted variation for the purest material considered in Figure 8.8(a)

sharp V-shaped minimum near $y = 0.6$. The high mobility in the ternary $Ga_{0.47}In_{0.53}As$ makes it potentially of particular importance in microwave semiconductors such as FETs and TEOs, and for this reason it has received the attention of several authors. The highest mobility in a sample grown by G. R. Antell (unpublished work) and measured by the present authors was 11 200 cm^2 V^{-1} s^{-1}. Although its temperature dependence indicated minimal ionized-impurity scattering at room temperature, greater mobilities have been reported, viz. 11 500 cm^2 V^{-1} s^{-1} by Pearsall *et al.*[10] and 13 800 cm^2 V^{-1} s^{-1} by Oliver and Eastman.[32] Of particular interest is that Pearsall *et al.* have also examined samples at $y = 1$ grown by VPE and metalorganic chemical vapour deposition (MOCVD) and obtained mobility values (10 500 and 9 500 cm^2 V^{-1} s^{-1} respectively) almost as high as for LPE material.

8.3.2. Temperature dependence of μ_e

In order to understand the scattering mechanisms which so drastically limit the mobility of samples in the centre of the alloy range it is useful to investigate the temperature dependence of the electron mobility. Much of the experimental data for the low-field electron mobility in binary compound semiconductors with direct gaps can be interpreted using a relatively simple theoretical model which considers only two scattering mechanisms, polar optical phonon (PO) scattering and ionized-impurity (II) scattering. The simplest way to combine these uses Matthiessen's rule that the total scattering rate is simply the sum of the scattering rates related to each mechanism, resulting in the following relationship for combining mobility limits μ_i into an overall mobility μ_t

$$\mu_t^{-1} = \sum \mu_i^{-1}$$

Rode[19] has discussed in some detail the errors caused by this assumption in some binary compound semiconductors. Pearsall *et al.*[10] concluded, however, that Matthiessen's rule could be used to interpret mobility data for the ternary alloy $Ga_{0.47}In_{0.53}As$ given by Takeda and Sasaki.[20] Marsh *et al.*[3] first determined the parameters for best fit using Matthiessen's rule and then used these in their Monte Carlo program to reconstitute the mobility variation with temperature. Agreement between these results and the experimentally measured mobilities was 'invariably close'. It therefore seems reasonable at this early stage to use Matthiessen's rule as the simplest analytical technique. It has the considerable advantage that it allows clear graphical displays of the mobility associated with each individual scattering mechanism leading to ease of comparison between approaches by different workers. It will be shown (for example in Figure 8.3 and 8.4) that experimentally determined mobility data covering wide ranges of temperature and pressure agree closely with theoretical predictions which are based on accepted scattering mechanisms and the use of Matthiessen's rule. On the other hand, the validity of Matthiessen's rule for alloy scattering is disputed in Chapter 9.

The theory of polar optical phonon scattering was developed by Ehrenreich,[21] and subsequently by Fortini *et al.*[22] The electron mobility limited by PO scattering alone as given by Ehrenreich and used here may be written

$$\mu_{PO} = 2.6 \times 10^{51} \left(\frac{T}{300}\right)^{1/2} \left(\frac{e}{e^*}\right)^2 \left(\frac{m_0}{m^*}\right)^{3/2} M\Omega\theta(e^z - 1)G(z) \tag{8.1}$$

The mobility limit μ_{PO} is in square centimetres per volt-second, T is the temperature, e is the electronic charge and e^* the Callan effective charge, M is the reduced mass of the cell in kilograms, Ω is the volume of the primitive cell, remaining constant at 5.05×10^{-29} m^3 throughout the alloy composition range, θ is the polar phonon temperature, and z is θ/T. $G(z)$ is a screening factor as given by Ehrenreich.[21] The values for e^*, M, and θ for the constituent binary compounds are given in Table 8.1. The value of m^* for electrons, m_e^*, was taken from Nicholas *et al.*[12] The following interpolation formula was used to calculate values of a parameter A_Q for the various quaternary alloy compositions from the values A_1, A_2, A_3, and A_4 corresponding to the respective binary compounds InAs, InP,

Table 8.1 Values of physical parameters of binary compounds used as a basis for interpolated estimates of the parameters in the quaternary alloy. The symbols are defined in section 8.3.2

	InAs	InP	GaAs	GaP
e^*/e	0.2	0.27	0.20	0.24
m^*_h/m_0	0.60	0.85	0.62	0.79
m^*_l/m_0	0.027	0.089	0.074	0.14
$\theta(K)$	350	498	421	582
$M \times 10^{26}$ (kg)	7.4	3.99	5.92	3.515
k_s	14.55	12.35	12.9	11.1
$\bar{u}$ ($\mathrm{m\,s^{-1}}$)	3090	3810	3900	4760
ρ ($\mathrm{kg\,m^{-3}}$)	5667	4787	5307	4130
E_{AC} (eV)	3.2	3.6	3.5	3.5
E_{NPO} (eV)	5.7	6.3	6.5	6.7
Δ_0 (eV)	0.38	0.13	0.34	0.08

The units used permit the parameters to be applied directly to the equations given in the text for calculations of mobilities in $\mathrm{cm^2\,V^{-1}\,s^{-1}}$. Where appropriate, the equations have been modified from the form in which they were originally published so that a consistent set of units can be applied.

GaAs, and GaP:

$$A_Q = (1-x)yA_1 + (1-x)(1-y)A_2 + xyA_3 + x(1-y)A_4$$

The mobility limit imposed by ionized-impurity scattering in the Γ_{1C} minimum may be written approximately according to the Brooks–Herring formula

$$\mu_{II} = 3.28 \times 10^{15} \frac{k_s^2 T^{3/2}}{(m^*/m_0)^{1/2} N_i} \left(\ln(1+b) - \frac{b}{1+b} \right)^{-1} \tag{8.2}$$

where

$$b = \frac{1.29 \times 10^{14} k_s m^* T^2}{m_0 n'}$$

$$n' = n + (N_D - N_A - n)(n + N_A)/N_D.$$

and

$$N_i = n + 2N_A$$

where n is the electron density, k_s is the static dielectric constant as given in Table 8.1, and N_D and N_A ($\mathrm{cm^{-3}}$) are the density of donors and acceptors respectively.

Extensive treatments of electron mobility in compound semiconductors have considered additional scattering mechanisms. Deformation-potential scattering was analysed by Wolfe *et al.*[23] and a discussion of piezoelectric scattering theory has been given by Rode.[19] These two scattering mechanisms are, however, not generally considered to play a significant part in determining the mobilities in quaternary alloys, although Marsh *et al.*[3] included deformation-potential scattering in their analysis.

Two further scattering mechanisms become significant when ternary and, more particularly, quaternary alloys are considered. Weisberg[24] considered the effect of large regions of space charge caused by inhomogeneity in the material. The space-charge-limited mobility may be written in the general form

$$\mu_{SC} = \frac{10^4 e}{\sqrt{2} N_S A k^{1/2}} m^{*-1/2} T^{-1/2} \tag{8.3}$$

where N_S is the density (in m^{-3}) of space-charge scattering regions, A (in m^2) is their effective scattering cross-sectional area, k is the Boltzmann constant, and m^* is the effective mass (in kg).

Alloy scattering is thought to arise from non-periodic potential fluctuations caused by the distribution of, for example, Ga and In atoms on the group III sublattice. This occurs in the ternary $Ga_{0.47}In_{0.53}As$ whereas in the quaternary there is also the distribution of the As and P atoms on the group V sublattice to be considered. Littlejohn *et al.* derived an expression for the mobility limit imposed by alloy scattering of the form

$$\mu_{AL} = \frac{8 \times 10^4 \sqrt{2} \hbar^4}{3\pi \Omega e k^{1/2} S(\alpha)(\Delta U)^2} m^{*-5/2} T^{-1/2} \tag{8.4}$$

where $S(\alpha)$ refers to the degree of randomness, being unity when there is total disorder but becoming zero in a perfectly ordered structure such as a superlattice. ΔU is called the alloy scattering potential (in eV) and is a measure of the magnitude of the fluctuations caused by the atomic distribution variations. Of particular importance in this section is the fact that both μ_{SC} and μ_{AL} may have the same temperature variation. Their different m^* dependences are discussed in section 8.3.3.

The experimental results for representative samples with electron concentrations in the low 10^{16} cm^{-3} range are shown in Figure 8.3. Their analysis in terms of Matthiessen's rule is also shown. The curve for the polar optical phonon scattering-limited mobility μ_{PO} was calculated using equation (8.1) and values obtained from Table 8.1 using the interpolation procedures described. The ionized-impurity scattering mobility μ_{II} was then added using equation (8.2). The value of n was determined experimentally from the Hall constant, and the total density of ionized-impurity centres $N_A + N_D$ was taken as an adjustable parameter. No value of $N_A + N_D$ enabled a fit to be obtained across the whole temperature range using a combination of equations (8.1) and (8.2) only[1] since the very large value of $N_A + N_D$ required to obtain the measured room-temperature mobility implied also that the mobility should decrease with decreasing temperature. The addition of another scattering mechanism (either alloy or space-charge scattering) with an effect on mobility proportional to $T^{-1/2}$ permitted good fits to be obtained throughout the temperature range as shown in Figure 8.3 with plausible values of ionized-impurity concentration, $N_A + N_D$.

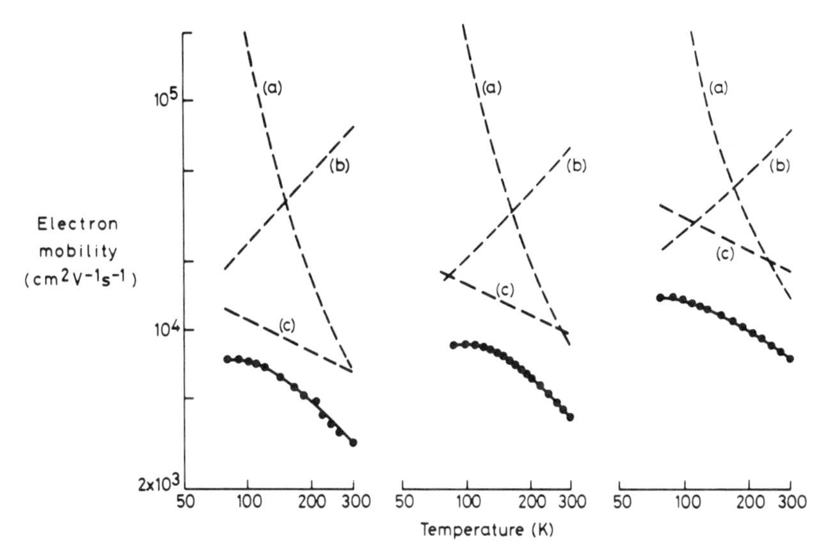

Figure 8.3 Electron mobility in $Ga_xIn_{1-x}As_yP_{1-y}$ alloys as a function of temperature for samples with the composition parameter y equal to 0.31 (*left*), 0.50 (*centre*), and 0.91 (*right*). The experimental data shown as points can be matched by a combination, shown as a full curve, of the mobility limits imposed by (a) polar optical phonon scattering, (b) ionized-impurity scattering, and (c) alloy scattering

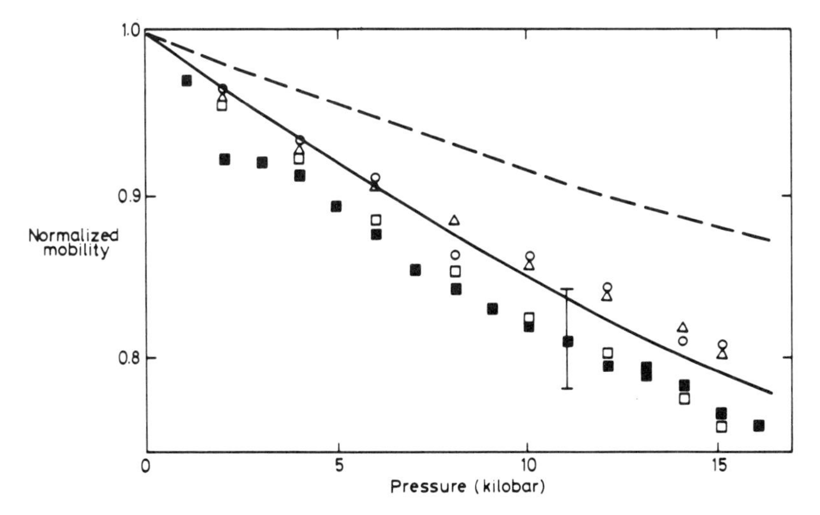

Figure 8.4 Room-temperature electron mobility in various alloy samples with y about 0.5 as a function of hydrostatic pressure. The mobility axis is normalized to unity for 1. Theoretical predictions using alloy scattering and space-charge scattering are shown as full and broken curves respectively

Equally good fits were obtained for many samples over the whole alloy composition range. Similar conclusions have also been reached by Leheny *et al.*[2] and by Ku *et al.*[25] It can therefore be concluded that the mobility variation with temperature of n-type quaternary alloys can be explained within experimental error using the scattering mechanisms described above. The addition of deformation-potential and piezoelectric scattering would not destroy the agreement but simply add some further refinement which at present is within the experimental uncertainty of the measured mobility, the parameters in Table 8.1, and the degree of validity of the analytical expressions.

In the following chapter Takeda discusses in detail the theoretical magnitude of the alloy scattering mobility μ_{AL}. However, in equation (8.3) for the space-charge scattering mobility, extremely large variations in the product $N_S A$ are physically possible, allowing a fit to be obtained with the experimental results shown in Figure 8.4. The reason why the $T^{-1/2}$ mobility curve has been designated μ_{AL} is discussed in the following section (8.3.3). It should be mentioned that in all the fits to lightly doped material, ionized-impurity scattering only starts to play a dominant role below 100 K and can, to a first approximation, be ignored at room temperature where μ_{PO} and μ_{AL} are roughly equal for mid-range alloys as shown in Figure 8.7.

8.3.3 Pressure dependence of μ_e

Equations (8.3) and (8.4) show that both space-charge and alloy scattering may have the desired $T^{-1/2}$ dependence which enables the results in Figure 8.4 to be fitted within experimental accuracy. A distinguishing feature is, however, their effective-mass dependence, and this can be investigated experimentally by high-pressure measurements.

The effective mass m_e^* of the electrons at the Γ_{1c} minimum may be simply related to the direct bandgap E_0 using the **k.p** theory of Kane[26]

$$\frac{m_0}{m_e^*} = 1 + \frac{Q^2}{3}\left(\frac{2}{E_0} + \frac{1}{E_0 + \Delta_0}\right) \tag{8.5}$$

where Δ_0 is the spin–orbit splitting of the valence band at the Γ point. The values of Δ_0 are listed in Table 8.1 and the value of Q^2 can therefore be determined for any alloy composition from equation (8.5) using the values of m_e^* determined by Nicholas *et al.*[12] and the value of the direct bandgap E_0. When hydrostatic pressure P is applied to a sample, E_0 increases quite considerably while Q and Δ_0 are relatively independent of pressure.[27] Therefore, the mobility dependence on m_e^* can be determined from its pressure dependence provided that the variation of E_0 with pressure is known.

An estimate of the pressure dependence of the direct bandgap can simply be made from the fact that for all four binary alloys the figure is around

0 μeV bar^{-1}, but direct measurements on material grown as described in section .2 have also been made by D. Patel (unpublished work). The variation of the Hall mobility μ_e with P for several samples[6] close to $y = 0.5$ is shown in Figure 8.4. Taking the magnitudes of μ_{PO}, μ_{II}, and μ_{AL} as determined from the variation of μ_e with T, as indicated in Figure 8.3 and the dependences on m_e^* of μ_{PO}, μ_{II}, and μ_{AL} given by equations (8.1), (8.2), and (8.4) respectively, it is possible to calculate the expected pressure variation of μ_e as shown by the full curve in Figure 8.4. When, however, it is assumed that alloy scattering is replaced by space-charge scattering with its relatively weak m^* dependence, as given by equation (8.3), then the broken curve in Figure 8.4 is predicted. Further data of this type are given elsewhere.[6]

It is clear that the results strongly support an interpretation of μ_e in terms of alloy scattering with its functional dependence $m^{*-5/2}T^{-1/2}$ as described by Littlejohn *et al.* It is interesting to note that the pressure variation of μ_e for the better samples showed a small but distinct bowing as predicted by theory, showing that, for an individual sample, the experimental error was better than the limits indicated and that the change was associated with a change in m^* as assumed.

8.3.4 Alloy scattering potential for electrons

Having established that the extra scattering occurring in $Ga_xIn_{1-x}As_yP_{1-y}$ appears to be of the form described by Littlejohn *et al.*, it is possible to use the magnitude of the alloy scattering at the various compositions (as shown for example in Figure 8.3) to deduce the magnitude of the parameter ΔU_e as a function of y, provided that the disorder parameter $S(\alpha)$ is known. The configurational entropy contribution to the Gibbs free energy of the quaternary alloy is proportional to the temperature. It is therefore possible that reduction of temperature may eventually result in single-phase quaternary alloys becoming unstable. However, segregation of $Ga_xIn_{1-x}As_yP_{1-y}$ into two (or more) separate phases appears to be negligibly slow in practice. On the other hand, calculations by de Cremoux *et al.*[28] have suggested that some quaternary alloy compositions may be unstable at temperatures used for liquid-phase epitaxial growth. If this is so, the ability to grow a full range of lattice-matched quaternary alloys under near-equilibrium conditions may depend on the stabilizing influence of the substrate. However, the accuracy with which the interaction parameters for the solid solutions can be estimated seems at present inadequate for useful predictions of the extent of ordering of atoms on either sublattice.

Experimental evidence concerning the degree of disorder in the material has been provided by the investigation of the reflectivity of a range of quaternary alloys. Pickering[14] was able to interpret his reflectivity data on the basis of four types of phonon, possessing frequencies corresponding to phonons present in the four binary compounds, InP, InAs, GaAs, and GaP. The relative contributions of

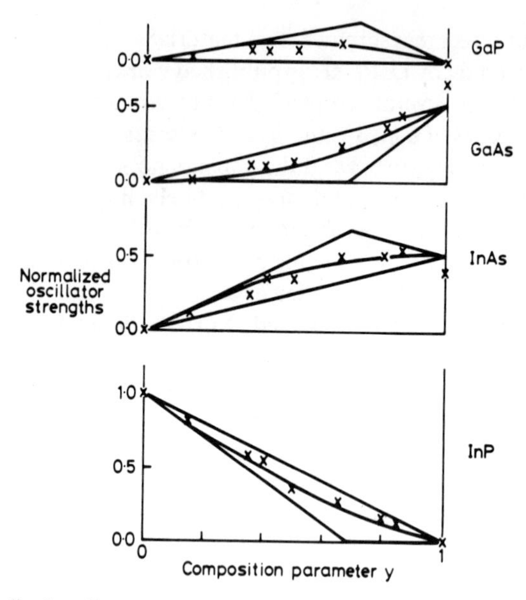

Figure 8.5 Normalized oscillator strengths of phonons related to those in GaP, GaAs, InAs, and InP as a function of the quaternary alloy composition parameter y. The experimental points are from Pickering.[14] The theoretical predictions for strong preferential pairing of Ga and As, strong preferential pairing of Ga and P, and random pairing take the form of the straight, angled and curved lines respectively

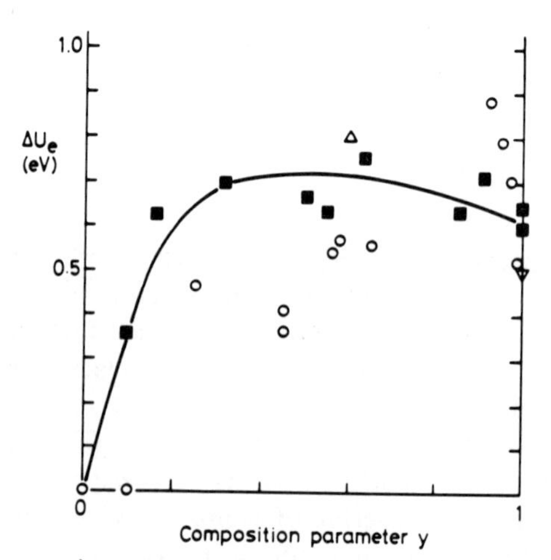

Figure 8.6 The alloy scattering parameter ΔU_e for electrons as a function of composition: ■, present work obtained by the procedure as illustrated in Figure 8.3; O, Marsh *et al.*[3]; Δ, Littlejohn *et al.*[8, 15]; ∇, derived from Oliver and Eastman.[32] The full curve is fitted to the authors' own experimental points only

he four types of phonon varied with the alloy composition, in a manner which is nconsistent with a high degree of ordering. In Figure 8.5 the effect of phonon tensity of strong preferences for either Ga–As and In–P pairings or Ga–P and n–As pairings is shown. The experimental data clearly lie between these two xtremes, close to the line for random pairings. Furthermore it is shown in Chapter 12 that the assumption of random atomic distribution on each sublattice eads to a correct prediction of the bowing in the variation of bandgap with alloy omposition. It is therefore concluded that the disorder parameter $S(\alpha)$ should be lose to unity.

The results obtained by the present authors for ΔU_e shown in Figure 8.6 ndicate a swift rise from zero at $y = 0$ to a broad maximum at the centre of the lloy system where ΔU_e is approximately 0.7 eV then a slow decrease towards $y = 1$. The full curve was drawn on the basis of these results only. The trend seems o be supported by the results of Littlejohn *et al.* and can also be deduced from the esults of Leheny *et al.*[2] Sharp peaks in the variation of ΔU_e with y were reported y Bhattacharya *et al.*[11] and by Marsh *et al.*,[3] but located at different values of y,).5 and 0.95 respectively. The values of ΔU_e deduced by Marsh *et al.* are lower at ow y. This is partly due to the fact that they included a significant amount of leformation-potential scattering near $y = 0$ which would affect particularly their oints close to $y = 0.1$ and $y = 0.25$. The use of strong deformation-potential cattering close to $y = 0$ appears to be supported by fits to the temperature lependence of μ_e in InP.[29] On the other hand, the measured pressure coefficients, vhich might be expected to provide a reasonable guide to the behaviour of lectrons in a Γ_{1c} minimum, suggest that the deformation potential scattering vould be weak in InP as in other similar binary compound semiconductors.

The magnitude of the mobility at $y = 1$ is of particular importance for device erformance. A calculation of ΔU_e for $y = 1$ based on data from Oliver and Eastman[32] for a high-mobility ternary sample gave a value of about 0.5 eV. The nfluence of alloy scattering decreases towards $y = 1$ not only because ΔU_e is lecreasing, but also because of the reduction in effective mass m_e^*. Since the nobility limit imposed by alloy scattering varies as $m_e^{*-5/2}$, and the limit imposed y polar optical phonon scattering varies as $m_e^{*-3/2}$, the relative importance of lloy scattering declines as m_e^* approaches its minimum value at the composition $y = 1$. This is illustrated in Figure 8.7 which shows how μ_{PO} and μ_{AL} vary with the omposition parameter y. Figure 8.7 is based on the full curve of Figure 8.6, which night be regarded as giving a pessimistic (high) value for alloy scattering at $y = 1$. A somewhat similar curve was given by Leheny *et al.*[2]

The full curve in Figure 8.7 shows the predicted mobility at room temperature or a sample without ionized-impurity scattering but with polar optical phonon cattering and alloy scattering based on the full curve of Figure 8.6. (The full curve of Figure 8.2 is almost identical, but includes a small contribution from ionized-mpurity scattering.) Using the data for μ_{AL} and μ_{PO} shown in Figure 8.7, it is

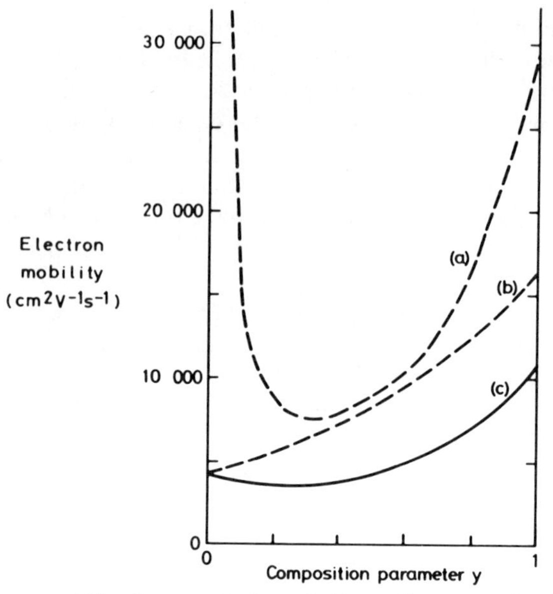

Figure 8.7 Electron mobility in pure $Ga_xIn_{1-x}As_yP_{1-y}$ alloys as a function of the alloy composition parameter y. The contributions of the mobility limits imposed by (a) alloy scattering and (b) polar optical phonon scattering are shown separately and also combined together (c)

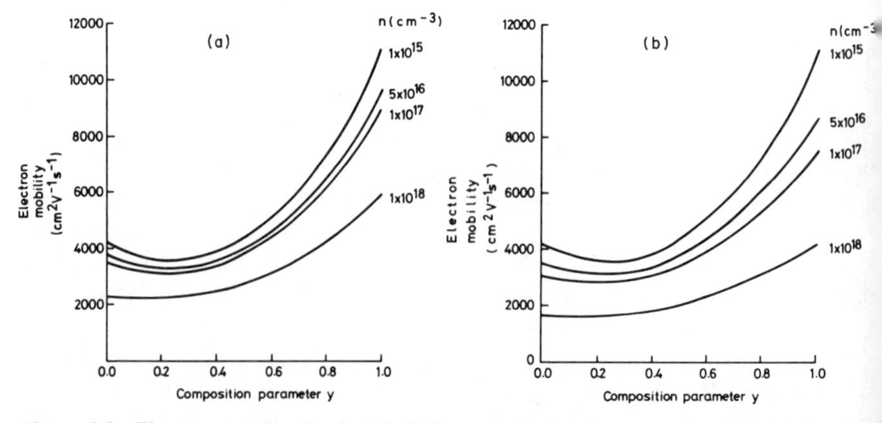

Figure 8.8 Electron mobility in doped $Ga_xIn_{1-x}As_yP_{1-y}$ alloys as a function of the alloy composition parameter y for various carrier concentrations: (a) uncompensated material ($n = N_D - N_A = 0$); (b) compensated material ($n = N_D - N_A = N_D/2$)

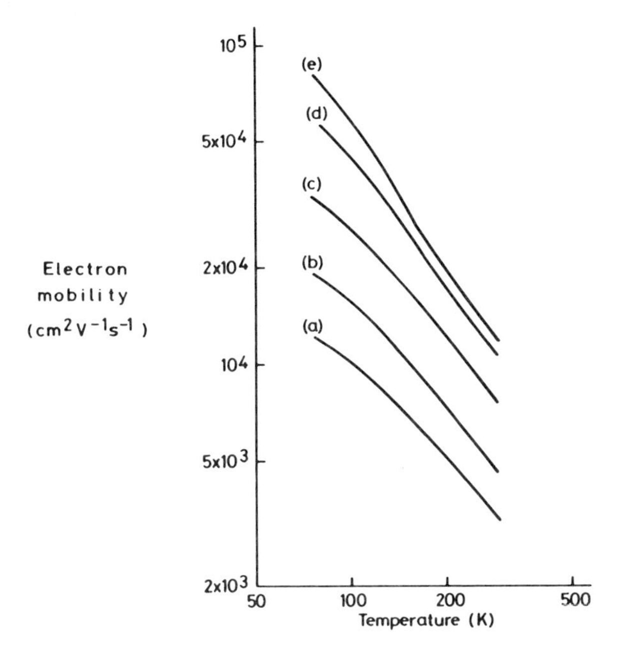

Figure 8.9 Predicted electron mobility as a function of temperature in the absence of ionized-impurity scattering for three alloy samples with the composition parameter y equal to (a) 0.30, (b) 0.58, (c) 0.91, and (d) 1.00. The values of ΔU_e correspond to the curve drawn in Figure 8.6. Curve (e) shows the mobility of a pure ternary alloy with $\Delta U_e = 0.5$ eV, derived from data given by Oliver and Eastman[32]

possible to predict the variation of the net mobility as a function of y and of the concentrations of donors N_D and compensating acceptors N_A. Examples are presented in Figure 8.8 for various values of $(N_D - N_A)$ and (N_A/N_D). Three assumptions have been made in these calculations; that Matthiessen's rule can be applied, that ΔU_e varies with y but is independent of N_D and N_A, and that the Brooks–Herring relationship holds for quite large values of $(N_D - N_A)$ and $(N_D + N_A)$. Further evidence for the validity of such assumptions is presented in section 8.4.4 on the mobility of holes.

As can be seen in Figure 8.3, the influence of alloy scattering becomes more pronounced as the temperature is lowered. Mobilities of just over 200 000 $cm^2 V^{-1} s^{-1}$ at liquid-nitrogen temperatures have been observed in very pure GaAs. This cannot be expected in the quaternary alloy; instead the alloy scattering potential of Figure 8.6 would predict that the 77 K mobility is less than 100 000 $cm^2 V^{-1} s^{-1}$ for all compositions as shown in Figure 8.9. At low temperatures the magnitude of ΔU_e is more significant even at $y = 1$ and a curve for $\Delta U_e = 0.5$ eV at $y = 1$ is also shown for comparison.

8.4 HOLE MOBILITY

8.4.1 Composition dependence of μ_p

The room-temperature Hall Mobility of holes, μ_p, was measured by the present authors on many samples grown and prepared in the manner described in section 2. The results for hole concentrations in the low 10^{16} cm^{-3} range, and in the range (1–2) $\times 10^{18}$ cm^{-3} are plotted against alloy composition in Figure 8.10. As can be seen, the downward bowing towards the centre of the alloy range is even more pronounced than that of μ_e. The μ_p value of about 140 cm^2 V^{-1} s^{-1} at $y = 0$ for the lightly doped sample is in agreement with accepted values[30] for InP. The hole mobility at $y = 1$ has not received as much attention as the electron mobility, but Pearsall *et al.*[10] reported a value of 300 cm^2 V^{-1} s^{-1}. For holes, as for electrons, the highest mobility is found at the ternary limit of the composition range.

Arguments similar to those used for μ_e apply to μ_p. What little direct experimental evidence there is indicates that the phonon spectrum is well-behaved. The heavy hole effective masses m_h of the constituent binary compounds are listed in Table 8.1. As can be seen, they do not vary widely and the error involved in their linear interpolation cannot explain the downward bowing, so an additional scattering mechanism must be considered. In the following two sections, measurements of μ_p as functions of temperature and pressure are presented, and it is again concluded that alloy scattering makes a significant contribution.

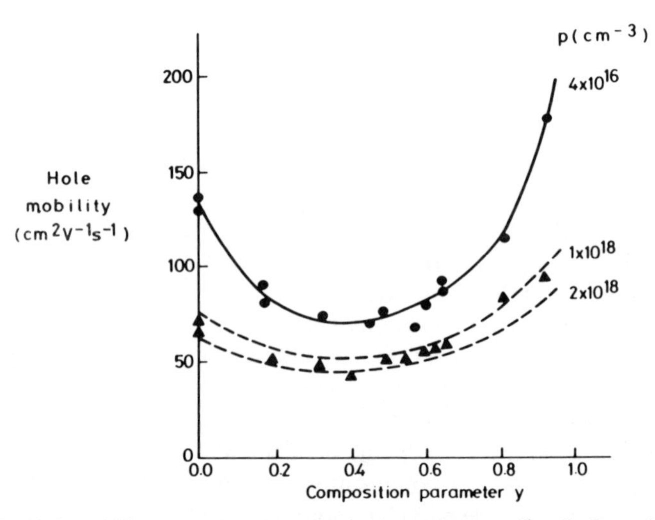

Figure 8.10 Hole mobility at room temperature in the quaternary alloy $Ga_xIn_{1-x}As_yP_{1-y}$ as a function of the alloy composition parameter y. The experimental points correspond to samples with free hole concentrations around 4×10^{16} cm^{-3} (●), and 2×10^{18} cm^{-3} (▲)

8.4.2 Temperature dependence of μ_p

The temperature variation of μ_p for a representative, lightly doped sample is shown in Figure 8.11. In all lightly doped samples investigated, μ_p continued to rise from room temperature down to almost 100 K, indicating that ionized-impurity scattering is not dominant at room temperature. Analysis of the hole mobility is more complex than that of electrons. It has, however, been dealt with in an excellent article by Wiley[30] and here the same approach using Matthiessen's rule is followed. Three factors additional to those mentioned for electrons in section 8.3 must be taken into account. Firstly, because the light and heavy hole bands are degenerate at the valence band maximum, both types of carrier must be taken into account. Owing to the strong light-hole-band to heavy-hole-band scattering, light holes do not have such a large effect on the transport as would otherwise be expected and, following Wiley, they can be taken into account in each scattering mechanism by multiplying equations (8.1), (8.2), (8.3), and (8.4) by an appropriate function of r, where $r = m_h^*/m_l^*$ the ratio of the effective masses of heavy and light holes.[33, 34] Additionally, the surfaces of constant energy for holes are not spherical as for electrons, but have strong p-like symmetry. Taking these factors into account, the equations for the mobility limits imposed by polar

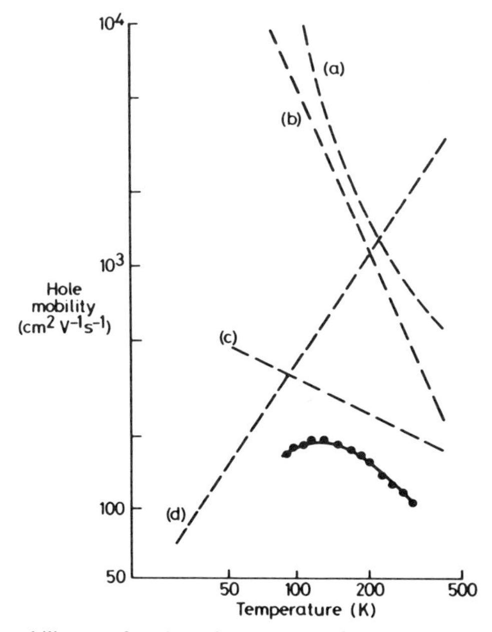

Figure 8.11 Hole mobility as a function of temperature for a quaternary alloy with $y = 0.8$. The experimental points are matched to a curve representing a mobility derived by combining the mobility limits imposed by (a) polar optical phonon scattering, (b) non-polar optical and acoustic scattering, (c) alloy scattering, and (d) ionized-impurity scattering

optical phonon scattering μ_{PO}^{p}, ionized-impurity scattering μ_{II}^{p}, alloy scattering μ_{AL}^{p}, and space-charge scattering μ_{SC}^{p} then become

$$\mu_{PO}^{p} = 2K\mu_{PO} \tag{8.6}$$

$$\mu_{II}^{p} = 1.5\left(\frac{r^{1/2}+r^{3/2}}{1+r^{3/2}}\right)\mu_{II} \tag{8.7}$$

$$\mu_{SC}^{p} = \frac{r^{5/2}(1+r^{1/2})}{(1+r^{3/2})^2}\mu_{SC} \tag{8.8}$$

$$\mu_{AL}^{p} = \frac{r^{5/2}(1+r^{1/2})}{(1+r^{3/2})^2}\mu_{AL} \tag{8.9}$$

where μ_{PO}, μ_{II}, μ_{SC}, and μ_{AL} are obtained from equations (8.1), (8.2), (8.3), and (8.4) respectively using the heavy hole mass m_h^*.

The factor 2 in equation (8.6) and the factor 1.5 in equation (8.7) take into account the p-like symmetry of the hole wavefunctions, and K is a correction factor to take into account the contribution from light holes and is shown graphically by Wiley[30] whose paper provides a detailed discussion. Finally, when considering hole mobilities, acoustic deformation-potential scattering and also non-polar optical phonon scattering must be included. The two mechanisms may be combined to give a mobility limit

$$\mu_{AC+NPO} = 3.17\times10^{-4}\frac{r^{5/2}(1+r^{1/2})}{(1+r^{3/2})^2}\frac{\rho\bar{u}^2}{(m_h^*/m_0)^{5/2}}\frac{S(\theta,\eta,T)}{E_{AC}^2}T^{-3/2} \tag{8.10}$$

where ρ, $\bar{u}$, m_h^*, and E_{AC} are the density, average sound velocity, heavy hole effective mass, and hole deformation potential respectively and can be obtained from Table 8.1, $\eta = (E_{NPO}/E_{AC})^2$ where E_{NPO} is the optical phonon deformation potential also available from Table 8.1 and S is a function given graphically by Wiley.

Analysis of the μ_p versus T curves using this approach is shown graphically in Figure 8.11. A very good fit can be achieved, well within experimental error, provided that a scattering mechanism with a mobility limit proportional to $T^{-1/2}$ is included corresponding to alloy or space-charge scattering. This ambiguity can again be resolved by high-pressure measurement.

8.4.3 Pressure dependence of μ_p

Hydrostatic pressure variation causes a change in the hole effective mass m_h^* and thereby a change in mobility. Since alloy and space-charge scattering are different functions of m_h^* they can be distinguished by this means. Unfortunately the calculation of the hole effective mass with pressure is more complex than that for electrons since more bands must be taken into account in determining an accurate

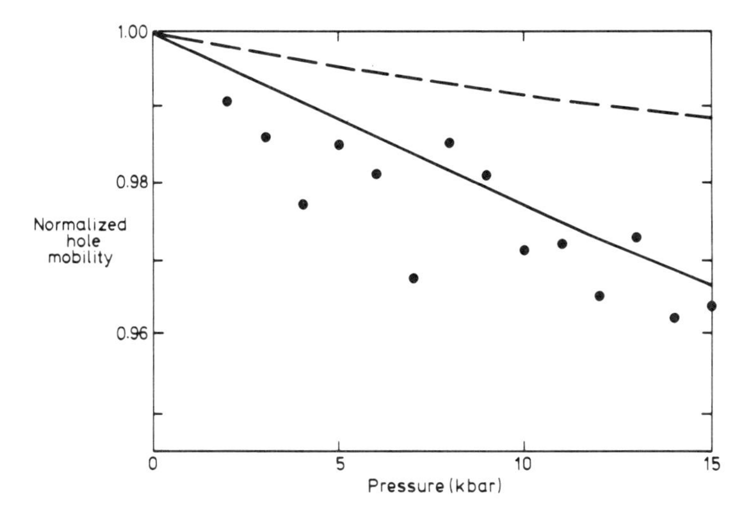

Figure 8.12 Hole mobility in one alloy sample with $y = 0.64$ as a function of hydrostatic pressure. The mobility axis is normalized to unity at 1 atm. The experimental points lie closer to a theoretical prediction (full curve) which assumes the presence of an alloy scattering component than to a prediction (broken curve) with space-charge scattering instead

value of m_h^*. The details, which are based on the analysis of Lawaetz,[31] are beyond the scope of the present work and will be published elsewhere.

The measured variation of μ_P with P for a sample of $y = 0.64$ is shown in Figure 8.12. The scatter appears large but the total change is only about 3% to 15 kbar. Taking into account the effective-mass dependence of μ_{PO}, μ_{AC+NPO}, μ_{II}, and μ_{AL} in the appropriate ratios indicated by the temperature dependence of μ_p at $y = 0.64$, the full curve in Figure 8.12 was calculated. When μ_{AL} was replaced by μ_{SC} with its weaker m_h^* dependence, the broken curve was obtained. The results indicate that interpretation in terms of alloy scattering is to be preferred. It should be mentioned that the results shown in Figure 8.12 further support the interpretation of those shown in Figure 8.4. If for example the change in μ_e in Figure 8.4 were due to a change with pressure of A, the space-charge scattering cross-section, then a similar change might be expected in μ_p with P. In fact it is almost an order of magnitude smaller as predicted by changes in m^* alone.

8.4.4 Alloy scattering potentials for holes

All the lightly doped p-type samples for which the temperature dependence of hole mobility was investigated (see Figure 8.11) showed that ionized-impurity scattering could be ignored at room temperature. It is therefore possible to calculate $(\Delta U_p)^2$ directly from the measured values of μ_p for the lightly doped samples shown in Figure 8.10 without recourse each time to the tedious temperature variation measurements. For the reasons discussed in section 8.3.4,

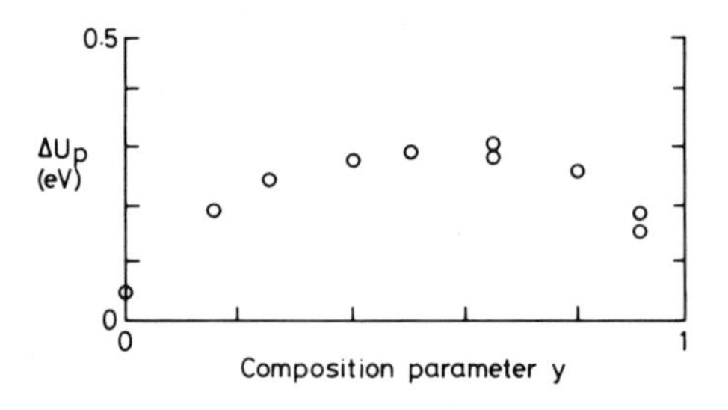

Figure 8.13 The alloy scattering parameter ΔU_p for holes as a function of composition

$S(\alpha)$ is taken as unity and the resulting values for ΔU_p are shown as a function of y in Figure 8.13.

As can be seen, ΔU_p goes through a broad maximum near the centre of the alloy range. At $y = 0.5$, ΔU_p is approximately 0.3 eV. Again, interpretation of the relative influence of alloy scattering, based on the values of ΔU_p shown in Figure 8.13, must be tempered by its effective-mass dependence. It is interesting to note for example that, despite the fact that ΔU_p is less than ΔU_e, alloy scattering near $y = 0.5$ has a much more drastic effect on holes than on electrons because of the greater hole mass and the strong dependence of μ_{AL} on m^*. To take into account

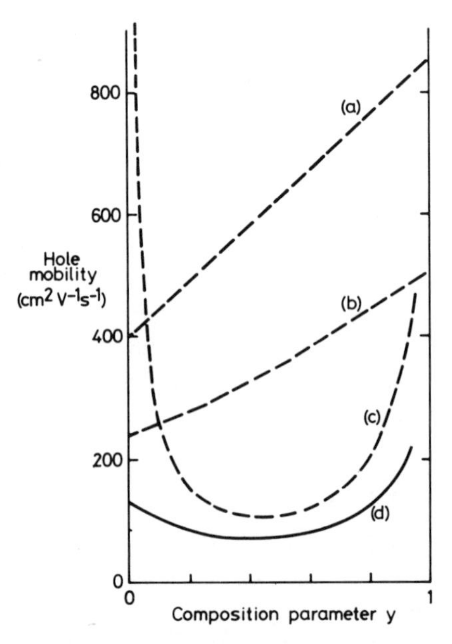

Figure 8.14 Hole mobility in pure $Ga_xIn_{1-x}As_yP_{1-y}$ alloys as a function of the alloy composition parameter y. The contributions of (a) polar optical phonon scattering, (b) non-polar optical and acoustic phonon scattering, and (c) alloy scattering to the overall mobility (d) are shown separately

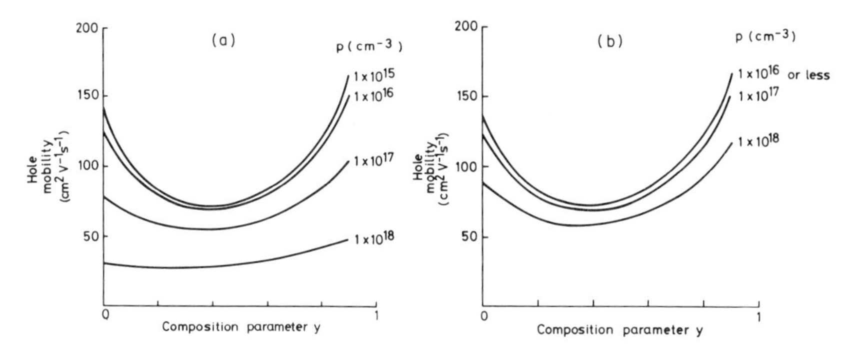

Figure 8.15 Hole mobility in doped $Ga_xIn_{1-x}As_yP_{1-y}$ alloys as a function of the alloy composition parameter y for various carrier concentrations: (a) compensated material ($p = N_A - N_D = 0.3N_A$); (b) uncompensated material ($p = N_A$, $N_D = 0$)

the variation of m_h^* with y, the room-temperature values of μ_{PO}, $\mu_{AC+NPO} + \mu_{AL}$ are shown in Figure 8.14. The μ_{AL} curve is based on experimental points in Figure 8.13 using equation (8.9) and the full curve in Figure 8.14 is the mobility resulting from adding the individual scattering mechanisms using Matthiessen's rule. This is also the full curve shown in Figure 8.10.

Ideally one might study the temperature dependence of μ_e or μ_p for different doping levels and analyse the results as shown in Figures 8.3 and 8.11. Unfortunately this is not so useful, since equation (8.2) is not valid at low temperatures at the high concentrations that would be of interest. However, by studying the variation of hole concentration with T in Mn-doped samples, it has been possible to show[18] that the compensation ratio N_D/N_A is about 0.7 in each case. On the assumption that this compensation ratio applies for every sample, it is possible to calculate the degree of ionized-impurity scattering with hole density at room temperature, where the Brooks–Herring relationship may be applied. The details are beyond the scope of the present discussion; however, Figure 8.10 shows the calculated mobility curves for $p = 1 \times 10^{18}$ and 2×10^{18} cm^{-3}, keeping the value of ΔU_p deduced from higher-purity material. The good agreement between theory and experiment suggests, to a first approximation, that the alloy scattering potential is independent of carrier concentration. This, it should be remembered, was used in the generation of the curves for the electron mobility shown in Figures 8.8(a) and (b). Figure 8.15(a) shows the variation of μ_p with y for different hole concentrations assuming $N_D/N_A = 0.7$ and Figure 8.15(b) shows the same curves for zero compensation.

8.5 CONCLUSIONS

The mobilities of electrons and holes in the quaternary alloy $Ga_xIn_{1-x}As_yP_{1-y}$ have been investigated experimentally for the full composition range that can be lattice-matched to InP substrates. Data have been obtained for the variation of mobility with both temperature and pressure. No significant dependence on the method of preparation has been noticed. The mobilities of both electrons and holes do not vary monotonically with composition but pass through a minimum. This minimum is an intrinsic property of the quaternary alloy. The mobility data for electrons can be satisfactorily interpreted by a combination of three scattering mechanisms: polar optical phonon scattering, alloy scattering, and ionized-impurity scattering. In the case of holes, acoustic and non-polar optical phonon scattering must also be considered.

Alloy scattering of the functional form proposed by Littlejohn *et al.* appears satisfactory provided that the alloy scattering parameter ΔU is regarded as an adjustable parameter, which varies with composition,[15] is different for electrons and holes, but is independent of the dopant concentration. The physical interpretation of ΔU remains unclear. It has been suggested[7] that the variation of ΔU as a function of y can be derived by considering fluctuations in bandgap (for ΔU_e) and ionization potential (for ΔU_p) providing these values are calculated for binary compounds with a hypothetical lattice parameter equal to that of InP. An alternative interpretation (for ΔU_e) in terms of the electronegativity difference is proposed in Chapter 9. A further feature of the alloy which may contribute to alloy scattering is that the differences in size of the constituent atoms implies that the mean position of each atom is irregular.

Alloy scattering is absent in the binary compound InP which forms one end of the composition range ($y = 0$) and becomes relatively unimportant at the other end ($y = 1$). Although the value of ΔU does not return to zero at $y = 1$, the decrease in effective mass as y approaches 1 leads to a reduction in the alloy scattering rate. The electron mobility in the pure lattice-matched ternary $Ga_{0.47}In_{0.53}As$ is about 12 000 $cm^2\ V^{-1}\ s^{-1}$ at room temperature, significantly greater than in GaAs. This makes the ternary alloy of potential interest for high-speed or high-frequency devices.

Because alloy scattering increases strongly with increasing effective mass, it has more influence on hole mobility than on electron mobility, although the ΔU values are lower for holes than for electrons. The maximum hole mobility in lattice-matched material is also to be found at the ternary end of the composition range where it can reach about 300 $cm^2\ V^{-1}\ s^{-1}$, a figure only about two-thirds of that in pure GaAs.

REFERENCES

1. P. D. Greene, S. A. Wheeler, A. R. Adams, A. N. El-Sabbahy, and C. N. Ahmad, 'Background carrier concentration and electron mobility in LPE $In_{1-x}Ga_xAs_yP_{1-y}$ layers', *Appl. Phys. Lett.*, **35**, 78–80, 1979.
2. R. F. Leheny, A. A. Ballman, J. C. DeWinter, R. E. Nahory, and M. A. Pollack, 'Compositional dependence of the electron mobility in $In_{1-x}Ga_xAs_yP_{1-y}$', *J.* Electron. Mater., **9**, 561–568, 1980.

3. J. H. Marsh, P. A. Houston, and P. N. Robson, 'Compositional dependence of the mobility, peak velocity and threshold field in $In_{1-x}Ga_xAs_yP_{1-y}$', *Gallium Arsenide and Related Compounds* 1980, Conf. Ser. no. 56, Institute of Physics, Bristol, 1981, pp. 621–630.
4. J. R. Hayes, A. R. Adams, and P. D. Greene, 'Mobility of holes in the quaternary alloy $In_{1-x}Ga_xAs_yP_{1-y}$', *Electron. Lett.*, **16**, 282–284. 1980.
5. J. R. Hayes, H. L. Tatham, A. R. Adams, and P. D. Greene, 'Pressure dependence of the hole mobility in $In_{1-x}Ga_xAs_yP_{1-y}$ and its relation to alloy scattering', *Electron. Lett.*, **17**, 230–232, 1981.
6. A. R. Adams, H. L. Tatham, J. R. Hayes, A. N. El-Sabbahy, and P. D. Greene, 'Evidence for alloy scattering from pressure-induced changes of electron mobility in $In_{1-x}Ga_xAs_yP_{1-y}$', *Electron. Lett.*, **16**, 560–562, 1980.
7. A. R. Adams, J. R. Hayes, and P. D. Greene. 'The influence of alloy scattering on electrons and holes in $In_{1-x}Ga_xAs_yP_{1-y}$', *Jpn. J. Appl. Phys.*, **19**, supp. 19–3, 315–320, 1980.
8. M. A. Littlejohn, R. A. Sadler, T. H. Glisson, and J. R. Hauser, 'Carrier compensation and alloy scattering in $Ga_{1-x}In_xP_{1-y}As_y$ grown by liquid-phase epitaxy', *Gallium Arsenide and Related Compounds 1978*, Conf. Ser. no. 45, Institute of Physics, Bristol, 1979, pp. 239–247.
9. M.Enda, 'Preparation and properties of VPE GaInAsP', *Jpn. J. Appl. Phys.*, **18**, 2167–2168, 1979.
10. T. P. Pearsall, G. Beuchet, J. P. Hirtz, N. Visentin, and M. Bonnet, 'Electron and hole mobilities in $Ga_{0.47}In_{0.53}As$', *Gallium Arsenide and Related Compounds 1980*, Conf. Ser. no. 56, Institute of Physics, Bristol, 1981, pp. 639–649.
11. P. K. Bhattacharya, J. W. Ku, S. J. T. Owen, G. H. Olsen, and S. H. Chias, 'LPE and VPE $In_{1-x}Ga_xAs_yP_{1-y}$/InP: transport properties, defects and device consideration', *IEEE J. Quantum Electron.*, **QE-17**, 150–160, 1981.
12. R. J. Nicholas, J. C. Portal, C. Houlbert, P. Perrier, and T. P. Pearsall, 'An experimental determination of the effective masses for $Ga_xIn_{1-x}As_yP_{1-y}$ alloys grown on InP', *Appl. Phys. Lett.*, **34**, 492–494, 1979.
13. J. B. Restorff, B. Houston, R. S. Allgaier, M. A. Littlejohn, and S. B. Phatak, 'The electron effective mass in $In_{1-x}Ga_xAs_yP_{1-y}$ from Shubnikov–deHaas measurements', *J. Appl. Phys.*, **51**, 2277–2278, 1980.
14. C. Pickering, 'Infra-red reflectivity studies of $Ga_xIn_{1-x}As_yP_{1-y}$ quaternary compounds', *J. Electron. Mater.*, **10**, 901–918, 1981.
15. M. A. Littlejohn, J. R. Hauser, T. H. Glisson, D. K. Ferry, and J. W. Harrison, 'Alloy scattering and high field transport in ternary and quaternary semiconductors', *Solid-State Electron.*, **21**, 107–114, 1978.
16. Y. Takeda and T. P. Pearsall, 'Failure of Matthiessen's rule in calculation of carrier mobility and alloy scattering effects in $Ga_{0.47}In_{0.53}As$', *Electron. Lett.*, **17**, 573–574, 1981.
17. M. Feng, L. W. Cook, M. M. Tashima, T. H. Windhorn, and G. E. Stillman, 'The influence of LPE growth techniques on the alloy composition of InGaAsP', *Appl. Phys. Lett.*, **34**, 292–294, 1979.
18. J. R. Hayes, A. R. Adams, B. Sealy, and P. D. Greene, unpublished data.
19. D. L. Rode, 'Low field electron transport', *Semiconductors and Semimetals*, eds R. K. Willardson and A. C. Beer, Academic Press, New York, 1975, pp. 1–89.
20. Y. Takeda, and A, Sasaki, 'Hall mobility and Hall factor of $In_{0.53}Ga_{0.47}As$', *Jpn. J. Appl. Phys.*, **19**, 383–384, 1980.
21. H. Ehrenreich, 'Screening effects in polar semiconductors', *J. Phys. Chem. Solids*, **8**, 130–135, 1959.
22. A. Fortini, D. Diguet, and J. Lugand, 'Analysis of polar optical scattering of electrons in GaAs', *J. Appl. Phys.*, **41**, 3121–3127, 1970.

23. C. M. Wolfe, G. E. Stillman, and W. T. Lindley, 'Electron mobility in high purity GaAs', *J. Appl. Phys.* **41**, 3088–3091.
24. L. R. Weisberg, 'Anomalous mobility effects in some semiconductors and insulators', *J. Appl. Phys.*, **33**, 1812–1817, 1962.
25. J. W. Ku, P. K. Bhattacharya, K. K. Smith, and S. H. Chiao, 'Electrical and optical properties of epitaxial $In_{1-x}Ga_xAs_yP_{1-y}$', private communication.
26. E. O. Kane, 'The **k. p** method', *Semiconductors and Semimetals*, vol. 1, eds R. K. Willardson and A. C. Beer, Academic Press, New York, 1966, pp. 75–100.
27. G. D. Pitt, J. Lees, R. A. Hoult, and R. A. Stradling, 'Magnetophonon effect in GaAs and InP to high pressures', *J. Phys. C: Solid State Phys.*, **6**, 3282–3294, 1973.
28. B. de Cremoux, P. Hirtz, and J. Ricciardi, 'On the presence of a solid immiscibility domain in the GaInAsP phase diagram', *Gallium Arsenide and Related Compounds 1980*, Conf. Ser. no. 56, Institute of Physics, Bristol, 1981, pp. 115–124.
29. B. R. Nag, and G. M. Dutta, 'Electron mobility in InP', *J. Phys. C: Solid State Phys.*, **11**, 119–123, 1978.
30. J. D. Wiley, 'Mobility of holes in III–V compounds', *Semiconductors and Semimetals*, vol. 10, eds R. K. Willardson and A. C. Beer, Academic Press, New York, 1975, pp. 91–174.
31. P. Lawaetz, 'Valence band parameters in cubic semiconductors', *Phys. Rev.*, B, **4**, 3460–3467, 1971.
32. J. D. Oliver, and L. F. Eastman, 'Liquid phase epitaxial growth and characterization of high purity lattice matched $Ga_xIn_{1-x}As$ on $\langle 111 \rangle$B InP', *J. Electron. Mater.*, **9**, 693–712, 1980.
33. K. Alavi, R. L. Aggarwal, and S. H. Groves, 'Interband magneto-absorption of $In_{0.53}Ga_{0.47}As$', *Phys. Rev.*, B, **21**, 1311–1315, 1980.
34. C. Hermann and T. P. Pearsall, 'Optical pumping and the valence-band light-hole effective mass in $Ga_xIn_{1-x}As_yP_{1-y}$ ($y \simeq 2.2x$)', *Appl. Phys. Lett.*, **38**, 450–452, 1981.

GaInAsP Alloy Semiconductors
Edited by T. P. Pearsall

Chapter 9

Low-field Transport Calculations

YOSHIKAZU TAKEDA*

Department of Electrical Engineering, North Carolina State University, Raleigh, NC 27650, USA

9.1 INTRODUCTION

The performance of electronic devices which utilize electron transport in semiconductors depends critically on low-field and/or high-field electron transport properties. Since these transport properties are substantially determined by the band structure and the electron scattering mechanisms in the semiconductor, it is very important to analyse the experimental transport properties carefully in terms of the band structure and scattering mechanisms and to have reasonable models to predict device performance. The important factor in the scattering mechanisms in alloy semiconductors is the extent of the role of 'alloy scattering'. In addition to the scattering mechanisms usually considered in compound semiconductors, such as longitudinal polar optical phonon scattering, piezoelectric scattering, acoustic deformation potential scattering, and ionized-impurity scattering, 'alloy scattering' due to the aperiodic arrangement of constituent atoms on lattice sites may limit the electron transport to an extent determined by material parameters of alloy semiconductors. While a very detailed discussion of the scattering mechanisms and the scattering rate would require more precise material parameters and mobility measurements, both the temperature dependence,[1-8] the composition dependence,[3,4,7] and also the pressure dependence[9] of the electron mobility in $Ga_xIn_{1-x}As_yP_{1-y}$ appear to support the existence of alloy scattering in these alloys.

Without using such simplifying assumptions as Matthiessen's rule and a relaxation-time approximation for polar optical phonon scattering, both the iterative technique[10-13] and the Monte Carlo technique[14,15] can be used to calculate accurately the electron mobility in the presence of several scattering mechanisms. Since most of the experimentally measured mobilities are obtained using the Hall effect, theoretical Hall mobility is also calculated by the iterative technique. All calculations here are based on the classical Boltzmann transport equation.

* Present address: The Department of Electrical Engineering, Kyoto University, Kyoto 606, Japan.

After describing the calculation techniques and the scattering mechanism formulation in section 9.2, the electron mobility in $Ga_{0.47}In_{0.53}As$ is discussed first in section 9.3 because the electrical properties of this ternary alloy have been most extensively studied[1, 5, 6, 8, 16–21] and high-purity crystals have been grown.[5] The temperature dependence of the electron mobility is used to study the scattering mechanisms which limit mobility. The conclusions obtained for $Ga_{0.47}In_{0.53}As$ in section 9.3 are extended to $Ga_xIn_{1-x}As_yP_{1-y}$ in section 9.4 where the composition dependence of the electron mobility is of major interest. Detailed comparisons with experimental and theoretical mobilities are presented in both sections.

9.2 LOW-FIELD MOBILITY CALCULATION

9.2.1 The iterative technique and the Monte Carlo technique

The iterative technique and the Monte Carlo technique are solution methods of the Boltzmann equation which can be used to calculate the low-field mobility without using the relaxation-time approximation,[22] Matthiessen's rule, or complicated mathematical expressions.[23–25] A combination of any kinds of scattering mechanisms is allowed in both calculational techniques. The iterative technique, in which the Boltzmann equation is solved by numerical iteration, has been described by Rode.[10–13] The calculation can be carried to any arbitrary degree of convergence. Starting from the first iteration which gives the usual result for a relaxation-time approximation, the contribution of the scattered-in electrons to the state of momentum **k** by LO (longitudinal polar optical) phonon absorption and emission is accounted for in the higher iteration. The number of the phonon absorption and emission sequence is increased with increasing iterations. To calculate to less than 1% error, five iterations are required for the drift mobility and seven iterations are necessary for the Hall mobility in $Ga_{0.47}In_{0.53}As$ at 300 K. The error is less at lower temperatures with the same iterations. The iterative technique has been used for all calculations in this chapter.

The Monte Carlo technique for hot carrier calculations in p-Ge was first introduced by Kurosawa[26] and extended to calculate the high-field electron transport properties in GaAs,[27] but the technique was supposed to be too slowly convergent to calculate the low-field mobility. However, with the technique of using statistical estimators for the diffusion coefficient,[14] the convergence has been much improved in the drift mobility calculation. This has been confirmed by comparing the results of this technique and the iterative technique as discussed in Chapter 10, where the Monte Carlo technique and its application to high-field transport are described in detail.

Both calculation techniques are based on classical transport theory, but must be used carefully in apparently random alloy materials, where multiple scattering and cluster effects which are not treatable in the above methods could occur. These effects can be treated with the coherent potential approximation[28, 29] or with the average *t*-matrix approximation.[30, 31] However, in the calculations in $Ga_xIn_{1-x}As$ and $Al_xGa_{1-x}As$ using the self-consistent Green's function method within the coherent potential approximation, Nishinaga *et al.* have shown that the multiple scattering and the cluster effects are negligibly small and that Brook's formula for alloy scattering[32, 33] is a good approximation when the effective mass and the alloy scattering potential are small.[34, 35]

9.2.2 Drift mobility and Hall mobility

The mobility calculated theoretically is usually the drift mobility μ_D. This is easier to deal with because the electron distribution is perturbed only by the electric field. On the other hand, most of the experimental mobility data are obtained by the Hall effect (μ_H) measured under the influence of both an electric field and a magnetic field.[36, 37] The electron distribution is perturbed by both fields. Although it might be possible to correlate the drift mobility and the Hall mobility through the Hall factor $r_H(\equiv \mu_H/\mu_D)$ by somehow estimating its value ,[11] the Hall factor for combined scattering is very complicated.[12, 13, 18, 38, 39] The calculated value of the Hall factor at room temperature is 1.10–1.20 in high-purity material.[12, 13, 18, 38] If the Hall factor is approximated to be unity, the contribution of alloy scattering to the total mobility is not accurately estimated because its contribution is not more than about 20% in GaInAsP. Thus, a theoretical calculation of the Hall mobility is required when analysing the experimental Hall mobility, especially in a purer sample. Knowledge of the drift mobility is also necessary to understand the electronic conduction in semiconductors as observed in many semiconductor devices. Both mobilities are presented in this chapter.

If the perturbation distribution functions are obtained, the drift mobility, the Hall mobility, and the Hall factor are calculated as shown below. The perturbation distribution is determined by the scattering mechanisms and the applied external forces and is calculated by the iterative technique as described in section 9.2.3.

The distribution function in the presence of a small electric field **F** is given by

$$f(\mathbf{k}) = f_0(k) + xg(k) \tag{9.1}$$

where x is the cosine of the angle between **k** and **F**, $f_0(k)$ is the equilibrium distribution, and $g(k)$ is the perturbation distribution. The drift mobility μ_D is by definition,

$$\mu_D = \frac{\mathbf{j}}{en\mathbf{F}} = \frac{1}{n}\frac{V}{3\pi^2}\int_0^\infty \frac{\hbar k^3}{Fm_0 d} g(k)\mathrm{d}k \tag{9.2}$$

where d is the augmented density of states and related to α of equation (9.12) as $1/d = 1 + (m_0 - m^*)/m^*\alpha$.

The electron concentration is:

$$n = \frac{V}{4\pi^3}\int_0^\infty f(\mathbf{k})\mathrm{d}\mathbf{k} = \frac{V}{\pi^2}\int_{k^2 0}^\infty f_0(k)\mathrm{d}k \tag{9.3}$$

Introducing brackets $\langle\square\rangle \equiv \int_0^\infty (k^3/d)\mathrm{d}k$, and $\langle\langle\square\rangle\rangle \equiv \int_0^\infty k^2\mathrm{d}k$, μ_D is expressed as

$$\mu_\mathrm{D} = \frac{\hbar}{3m_0 F}\frac{\langle g\rangle}{\langle\langle f_0\rangle\rangle} \tag{9.4}$$

The distribution function under the influence of both a small electric field and magnetic field **B** is given by

$$f(\mathbf{k}) = f_0(k) + xg(k) + yh(k) \tag{9.5}$$

where y is the direction cosine from $\mathbf{B}\times\mathbf{F}$ to $\mathbf{k}$. The x-component j_x and y-component j_y of current $\boldsymbol{j}$ are expressed using $g(k)$ and $h(k)$ as follows:

$$j_x = \int ev_x f(\mathbf{k})\mathrm{d}\mathbf{k} = \frac{V}{3\pi^3}\frac{e\hbar}{m_0 F}(\langle g\rangle F_x - \langle h\rangle F_y) \tag{9.6}$$

$$j_y = \int ev_y f(\mathbf{k})\mathrm{d}\mathbf{k} = \frac{V}{3\pi^3}\frac{e\hbar}{m_0 F}(\langle g\rangle F_y + \langle h\rangle F_x) \tag{9.7}$$

With the condition $j_y = 0$ and by definition, the Hall coefficient is

$$R_\mathrm{H} \equiv \frac{F_y}{j_x B} = -\frac{3\pi^3 m_0 F}{e\hbar V}\frac{\langle h\rangle}{\langle g\rangle^2 + \langle h\rangle^2}\frac{1}{B} \tag{9.8}$$

From (9.6) $j_x = \sigma F_x$ where σ is the conductivity. Using R_H and σ the Hall mobility is

$$\mu_\mathrm{H} \equiv R_\mathrm{H}\sigma = -\frac{\langle h\rangle}{\langle g\rangle}\frac{1}{B} \tag{9.9}$$

and the Hall factor is

$$r_\mathrm{H} \equiv \mu_\mathrm{H}/\mu_\mathrm{D} \tag{9.10}$$

9.2.3 The Boltzmann equation and scattering mechanisms

$Ga_xIn_{1-x}As_yP_{1-y}$ lattice-matched to InP is a direct-gap semiconductor over the whole composition range y, and in Chapter 12 this alloy is shown to have a well-behaved band structure which is expressed by the $\mathbf{k}\cdot\mathbf{p}$ theory.[40] It is assumed that Kane's expression[41] for the energy dispersion relation near the conduction band edge at $\mathbf{k}$ = (0,0,0) is applicable to this quaternary alloy. With the approximation that the spin–orbit splitting energy $\Delta = 0$ the energy dispersion is written as[10]

$$E = E_0 + (\alpha - 1)\tfrac{1}{2}E_g \tag{9.11}$$

$$\alpha^2 = 1 + 4E_0(m_0 - m^*)/m^* E_g \tag{9.12}$$

where $E_0 = \hbar^2 \mathbf{k}^2/2m_0$. m_0 is the free-electron mass, E_g is the effective-mass energy gap,[24] and m^* is the electron effective mass at the conduction band edge.

The electron wavefunction has a p-function admixture near $\mathbf{k} = (0,0,0)$. If a and c are the coefficients for the s-function and the p-function, respectively, they are related to α of (9.12) by

$$2a^2 = 1 + 1/\alpha \tag{9.13}$$

$$a^2 + c^2 = 1 \tag{9.14}$$

Because of this p-function admixture, the matrix element of the electron scattering from the momentum state $\mathbf{k}$ to $\mathbf{k}'$ for the s-function is modified and multiplied by

$$G(\mathbf{k}, \mathbf{k}') = (aa' + cc'\xi)^2$$

where ξ is the cosine of the angle between $\mathbf{k}$ and $\mathbf{k}'$.[10, 42] a and c are evaluated at $\mathbf{k}$ while a' and c' are evaluated at $\mathbf{k}'$.

The Boltzmann equation for a non-degenerate semiconductor is expressed as

$$\frac{e}{\hbar}\mathbf{F}\cdot\nabla_k f(\mathbf{k}) = \int [f(\mathbf{k}')s(\mathbf{k}', \mathbf{k}) - f(\mathbf{k})s(\mathbf{k}, \mathbf{k}')]\mathrm{d}\mathbf{k}' \tag{9.15}$$

where $s(\mathbf{k}', \mathbf{k})$ is the probability per unit time that an electron in the momentum state $\mathbf{k}'$ will make a transition into the momentum state $\mathbf{k}$. Substituting $f(\mathbf{k})$ in (9.15) by (9.1), multiplying both sides of (9.15) by the first-order Legendre polynomial x, and integrating over x, we will get the equation

$$\frac{eF}{\hbar}\frac{\partial f_0(k)}{\partial k} = \frac{v_i^{(1)} - v_o^{(1)}}{x} \tag{9.16}$$

where the v are the scattering probability fluxes and

$$v_i^{(1)} = \int [x' g(k') s(\mathbf{k}', \mathbf{k})]\mathrm{d}\mathbf{k}' \tag{9.17}$$

$$v_o^{(1)} = xg(k)\int s(\mathbf{k}, \mathbf{k}')\mathrm{d}\mathbf{k}' \tag{9.18}$$

Since the conduction band is isotropic and the differential scattering rate s depends on only k, k', and ξ, the angle between $\mathbf{k}$ and $\mathbf{k}'$, $v_i^{(1)}$ is easily shown to be proportional to x.[13] Subscripts o and i denote scattering-*out* and scattering-*in*, respectively. Since it is assumed that scattering events are mutually independent, the total differential scattering rate is the sum of the differential scattering rates for each scattering mechanism, and thus the total scattering probability flux is the sum of the scattering probability fluxes as is clear from equations (9.17) and (9.18).

Table 9.1 Scattering rates for five scattering mechanisms

Longitudinal polar optical phonon scattering

$$S^*_{\text{oPO}} = \lambda^*_{\text{o}} \frac{m_0 d^*}{\hbar k}\left(A^{*2} \ln\left|\frac{k^* + k}{k^* - k}\right| - A^* cc^* - aca^* c^*\right)$$

$$S^*_{\text{iPO}} = \lambda^*_{i} \frac{m_0 d^*}{\hbar k}\left((k^{*2} + k^2)A^{*2} \ln\left|\frac{k^* + k}{k^* - k}\right| \Big/ 2k^* k - A^{*2} - \tfrac{1}{3}c^2 c^{*2}\right)$$

$$A^* = aa^* + (k^{*2} + k^2)cc^*/2k^* k$$

$$\lambda^*_{\text{o}} = \frac{e^2 \omega_{\text{PO}}}{4\pi\hbar}\left(\frac{1}{\varepsilon_\infty} - \frac{1}{\varepsilon_0}\right)[N_{\text{PO}} + \tfrac{1}{2} * (-\tfrac{1}{2})]$$

$$\lambda^*_{\text{i}} = \frac{e^2 \omega_{\text{PO}}}{4\pi\hbar}\left(\frac{1}{\varepsilon_\infty} - \frac{1}{\varepsilon_0}\right)[N_{\text{PO}} + \tfrac{1}{2} * (\tfrac{1}{2})]$$

$$N_{\text{PO}} = \frac{1}{\exp(\hbar\omega_{\text{PO}}/k_{\text{B}}T) - 1}$$

The asterisk $*$ takes on the values + and −, corresponding to phonon absorption and emission, respectively.

Piezoelectric scattering

$$S_{\text{PE}} = \lambda_{\text{PE}} \frac{m_0 d}{\hbar k} \tfrac{1}{3}(3a^4 + c^4)$$

$$\lambda_{\text{PE}} = \frac{e^2 e_{14}{}^2 k_{\text{B}} T}{2\pi\varepsilon_0^2 \hbar^2 \rho u_{\text{P}}{}^2}$$

note: $$\frac{1}{\rho u_{\text{P}}^2} = \tfrac{1}{35}(12/c_l + 16/c_t),$$

$$c_l = \tfrac{1}{5}(3c_{11} + 2c_{12} + 4c_{44})$$

$$c_t = \tfrac{1}{5}(c_{11} - c_{12} + 3c_{44})$$

Acoustic deformation-potential scattering

$$S_{\text{AC}} = \lambda_{\text{AC}} \frac{m_0 dk}{\hbar} \tfrac{1}{3}(3a^4 + c^4 - 2a^2 c^2)$$

$$\lambda_{\text{AC}} = \frac{e^2 E_1^2 k_{\text{B}} T}{\pi \hbar^2 \rho u_{\text{a}}^2}$$

$$\rho u_{\text{a}}{}^2 = c_l$$

Ionized-impurity scattering

$$S_{\text{II}} = \frac{e^4 N m_0 d}{8\pi\varepsilon_0 \hbar^3 k^3}\left[D \ln\left(1 + \frac{4k^2}{\beta^2}\right) - B\right]$$

$N = N^+ + N^-$
β = inverse of the screening length of the Coulomb potential
$D = 1 + 2\beta^2 c^2/k^2 + 3\beta^4 c^4/4k^4$
$B = [4k^2 + 8(\beta^2 + 2k^2)c^2 + (3\beta^4 + 6\beta^2 k^2 - 8k^4)c^4/k^2]/(\beta^2 + 4k^2)$

Alloy scattering

$$S_{\text{AL}} = \lambda_{\text{AL}} \frac{m_0 dk}{\hbar} \tfrac{1}{3}(3a^4 + c^4 - 2a^2 c^2)$$

$$\lambda_{\text{AL}} = \frac{3\pi}{64} \frac{l^3}{\hbar^2} |\Delta U_{\text{AL}}|^2$$

Therefore

$$v_{\rm i}^{(1)} - v_{\rm o}^{(1)} = (v_{\rm iPO}^{(1)} - v_{\rm oPO}^{(1)}) + (v_{\rm iPE}^{(1)} - v_{\rm oPE}^{(1)}) + (v_{\rm iAC}^{(1)} + v_{\rm oAC}^{(1)}) + (v_{\rm iII}^{(1)} - v_{\rm oII}^{(1)}) + (v_{\rm iAL}^{(1)} - v_{\rm oAL}^{(1)}) \quad (9.19)$$

where PO, PE, AC, II, and AL denote longitudinal polar optical phonon scattering, piezoelectric scattering, acoustic deformation potential scattering, ionized-impurity scattering, and alloy scattering, respectively. These five scattering mechanisms are considered as the principal scattering mechanisms in $Ga_xIn_{1-x}As_yP_{1-y}$ alloys. Since the v are proportional to x, it may be convenient to define the scattering rate S for each scattering. This scattering rate is exactly equal to $1/\tau$ if the relaxation time τ is defined. In the inelastic scattering process of polar optical phonon scattering, the scattering probability fluxes must be evaluated at k^+ or k^-, corresponding to phonon absorption or phonon emission. $k^\pm$ are the momenta at the energy corresponding to k plus or minus the energy $\hbar\omega_{\rm PO}$ of a longitudinal polar optical phonon. Thus,

$$v_{\rm oPO}^{(1)} = S_{\rm oPO}^{+} xg(k) + S_{\rm oPO}^{-} xg(k) \equiv S_{\rm oPO}\, xg(k) \quad (9.20)$$

$$v_{\rm iPO}^{(1)} = S_{\rm iPO}^{+} xg(k^+) + S_{\rm iPO}^{-} xg(k^-) \quad (9.21)$$

In the elastic scattering process there is no change in the magnitude of the momentum, so that

$$v_{\rm oPE}^{(1)} - v_{\rm iPE}^{(1)} = S_{\rm PE}\, xg(k) \quad (9.22)$$

$$v_{\rm oAC}^{(1)} - v_{\rm iAC}^{(1)} = S_{\rm AC}\, xg(k) \quad (9.23)$$

$$v_{\rm oII}^{(1)} - v_{\rm iII}^{(1)} = S_{\rm II}\, xg(k) \quad (9.24)$$

$$v_{\rm oAL}^{(1)} - v_{\rm iAL}^{(1)} = S_{\rm AL}\, xg(k) \quad (9.25)$$

These scattering rates are calculated from the matrix elements for the scattering mechanisms, available in the literature,[22, 24, 43] with modification by the p-function admixture. All these scattering rates are listed in Table 9.1.

Using $S_{\rm o} = S_{\rm oPO} + S_{\rm PE} + S_{\rm AC} + S_{\rm II} + S_{\rm AL}$, equation (9.16) is reduced as follows:

$$g(k) = \frac{S_{\rm iPO}^{+} g(k^+) + S_{\rm iPO}^{-} g(k^-) - (eF/\hbar)(\partial f(k)/\partial k)}{S_{\rm o}} \quad (9.26)$$

When all the scattering-in terms are neglected

$$g(k) = -\frac{eF}{\hbar S_{\rm o}} \frac{\partial f_0(k)}{\partial k} \quad (9.27)$$

Table 9.1, cont'd.

The constants are in appearing order; e = electron charge, $\omega_{\rm PO}$ = longitudinal optical phonon frequency, $\hbar$ = Planck's constant, $k_{\rm B}$ = Boltzmann constant, ε_∞ = optical dielectric constant, e_{14} = piezoelectric constant, ρ = material density, c_{ii} = elastic constants, E_1 = acoustic deformation potential, l = lattice parameter, $\Delta U_{\rm AL}$ = alloy scattering potential including composition-dependent terms.

This relaxation-time formula is the first iteration, then the ith iteration is

$$g_{i+1}(k) = \frac{S_{\mathrm{iPO}}^{+} g_i(k^+) + S_{\mathrm{iPO}}^{-} g_i(k^-) - (eF/\hbar)(\partial f_0(k)/\partial k)}{S_o} \tag{9.28}$$

Similarly, $g(k)$ and $h(k)$ for (9.5) are derived as

$$g(k) = \frac{S_{\mathrm{iPO}}^{+} g(k^+) + S_{\mathrm{iPO}}^{-} g(k^-) - (eF/\hbar)(\partial f_0(k)/\partial k) + \beta S_o h(k)}{S_o} \tag{9.29}$$

$$h(k) = \frac{S_{\mathrm{iPO}}^{+} h(k^+) + S_{\mathrm{iPO}}^{-} h(k^-) - \beta S_o g(k)}{S_o} \tag{9.30}$$

where

$$\beta = \frac{eB}{\hbar^2}\frac{\partial E}{\partial k}\frac{1}{kS_o}.$$

The iterative method is used again for both $g(k)$ and $h(k)$ to obtain numerical results.

Though all these perturbation functions (9.26), (9.29), and (9.30) are derived here for a non-degenerate semiconductor, those for a degenerate case are easily derived similarly.[13]

Two expressions are proposed for the alloy scattering rate,[32, 33, 43] and there is a slight difference in the coefficients due to the different definitions of the extent of the alloy scattering potential. The alloy scattering rate in Table 9.1 is that derived in reference 43 and the alloy scattering potential is assumed to have a radius equal to the nearest neighbour distance. The alloy scattering potential will be derived in section 9.3.1 and section 9.4.1.

9.2.4 Material parameters

Many material parameters are required for the mobility calculations, as can be found in the preceding section, but not all are available for $Ga_xIn_{1-x}As_yP_{1-y}$ alloys. Some of the important parameters have been determined experimentally. The effective-mass ratio[44]

$$m^*/m_0 = 0.08 - 0.039y$$

is used because this linear approximation is, from the point of view of the $\mathbf{k}\cdot\mathbf{p}$ theory,[40, 41] close to the very small bowing of the energy gap with composition y.[45, 46] Vegard's law is confirmed in the lattice-matched $Ga_xIn_{1-x}As_yP_{1-y}$ to InP.[45, 46] Since the lattice parameter is equal to that of InP, the composition is determined uniquely through the equation

$$x(1.032 - 0.032y) = 0.47y$$

or it could be approximated by $y \simeq 2.2x$. The energy gaps at 0 K are 1.42 eV for

Table 9.2 Material parameters of InP and $Ga_{0.47}In_{0.53}As$ used in the calculations. Linear interpolations between the two would be a good approximation for material parameters of GaInAsP

	InP	$Ga_{0.47}In_{0.53}As$
Effective mass ratio m^*/m_0	0.08	0.041
Static dielectric constant ε_0	12.35	13.77
Optical dielectric constant ε_∞	9.52	11.38
Optical phonon energy $\hbar\omega_{PO}$ (meV)	43.43	34.12
Energy gap at 0 K (eV)	1.42	0.812
Acoustic deformation potential E_1 (eV)	6.5	5.89
Piezoelectric constant e_{14} (C m^{-2})	0.035	0.099

InP[47] and 0.812 eV for $Ga_{0.47}In_{0.53}As$[21] and interpolated between these two values for GaInAsP. Measured LO phonon frequencies are available for GaInAsP.[48, 49] At least two LO-phonon frequencies are clearly resolved.[48] For a one-LO-phonon model in GaInAsP, the frequency is linearly interpolated between 350 cm^{-1} (InP) and 275 cm^{-1} ($Ga_{0.47}In_{0.53}As$).[48] Other parameters such as dielectric constants, sound of speed in the material, piezoelectric constant, and acoustic deformation potential, which are not experimentally available, are interpolated from those parameters of the constituent binary materials. Among these parameters, LO phonon frequency, dielectric constants, and effective mass are the most influential on the mobility calculation because these parameters appear in the three predominant scattering mechanisms, i.e. longitudinal optical phonon scattering, ionized-impurity scattering, and alloy scattering. Material parameters of InP and $Ga_{0.47}In_{0.53}As$ used in the calculations are listed in Table 9.2.

9.3 HALL MOBILITY AND DRIFT MOBILITY IN $Ga_{0.47}In_{0.53}As$

9.3.1 Alloy scattering potential

The functional dependence of alloy scattering on material and experimental parameters was first given by Nordheim.[50] Many models have been proposed for the alloy scattering potential.[32, 33, 51-55] Among those, potentials due to energy

Table 9.3 Alloy scattering potentials (in eV) for ternary III–V compounds relevant to GaInAsP for the energy gap difference ΔU_{EG}, the electron affinity difference ΔU_{EA}, and the Phillips' electronegativity difference ΔU_{EN}

Material	ΔU_{EG}	ΔU_{EA}	$\Delta U_{EN}(x = 0.5)$
$Ga_xIn_{1-x}As$	1.08	0.83	0.529
$Ga_xIn_{1-x}P$	1.38	0.40	0.559
$GaAs_yP_{1-y}$	1.30	0.07	0.637
$InAs_yP_{1-y}$	0.98	0.50	0.581

gap difference (ΔU_{EG}),[32, 33] electron affinity difference (ΔU_{EA}),[52] and Phillips' electronegativity difference (ΔU_{EN})[51, 53–55] are mainly considered here because they have been the most debated models, and they do not have an adjustable parameter in their formulation. The alloy scattering potentials for several ternary III–V compounds for ΔU_{EG}, ΔU_{EA}, and ΔU_{EN} have been tabulated by Littlejohn *et al.*[55] These potentials for four ternary III–V compounds relevant to GaInAsP are shown in Table 9.3. ΔU_{EN} has a slight functional dependence on the alloy composition.[51, 55] The alloy scattering potential for the scattering rate in Table 9.1 is expressed as

$$|\Delta U_{AL}|^2 = x(1-x)|\Delta U|^2 \tag{9.31}$$

for ternary alloys and $x = 0.47$ for $Ga_{0.47}In_{0.53}As$.

9.3.2 Comparison with experiments

$Ga_{0.47}In_{0.53}As$ has shown the highest mobility among the GaInAsP alloys[3, 4, 7] and also among the semiconductors which have a larger energy gap than about 0.7eV, suitable for room-temperature device applications.[1,16,20] However, there is a large scatter in the measured mobility of $Ga_{0.47}In_{0.53}As$ at the same carrier concentration because of many factors such as the carrier compensation, lattice mismatch effects, composition grading and fluctuation, and perhaps experimental errors.

Table 9.4 The calculated Hall mobilities compared with the experimental data at 300 K and 77 K for three high-purity samples. Hall mobilities were calculated for three different alloy scattering potentials with the same carrier concentration as each experimental datum assuming carrier compensation ratio 1. Samples a and b are from reference 5 and sample c from reference 8

	$n\,10^{15}\,cm^{-3}$, 300 K	$\mu_H\,(cm^2\,V^{-1}\,s^{-1})$ 300 K	77 K
Sample a	0.346	12 000	70 000
ΔU_{EG}		10 260	42 260
ΔU_{EA}		12 170	63 720
ΔU_{EN}		14 570	119 320
Sample b	1.9	13 800	43 000
ΔU_{EG}		9 800	32 940
ΔU_{EA}		11 600	47 210
ΔU_{EN}		13 840	80 050
Sample c	15	10 400	26 500
ΔU_{EG}		8 330	20 750
ΔU_{EA}		9 750	27 070
ΔU_{EN}		11 510	38 150

Those samples of lower carrier concentration and higher electron mobility both at 300 K and 77 K are selected to test the calculated Hall mobility with various alloy scattering potentials. The electron Hall mobility is calculated at the same carrier concentration as the experimental data assuming the carrier compensation ratio $(N_D + N_A)/n = 1$ (n is the carrier concentration, N_D the donor concentration, and N_A the acceptor concentration). The impurities are assumed to be fully ionized and $n = N_D - N_A$. In this ideal case, $N_A = 0$ so that no impurity compensation exists. Therefore, the calculated Hall mobility in this ideal case should be higher than the experimental data because the impurities are somewhat compensated in the grown materials. The results are listed and compared in Table 9.4. The high-purity samples are taken from references 5 (sample a and b), and 8 (sample c). It is clear that both the energy gap difference and the electron affinity difference fail to explain these high mobilities at 300 K and 77 K for all three samples. Only the Phillips' electronegativity difference is possible for the alloy scattering potential among the three possibilities. To confirm this, data fitting has been done to the sample c. Its temperature dependence of mobility is well explained over a wide temperature range 60 K $\leq T \leq$ 295 K with the Phillips' electronegativity difference for the alloy scattering potential and with a carrier compensation ratio 1.67 which is reasonable for LPE-grown semiconductors.[56]

9.3.3 Two LO phonon modes

From the analysis of infrared reflectivity spectra, $Ga_xIn_{1-x}As$ has been shown to have two LO-phonon modes over a part of the composition $0.2 \lesssim x < 1.0$.[57–59] On the other hand, Raman scattering spectra show only one strong LO-phonon peak[48] which has a frequency close to the interpolated value between GaAs and InAs. In GaInAsP alloys two clearly resolved LO phonons are observed over the composition range $0 < y < 1$ and their intensities are found to be proportional to the composition.[48] This is not the case for $Ga_xIn_{1-x}As$ from the Raman scattering experiments.[48] Though the strength of the LO phonon peak is not necessarily a reliable measure of the electron–phonon scattering rate, it is possible from the modified random-element-isodisplacement model to show that the polarization of two LO-phonon modes is proportional to the composition.[60] As

Table 9.5 Calculated mobilities in $Ga_{0.47}In_{0.53}As$ for one-LO-phonon model and for two-LO-phonon model. Electron concentration is 1×10^{16} cm^{-3} and carrier compensation ratio is 1

	300 K	77 K
One LO μ_H (cm^2 V^{-1} s^{-1})	12 100	44 970
Two LO μ_H (cm^2 V^{-1} s^{-1})	11 200	42 430

an extreme case it would be possible to assume the composition-proportional rate of electron–phonon scattering. This has been done and the calculated mobilities are listed in Table 9.5. Phonon Debye temperatures of 394 K and 334 K[57] are used for the two LO phonons at $x = 0.47$. The carrier concentration is 1×10^{16} cm^{-3} and carrier compensation ratio is 1. The mobilities for the two-LO-phonon model is lower by several per cent.

However, because of the lack of experimental data for the scattering intensities by the two LO-phonon modes in $Ga_xIn_{1-x}As$, the two-LO-phonon effect is not discussed further here but is considered in the next section for GaInAsP.

9.3.4 Hall factor

Hall factor relates the Hall mobility μ_H to the drift mobility μ_D through the equation $r_H = \mu_H/\mu_D$ as shown in (9.10). It is also the correction factor to the carrier concentration deduced from a Hall effect measurement, $n = r_H/eR_H$ which is easily derived from equation (9.9). Usually, the Hall factor is assumed to be 1 in this carrier concentration calculation, but it underestimates the real concentration. Temperature dependences of Hall factors for several combinations of scattering mechanisms and ionized-impurity concentration are shown in Figure 9.1 for the one-LO-phonon model in $Ga_{0.47}In_{0.53}As$. The Hall factor for intrinsic $Ga_{0.47}In_{0.53}As$ with no alloy scattering is very similar to that for

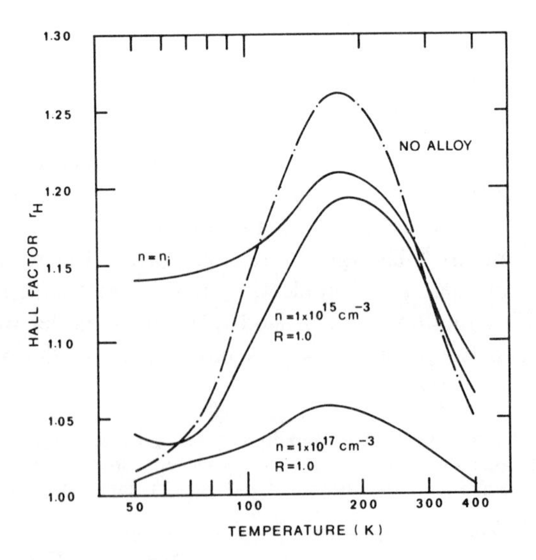

Figure 9.1 Temperature dependence of Hall factors in $Ga_{0.47}In_{0.53}As$ for combinations of several scattering mechanisms. The chain curve is for intrinsic $Ga_{0.47}In_{0.53}As$ without alloy scattering. Full curves are the Hall factor at $n = n_i$ (intrinsic), 1×10^{15} cm^{-3}, and 1×10^{17} cm^{-3} with alloy scattering. Carrier compensation ratio is 1

GaAs. When alloy scattering is included, the Hall factor flattens at lower temperatures because the alloy scattering is dominant here and it has a constant Hall factor approximately equal to 1.18.[18] For these two, a non-degenerate distribution function was used in the calculation. When the ionized-impurity scattering is also included and the degenerate distribution function (Fermi–Dirac function) is used, the Hall factor is closer to 1 partly because of a weakly momentum-dependent scattering rate in a combination of ionized-impurity and LO phonon scattering and partly because of the degeneracy of the carriers. These effects are more prominent at a higher carrier concentration such as $1 \times 10^{17}\ \text{cm}^{-3}$.

There are two reports concerning the value of r_H in GaInAsP quaternary alloys.[3,61] From the comparison of Hall mobilities at low and high magnetic fields, one paper reports that r_H is in the range of 0.98–1.00,[3] and other paper reports that it is between 1.00 and 1.03.[61] The calculated value of r_H in $Ga_{0.47}In_{0.53}As$ ternary alloy at 300 K has dependences of both carrier concentration n and carrier compensation ratio R.[38] In these reports both n and R are not specified, but if $R = 2.0$ is assumed the calculated r_H is 1.09 at $n = 1 \times 10^{16}\ \text{cm}^{-3}$ and it is 1.05 at $n = 1 \times 10^{17}\ \text{cm}^{-3}$. The value of r_H could be close to 1 in this carrier concentration range. However, calculated values for r_H are as high as 1.17 in a very pure sample.[38]

9.3.5 Carrier concentration dependence of Hall mobility and drift mobility in $Ga_{0.47}In_{0.53}As$

Theoretically calculated Hall mobilities at 300 K and 77 K are shown in Figures 9.2 and 9.3 for a wide range of free-electron concentrations and carrier compensation ratio of 1, 2, 5, and 10. The top curves are for the Hall mobility with no alloy scattering at carrier compensation ratio 1. All are calculated with the one-LO-phonon model and with the Phillips' electronegativity difference taken for the alloy scattering potential. Experimental Hall mobility data from various papers[4,5,8,16,19] are also shown in these figures.

Here, sources of error in the theoretical calculation and experiments are discussed. An error of $\pm 2\%$ in the electron effective mass affects the electron mobility by $\mp 3.4\%$ at 300 K. The same error in the static dielectric constant varies the electron mobility by $\mp 6.9\%$ at 300 K. These results indicate that further accurate measurements of material parameters are essential for a more detailed discussion of the electron mobility. In the theoretical calculations the alloy is assumed to be completely random, homogeneous, and free from strain. Because of heteroepitaxy, lattice mismatch and difference of thermal expansion coefficient[62] between the substrate InP and $Ga_{0.47}In_{0.53}As$ would affect the mobility and its temperature dependence. The effect of lattice mismatch on crystal perfection has been demonstrated.[63–65] Composition grading and fluctuation may cause an additional scattering such as space-charge scattering[66] or

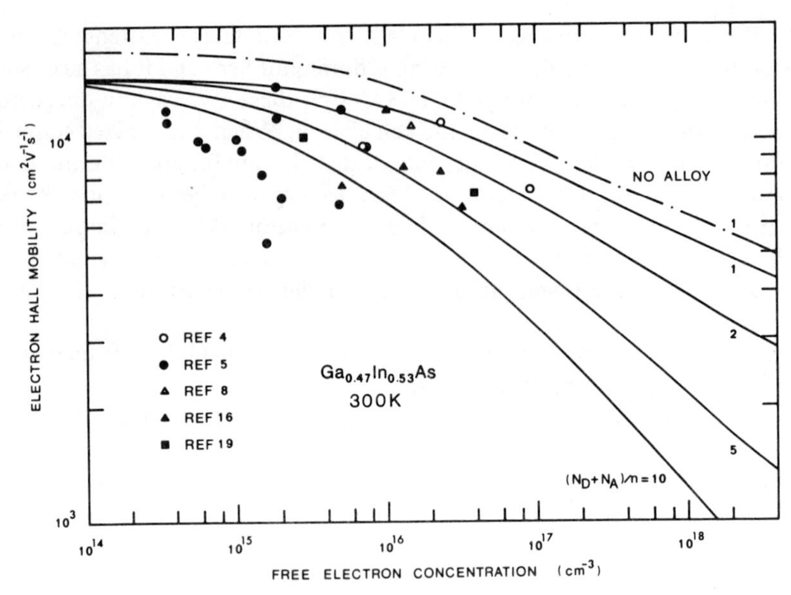

Figure 9.2 Electron Hall mobilities in $Ga_{0.47}In_{0.53}As$ at 300 K calculated by the iterative technique with one-LO-phonon model and Phillips' electronegativity difference for alloy scattering potential. The chain curve is the mobility with no alloy scattering. The numbers 1, 2, 5, and 10 at the right edge are the carrier compensation ratios, $(N_D + N_A)/n$. All the donors and acceptors are assumed to be fully ionized. Experimental data are well covered by these calculated mobilities

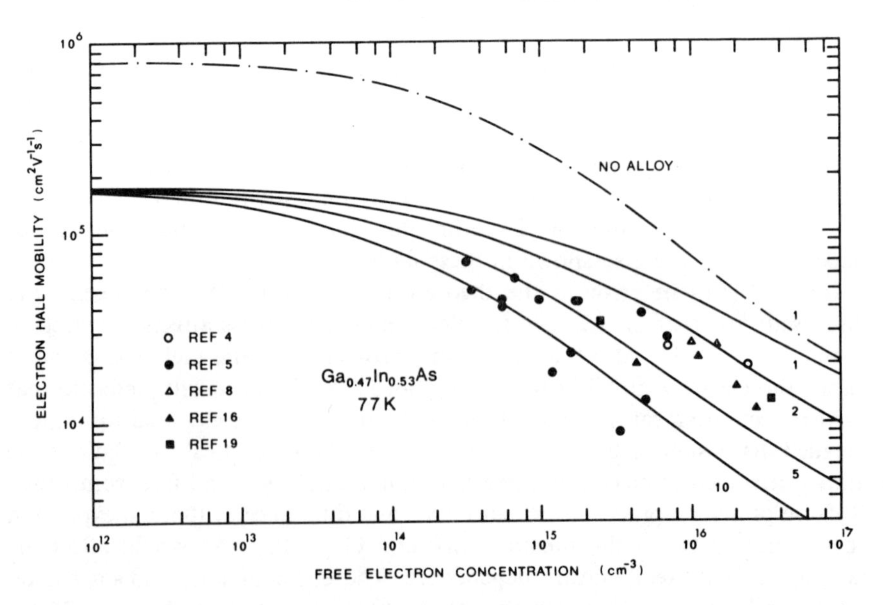

Figure 9.3 Electron Hall mobilities in $Ga_{0.47}In_{0.53}As$ at 77 K with the same model as for Figure 9.2. Effect of alloy scattering on mobility is prominent at 77 K

potential cluster effects.[53,59] Mobility limited by these effects is known to have the same temperature dependence ($T^{-1/2}$) as the alloy-scattering-limited mobility. Neutral impurity scattering,[67] if present, will be effective at lower temperatures in $Ga_{0.47}In_{0.53}As$ because its effective mass is smaller and the dielectric constant is larger when compared with those of GaAs and InP. This scattering mechanism flattens the temperature dependence of the total mobility. Though these additional scattering mechanisms make the total mobility lower, there is a mechanism which will apparently enhance the Hall mobility in inhomogeneous semiconductors.[68] This enhancement in $Ga_{0.47}In_{0.53}As$ was suggested in relatively highly doped ($> 1 \times 10^{17}\,cm^{-3}$) VPE materials.[69] It should be noticed that the Hall factor has a magnetic field dependence.[12,70] When the product of μB (mobility and magnetic field intensity) is much greater than 1, the Hall factor is close to 1. The magnetic field strength in the Hall effect measurement should be kept low enough to satisfy $\mu B \ll 1$; 5 kG is already high in this high mobility material ($\mu_H \gtrsim 10\,000\,cm^2\,V^{-1}\,s^{-1}$). It is far beyond the scope of this chapter to examine all these possible effects in all of the experimental data. However, from the present status of the experiments and even with unknown material parameters and their temperature dependences, the calculated Hall mobilities with the Phillips' electronegativity difference as the alloy

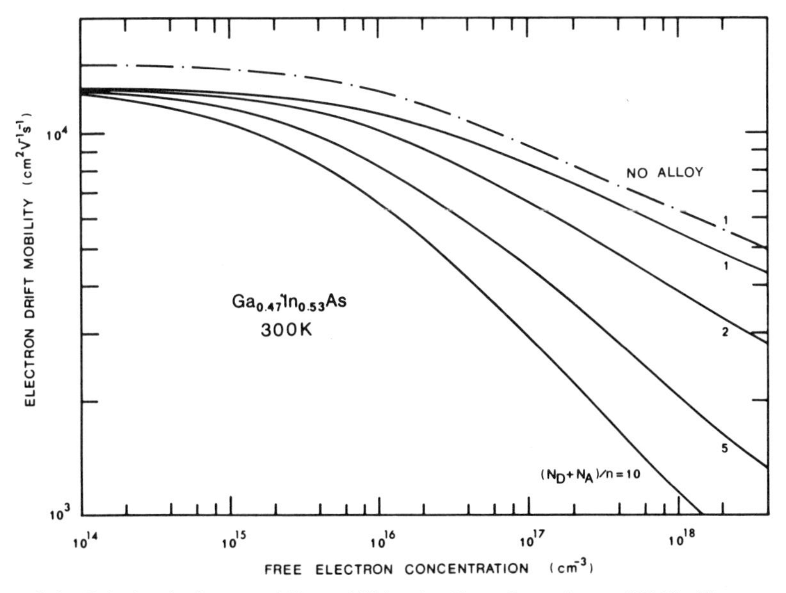

Figure 9.4 Calculated electron drift mobilities in $Ga_{0.47}In_{0.53}As$ at 300 K. The scattering mechanisms and material parameters are exactly the same as for Hall mobility calculations. The chain curve is the mobility with no alloy scattering. The numbers at the right edge are the carrier compensation ratios, $(N_D + N_A)/n$. The ratio of mobility value in Figure 9.2 and Figure 9.4 is the Hall factor. The drift mobility in $Ga_{0.47}In_{0.53}As$ is always higher than that of GaAs at 300 K

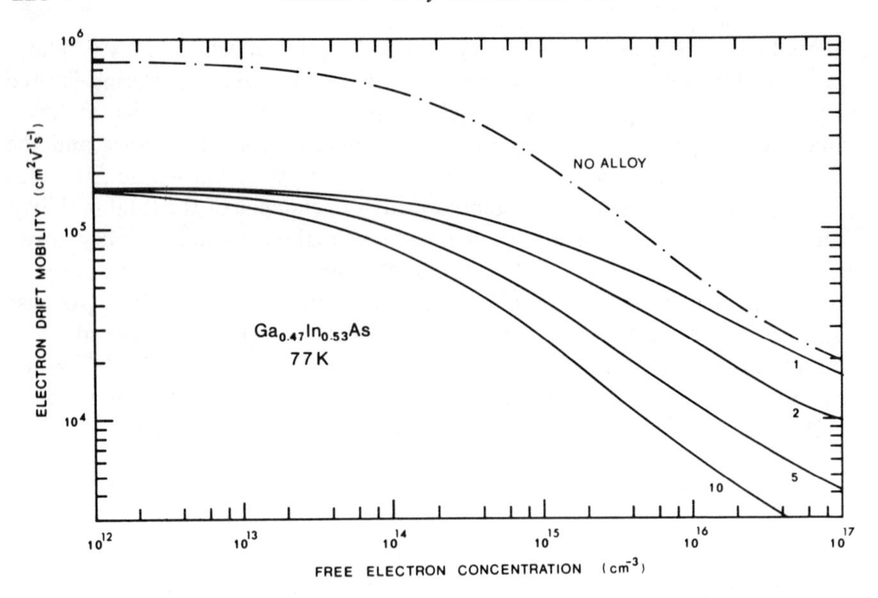

Figure 9.5 Calculated electron drift mobilities in $Ga_{0.47}In_{0.53}As$ at 77 K. At this temperature the drift mobility in $Ga_{0.47}In_{0.53}As$ is higher only at carrier concentration $n \gtrsim 1 \times 10^{15}$ cm^{-3} because of the alloy scattering effects in this alloy

scattering potential are seen to be in quite reasonable agreement with the experimental data in $Ga_{0.47}In_{0.53}As$.

It is also very important to point out here that Matthiessen's rule is an unacceptable approximation to analyse the scattering mechanisms in semiconductors especially when we deal with alloy scattering problems. Normally, Matthiessen's rule leads to an overestimation of the mobility as shown in GaAs[11] and $Ga_{0.47}In_{0.53}As$.[71] For example, from the calculation in section 9.3.2 for the sample c it was found that the alloy scattering potential due to the Phillips' electronegativity difference and a carrier compensation ratio 1.67 explain the temperature dependence of the mobility very well. However, if Matthiessen's rule is applied to analyse this mobility, an anomalous temperature dependence of the alloy scattering limited mobility is required.[8,71]

Figure 9.4 and 9.5 are the calculated *drift* mobilities in $Ga_{0.47}In_{0.53}As$ at 300 K and 77 K. The scattering mechanisms and material parameters are exactly the same as for the Hall mobility calculation. The top curves are for the drift mobility with no alloy scattering and at a carrier compensation ratio of 1. The one-LO-phonon model is employed for all calculations. The drift mobility in $Ga_{0.47}In_{0.53}As$ at 300 K is higher than that of GaAs[13] at any carrier concentration and carrier compensation ratio. However, the drift mobility in $Ga_{0.47}In_{0.53}As$ at 77 K is lower than that of GaAs[13] at carrier concentrations less than about 1×10^{15} cm^{-3} because of the increasing effect of alloy scattering in $Ga_{0.47}In_{0.53}As$ at lower temperature and lower carrier concentration.

9.4 HALL MOBILITY AND DRIFT MOBILITY IN GaInAsP

9.4.1 Extension of models from ternary to quaternary

As discussed in section 9.2.4 the material parameters of GaInAsP lattice-matched to InP have a compositional variation which is simpler than a sophisticated interpolation scheme would suggest. Linear interpolation between the two end materials, InP and $Ga_{0.47}In_{0.53}As$ appears to be the appropriate approximation for many material parameters. As for the LO phonon frequency, at least two clear peaks are observed in Raman scattering[48] and it will be necessary to deal with multi-LO-phonon models in order to discuss phonon scattering realistically.[72] However, even one LO phonon is shown in section 9.4.3 to be a reasonable approximation. An expression for the quaternary alloy scattering potential $\Delta U_Q(x, y)$ has been proposed by Littlejohn *et al.*[55] as

$$|\Delta U_Q(x, y)|^2 = x(1-x)y^2|\Delta U_{GIA}|^2 + x(1-x)(1-y)^2|\Delta U_{GIP}|^2 + y(1-y)x^2|\Delta U_{GAP}|^2 + y(1-y)(1-x)^2|\Delta U_{IAP}|^2 \quad (9.32)$$

where ΔU's on the right-hand side of (9.32) are the ternary alloy scattering potentials. For example, ΔU_{GIA} is the scattering potential of the ternary $Ga_xIn_{1-x}As$ and ΔU_{GAP} is the scattering potential of the ternary $GaAs_yP_{1-y}$. This expression was derived assuming interference among four atomic potentials. On the other hand, only interference between Ga and In on group III sites and between As and P on group V sites was proposed by Pearsall[40] in the expression of bowing parameter of energy gap using the Phillips' electronegativity difference as the scattering potential. In this case the alloy scattering potential will be expressed as

$$|\Delta U_Q(x, y)|^2 = x(1-x)C^2_{Ga\text{–}In} + y(1-y)C^2_{As\text{–}P} \quad (9.33)$$

where $C_{A\text{–}B}$ is the Phillips' electronegativity difference between the elements A and B. Effects of these two different models on electron mobility will be discussed in section 9.4.2.

9.4.2 Comparison with experiments

Three alloy scattering potentials are tested here again and only the Phillips' electronegativity difference gives satisfactory results over the whole composition range both at 300 K and 77 K. Electron Hall mobilities at carrier concentrations $1 \times 10^{16}\,cm^{-3}$ and $1 \times 10^{17}\,cm^{-3}$ with carrier compensation ratio 2 were calculated for the one-LO-phonon model at 300 K and 77 K. The LO phonon frequency was linearly interpolated between $350\,cm^{-1}$ (InP) and $275\,cm^{-1}$ ($Ga_{0.47}In_{0.53}As$).[48] The expression of equation (9.32) was used for the quaternary alloy scattering potential. The results are compared with experimental results in Figures 9.6 and 9.7. The experimental data are from references 1, 3, 4, 8, 16, 17, 19,

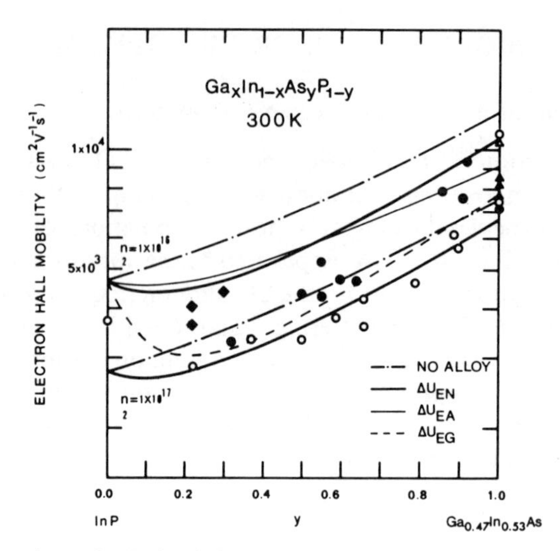

Figure 9.6 Comparison of calculated electron Hall mobilities with experimental data over the composition y at 300 K. Three alloy scattering potentials are tested at $n = 1 \times 10^{16}\ cm^{-3}$: thick full curve, Phillips' electronegativity difference (ΔU_{EN}); thin full curve, electron affinity difference (ΔU_{EA}); broken curve, energy gap difference (ΔU_{EG}). At $n = 1 \times 10^{17}\ cm^{-3}$, the alloy scattering potential (thick full curve) is ΔU_{EN}. In the calculation carrier compensation of 2 is assumed and the one-LO-phonon model for phonon scattering of (9.32) for alloy scattering potential is used. The chain curves are for the mobility without alloy scattering. The experimental data for which carrier concentrations are between $1 \times 10^{16}\ cm^{-3}$ and $1 \times 10^{17}\ cm^{-3}$ are plotted. Their sources are the same as in Figure 9.7. Only ΔU_{EN} can cover the experimental data over the whole composition range

and 73–75 and those data which have the carrier concentration between 1×10^{16} and $1 \times 10^{17}\ cm^{-3}$ are plotted. Thus, the theoretical curve with the correct alloy scattering potential at $n = 1 \times 10^{16}\ cm^{-3}$ should be higher than the experimental data both at 300 K and 77 K. The mobility calculated with the electron affinity difference as the alloy scattering potential fails to cover the higher composition side close to $Ga_{0.47}In_{0.53}As$. The mobility with the energy gap difference for the alloy scattering potential is considerably lower over the whole composition range except at $y = 0$, InP, where there is no alloy scattering. The Phillips' electronegativity difference again explains very well these experimental data over the whole composition range both at 300 K and 77 K; most of the experimental data are between the two thick full curves. Some data below the curve at $n = 1 \times 10^{17}\ cm^{-3}$ would have higher carrier compensation ratios.

An earlier theoretical calculation in this alloy predicted an upward bowing of the drift mobility along the composition y.[15] This was because interpolated effective-mass values used were significantly lower than recently measured values.[44] An effective mass varying linearly with composition makes the calculated mobility much closer to the experimental results.[72] Very recent work on the pressure dependence of the electron mobility in samples with $y = 0.5$ and $y = 0.91$ seems to confirm the existence of the alloy scattering mechanism in

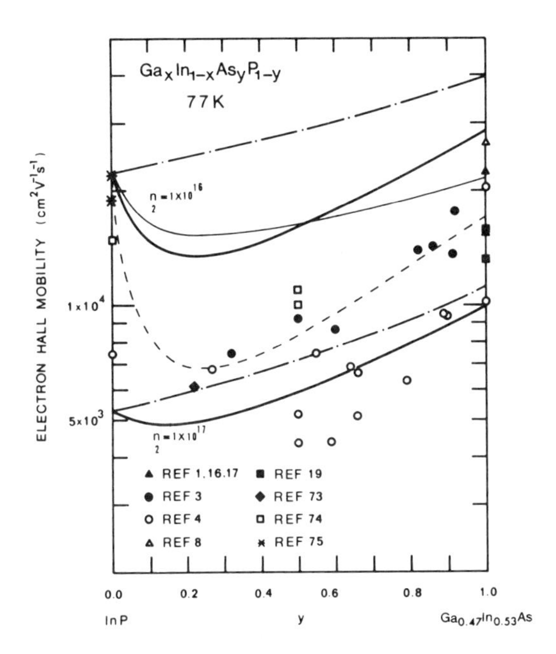

Figure 9.7 Comparison of calculated electron Hall mobilities with experimental data at 77 K. Models and alloy scattering potentials are the same as in Figure 9.6. At 77 K only ΔU_{EN} covers the experimental data over the whole composition range, as at 300 K

GaInAsP alloys.[9] However, the value and the composition dependence of the alloy scattering potential are not yet clear.

The alloy-scattering-limited Hall mobility and the total Hall mobility at $n = 1 \times 10^{15}\,cm^{-3}$ and $R = 1.0$ were calculated for the two expressions of the alloy scattering potential, (9.32) and (9.33). The Phillips' electronegativity difference is the alloy scattering potential and the one-LO-phonon model is employed for phonon scattering. The results at 300 K are shown in Figure 9.8. The alloy-scattering-limited mobilities have a noticeable difference when $y \lesssim 0.6$, but the total mobilities are almost the same over the whole composition range even at this low carrier concentration. On the other hand, at 77 K the difference between the two total mobilities is as large as $\sim 40\,\%$ at $y = 0.2$ because alloy scattering is the dominant scattering mechanism at this temperature in a high-purity alloy ($n = 1 \times 10^{15}\,\text{cm}^{-3}$). The theoretical mobility in Figure 9.7 at 77 K was calculated with the expression of equation (9.32), and to fit the experimental data further downward bowing in the theoretical curve is required. This is the opposite tendency of that of the mobility calculated with the expression (9.33). However, it is difficult to discuss the details of the alloy scattering potential from the mobility values in this y composition range because, as it is suggested in section 9.2.1 and in reference 35, the multiple scattering and clustering effects will be more influential in this composition range due to the heavier electron effective mass. From this

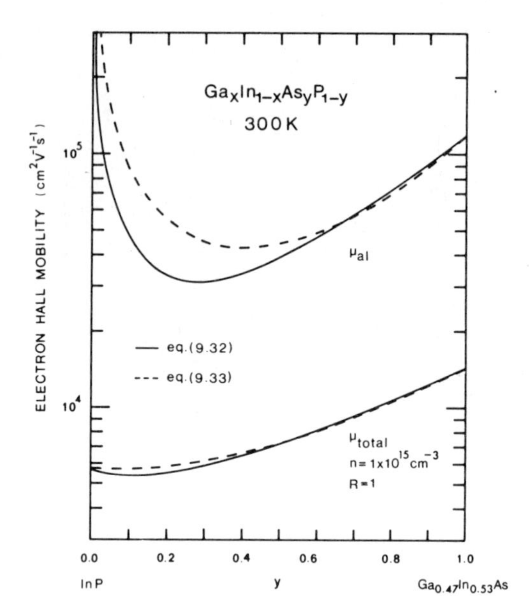

Figure 9.8 The upper two curves are the alloy-scattering-limited Hall mobility and the lower two curves represent the total Hall mobility at electron concentration $n = 1 \times 10^{15}$ cm^{-3} and with carrier compensation ratio $R = 1$ at 300 K. Full curves are for the alloy scattering potential (9.32) and broken curves are for the potential (9.33). The differences in the total mobilities are very small at 300 K but noticeable at 77 K because of the predominance of alloy scattering at lower temperature

physical point of view the alloys with $0 < y < 0.5$ are more interesting than those with $0.5 < y \leq 1.0$ which is the region which may be more useful for high-speed logic device applications.

9.4.3 Two LO-phonon modes

In GaInAsP Raman backscattering measurements resolved at least two LO-phonon modes[48] and far-infrared reflectivity measurements showed substantial agreement with the Raman results but one additional mode was observed.[49] Here a simplified two-LO-phonon model is used for the mobility calculation and effects of two LO-phonon modes on electron scattering are briefly discussed. As observed in the Raman scattering, the InP-like phonon and the GaInAs-like phonon are assumed. The phonon scattering rate S^*_{0PO} and S^*_{1PO} in Table 9.1 may be separated into the InP-like and the GaInAs-like phonon scattering rate. As a reasonable model the scattering rate by the InP-like phonon will be weighted by $(1-x)(1-y)$ which is the InP concentration in the quaternary, and the scattering rate by the GaInAs-like phonon is weighted by y which is the GaInAs concentration in the quaternary. This weighting by concentration was found to be a good approximation for the Raman scattering intensity ratio.[48] If the

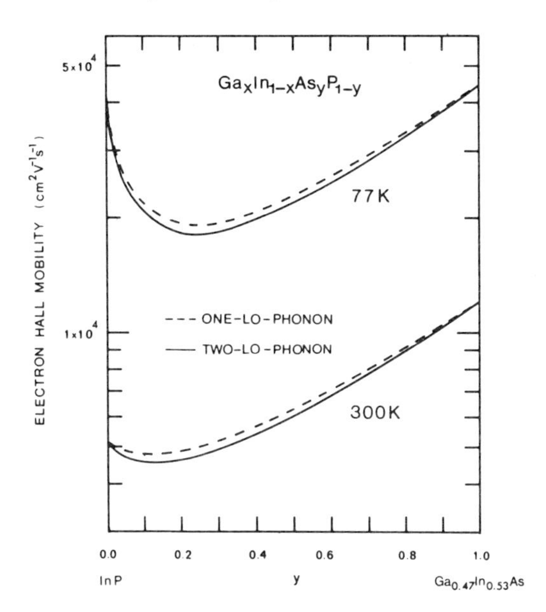

Figure 9.9 Electron Hall mobilities calculated with one-LO-phonon model (broken curves) and two-LO-phonon model (full curves) are compared at 300 K and 77 K. Electron concentration is 1×10^{16} cm^{-3} and carrier compensation ratio is 1. Greater bowing of the mobility, though it is small, is observed for the two-LO-phonon model

oscillator strength for each mode has a linear variation with concentration, it can be shown that the polarizability has the same concentration dependence.[60] The mode frequency of the InP-like phonon varies linearly from 350 cm^{-1} to about 328 cm^{-1} and that of the GaInAs-like phonon increases linearly from about 200 cm^{-1} to 275 cm^{-1} with increasing composition y.

In Figure 9.9 the results are compared with Hall mobilities for the one-LO-phonon model: the electron concentration $n = 1 \times 10^{16}$ cm^{-3} and carrier compensation ration $R = 1$ are assumed in both calculations. The quaternary alloy scattering potential of (9.32) is used. The two-LO-phonon model enhances the downward bowing of the mobility curve. This is caused by the incorporation of the lower-energy (GaInAs-like) phonon to the phonon scattering process. An electron which has energy equal to the phonon energy can emit the phonon and lose its energy. This process causes a sharp drop in the perturbation distribution of electrons.[10] In the two-LO-phonon model this drop occurs at a lower energy (GaInAs-like phonon energy) and thus the difference in the perturbation distribution makes the mobility smaller.[76] However, the difference between the calculated mobilities are not large both at 300 K and 77 K. From Figure 9.9, even the one-LO-phonon model appears to be a fairly good approximation to the experimental data so far reported. For further discussion, consideration of a more realistic estimation of the electron–phonon coupling for multiphonon modes will be necessary.

9.4.4 Carrier concentration dependence of Hall mobility and drift mobility in GaInAsP

Electron Hall mobility and drift mobility in GaInAsP are calculated for the one-LO-phonon model at various carrier concentrations. The one-LO-phonon model is used for simplicity and for its satisfactory results as discussed in the previous section. The expression for the quaternary alloy scattering potential is equation (9.32) in which interference among all four atomic potentials is assumed. The alloy scattering potential is the Phillips' electronegativity difference. Electron Hall mobilities calculated with these models explain fairly well the experimental data so far reported, and thus are extended to the wider carrier concentration and to the drift mobility calculation. Electron Hall mobilities at 300 K and 77 K are

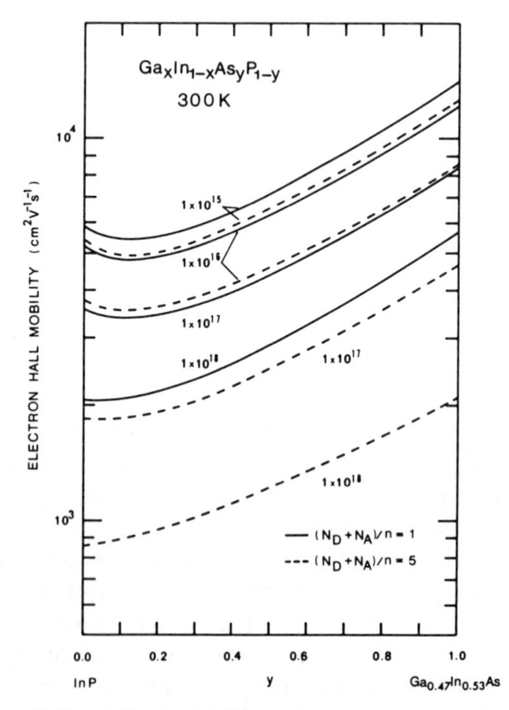

Figure 9.10 Electron Hall mobility in GaInAsP lattice-matched to InP calculated at 300 K by the iterative technique with the one-LO-phonon model and with the Phillips' electronegativity difference as the alloy scattering potential. Full curves are for the Hall mobility with carrier compensation 1, and broken curves represent the Hall mobility with carrier compensation 5. In both cases electron concentration is, from the top, 1×10^{15} cm^{-3}, 1×10^{16} cm^{-3}, 1×10^{17} cm^{-3}, and 1×10^{18} cm^{-3}

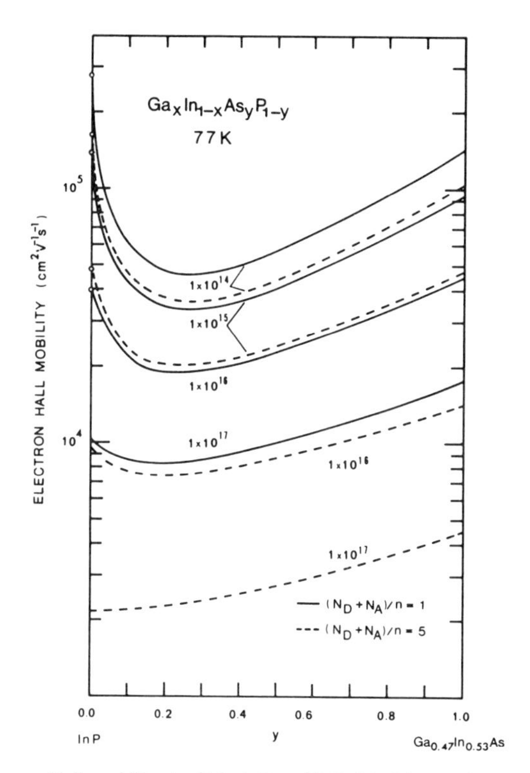

Figure 9.11 Electron Hall mobility in GaInAsP at 77 K. Models are the same as for Figure 9.10. Carrier concentration is, from the top 1×10^{14} cm^{-3}, 1×10^{15} cm^{-3}, 1×10^{16} cm^{-3}, and 1×10^{17} cm^{-3} for both full curves ($R = 1$) and broken curves ($R = 5$)

shown in Figures 9.10 and 9.11, respectively. Electron drift mobilities at 300 K and 77 K are shown in Figures 9.12 and 9.13, respectively. In these four figures the full curves are for the mobility calculated at the carrier compensation of 1 and the broken curves represent the mobility with the carrier compensation 5. As a general tendency in purer materials ($n < 1 \times 10^{16}$ cm^{-3}) at 300 K the value of r_H ($= \mu_H/\mu_D$) decreases with increasing y. In the temperature dependence curve of the Hall factor, the maximum is at approximately $\theta/2$ where θ is the LO-phonon Debye temperature.[12] Since in the one-LO-phonon model the phonon Debye temperature decreases with increasing y, the intercept at 300 K of the right side slope of the Hall factor curve (see Figure 9.1) decreases with increasing y. At a higher carrier concentration the value of r_H is influenced more by a combination of scattering mechanisms. At $n = 1 \times 10^{17}$ cm^{-3} and $R = 2.0$, r_H has the lowest value of 1.02 at $y \simeq 0.5$.

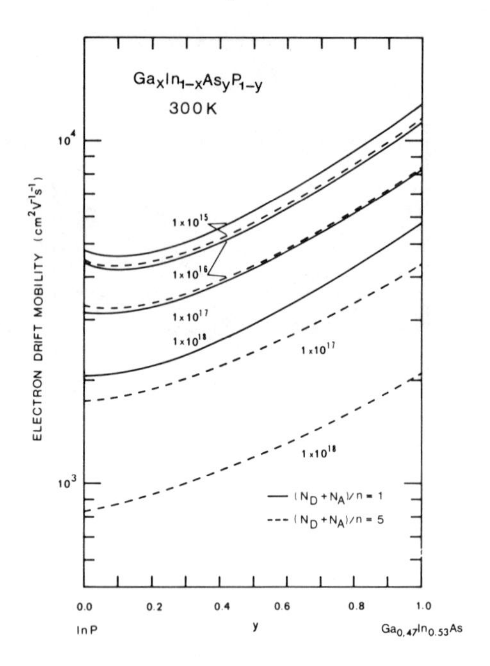

Figure 9.12 Electron drift mobility in GaInAsP at 300 K. Models and calculation technique are the same as for Figure 9.10. Carrier concentration is, from the top, 1×10^{15} cm^{-3}, 1×10^{16} cm^{-3}, 1×10^{17} cm^{-3}, and 1×10^{18} cm^{-3} for both full curves ($R = 1$) and broken curves ($R = 5$)

9.5 SUMMARY

The electron Hall mobility and drift mobility in $Ga_{0.47}In_{0.53}As$ and $Ga_xIn_{1-x}As_yP_{1-y}$ lattice-matched to InP have been calculated by the iterative technique. With an accurate model of scattering mechanisms and material parameters, the iterative technique has been shown to be a very powerful calculation technique for the electron Hall mobility and drift mobility over a wide range of temperature and carrier concentration. Using those calculations and available experimental data, low-field electron transport in $Ga_{0.47}In_{0.53}As$ and $Ga_xIn_{1-x}As_yP_{1-y}$ have been analysed. Among several alloy scattering potentials, the Phillips' electronegativity difference was found to give the best fit to the experimental data. Although the author does not eliminate the possibility of another scattering model for alloy scattering, with the classic transport theory and currently available experimental data the Phillips' electronegativity difference was found to be the best-suited alloy scattering potential to explain the electron Hall mobility both in $Ga_{0.47}In_{0.53}As$ and $Ga_xIn_{1-x}As_yP_{1-y}$ over a wide range of temperature and carrier concentration. It has been pointed out that

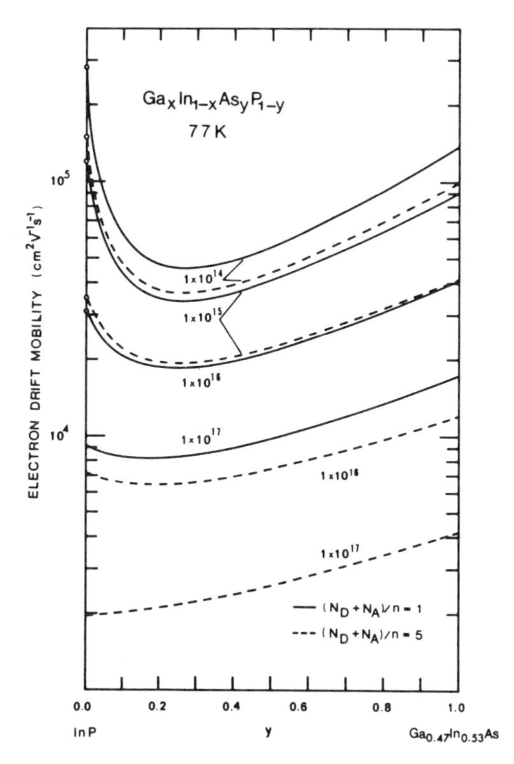

Figure 9.13 Electron drift mobility in GaInAsP at 77 K. Models and calculation technique are the same as for Figure 9.10. Carrier concentration is, from the top, 1×10^{14} cm^{-3}, 1×10^{15} cm^{-3}, 1×10^{16} cm^{-3}, and 1×10^{17} cm^{-3} for both full curves ($R = 1$) and broken curves ($R = 5$)

precise determination of material parameters and careful mobility measurements are necessary for a further detailed discussion on scattering mechanisms and multiphonon frequency effects in these alloys. Theoretical calculations both of the Hall mobility and the drift mobility have been extended to a wide range of carrier concentrations and thus enable us to have a wide systematic view of the mobility variations in $Ga_{0.47}In_{0.53}As$ and $Ga_xIn_{1-x}As_yP_{1-y}$ alloys lattice-matched to InP.

ACKNOWLEDGMENTS

I would like to acknowledge many useful discussions with and support from Dr M. A. Littlejohn and support and encouragement from Dr J. R. Hauser at North Carolina State University throughout this work. I am very grateful to Professor A. Sasaki at Kyoto University who kindly allowed me a one-year leave from Kyoto University without which this mutually stimulating work at North Carolina State University was not possible. I am also grateful to Dr T. P. Pearsall at Bell Laboratories (Murray Hill) for his continuous encouragement and enlightening discussions. I would like to thank Ms N. K. Tyson for her cheerful help in correcting and typing of the manuscript. The assistance of Mr R. Sultan in computation is also acknowledged.

REFERENCES

1. Y. Takeda, A. Sasaki, Y. Imamura, and T. Takagi, 'Electron mobility and energy gap of $In_{0.53}Ga_{0.47}As$ on InP substrate', *J. Appl. Phys.*, **47**, 5405–5408, 1976.
2. M. A. Littlejohn, R. A. Sadler, T. H. Glisson, and J. R. Hauser, 'Carrier compensation and alloy scattering in $Ga_{1-x}In_xP_{1-y}As_y$ grown by liquid-phase epitaxy', *Gallium Arsenide and Related Compounds 1978*, Conf. Ser. no. 45, Institute of Physics, Bristol, 1979, pp. 239–247.
3. P. D. Greene and S. A. Wheeler, 'Background carrier concentration and electron mobility in LPE $In_{1-x}Ga_xAs_yP_{1-y}$ layers', *Appl. Phys. Lett.*, **35**, 78–80, 1979.
4. R. F. Leheny, A. A. Ballman, J. C. DeWinter, R. E. Nahory, and M. A. Pollack, 'Compositional dependence of the electron mobility in $In_{1-x}Ga_xAs_yP_{1-y}$', *J. Electron. Mater.*, **9**, 561–568, 1980.
5. J. D. Oliver, Jr and L. F. Eastman, 'Liquid phase epitaxial growth and characterization of high purity lattice matched $Ga_xIn_{1-x}As$ on <111> B InP', *J. Electron. Mater.*, **9**, 693–712, 1980.
6. R. F. Leheny, J. Shah, J. Degani, R. E. Nahory, and M. A. Pollack, 'Electron transport in $In_{0.53}Ga_{0.47}As$', *Gallium Arsenide and Related Compounds* 1980, Conf. Ser. no. 56, Institute of Physics, Bristol, 1981, pp. 511–517.
7. J. H. Marsh, P. A. Houston, and P. N. Robson, 'Compositional dependence of the mobility, peak velocity and threshold field in $In_{1-x}Ga_xAs_yP_{1-y}$', *Gallium Arsenide and Related Compounds 1980*, Conf Ser. no. 56, Institute of Physics, Bristol, 1981, pp. 621–630.
8. T. P. Pearsall, G. Beuchet, J. P. Hirtz, N. Visentin, M. Bonnet, and A. Roizes, 'Electron and hole mobilities in $Ga_{0.47}In_{0.53}As$', *Gallium Arsenide and Related Compounds 1980*, Conf. Ser. no. 56, Institute of Physics, 1981, pp. 639–649.
9. A. R. Adams, H. L. Tatham, J. R. Hayes, and A. N. El-Sabbahy, 'Evidence for alloy scattering from pressure-induced changes of electron mobility in $In_{1-x}Ga_xAs_yP_{1-y}$', *Electron. Lett.*, **16**, 560–562, 1980.
10. D. L. Rode, 'Electron mobility in direct-gap polar semiconductors', *Phys. Rev.*, B, **2**, 1012–1024, 1970.
11. D. L. Rode, 'Electron transport in GaAs', *Phys. Rev.*, B, **3**, 2534–2541, 1971.
12. D. L. Rode, 'Theory of electron galvanomagnetics in crystals: Hall effect in semiconductors and semimetals', *Phys. Stat. Sol.*, b, **55**, 687–696, 1973.
13. D. L. Rode, 'Low-field electron transport', *Semiconductors and Semimetals*, vol. 10, eds. R. K. Willardson and A. C. Beer, Academic Press, New York, 1975, pp. 1–89.
14. C. Canali, C. Jacoboni, F. Nava, G. Ottaviani, and A. Alberigi-Quaranta, 'Electron drift velocity in silicon', *Phys. Rev.*, B, **12**, 2265–2284, 1975.
15. M. A. Littlejohn, J. R. Hauser, and T. H. Glisson, 'Velocity-field characteristics of $Ga_{1-x}In_xP_{1-y}As_y$ quaternary alloys', *Appl. Phys. Lett.*, **30**, 242–244, 1977.
16. A. Sasaki, Y. Takeda, N. Shikagawa, and T. Takagi, 'Liquid phase epitaxial growth, electron mobility, and maximum drift velocity of $In_{1-x}Ga_xAs$ ($x \simeq 0.5$) for microwave devices', *Proc. 8th Conf. on Solid State Devices*, Tokyo, 1976; *Jpn. J. Appl. Phys.*, **16**, *Suppl.* 16-1, 239–243, 1977.

17. Y. Takeda, A. Sasaki, Y. Imamura, and T. Takagi, 'Properties of liquid phase epitaxial $In_{1-x}Ga_xAs$ ($x \simeq 0.5$) on InP substrate', *J. Electrochem. Soc.*, **125**, 130–135, 1978.
18. Y. Takeda and A. Sasaki, 'Hall mobility and Hall factor of $In_{0.53}Ga_{0.47}As$', *Jpn. J. Appl. Phys.*, **19**, 383–384, 1980.
19. R. Sankaran, R. L. Moon, and G. A. Antypas, 'Liquid phase epitaxial growth of InGaAs on InP', *J. Cryst. Growth*, **33**, 271–280, 1976.
20. T. P. Pearsall, R. Bisaro, P. Merenda, G. Laurencin, R. Ansel, J. C. Portal, C. Houlbert, and M. Quillec, 'The characterization of $Ga_{0.47}In_{0.53}As$ grown lattice-matched on InP substrates', *Gallium Arsenide and Related Compounds 1978*, Conf. Ser. no. 45, Institute of Physics, Bristol, 1979, pp. 94–102.
21. T. P. Pearsall, '$Ga_{0.47}In_{0.53}As$: a ternary semiconductor for photodetector applications', *IEEE J. Quantum Electron.*, **QE-16**, 709–720, 1980.
22. E. M. Conwell, 'High field transport in semiconductors', *Solid State Physics, Suppl.* 9, eds. F. Seitz, D. Turnbull, and H. Ehrenreich, Academic Press, New York, 1967, chap. 3.
23. D. J. Howarth and E. H. Sondheimer, 'The theory of electronic conduction in polar semiconductor', *Proc. R. Soc. London*, A, **219**, 53–74, 1953.
24. H. Ehrenreich, 'Electron scattering in InSb', *J. Phys. Chem. Solids*, **2**, 131–149, 1957.
25. H. Ehrenreich, 'Transport of electrons in intrinsic InSb', *J. Phys. Chem. Solids*, **9**, 129–148, 1959.
26. T. Kurosawa, 'Monte Carlo calculation of hot electron problems', *Proc. Int. Conf on the Physics of Semiconductors*, Kyoto, 1966; *J. Phys. Soc. Jpn*, **21**, *Suppl.*, 424–426, 1966.
27. W. Fawcett, A. D. Boardman, and S. Swain, 'Monte Carlo determination of electron transport properties in gallium arsenide', *J. Phys. Chem. Solids*, **31**, 1963–1990, 1970.
28. F. Yonezawa, 'A note on electronic state of random lattice', *Prog. Theor. Phys.*, **31**, 357–377, 1964.
29. F. Yonezawa, 'A systematic approach to the problem of random lattice. I', *Prog. Theor. Phys.*, **40**, 734–757, 1968.
30. P. N. Argyres, 'Theory of the self-energy and its single-site approximations for disordered alloys', *Phys. Rev.*, B, **23**, 2447–2454, 1981.
31. P. N. Argyres and S. C. Papadopoulos, 'Theory of electrical conductivity of random binary alloys in the average t-matrix approximation', *Phys. Rev.*, B, **23**, 2455–2470, 1981.
32. Suggested by H. Brooks, 'Theory of the electrical properties of germanium and silicon', *Advances in Electronics and Electron Physics*, vol. 7 ed. L. Marton, Academic Press, New York, 1955, pp. 85–182. The expression for the formula is found in ref. 33.
33. L. Makowski and M. Glicksman, 'Disorder scattering in solid solutions of III–V semiconducting compounds', *J. Phys. Chem. Solids*, **34**, 487–492, 1973.
34. T. Nishinaga, O. Hori, and S. Uchiyama, 'Theoretical studies of disorder scattering in compound semiconductor alloys', *J. Phys. Soc. Jpn*, **41**, 1603–1610, 1976.
35. T. Nishinaga and K. Hiramatsu, 'Studies of disorder scattering in $(Ga_{1-x}Al_x)As$', Private communication.
36. See standard textbook on electronic conduction.
37. L. J. van der Pauw, 'A method of measuring specific resistivity and Hall effect of discs of arbitrary shape', *Philips Res. Rep.*, **13**, 1–9, 1958.
38. Y. Takeda, M. A. Littlejohn, and J. R. Hauser, 'Carrier concentration dependence of Hall factor in $In_{0.53}Ga_{0.47}As$', *Appl. Phys. Lett.*, **40**, 000–000, 1982.
39. R. T. Bate, R. K. Willardson, and A. C. Beer, 'Transverse magnetoresistance and Hall effect in n-type InSb', *J. Phys. Chem. Solids*, **9**, 119–128, 1959.
40. T. P. Pearsall, Chapter 12 in this volume.

41. E. O. Kane, 'Band structures of indium antimonide', *J. Phys. Chem. Solids*, **1**, 249–261, 1957.
42. D. Matz, 'Effects of nonparabolicity on non-ohmic transport in InSb and InAs', *Phys. Rev.*, **168**, 843–849, 1968.
43. J. W. Harrison and J. R. Hauser, 'Alloy scattering in ternary III–V compounds', *Phys. Rev.*, B, **13**, 5347–5350, 1976.
44. R. J. Nicholas, S. J. Sessions, and J. C. Portal, 'Cyclotron resonance and the magnetophonon effect in $Ga_xIn_{1-x}As_yP_{1-y}$', *Appl. Phys. Lett.*, **37**, 178–180, 1980.
45. R. E. Nahory, M. A. Pollack, W. D. Johnston, Jr, and R. L. Barns, 'Band gap versus composition and demonstration of Vegard's law for $In_{1-x}Ga_xAs_yP_{1-y}$ lattice matched to InP', *Appl. Phys. Lett.*, **33**, 659–661, 1978.
46. K. Nakajima, A. Yamaguchi, K. Akita, and T. Kotani, 'Composition dependence of the band gaps of $In_{1-x}Ga_xAs_{1-y}P_y$ quaternary solids lattice matched on InP substrates', *J. Appl. Phys.*, **49**, 5944–5950, 1978.
47. M. Neuberger, 'III–V ternary semiconducting compounds—data tables', *Handbook of Electronic Materials*, vol. 7, IFI/Plenum, New York, 1972.
48. A. Pinczuk, J. M. Worlock, R. E. Nahory, and M. A. Pollack, 'Lattice vibrations of $In_{1-x}Ga_xAs_yP_{1-y}$ quaternary compounds', *Appl. Phys. Lett.*, **33**, 461–463, 1978.
49. P. M. Amirtharaj, G. D. Holak, and S. Perkowitz, 'Far-infrared spectroscopic study of $In_{1-x}Ga_xAs_yP_{1-y}$', *Phys. Rev.*, B, **21**, 5656–5661, 1980.
50. L. Nordheim, 'Zur Elektronen Theorie der Metalle. II', *Ann. Phys. Lpz.*, **9**, 641–678, 1931.
51. J. A. Van Vechten and T. K. Bergstresser, 'Electronic structure of semiconductor alloys', *Phys. Rev.*, B, **1**, 3351–3358, 1970.
52. J. W. Harrison and J. R. Hauser, 'Theoretical calculations of electron mobility in ternary III–V compounds', *J. Appl. Phys.* **47**, 292–300, 1976.
53. F. Oosaka, T. Sugano, Y. Okabe, and Y. Okada, 'Scattering of electrons by potential clusters in ternary alloy semiconductor', *Jpn. J. Appl. Phys.*, **15**, 2371–2380, 1976.
54. A. Sasaki, 'Theory for electron mobility in ternary mixed semiconductors', *Proc. 9th Conf. on Solid State Devices*, Tokyo, 1977; *Jpn. J. Appl. Phys.*, **17**, *Suppl.* 17-1, 161–166, 1978.
55. M. A. Littlejohn, J. R. Hauser, T. H. Glisson, D. K. Ferry, and J. W. Harrison, 'Alloy scattering and high field transport in ternary and quaternary III–V semiconductors', *Solid-State Electron.*, **21**, 107–114, 1978.
56. Y. Takeda, M. A. Littlejohn, and J. R. Hauser, 'Electron Hall mobility calculations and alloy scattering in $Ga_{0.47}In_{0.53}As$', *Electron. Lett.*, **17**, 377–379, 1981.
57. M. H. Brodsky and G. Lucovsky, 'Infrared reflection spectra of $Ga_{1-x}In_xAs$: A new type of mixed-crystal behavior', *Phys. Rev. Lett.*, **30**, 990–993, 1968.
58. G. Lucovsky and M. F. Chen, 'Long wave optical phonons in the alloy system: $Ga_{1-x}In_xAs$, $GaAs_{1-x}Sb_x$, and $InAs_{1-x}Sb_x$', *Solid State Commun.*, **8**, 1397–1401, 1970.
59. S. Yamazaki, A. Ushirokawa, and T. Katoda, 'Effect of clusters on long-wavelength optical phonons in $Ga_{1-x}In_xAs$', *J. Appl. Phys.*, **51**, 3722–3729, 1980.
60. I. F. Chang and S. S. Mitra, 'Application of a modified random-element isodisplacement model to long-wavelength optic phonons of mixed crystals', *Phys. Rev.*, **172**, 924–933, 1968.
61. P. K. Bhattacharya, J. W. Ku, S. J. T. Owen, G. H. Olsen, and S. H. Chiao, 'LPE and VPE $In_{1-x}Ga_xAs_yP_{1-y}$/InP: transport properties, defects, and device considerations', *IEEE J. Quantum Electron.*, **QE-17**, 150–161, 1981.
62. R. Bisaro, P. Merenda, and T. P. Pearsall, 'The thermal-expansion parameters of some $Ga_xIn_{1-x}As_yP_{1-y}$ alloys', *Appl. Phys. Lett.*, **34**, 100–102, 1979.

63. Y. Takeda and A. Sasaki, 'Composition latching phenomenon and lattice mismatch effects in LPE-grown $In_{1-x}Ga_xAs$ on InP substrate', *J. Cryst. Growth*, **45**, 257–261, 1978.
64. K. Nakajima, S. Komiya, K. Akita, T. Yamaoka, and O. Ryuzan, 'LPE growth of misfit dislocation-free thick $In_{1-x}Ga_xAs$ layers on InP', *J. Electrochem. Soc.*, **127**, 1568–1572, 1980.
65. Y. Takeda and A. Sasaki, 'Low EPD and high avalanche multiplication of lattice-mismatched $In_{0.52}Ga_{0.48}As$ on InP substrate', *Proc. 12th Conf. on Solid State Devices*, Tokyo, 1980; *Jpn. J. Appl. Phys.*, **20**, *Suppl.* 20-1, 189–192, 1981.
66. T. Katoda, F. Osaka, and T. Sugano, 'Electron mobility in $In_{1-x}Ga_xAs$ epitaxial layer', *Jpn. J. Appl. Phys.*, **13**, 561–562, 1974.
67. C. Erginsoy, 'Neutral impurity scattering in semiconductors', *Phys. Rev.*, **79**, 1013–1014, 1950.
68. C. M. Wolfe and G. E. Stillman, 'Apparent mobility enhancement in inhomogeneous crystals', *Semiconductors and Semimetals*, vol. 10, eds. R. K. Willardson and A. C. Beer, Academic Press, New York, 1975, pp. 175–220.
69. G. Beuchet, M. Bonnet, P. Thébault, and J. P. Duchemin, 'GaInAs/InP heterostructures grown in a four-barrel reactor by the hydride method', *Gallium Arsenide and Related Compounds 1980*, Conf. Ser. no. 56, Institute of Physics, Bristol, 1981, pp. 37–44.
70. G. E. Stillman, C. M. Wolfe, and J. O. Dimmock, 'Hall coefficient factor for polar mode scattering in n-type GaAs', *J. Phys. Chem. Solids*, **31**, 1199–1204, 1970.
71. Y. Takeda and T. P. Pearsall, 'The failure of Matthiessen's rule in the calculation of carrier mobility and alloy scattering effects in $Ga_{0.47}In_{0.53}As$', *Electron. Lett.*, **17**, 573–574, 1981.
72. Y. Takeda, M. A. Littlejohn and J. R. Hauser, 'Electron Hall mobility in $Ga_xIn_{1-x}As_yP_{1-y}$ calculated with two-LO-phonon model', *Appl.Phys. Lett.*, **39**, 620–621, 1981.
73. B. Houston, J. B. Restorff, R. S. Allgaier, J. R. Burke, D. K. Ferry, and G. A. Antypas, 'Hot electron and magneto-transport properties of $In_{1-x}Ga_xP_{1-y}P_y$ liquid phase epitaxial films', *Solid-State Electron.*, **21**, 91–94, 1978.
74. S. H. Groves and M. C. Plonko, 'LPE growth of nominally undoped InP and $In_{0.8}Ga_{0.2}As_{0.5}P_{0.5}$ alloys', *Gallium Arsenide and Related Compounds 1978*, Conf. Ser. no. 45, Institute of Physics, Bristol, 1979, pp. 71–77.
75. V. L. Wrick, G. J. Scilla, L. F. Eastman, R. L. Henry, and E. M. Swiggard, 'The effects of baking time on LPE: purity and morphology', *Gallium Arsenide and Related Compounds (Edinburgh) 1976*, Conf. Ser. no. 33a Institute of Physics, Bristol, 1977, pp. 35–40.
76. Y. Takeda, M. A. Littlejohn, and J. R. Hauser, 'Effects of two LO-phonon modes on electron distribution in GaInAsP', *Solid State Comm.*

GaInAsP Alloy Semiconductors
Edited by T. P. Pearsall

Chapter 10

Hot Electron Transport in n-Type $Ga_{1-x}In_xAs_yP_{1-y}$ Alloys Lattice-matched to InP

M. A. LITTLEJOHN, T. H. GLISSON, AND J. R. HAUSER

Department of Electrical Engineering, North Carolina State University, Raleigh, NC 27650, USA

10.1 INTRODUCTION

This chapter presents results of Monte Carlo calculations on hot electron transport in $Ga_{1-x}In_xAs_yP_{1-y}$ quaternary alloys. The influence of material properties on hot electron transport is examined in detail. Wherever possible, we use experimentally determined values for material parameters, e.g. recently measured values of central-valley effective mass are used in the calculations. However, relatively few material parameters of this quaternary alloy have been determined experimentally. A range of parameters is used for unknown material parameters. Emphasis is placed on the effect of satellite-valley material parameter variations on the results.

Also, the effects on hot electron transport of a pseudo-two-mode longitudinal optical phonon spectrum recently observed by Raman spectroscopy are considered. Existing discrepancies between experimental and theoretical ohmic mobilities and their effects on hot electron transport are examined.

The alloy semiconductor $Ga_{1-x}In_xAs_yP_{1-y}$ has been of interest for device applications since 1972.[1] Most studies of $Ga_{1-x}In_xAs_{1-y}P_y$ have been directed towards optical devices, and especially towards sources and detectors for optical communications. The development of silica optical fibres having low transmission loss and low dispersion in the region 1.0 μm to 1.7 μm[2] has promoted much interest in $Ga_{1-x}In_xAs_yP_{1-y}$.[2] This alloy has several properties that make it attractive for optical applications in other spectral regions as well. It can be lattice-matched to both InP and GaAs, and most of these lattice-matched alloy compositions are direct-bandgap materials.[2] When lattice-matched to InP, the quaternary has a direct energy gap that varies at room temperature from 1.35 eV

(0.92 μm) to 0.75 eV (1.65 μm) as y varies from 0 to 1 according to[3]

$$y = 2.065(1-x)+0.0642xy \tag{10.1}$$

A linear approximation to equation (10.1) is

$$y = 2.13(1-x) \tag{10.2}$$

This equation gives $y = 0$ when $x = 1$ (i.e. InP) and $x = 0.53$ when $y = 1$ (i.e. $Ga_{0.47}In_{0.53}As$).

Another property of this quaternary that is attractive for fibre optical communications is the ability to grow well-controlled n- and p-type thin films with excellent surface morphology by liquid-phase epitaxy (LPE),[4] vapour-phase (VPE),[5] and molecular-beam epitaxy (MBE).[6] Double-heterostructure lasers with room-temperature operating lifetimes in excess of 5000 h have been built. Also, multiple quantum well laser structures have also been fabricated and tested.[4–6] The factors that make $Ga_{1-x}In_xAs_yP_{1-y}$ attractive for optoelectronic devices also make this material attractive for high-speed digital and analogue microwave devices. Specifically, this quaternary meets many of the requirements for compound semiconductor field-effect transistors.[7] The 1 eV direct bandgap, the small central-valley effective mass, and the large predicted inter-valley energy separation are consistent with requirements for such devices.[8, 9] The use of the quaternary in FETs depends on the ability to synthesize the material while controlling its properties and on modelling transport in the material. In a practical sense, the realization of FETs in $Ga_{1-x}In_xAs_yP_{1-y}$ depends on technological mastery of such processes as ion implantation, Schottky barrier formation, formation of gate insulators, and formation of ohmic contacts, on understanding of interfacial properties, and on substrate availability.[10, 11]

The trend today is towards reduced dimensions and novel structures for compound semiconductor FETs.[7] The performance of such devices depends critically on the transport properties of the material. In particular, device performance depends on both low-field ohmic transport and high-field (hot electron) transport to extents determined by device structure and material properties.[7] Thus, it is important to have models that accurately predict low-field and high-field transport if these materials are to be used for FETs and other high-speed devices.

The Monte Carlo method is well-established for simulation of hot electron and low-field transport in semiconductor materials.[12–18] It has been applied widely to GaAs and to InP.[12–16] It is probably the most complete and exact way to study transport in semiconductors because it can include all known scattering mechanisms in the simulations.

The Monte Carlo method was first used to study electron transport in $Ga_{1-x}In_xAs_yP_{1-y}$ in 1977.[8] The main purpose of that study was to examine the effects of alloy scattering on transport in ternary and quaternary alloys and to compare these materials with GaAs and InP. The effects of alloy scattering were later studied in more detail.[19] The calculations reported at that time used a simple interpolation to estimate many material parameters for the alloys because of lack of experimental data. However, available experimental data were used. This interpolation is based on a knowledge of material parameters for the binary constituents. Hot electron transport properties (velocity-field characteristics) and low-field drift mobility were calculated.[8] Since the original calculations for $Ga_{1-x}In_xAs_yP_{1-y}$ were made, little experimental data on hot electron transport have been reported. However, there have been several Hall effect studies of the low-field electron mobility. Figure 10.1 shows the available experimental data, and also shows the original calculated drift mobility for comparison. The experimental data are extracted from publications by Wieder,[10] Leheny *et al.*,[20] Greene *et al.*,[21] Adams *et al.*,[22] Oliver and Eastman,[23] and Houston *et al.*[24] Calculated mobilities in Figure 10.1 are for an uncompensated semiconductor with a net ionized-impurity density of $1 \times 10^{16}\,cm^{-3}$. The experimental Hall mobilities were obtained on samples with electron concentrations ranging from $1 \times 10^{16}\,cm^{-3}$ to $1 \times 10^{17}\,cm^{-3}$.[20] Figure 10.1 also shows a plot of Hall mobility obtained by multiplying the calculated drift mobility by the Hall factor for $Ga_{0.47}In_{0.53}As$ obtained by Takeda and Sasaki,[25] and the mobility calculated

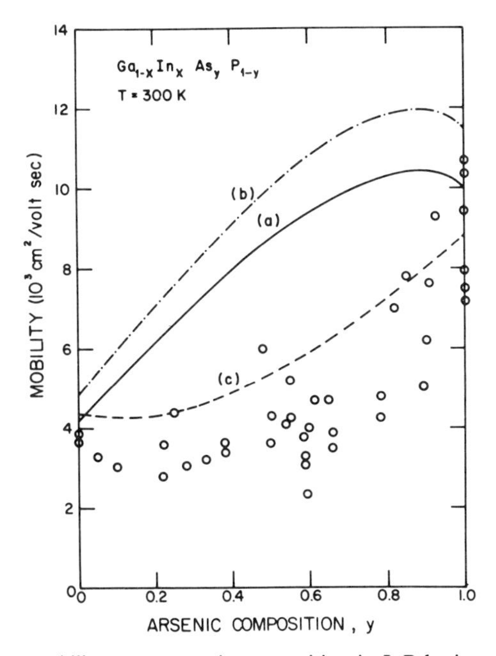

Figure 10.1 Electron mobility versus arsenic composition in InP lattice-matched alloys. Curve (a) shows our original calculations[8] for a doping level of $10^{16}\,cm^{-3}$. Curve (b) is a plot of Hall mobility using a representative Hall factor of 1.15.[25] Curve (c) is calculated using a linear effective-mass equation (see section 10.2.5). The circles reprsent experimental data

using a linear effective-mass variation with composition. This curve is discussed in subsequent sections.

The experimental data shown in Figure 10.1 represent the state of electron transport measurements in $Ga_{1-x}In_xAs_yP_{1-y}$. As can be seen, experimental values are in rather serious disagreement with previous calculations. This provides a point of departure for our discussion of hot electron transport in this quaternary. We focus on compositions that are lattice-matched to InP according to equation (10.1). Chapter 9 in this volume focuses on low-field transport. We consider low-field transport only to the extent that it influences hot electron transport, for example through the central-valley effective mass and its variation with composition. The present paper can be considered as a re-evaluation of the hot electron transport properties of $Ga_{1-x}In_xAs_yP_{1-y}$ and a reassessment of this material in light of experimental data. It is shown that much of the disagreement between experiment and theory can be explained. However, much more experimental work is needed before a complete description of hot electron transport in $Ga_{1-x}In_xAs_yP_{1-y}$ can be achieved.

10.2 THE MONTE CARLO METHOD

The Monte Carlo method is a general technique for simulating random processes which can be applied to electron transport in semiconductors.[12, 26] Application of Monte Carlo methods to transport calculations has been reviewed recently by Nag.[27] A Monte Carlo transport calculation consists of modelling the motion of an electron as a sequence of free flights (in a constant applied field) interrupted by collisions (scattering events). Between collisions, the electron obeys classical laws of motion determined from the band structure for the material. Band structure is usually taken into account through a band-edge electron effective mass and energy band non-parabolicity. The collisions are regarded as random events whose probabilities are known functions of energy. The duration of a free flight, the type of scattering processes, and the change in momentum produced by a scattering are determined using numbers produced by a pseudorandom number generator. In effect, this method produces a numerical solution to the Boltzmann transport equation for a specified set of scattering processes and material parameters.[12, 26, 27]

Figure 10.2 shows the main steps in a Monte Carlo simulation. A carrier is initialized by selecting its three components of momentum from a Maxwellian distribution for lattice temperature T. The carrier then accelerates freely under an applied field E until it is determined that an interaction has occurred; this determination is based on total scattering rate as a function of carrier energy and on a pseudorandom number selected at the beginning of a free flight. After a free flight, another pseudorandom number is selected; this number and the individual scattering rates determine which of various possible interactions occurs. Once a scattering mechanism has been selected, additional pseudorandom numbers

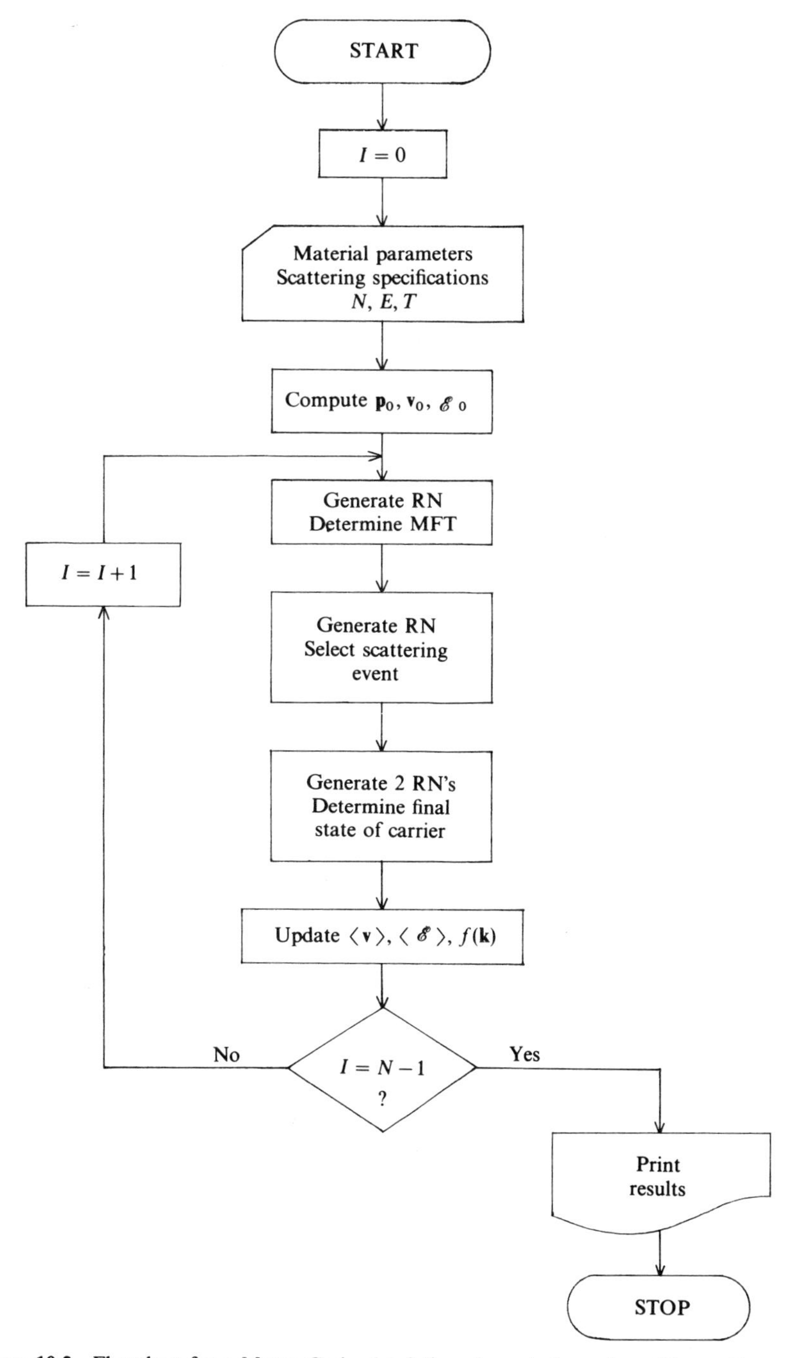

Figure 10.2 Flowchart for a Monte Carlo simulation. *Legend:* I, number of interactions completed; N, number of interactions specified; RN, random number; MFT, mean free time; E, electrical field intensity; $\mathscr{E}$, energy; $\mathbf{p}$, momentum; $\mathbf{v}$, velocity; T, temperature; $\mathbf{k}$, electron wave vector; $f\{\mathbf{k}\}$, distribution function

determine the state (momentum) of the carrier after the interaction; this state becomes the initial state of the carrier for the next free flight. These steps (free flight ended by a scattering) are repeated until a specified number (N) of interactions has occurred or until a specified time has elapsed. This process is described in more detail in the following sections.

10.2.1 Energy bands and equations of motion

Energy band structure is incorporated into the Monte Carlo program used in this work through a three-valley model with spherical symmetry in each valley.[12, 16, 26, 27] The three valleys are centred at specific points in the first Brillouin zone, and are referred to as the Γ band (000 direction), L band (111 direction), and X band (100 direction). In addition to being separated in momentum space, each valley is also separated in energy from its two neighbour valleys. Thus, $\Delta\mathscr{E}_{\Gamma L}$ is the energy separation between the Γ and L valleys. For GaAs, $\Delta\mathscr{E}_{\Gamma L} = 0.33$ eV.[16] Each valley includes band non-parabolicity by relating the electron energy $\mathscr{E}$ to the wavevector $\mathbf{k}$ by

$$\hbar^2 k^2 = 2m^* \gamma(\mathscr{E}) \tag{10.3}$$

where $\hbar$ is Planck's constant, k is the magnitude of $\mathbf{k}$, m^* is the band-edge effective mass, and $\gamma(\mathscr{E})$ describes the band strucuture for the material. Here $\mathscr{E}$ is electron kinetic energy measured from the band edge. Each non-parabolic conduction band is described by a function of the form

$$\gamma(\mathscr{E}) = \mathscr{E}(1 + \alpha\mathscr{E}) \tag{10.4}$$

where α is a 'non-parabolicity parameter'. In the Γ valley, α is given by

$$\alpha = \frac{1}{\mathscr{E}_g}(1 - m^*/m_0)^2 \tag{10.5}$$

where $\mathscr{E}_g$ is the bandgap energy and m_0 is the free-electron mass.[12] Equation (10.5) is used to estimate α for the L and X valleys.

During a free flight (between scatterings), motion of an electron is governed by

$$\hbar\frac{d\mathbf{k}}{dt} = q\mathbf{E} \tag{10.6}$$

where q is electronic charge and $\mathbf{E}$ is electric field strength. This equation has been termed the 'quasiclassical approximation'.[26] The constant applied electric field changes the electron wavevector linearly with time during a free flight. The electron energy in a given valley then increases with time according to equation (10.3) until it loses (or gains) energy by the scattering event. Equations (10.3) and (10.6) describe the free flight kinetics.

The form of band structure described by equations (10.4) and (10.5) is a simple approximation to the actual band structure. Recently, Shichijo has applied detailed band structure calculated from pseudopotential theory in the Monte Carlo method to study avalanche effects in GaAs.[28] He found that it is important to include these band structure details at large electron energies above about 1.5 eV, which requires field strengths above 50 kV cm^{-1}. For field strengths between 1 kV cm^{-1} and 20 kV cm^{-1} the simple band structure can be expected to yield accurate hot electron calculations. The work of Shichijo should provide a basis for more accurate hot electron calculations as device dimensions decrease.

10.2.2 Scattering mechanisms and scattering rates

In a semiconductor, there are a number of physical processes which can cause an electron moving through the lattice to be scattered.[27] The scattering mechanisms included in this work are listed in Table 10.1. The relative importance of each process in a given material depends on electric field strength (average energy) and material parameters.

When an electron is scattered, the wavevector of the electron is changed from $\mathbf{k}$ to $\mathbf{k}'$. The time required for a scattering process to change the wavevector from $\mathbf{k}$ to $\mathbf{k}'$ is called the collision duration τ_c. In this work collisions are assumed to occur with $\tau_c = 0$. While this assumption is usually valid, there is some concern that it may be violated in submicrometre semiconductor devices.[29]

The inclusion of a scattering rate in the Monte Carlo method requires the formulation of the total scattering rate and the angular dependence of the scattering between state $\mathbf{k}$ and state $\mathbf{k}'$. The transition rate $S(\mathbf{k}, \mathbf{k}')$ from state $\mathbf{k}$ to state $\mathbf{k}'$ is given by Fermi's golden rule[30]

$$S(\mathbf{k}, \mathbf{k}') = \frac{2\pi}{\hbar} |\langle \mathbf{k}' | H^{(1)} | \mathbf{k} \rangle|^2 \delta[\mathscr{E}(\mathbf{k}') - \mathscr{E}(\mathbf{k}) \pm \Delta\mathscr{E}] \qquad (10.7)$$

where $\langle \mathbf{k}' | H^{(1)} | \mathbf{k} \rangle$ is the matrix element of the perturbing (scattering) potential $H^{(1)}$ between the initial state ($\mathbf{k}$) and the final state ($\mathbf{k}'$) and $\Delta\mathscr{E}$ is the energy loss (+) or gain (−) during the transition. The angular dependence of the scattering is

Table 10.1 Scattering Mechanisms included in Monte Carlo Calculation

1. Acoustic mode (deformation potential) scattering
2. Polar optical scattering
3. Piezoelectric scattering
4. Non-polar optical scattering
5. Equivalent inter-valley scattering (e.g. L–L)
6. Non-equivalent inter-valley scattering (e.g. Γ–L)
7. Ionized-impurity (Coulomb) scattering
8. Alloy (random potential) scattering

then obtained from $S(\mathbf{k}, \mathbf{k}')$ for specific values of $\mathbf{k}$ and $\mathbf{k}'$ as

$$P(\beta)\,d\beta \sim S(\mathbf{k}, \mathbf{k}')\sin\beta\,d\beta \tag{10.8}$$

where β is the angle between $\mathbf{k}$ and $\mathbf{k}'$. Here, $P(\beta)$ is the angular probability density, and the proportionality constant in equation (10.8) is obtained by normalization of $P(\beta)$; i.e.

$$\int_0^{\pi} P(\beta)\,d\beta = 1$$

The total scattering rate out of the state $\mathbf{k}$ is obtained by summation of equation (10.7) over all allowed final states. Thus

$$\frac{1}{\tau(\mathbf{k})} = \sum_{\mathbf{k}'} S(\mathbf{k}, \mathbf{k}') \tag{10.9}$$

where $\tau(\mathbf{k})^{-1}$ is the total scattering rate. The summation over final states can be approximated as an integral by the relation[12, 27, 30]

$$\sum_{\mathbf{k}} S(\mathbf{k}, \mathbf{k}') = \frac{V}{(2\pi)^3}\int_{\mathbf{k}'} S(\mathbf{k}, \mathbf{k}')\,d\mathbf{k}' \tag{10.10}$$

which in spherical coordinates becomes

$$\frac{1}{\tau(\mathbf{k})} = \frac{V}{(2\pi)^3}\int_0^{\infty}\int_0^{\pi}\int_0^{\pi} S(\mathbf{k}, \mathbf{k}')\,k'^2 \sin\beta\,d\phi\,d\beta\,dk' \tag{10.11}$$

where V is the volume of the crystal and ϕ is the azimuthal angle. The total scattering rate can be expressed as a function of energy $\mathscr{E}$ by using equation (10.3).

Equations for the total scattering rates of the scattering mechanisms listed in Table 10.1 are given in several publications.[12–15, 19, 26–28, 30] A good discussion of scattering rates used in Monte Carlo simulations is given by Shichijo.[31] The scattering rates for the mechanisms used in this work are given in Appendix 10.1 for completeness.

The overall electron scattering rate $\lambda(\mathscr{E})$ is an important relation for Monte Carlo calculations. This is given by

$$\lambda(\mathscr{E}) = \sum_{i=1}^{M} \frac{1}{\tau_i(\mathscr{E})} \tag{10.12}$$

where τ_i^{-1} is the scattering rate for the ith scattering mechanism (equation (10.11)) listed in Table 10.1 as a function of electron energy $\mathscr{E}$. The sum is over all scattering events listed in Table 10.1. Figure 10.3 shows the overall scattering rate for $Ga_{0.47}In_{0.53}As$ for doping levels of $1 \times 10^{14}\,cm^{-3}$ and $1 \times 10^{17}\,cm^{-3}$. The scattering rate for GaAs at a doping level of $1 \times 10^{14}\,cm^{-3}$ is shown for comparison.[28] The scattering rates for electron energies up to 1.0 eV are shown in Figure 10.3. During a simulation, an electron can make excursions to high energies. For example, in GaAs with fields of $500\,kV\,cm^{-1}$, the energy excursions

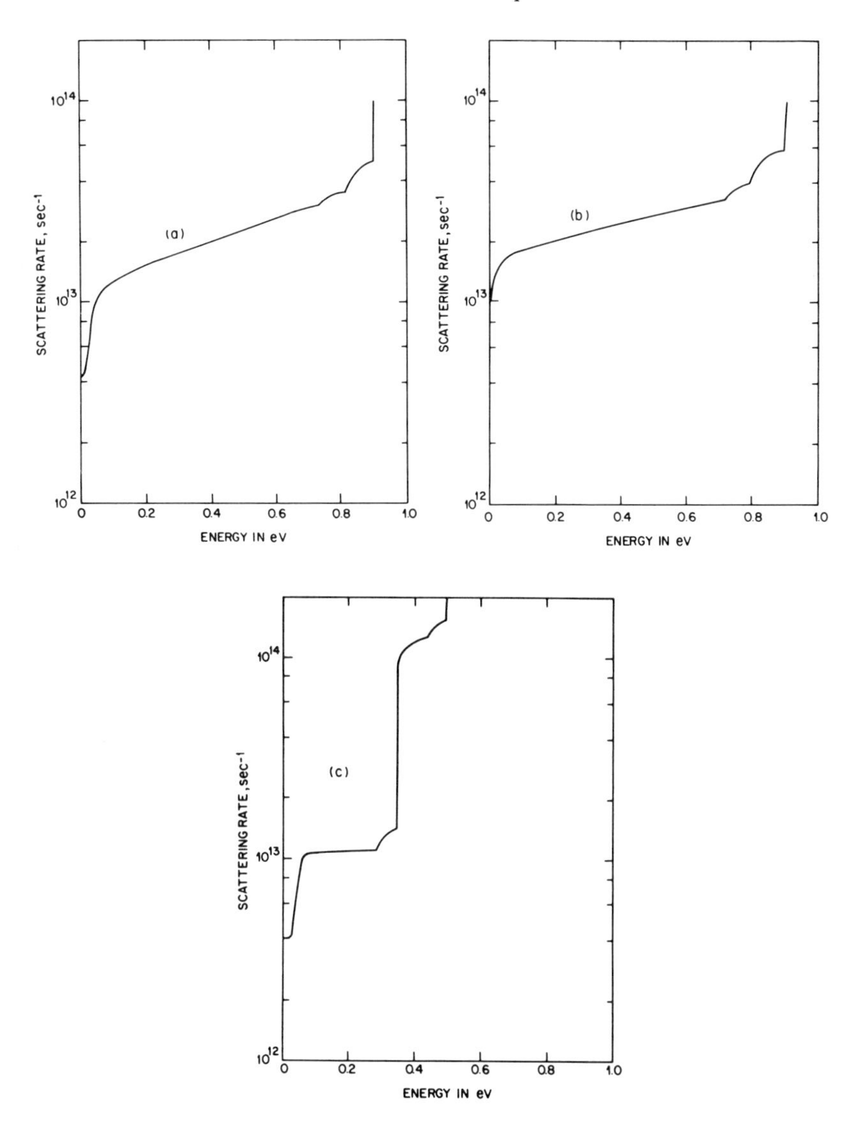

Figure 10.3 Total electron scattering rate versus energy: curve (a), $Ga_{0.47}In_{0.53}As$, $N_D = 10^{14}\ cm^{-3}$; curve (b), $Ga_{0.47}In_{0.53}As$, $N_D = 10^{17}\ cm^{-3}$; curve (c), GaAs, $N_D = 10^{14}\ cm^{-3}$

range from 0.3 eV to 2.0 eV while the average energy is 0.85 eV.[28] For $100\ kV\ cm^{-1}$ field strengths, the excursions are from 0.15 eV to 0.8 eV while the average energy is 0.5 eV.[28] Band structure and avalanche effects become critical for high electron energies, and one must question the simple band structure described in equation (10.4). For the work described here, field strengths of less than 20 kV cm^{-1} will be used. The average energies and the energy excursions should be such that the simple band structure will be a reasonable approximation.

10.2.3 The use of random numbers in the Monte Carlo method

Random numbers from a pseudorandom number generator are used at several points in the Monte Carlo simulation. They are used to select initial conditions for the particle, the time of free flight, the scattering process selected after a free flight, and directional aspects of the final state after conservation of energy and momentum are satisfied. These will be briefly discussed below.

Initial conditions

In the Monte Carlo method, the electron is released with an energy equal to the average energy of a Maxwell–Boltzmann distribution with lattice temperature T. For a single parabolic band

$$\mathscr{E}_0 = \frac{3}{2} k_B T \tag{10.13}$$

where $\mathscr{E}_0$ is the initial energy and k_B is Boltzmann's constant. The momentum $p = \hbar k$ can be calculated from equation (10.3). The momentum p and the lattice temperature T establish the distribution function $f(v_i)$, $i = x, y, z$, for each component of the velocity. The velocity components v_i are then selected from an equation of the form

$$R_n = \int_{-\infty}^{v_i} f(v_i')\,dv_i' \tag{10.14}$$

where R_n is a uniformly distributed random number, $0 \leq R_n \leq 1$. A different random number is used for each velocity component.

The electron is started in the lowest valley although an alternative starting valley can be specified. The initial conditions are not critical for the 'single particle' Monte Carlo method provided the electron undergoes many interactions. Usually we take $N = 25\,000$ and tests are made with $N = 250\,000$ to ensure good statistical estimates. Reasonable initial conditions will reduce the number of interactions required for 'good' results.

Selection of time of free flight

The probability per unit time P_d that an electron will drift for a time t_f and then be scattered is given by[26]

$$P_d(t_f) = \lambda(t_f)\exp\left(-\int_0^{t_f} \lambda(t)\,dt\right), \tag{10.15}$$

where $\lambda(t_f)$ is the overall scattering rate for the particle, i.e. $\lambda(t_f) = \lambda[\mathscr{E}(t_f)]$ from equations (10.3), (10.6), and (10.12). The procedure used to generate random free flight times t_f with a probability distribution given by equation (10.15) from a random number R_n with uniform probability distribution $P(R_n)$ between 0 and 1 is as follows. Since

$$P(R_n)\mathrm{d}R_n = P_d(t)\mathrm{d}t \tag{10.16}$$

then

$$\int_0^{R_n} P(R_n')\mathrm{d}R_n' = \int_0^{t_f} P_d(t)\mathrm{d}t \tag{10.17}$$

which, upon integration, gives

$$R_n = 1 - \exp\left(-\int_0^{t_f} \lambda(t)\mathrm{d}t\right) \tag{10.18}$$

This can be written as

$$-\ln(1 - R_n) = \int_0^{t_f} \lambda(t)\mathrm{d}t \tag{10.19}$$

In this work, equation (10.19) is numerically integrated by a trapezoidal rule to solve for t_f once R_n is generated. Extensive use of look-up tables for $\lambda(\mathscr{E})$ is employed in the numerical integration. We have found this to be more computationally efficient than the use of a 'self-scattering' mechanism employed by many others.[26]

Selection of scattering processes

Once the free flight time is chosen, the scattering mechanism is selected randomly. From the scattering rates, the relative probability P_i of each scattering mechanism is calculated from

$$P_i(\mathscr{E}) = \frac{\lambda_i(\mathscr{E})}{\lambda(\mathscr{E})} \tag{10.20}$$

The relative probabilities are then converted into a scattering probability table by defining M functions as follows.

$$f_k(\mathscr{E}) = \sum_{i=1}^{k} P_i(\mathscr{E}) \qquad k = 1, 2, \ldots, M \tag{10.21}$$

Here M is the total number of scattering mechanisms. Using the scattering probability table stored in memory and generating another uniformly distributed random number R_n $(0 \leq R_n \leq 1)$, the scattering mechanism is selected by solving the equation

$$R_n < \sum_{k=1}^{j} f_k(\mathscr{E}) \tag{10.22}$$

for the minimum value of j. The jth scattering mechanism is thus chosen. The ordering of the scattering mechanisms is unimportant for this ergodic process.[26] According to the ergodic hypothesis,[26, 27] the ensemble average properties are equal to the single-particle time-averaged properties. Thus, estimates for steady-state physical quantities, such as drift velocity, energy, etc., can be obtained from time averages over a single-particle history.

Final state selection

In order to obtain the scattering angle between initial state **k** and final state **k**′, one can generate another random number R_n and solve the equation

$$R_n = \int_0^{\beta} P(\beta')\mathrm{d}\beta' \tag{10.23}$$

where $P(\beta)$ is defined in equation 10.8.[12] We have found it more computationally convenient to determine the scattering wavevector $\mathbf{q}_\mathrm{p}$ (phonon wavevector) in all electron–phonon interactions. The scattering wavevector is completely specified by determining its magnitude q_p, the azimuthal angle ϕ relative to **k**, and the angle Φ between **k** and $\mathbf{q}_\mathrm{p}$. The requirement of momentum conservation can be written as

$$q_0^2 = k^2 + k'^2 - 2kk' \cos\beta \tag{10.24}$$

Then, equation (10.23) can be reformulated as

$$R_n = \frac{\displaystyle\int_{q_{\min}}^{q_\mathrm{p}} S(\mathbf{k}, \mathbf{k}')q'_\mathrm{p}\mathrm{d}q'_\mathrm{p}}{\displaystyle\int_{q_{\min}}^{q_{\max}} S(\mathbf{k}, \mathbf{k}')q'_\mathrm{p}\mathrm{d}q'_\mathrm{p}} \tag{10.25}$$

which can be solved directly for q. The angle θ is then determined from momentum conservation as

$$k'^2 = q_\mathrm{p}^2 + k^2 \pm 2q_\mathrm{p}k\cos\theta \tag{10.26}$$

which gives

$$\cos\theta = \frac{|k'^2 - k^2| \pm q_\mathrm{p}^2}{2kq_\mathrm{p}} \tag{10.27}$$

Here, the upper and lower signs correspond to absorption and emission, respectively. The azimuthal angle ϕ is determined by randomly selecting a value between 0 and 2π radians. Thus

$$\phi = 2\pi R_n \tag{10.28}$$

This then determines the final state $\mathbf{k}'$ through conservation of energy.

10.2.4 Evaluation of material properties

A Monte Carlo calculation including all eight scattering processes of Table 10.1 requires a large number of material parameters. These parameters can be classified as bulk material parameters and band-dependent parameters. The bulk material parameters are listed in Table 10.2 and the band-dependent parameters are listed in Table 10.3. Units given are SI. The parameters listed in Table 10.3 are required for each energy band or each set of equivalent valleys. These are not well-known because they are difficult to measure. However, once these material properties are given and the scattering models are specified, the Monte Carlo method is straightforward.

An interpolation procedure has been used[8] to describe the compositional dependence of quaternary material parameters. This procedure uses the parameters listed in Tables 10.2 and 10.3 for the binary constituents GaAs, GaP, InAs, and InP. (Most material parameters are better known for GaAs and InP[12-16] than for InAs and GaP.[32-34]) From these, we obtain parameters as functions of composition for the four ternary alloys contained in the quaternary. We use either measured values or values calculated by the method described by Harrison and Hauser[35] if no experimental data are available for the ternaries. Often, this results

Table 10.2 Bulk material parameters

1. Lattice constant (m)
2. Electron affinity (J)
3. Piezoelectric constant ($\mathrm{C\,m^{-2}}$)
4. Longitudinal optical phonon energy (J)
5. Material density ($\mathrm{kg\,m^{-3}}$)
6. Acoustic wave velocity ($\mathrm{m\,s^{-1}}$)
7. Low-frequency dielectric constant
8. High-frequency dielectric constant

Table 10.3 Valley-dependent material parameters

1. Acoustic deformation potential (J)
2. Effective mass (kg)
3. Optical deformation potential ($\mathrm{J\,m^{-1}}$)
4. Optical phonon energy (J)
5. Band non-parabolicity parameter ($\mathrm{J^{-1}}$)
6. Inter-valley deformation potential ($\mathrm{J\,m^{-1}}$)
7. Inter-valley phonon energy (J)

in the use of Vegard's law.[36]. The final step in estimating the quaternary parameters is to use an interpolation formula that gives a quaternary material parameter for $Ga_{1-x}In_xAs_yP_{1-y}$, Q(x, y), as a function of (x, y). This formula is[8]

$$Q(x, y) = \frac{x(1-x)[(1-y)T_{12}(x)+yT_{43}(x)]+y(1-y)[(1-x)T_{14}(y)+xT_{23}(y)]}{x(1-x)+y(1-y)} \tag{10.29}$$

Here, T_{12} and T_{43} are parameters for the ternaries along the $y = 0$ and $y = 1$ boundaries of the compositional plane, which are $Ga_{1-x}In_xP$ and $Ga_{1-x}In_xAs$, respectively, and T_{14} and T_{23} are the parameters for the ternaries along the $x = 0$ and $x = 1$ boundaries, which are $GaAs_yP_{1-y}$ and $InAs_yP_{1-y}$, respectively. Equation (10.29) is *ad hoc*: however, it reduces to the ternary expression along the compositional plane boundaries, reduces to the correct expression for the binary materials in the appropriate limit and gives the average of the four binary material parameters for $x = y = 0.5$. When used to obtain parameters for compositions lattice-matched to InP, equation (10.29) depends critically on material parameters for InP, $InAs_{1-y}P_y$, and $Ga_{1-x}In_xAs$. Table 10.4 gives material parameters calculated from equation (10.29) for $Ga_{0.23}In_{0.77}As_{0.5}P_{0.5}$ and (in parentheses) for $Ga_{0.47}In_{0.53}As$.

10.2.5 Experimental material properties

There are few experimental values for the parameters given in Table 10.4. Some values obtained recently by Pearsall[37] for $Ga_{0.47}In_{0.53}As$ are in reasonably good agreement with those of Table 10.4. However, there would be no way to proceed with the hot electron calculations if the only recourse was to experimental data. Thus, we must utilize the interpolation described in section 10.2.4 if we are to continue. The main justifications for using this interpolation method are that the material parameters for the binary compounds are reasonably well-known, and the quaternary material parameters predicted by the interpolation procedure are generally smooth functions of composition. Experimental data measured on ternary and quaternary III–V materials manifest this property. The interpolation formula given by equation (10.29) has been shown to yield reasonable to excellent estimates when applied to energy bandgaps and lattice constants. Thus, we will proceed using the interpolation procedure while attempting to be conservative in predictions made from the calculations. In addition, two particular material parameters which have been studied experimentally are of most interest. These are the central-valley effective mass and the longitudinal optical (LO) phonon energies. These parameters will be discussed below.

Central-valley effective mass

In a previous paper[8] effective-mass values are obtained for $Ga_{1-x}In_xAs_yP_{1-y}$ from equation (10.29) for reciprocal effective mass—a procedure suggested by

Table 10.4 Material properties for $Ga_{0.23}In_{0.77}As_{0.5}P_{0.5}$ and (in parentheses) for $Ga_{0.47}In_{0.53}As$

(a) Bulk material properties

Lattice constant (Å)	5.869	(5.869)
Density (g cm^{-3})	5.18	(5.59)
Piezoelectric constant (C m^{-3})	0.0630	(0.0924)
LO phonon energy (eV)	0.0424	(0.0327)
Longitudinal sound velocity (cm s^{-1})	5.07×10^5	(4.81×10^5)
Optical dielectric constant	10.46	(11.36)
Static dielectric constant	13.03	(13.73)

(b) Valley-dependent material parameters

	Γ(000)		L(111)		X(100)	
Effective mass (m^*/m_0)	0.0605	(0.041)	0.328	(0.291)	0.549	(0.679)
Non-parabolicity (eV^{-1})	0.968	(1.307)	0.634	(0.691)	0.199	(0.202)
Energy bandgap relative to valence band edge (eV)	0.955	(0.717)	1.717	(1.555)	1.989	(1.997)
Acoustic deformation potential (eV)	6.81	(5.89)	8.77	(8.35)	9.41	(9.29)
Optical deformation potential (eV cm^{-1})	0	(0)	3×10^8	(3×10^8)	0	(0)
Optical phonon energy (eV)	–		0.0361	(0.03146)	–	
Number of equivalent valleys	1		4		3	
Intervalley deformation potential (eV cm^{-1})						
from 000	0	(0)	1.13×10^9	(7.35×10^8)	5.59×10^8	(7.35×10^8)
from 111	1.13×10^9	(7.35×10^8)	4.44×10^8	(7.35×10^8)	5.0×10^8	(5×10^8)
from 100	5.59×10^8	(7.35×10^8)	5.0×10^8	(5×10^8)	4.4×10^8	(5.94×10^8)
Inter-valley phonon energy (eV)						
from 000	0	(0)	0.0320	(0.0243)	0.0186	(0.0236)
from 111	0.0320	(0.0243)	0.0322	(0.0238)	0.0321	(0.0272)
from 000	0.0186	(0.0236)	0.0321	(0.0272)	0.0186	(0.0236)

Harrison and Hauser.[35] The results are shown as curve (a) in Figure 10.4. Also shown in Figure 10.4 are a linear interpolation between the InP and $Ga_{0.47}In_{0.53}As$ effective-mass values (curve (b)) and the result of a theoretical calculation (curve (c)) based on $\mathbf{k} \cdot \mathbf{p}$ approximation.[37] Experimental data from various sources are also shown.[37, 39–42]. The interpolated values obtained from equation (10.29) are generally lower than the experimental values by as much as 34 %. Also, the interpolated effective mass has a minimum at $y \simeq 0.9$ which leads to the maximum calculated drift mobility for $y \simeq 0.9$ shown in Figure 10.1. By using the linear effective-mass equation[37]

$$m^*/m_0 = 0.08 - 0.039y \tag{10.30}$$

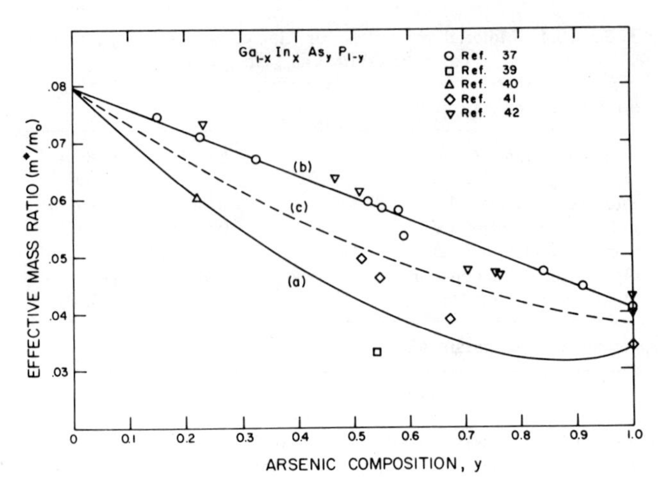

Figure 10.4 Central-valley electron effective-mass ratio versus arsenic composition in InP lattice-matched alloys: curve (a), obtained from equation (10.29); curve (b), linear interpolation, equation (10.30); curve (c), theoretical calculation

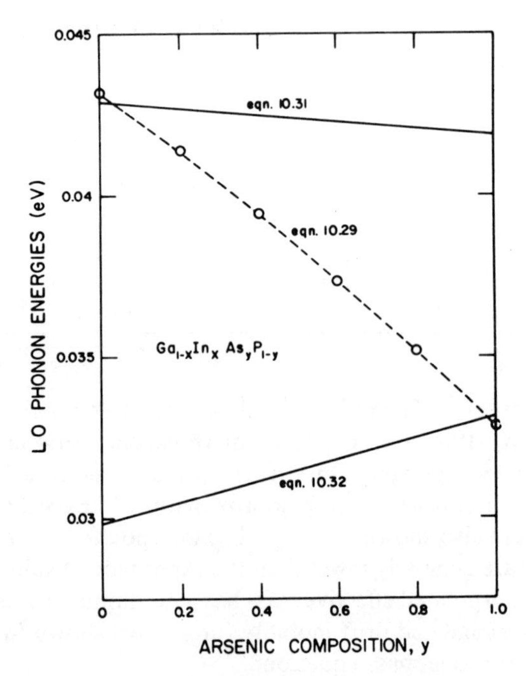

Figure 10.5 Longitudinal optical phonon energies versus arsenic composition in InP lattice-matched alloys

the curve (c) in Figure 10.1 is obtained for a doping density of 1×10^{16} cm^{-3}; all other parameters remain unchanged from the values used to obtain curve (a). A slight minimum at $y = 0.1$ is given by the Monte Carlo calculation. The trend of the experimental data is similar to that of the calculations.

Two-mode LO Phonon scattering

Recently, it has been shown that there are two longitudinal optical phonon modes that are prominent in Raman spectra for $Ga_{1-x}In_xAs_yP_{1-y}$.[43] These two modes are like those observed in InP and $Ga_{0.47}In_{0.53}As$, respectively, and it has been concluded that $Ga_{1-x}In_xAs_yP_{1-y}$ lattice-matched to InP has a pseudo-two-mode scattering mechanism.[43] In the quaternary the strengths of the modes are proportional to the concentrations of InP and GaInAs, and the phonon energies are functions of the fraction of arsenic along the lattice-matched line given by equation (10.2). The phonon energies can be expressed as (in eV)

$$\hbar\omega_1(\mathrm{InP}) = 0.04293 - 0.00107y \qquad (10.31)$$

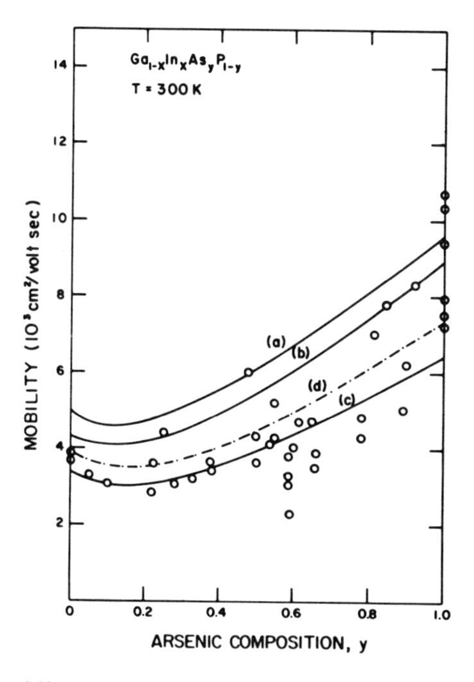

Figure 10.6 Electron drift mobilities versus arsenic composition in InP lattice-matched alloys using a linear effective-mass ratio and scattering by two LO phonons. Curve (a), $N_D = 10^{14}$ cm^{-3}; curve (b), $N_D = 10^{16}$ cm^{-3}; curve (c), $N_D = 10^{17}$ cm^{-3}; curve (d), using Hall factor of 1.15. The circles represent experimental data

and

$$\hbar\omega_2(\text{GaInAs}) = 0.02981 + 0.00338y \qquad (10.32)$$

Figure 10.5 shows a plot of these energies along with the single-mode energy obtained from equation (10.29). While the InP value ($y = 0$) and $Ga_{0.47}In_{0.53}As$ value ($y = 1$) obtained from equation (10.29) are in good agreement with experimental values,[43] it is obvious that the effects of single-mode scattering could differ considerably from those of two model LO scattering for the quaternary compositions.

Two-mode LO phonon scattering is included in the Monte Carlo calculations presented here. Figure 10.6 shows calculated ($100\ \text{V cm}^{-1}$) and measured low-field drift mobility for $Ga_{1-x}In_xAs_yP_{1-y}$. The calculated mobilities are in good agreement with the experimental data. The calculation is described in detail in the next section. The small differences between the measured and calculated values can be attributed to carrier compensation or alloy scattering. We conclude that revising effective-mass values to the linear variation of Figure 10.4 and using two-mode LO phonon scattering explains the measured low-field mobility of $Ga_{1-x}In_xAs_yP_{1-y}$. In Chapter 9, Dr Takeda uses the iterative integral to calculate low-field mobility. Our results are in excellent agreement with his. Figure 10.6 provides a point of departure for the discussion of hot electron effects in section 10.3.

10.2.6 Alloy scattering

Alloy scattering (random potential scattering) is an important physical mechanism which significantly influences transport in ternary and quaternary III–V compounds.[8,19,35,36] It is especially important in hot electron transport in $Ga_{1-x}In_xAs_yP_{1-y}$.[8] There has been no change in the status of theoretical considerations for alloy scattering since our earlier publication.[19] It is still not clear which alloy scattering potential model should be used to describe alloy scattering in a given material. There are three possible scattering potentials proposed, which are based on energy gap differences, electron affinity differences, and electronegativity differences of the binary constituents. In Chapter 9 of this volume, Takeda concludes that the model based on electronegativity differences gives the best fit between experiment and theory for low-field drift mobility in $Ga_{0.47}In_{0.53}As$.

In this chapter, we will not attempt to vary the alloy scattering potential. We will in general use the maximum amount of alloy scattering so that conservative estimates for velocity field characteristics can be made. In light of the uncertainty in material parameters it seems unwise to introduce another uncertainty into the simulations.

10.3 HIGH-FIELD CALCULATIONS FOR $Ga_{1-x}In_xAs_yP_{1-y}$

We have found that a linear variation of effective mass and two-mode LO phonon scattering bring the calculated low-field mobility of $Ga_{1-x}In_xAs_yP_{1-y}$ (lattice-matched to InP) into agreement with available experimental data. This gives us some confidence in the hot electron results given below. For example, Monte Carlo calculations for GaAs give results that agree with experimental data, even though many of the parameters of the scattering processes and the band structure have not been measured.[16] However, one must view these results critically until more experimental data (particularly material properties) become available. Meanwhile, these calculations for the quaternary provide useful information and some basis for comparison of future experimental data. The results presented in this section are obtained using the parameters listed in Table 10.4 and the effective masses and phonon energies described in section 10.2.5.

10.3.1 Two-Mode phonon scattering in $Ga_{1-x}In_xAs_yP_{1-y}$

The scattering rate (the number of scattering events per unit time) τ_{PO}^{-1} for electron–LO phonon interaction is given by[12–15,31]

$$\frac{1}{\tau_{PO}(\omega_{LO})} = \frac{qm^{*1/2}(\hbar\omega_{LO})}{4\pi\varepsilon_0\sqrt{2}\hbar^2}\left(\frac{1}{\varepsilon_\infty}-\frac{1}{\varepsilon_s}\right)\frac{d\gamma(\mathscr{E})/d\mathscr{E}}{[\gamma(\mathscr{E})]^{1/2}}F_0(\mathscr{E},\mathscr{E}') \times \begin{cases} N_0 & \text{absorption} \\ (N_0+1) & \text{emission} \end{cases} \tag{10.33}$$

where q is electronic charge, $\hbar$ is Planck's constant, m^* is band-edge effective mass, $\hbar\omega_{LO}$ is LO phonon energy, ε_0 is permittivity of free space, ε_∞ and ε_s are optical and static dielectric constants, respectively, $\gamma(\mathscr{E})$ is given by equation (10.3), and N_0 is average number of optical phonons given by

$$N_0 = \frac{1}{\exp(\hbar\omega_{LO}/k_BT)-1} \tag{10.34}$$

where k_B is Boltzmann's constant and T is absolute (lattice) temperature. The function $F_0(\mathscr{E},\mathscr{E}')$ accounts for admixture of p-type wavefunctions, called p-state mixing.[12] Here

$$\mathscr{E}' = \mathscr{E} \pm \hbar\omega_{LO} \tag{10.35}$$

where $\mathscr{E}'$ is energy of the final state, $\mathscr{E}$ is the energy of the initial state, and the $+(-)$ sign indicates absorption(emission) of an LO phonon.

To include two-mode LO-phonon scattering described by equations (10.31) and (10.32), accounting for compositional dependence of the mode strengths (43), two scattering rates are defined as follows:

$$\frac{1}{\tau_{P1}} = \frac{x(1-y)}{\tau_{PO}(\omega_1)} \tag{10.36}$$

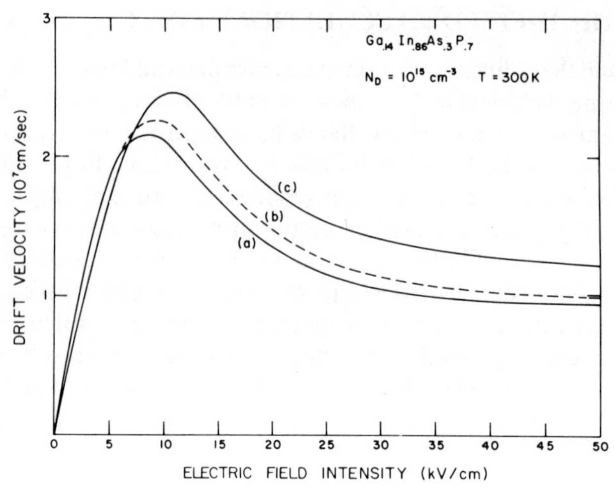

Figure 10.7 Electron drift velocity versus electric field intensity for $Ga_{0.14}In_{0.86}As_{0.3}P_{0.7}$. Curve (a), two LO phonons; Curve (b), one LO phonon; curve (c), InP. $N_D = 10^{15}\ cm^{-3}$ for each curve

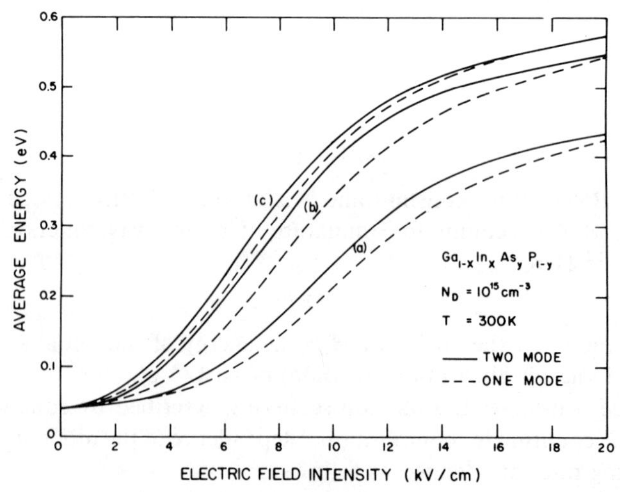

Figure 10.8 Average electron energy versus electric field intensity with two-mode (full curve) and one-mode (broken curve) LO phonon scattering. Curve (a), $y = 0.3$; curve (b), $y = 0.7$; curve (c), $y = 0.9$. $N_D = 10^{15}\ cm^{-3}$ for each curve

and

$$\frac{1}{\tau_{P2}} = \frac{y}{\tau_{PO}(\omega_2)} \tag{10.37}$$

where ω_1 (10.31) is used for InP-like LO-phonon scattering and ω_2 (10.32) is used for $Ga_{0.47}In_{0.53}As$-like LO-phonon scattering. The compositional dependences in equations (10.36) and (10.37) account for the variation with composition of the Raman intensities.[43]

These equations have been used in a Monte Carlo calculation of velocity–field (v–E) characteristics for $Ga_{1-x}In_xAs_yP_{1-y}$. Figure 10.7 shows results for $y = 0.3$, $x = 0.859$ with and without two-mode scattering. Also shown is a v–E characteristic for InP. Two-mode scattering generally decreases the drift velocity near and above threshold and decreases the threshold field as well. Low-field mobility is also reduced although this is not readily apparent from Figure 10.7 (see Figure 10.6).

Figure 10.8 shows the average energy in the Γ band for $y = 0.3, 0.7$, and 0.9. Two-mode scattering increases the average energy over that calculated using single-mode scattering until inter-valley scattering comes into play. This means that two-mode scattering has a large effect on the energy distribution. This is

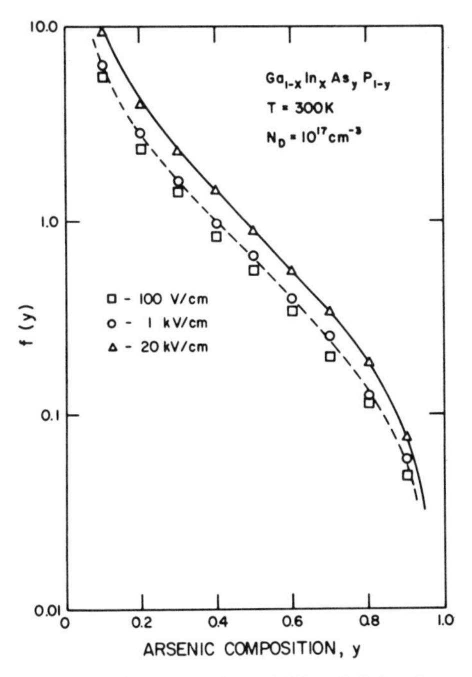

Figure 10.9 Monte Carlo data verifying equation (10.38) as defining the average ratio of scattering events for the two LO-phonon modes

expected, because for large energies, polar scattering strongly favours small scattering angles.[31]

The ratio of the scattering rates for the two LO modes is given by the ratio of equation (10.36) to equation (10.37). At high energies, $F_0(\mathscr{E}, \mathscr{E}')$ in equation (10.33) is nearly the same for both modes; in that case, the ratio of LO scattering rates is given by

$$f(y) = \frac{x(1-y)}{y} \frac{\omega_1(y)}{\omega_2(y)} \frac{\exp[\hbar\omega_2(y)/kT]-1}{\exp[\hbar\omega_1(y)/kT]-1} \tag{10.38}$$

where x and y are related by equation (10.2). Figure 10.9 shows this ratio and Monte Carlo calculations for the ratio of the percentage of scattering events in the InP mode to those in the $Ga_{0.47}In_{0.53}As$ mode (here the ratio is for polar optical emission). At low fields the functions $F_0(\mathscr{E}, \mathscr{E}')$ are not the same for the two modes; in that case the Monte Carlo data differ from equation (10.38) by approximately a constant factor. However, at high fields (high average energies) the scattering rates are nearly independent of the phonon energies and the ratio of the scattering events in the two LO modes is given by equation (10.38); this is essentially the ratio of the Raman intensities observed experimentally by Pinczuk *et al.*[43]

Finally, Figure 10.10 shows the velocity–field characteristics for $Ga_{1-x}In_xAs_yP_{1-y}$ for several other values of y. Also shown is a v–E characteristic for GaAs. It appears that two-mode LO-phonon scattering in $Ga_{1-x}In_xAs_yP_{1-y}$ reduces the low-field mobility, the peak velocity, the

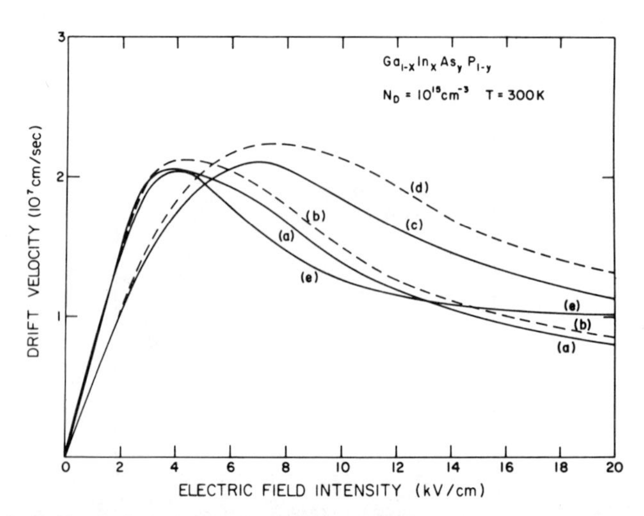

Figure 10.10 Drift velocity versus electric field intensity for InP lattice-matched alloys. Curve (a), $y = 0.9$, two LO modes; curve (b), $y = 0.9$, one LO mode; curve (c), $y = 0.5$, two LO modes; curve (d), $y = 0.5$, one LO mode; curve (e), GaAs, for comparison. $N_D = 10^{15}\ cm^{-3}$ for all curves

threshold field, and the saturated drift velocity. High-field drift velocities for $Ga_{1-x}In_xAs_yP_{1-y}$ including two LO-phonon modes and linear variation of effective mass are considerably lower than those predicted by our original calculations.[8].

10.3.2 High-field transport in $Ga_{0.47}In_{0.53}As$

It appears that $Ga_{0.47}In_{0.53}As$ has the largest low-field mobility of any composition of $Ga_{1-x}In_xAs_yP_{1-y}$ lattice-matched to InP. This section presents some new calculations for high-field transport properties of this ternary compound. The material parameters listed in Table 10.4 are used here, but trends are shown for ranges of uncertain material parameters.

Figure 10.11 shows the variation of the velocity–field characteristics for $Ga_{0.47}In_{0.53}As$ with central-valley effective mass. This parameter obviously has a major effect on both hot electron transport and low-field drift mobility. At the present time, $m_\Gamma^* = 0.041\,m_0$ appears to be the best experimental value.[37] Figure 10.12 shows the velocity–field characteristic for $Ga_{0.47}In_{0.53}As$ with $m_\Gamma^* = 0.041\,m_0$ with LO-phonon energy as a parameter. As the LO-phonon energy increases, the velocity increases at and above the threshold field; this is expected from a simple model for polar optical scattering.[30] Figure 10.13 shows the effect of doping concentration on the velocity–field characteristic for $Ga_{0.47}In_{0.53}As$ using $m_\Gamma^* = 0.041\,m_0$, $\hbar\omega_{LO} = 0.0327$ eV, and the other parameters given in Table 10.4. The influence of doping is small compared to that of m^* and $\hbar\omega_{LO}$.

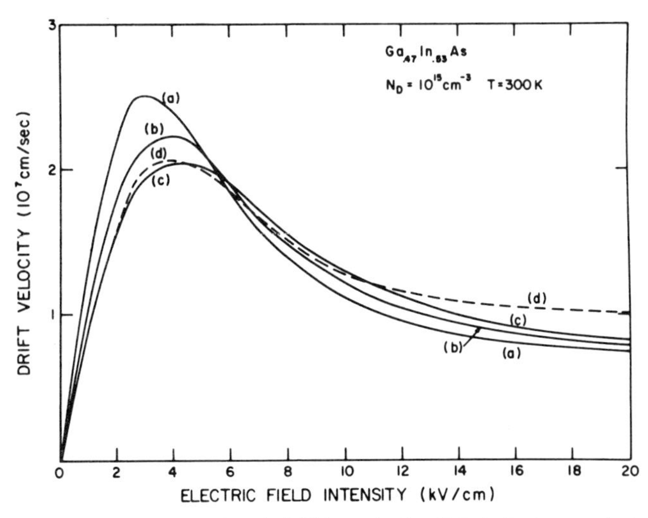

Figure 10.11 Drift velocity versus electric field intensity for GaInAs lattice-matched to InP. Curve (a), $m_\Gamma^* = 0.033m_0$; curve (b), $m_\Gamma^* = 0.037m_0$; curve (c), $m_\Gamma^* = 0.041m_0$; curve (d), GaAs, for comparison. $N_D = 10^{15}$ cm^{-3} for all curves

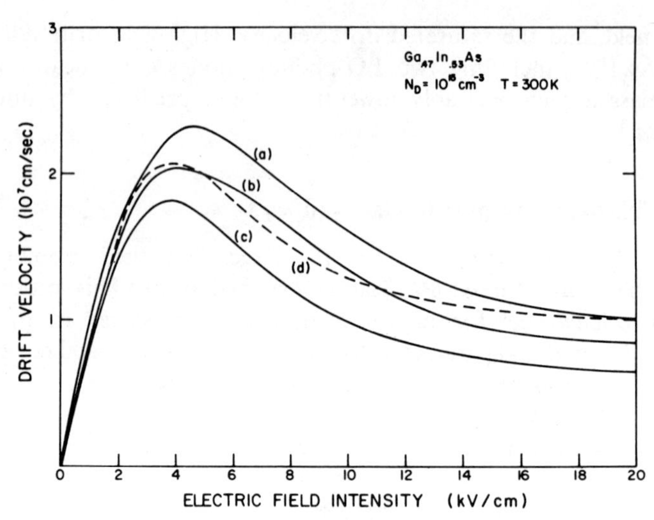

Figure 10.12 Drift velocity versus electric field intensity for GaInAs lattice-matched to InP. Curve (a), $\hbar\omega_{\mathrm{LO}} = 38$ meV; curve (b), $\hbar\omega_{\mathrm{LO}} = 33$ meV; curve (c), $\hbar\omega_{\mathrm{LO}} = 28$ meV; curve (d), GaAs for comparison. $N_{\mathrm{D}} = 10^{15}\ \mathrm{cm}^{-3}$ for all curves

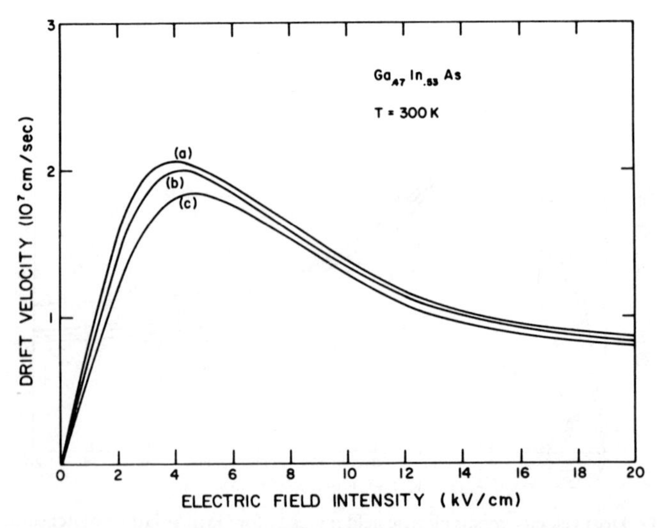

Figure 10.13 Drift velocity versus electric field intensity for GaInAs lattice-matched to InP. Curve (a), $N_{\mathrm{D}} = 0$; curve (b), $N_{\mathrm{D}} = 10^{16}\ \mathrm{cm}^{-3}$; curve (c), $N_{\mathrm{D}} = 10^{17}\ \mathrm{cm}^{-3}$

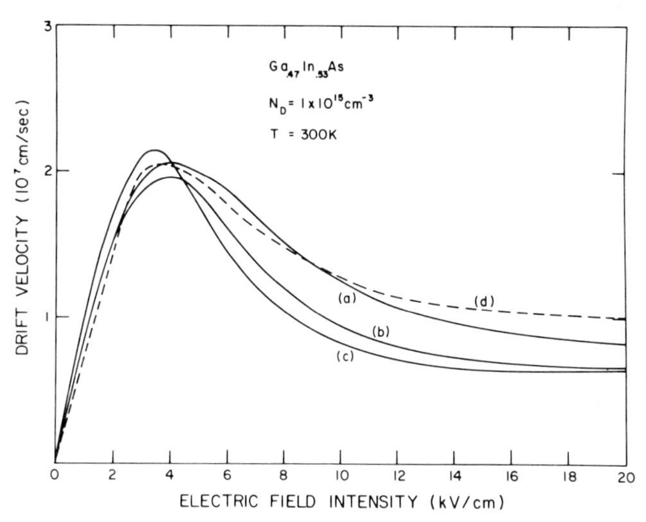

Figure 10.14 Drift velocity versus electric field intensity for GaInAs lattice-matched to InP. Curve (a), $m_\Gamma^* = 0.041 m_0$, $\Delta\mathscr{E}_{\Gamma\text{-L}} = 0.82$ eV; curve (b), $m_\Gamma^* = 0.041 m_0$, $\Delta\mathscr{E}_{\Gamma\text{-L}} = 0.48$ eV; curve (c), $m_\Gamma^* = 0.037 m_0$, $\Delta\mathscr{E}_{\Gamma\text{-L}} = 0.48$ eV; curve (d), GaAs, for comparison. $N_D = 10^{15}$ cm^{-3}

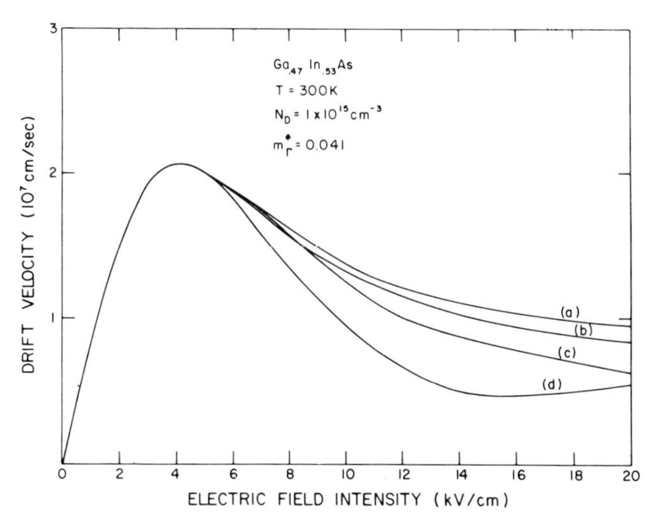

Figure 10.15 Drift velocity versus electric field intensity for GaInAs lattice-matched to InP. Curve (a), $m_L^* = 0.1\ m_0$, $D_{\Gamma L} = 2 \times 10^9$ eV cm^{-1}; curve (b), $m_L^* = 0.5\ m_0$, $D_{\Gamma L} = 2 \times 10^9$ eV cm^{-1}; curve (c), $m_L^* = 0.1\ m_0$, $D_{\Gamma L} = 2 \times 10^8$ eV cm^{-1}; curve (d), $m_L^* = 0.5\ m_0$, $D_{\Gamma L} = 2 \times 10^8$ eV cm^{-1}, $N_D = 10^{15}$ cm^{-3}

Table 10.5 Predicted transport properties for $Ga_{0.47}In_{0.53}As$ at a doping level (uncompensated) of 10^{15} cm^{-3}

1. Low-field drift mobility (cm^2 V^{-1} s^{-1})	9000–9500
2. Peak drift velocity (cm s^{-1})	$(2.0–2.2) \times 10^7$
3. Threshold field (V cm^{-1})	$(3–4) \times 10^3$
4. Saturated drift velocity (cm s^{-1}) (at $E = 20$ kV cm^{-1})	$(0.5–1.2) \times 10^7$

The energy separation $\Delta\mathscr{E}_{\Gamma\text{-L}}$ of the Γ and the L valleys in the ternary $Ga_{0.47}In_{0.53}As$ is an uncertain material parameter. Values from about 0.8 eV to 0.5 eV have been suggested and used by various authors.[8,10,44] Figure 10.14 shows the effect on the velocity–field characteristics of making $\Delta\mathscr{E}_{\Gamma\text{-L}} = 0.48$ eV. These results show a small effect on peak velocity and threshold field and a fairly large effect on saturated velocity.

Figure 10.15 shows the velocity–field characteristic with the L band effective mass and the coupling constant for Γ–L inter-valley scattering as parameters. These parameters largely determine the saturation velocity; the values used in Figure 10.15 are representative of those for many III–V compounds.[12–16]

From the results presented in Figures 10.11–10.15 for $Ga_{0.47}In_{0.53}As$, we obtain the values given in Table 10.5. The values in Table 10.5 appear to be reasonable values based on experience with GaAs and InP materials. We believe that these values indicate the ranges of values that could be measured for this compound. However, a final verdict must await measurements of material parameters for this alloy.

10.4 SUMMARY

This chapter has presented Monte Carlo calculations of hot electron transport in $Ga_{1-x}In_xAs_yP_{1-y}$ lattice-matched to InP. The required material parameters were obtained using an interpolation procedure, although experimentally determined material parameters were used when they were available. Central-valley effective-mass values were obtained from experiments reported in the literature. Experimental LO-phonon energies were used, and two dominant LO-phonon models observed in Raman spectra were included in the Monte Carlo calculations.

The low-field drift mobility was also studied with the Monte Carlo method. The calculated drift mobilities are in general agreement with experimental Hall mobilities when experimental central-valley effective-mass values and LO-phonon energies are used in the calculations. The theoretical and experimental mobilities reach their maximum values for the ternary lattice-matched composition $Ga_{0.47}In_{0.53}As$. Previous large discrepancies between experiment and theory for ohmic mobilities of $Ga_{1-x}In_xAs_yP_{1-y}$ are apparently resolved.

The material parameters which result in agreement between theoretical and experimental electron mobilities provide a basis for hot electron properties calculated in this article. The experimental effective-mass values and the two LO-phonon modes have a significant effect on the hot electron calculations. The use of these parameters results in the decrease of peak velocity, threshold field, and saturated drift velocity of $Ga_{1-x}In_xAs_yP_{1-y}$.

There are many other material parameters which affect hot electron transport in $Ga_{1-x}In_xAs_yP_{1-y}$. The parameters of the upper energy bands are essential for accurate transport calculations. These parameters for $Ga_{1-x}In_xAs_yP_{1-y}$ are unknown. In this article, we have used a range of parameters for the upper energy bands which are based on experience with GaAs and InP. From this range of material parameters, trends in the hot electron transport can be established. For example, for $Ga_{0.47}In_{0.53}As$ with a doping density of $10^{15}\,cm^{-3}$, we calculate a low-field drift mobility of 9000–9500 $cm^2\,V^{-1}\,s^{-1}$, a peak drift velocity of $(2\text{–}2.2) \times 10^7\,cm\,s^{-1}$, a threshold field strength of $(3\text{–}4) \times 10^3\,V\,cm^{-1}$, and a saturation velocity of $(0.5\text{–}1.2) \times 10^7\,cm\,s^{-1}$. The exactness of these predicted values must remain under careful scrutiny until further experimental measurements of material parameters and transport properties are made.

$Ga_{1-x}In_xAs_yP_{1-y}$ will play an important role in future electronic devices. The hot electron transport properties of this material will be a key factor in the performance of these devices. Improvements in materials technology for $Ga_{1-x}In_xAs_yP_{1-y}$ have been impressive. As material quality increases, key experimental data will become available which will allow future refinements of the hot electron calculations presented here.

ACKNOWLEDGEMENTS

The authors acknowledge the financial support of this work by the Office of Naval Research, Arlington, VA. The technical contributions by Mr Kenny Williams and the expert assistance in manuscript preparation by Ms Nancy Tyson are greatly appreciated.

REFERENCES

1. A. R. Clawson, 'Bibliography on the quaternary III–V semiconductor InGaAsP', *Technical Note* 830, Naval Ocean Systems Center, San Diego, CA 95152, February 1980.
2. H. C. Casey, Jr and M. B. Panish, *Heterostructure Lasers*, part B, *Materials and Operating Characteristics*, Academic Press, New York, 1978, p. 2.
3. R. E. Nahory, M. A. Pollack, W. D. Johnston, Jr and R. L. Barns, 'Band-gap versus composition and demonstration of Vegard's law for $In_{1-x}Ga_xAs_yP_{1-y}$ lattice matched to InP', *Appl. Phys. Lett.*, **33**, 659–661, 1978.
4. C. J. Neuse, 'III–V alloys for optoelectronic applications', *J. Electron. Mater.*, **6**, 253–293, 1977.
5. N. Holonyak, Jr, G. E. Stillman, and C. M. Wolfe, 75th Anniversary Review Series—

Compound Semiconductors, *J. Electrochem. Soc.*, **125**, 487C–499C, 1978.
6. A. Y. Cho, 'Recent Developments in molecular-beam epitaxy', *J. Vac. Sci. Technol.*, **16**, 275–284, 1979.
7. H. H. Wieder, 'Problems and prospects of compound semiconductor field-effect transistors', *J. Vac. Sci. Technol.*, **17**, 1009–1018, 1980.
8. M. A. Littlejohn, J. R. Hauser, and T. H. Glisson, 'Velocity-field characteristics of $Ga_{1-x}In_xP_{1-y}As_y$ quaternary alloys', *Appl. Phys. Lett.*, **30**, 242–244, 1977.
9. J. R. Hauser, T. H. Glisson, and M. A. Littlejohn, 'Negative resistance and peak velocity in the central (000) valley of III–V semiconductors', *Solid-State Electron.*, **22**, 487–493, 1979.
10. H. H. Wieder, 'Material options for field effect transistors', *J. Vac. Sci. Technol.*, **18**, 827–838, 1981.
11. Y. Shinada, M. Okamura, E. Yamaguchi, and T. Kobayashi, 'InGaAsP *n*-channel inversion-mode MOSFET', *Jpn. J. Appl. Phys.*, **19**, 2301–2302, 1980.
12. W. Fawcett, A. D. Boardman, and S. Swain, 'Monte-Carlo determination of electron-transport properties in gallium-arsenide', *J. Phys. Chem. Solids*, **31**, 1963–1990, 1970.
13. W. Fawcett and D. C. Herbert, 'High-field transport in gallium-arsenide and indium-phosphide', *J. Phys. C: Solid State Phys.*, **7**, 1641–1654, 1974.
14. D. C. Herbert, W. Fawcett, and C. Hilsum, 'High-field transport in indium-phosphide', *J. Phys. C: Solid State Phys.*, **9**, 3369–3975, 1976.
15. T. J. Maloney and J. Frey, 'Transient and steady-state electron-transport properties of GaAs and InP', *J. Appl. Phys.*, **48**, 781–787, 1977.
16. M. A. Littlejohn, J. R. Hauser, and T. H. Glisson, 'Velocity-field characteristics of GaAs with gamma–Γ_6^c–L_6^c–X_6^c conduction-band ordering', *J. Appl. Phys.*, **48**, 4587–4590, 1977.
17. G. Ottavian, L. Reggiani, C. Canali, F. Nava, and A. Alberigi Quartanta, 'Hole drift velocity in silicon', *Phys. Rev.*, B, **12**, 3318–3329, 1975.
18. M. A. Littlejohn, J. R. Hauser, and T. H. Glisson, 'Monte Carlo calculation of transport properties of GaN', *Appl. Phys. Lett.*, **26**, 625–627, 1975.
19. M. A. Littlejohn, J. R. Hauser, T. H. Glisson, D. K. Ferry, and J. W. Harrison, 'Alloy scattering and high field transport in ternary and quaternary III–V semiconductors', *Solid-State Electron.*, **21**, 107–114, 1978.
20. R. F. Leheny, R. E. Nahory, M. A. Pollack, A. A. Ballman, E. D. Beebe, J. C. DeWinter, and R. J. Martin, '$In_{0.53}Ga_{0.47}As$ junction field-effect transistor', *IEEE Electron. Device. Lett.*, **ED1**, 110–111, 1980.
21. P. D. Greene, S. A. Wheeler, A. R. Adams, A. N. El-Sabbahy, and C. N. Ahmad, 'Background carrier concentrations and electron-mobility in LPE $In_{1-x}Ga_xAs_yP_{1-y}$ layers', *Appl. Phys. Lett.*, **35**, 78–80, 1979.
22. A. R. Adams, H. L. Latham, J. R. Hayes, A. N. El-Sabbahy, and P. D. Greene, 'Evidence for alloy scattering from pressure-induced changes of electron mobility in $In_{1-x}Ga_xAs_yP_{1-y}$', *Electron. Lett.*, **16**, 560–561, 1980.
23. J. D. Oliver and L. F. Eastman, 'Liquid phase epitaxial growth and characterization of high purity lattice matched $Ga_xIn_{1-x}As$ on $\langle 111 \rangle$B InP', *J. Electron. Mater.*, **9**, 693–712, 1980.
24. B. Houston, J. B. Restorff, R. F. Allgaier, and J. R. Burke, 'Hot electron and magneto-transport properties of $In_{1-x}Ga_xP_{1-y}As_y$ liquid phase epitaxial films', *Solid-State Electron.*, **21**, 91–94, 1978.
25. Y. Takeda and A. Sasaki, 'Hall mobility and hall factor of $In_{0.53}Ga_{0.47}As$', *Jpn. J. Appl. Phys.*, **19**, 383–384, 1980.
26. W. Fawcett, *Electronic Transport in Crystalline Solids*, ed. A. Salam International Atomic Energy Agency, Vienna, 1973, pp. 531–618.

27. B. R. Nag, *Electron Transport in Compound Semiconductors, Springer Series in Solid-State Sciences*, vol. 11, Springer-Verlag, Berlin, 1980.
28. H. Shichijo and K. Hess, 'Band structure-dependent transport and impact ionization in GaAs', *Phys. Rev.*, **23**, 4197–4207, 1981.
29. J. R. Barker and D. K. Ferry, 'On the physics and modeling of small semiconductors', *Solid-State Electron.*, **23**, 519–530, 1980.
30. 30. E. M. Conwell, 'High field transport in semiconductors', *Solid-State Phys., Suppl.* **9**, Academic Press, New York, 1967.
31. H. Shichijo, 'Theoretical studies on high field transport in III–V semiconductors', *PhD Thesis*, University of Illinois at Champaign-Urbana, Urbana, Il, September 1980.
32. D. L. Rode, 'Electron mobility in Ge, Si, and GaP', *phys. Stat. Sol.*, B, **53**, 245–253, 1972.
33. D. K. Ferry, 'High-field transport in wide-gap semiconductors', *Phys. Rev.*, **12**, 2361–2369, 1975.
34. J. R. Hauser, M. A. Littlejohn, and T. H. Glisson, 'Velocity-field relationship of InAs–InP alloys including effects of alloy scattering', *Appl. Phys. Lett.*, **28**, 458–461, 1976.
35. J. W. Harrison and J. R. Hauser, 'Alloy scattering in ternary III–V compounds', *Phys. Rev.*, B, **13**, 5347–5350, 1976.
36. J. W. Harrison and J. R. Hauser, 'Theoretical calculations of electron-mobility in 3–5 compounds', *J. Appl. Phys.*, **47**, 292–300, 1976.
37. T. P. Pearsall, '$Ga_{0.47}In_{0.53}As$: a ternary semiconductor for photodetector applications', *IEEE J. Quantum Electron.*, **OE-16**, 709–720, 1980.
38. T. H. Glisson, J. R. Hauser, M. A. Littlejohn, and C. K. Williams, 'Energy bandgap and lattice constant contours of III–V quarternary alloys', *J. Electron. Mater.*, **7**, 1–15, 1978.
39. T. Nishino, Y. Yamazoe, and Y. Hamahawa, 'Electroreflectance of $In_{0.79}Ga_{0.21}As_{0.54}P_{0.46}$', *Appl. Phys. Lett.*, **33**, 861–862, 1978.
40. J. B. Restorff, B. Houston, J. R. Burke, and R. E. Hayes, 'Measurements of effective mass in $In_{0.9}Ga_{0.1}As_{0.22}P_{0.78}$ by Shubnikov–De Haas oscillations', *Appl. Phys. Lett.*, **32**, 189–190, 1978.
41. H. Brendecke, H. L. Störmer, and R. J. Nelson, 'Cyclotron-resonance in n-type $In_{1-x}Ga_xAs_yP_{1-y}$', *Appl. Phys. Lett.*, **35**, 772–774, 1979.
42. J. B. Restorff, B. Houston, R. S. Allgaier, M. A. Littlejohn, and S. B. Phatak, 'The electron effective mass in $In_{1-x}Ga_xAs_yP_{1-y}$ from Shubnikov–DeHaas measurements', *Appl. Phys. Lett.*, **51**, 2277–2278, 1980.
43. A. Pinczuk, J. M. Worlock, R. E. Nahory, and M. A. Pollack, 'Lattice vibrations of $In_{1-x}Ga_xAs_yP_{1-y}$ quaternary compounds', *Appl. Phys. Lett.*, **33**, 461–463, 1978.
44. D. K. Ferry, 'Material considerations for advances in submicron VLSI', *Advances in Electronics and Electron Physics*, Academic Press, New York, 1982, pp. 311–390.

APPENDIX 10.1 SCATTERING RATES FOR MONTE CARLO CALCULATIONS

This appendix lists the scattering rates used in the Monte Carlo method discussed in this chapter. For more detailed discussion of these scattering rates, the reader is referred to references 12, 19, 26, 27, 30, 31.

A. Acoustic phonon scattering

The acoustic phonon scattering rate is givenby

$$\frac{1}{\tau_{AC}} = \frac{Z_1^2 k_B T (m^*)^{3/2}}{\sqrt{2}\pi\hbar^4 \rho v_s^2} [\gamma(\mathscr{E})]^{1/2} \frac{d\gamma(\mathscr{E})}{d\mathscr{E}} F_a(\mathscr{E}) \tag{A.1}$$

where

$$F_a(\mathscr{E}) = \frac{(1+\alpha\mathscr{E})^2 + \frac{1}{3}(\alpha\mathscr{E})^2}{(1+2\alpha\mathscr{E})^2} \tag{A.2}$$

In this equation, Z_1 is the acoustic deformation potential, ρ is material density, and v_s is the velocity of sound in the medium.

B. Non-polar optical phonon scattering

The non-polar optical phonon scattering rate is given by

$$\frac{1}{\tau_{NPO}} = \frac{D_0^2 (m^*)^{3/2}}{\sqrt{2}\pi\hbar^2 \rho(\hbar\omega_0)} [\gamma(\mathscr{E}')]^{1/2} \frac{d\gamma(\mathscr{E}')}{d(\mathscr{E}')} \times \begin{cases} N_0 & \text{absorption} \\ (N_0+1) & \text{emission} \end{cases} \tag{A.3}$$

when D_0 is the optical deformation potential, $\hbar\omega_0$ is the optical phonon energy, and N_0 is the optical phonon occupaion number. In equation (A.3) $\mathscr{E}'$ is the final state energy given by

$$\mathscr{E}' = \begin{cases} \mathscr{E} + \hbar\omega_0 & \text{absorption} \\ \mathscr{E} - \hbar\omega_0 & \text{emission} \end{cases} \tag{A.4}$$

C. Polar optical phonon scattering

The polar optical phonon&scattering rate is given by

$$\frac{1}{\tau_{PO}} = \frac{q^2 \omega_{LO} (m^*)^{1/2}}{4\pi\sqrt{2}\varepsilon_0 \hbar} \left(\frac{1}{\varepsilon_\infty} - \frac{1}{\varepsilon_s}\right) \frac{d\gamma(\mathscr{E})/d\mathscr{E}}{[\gamma(\mathscr{E})]^{1/2}} F_0(\mathscr{E}, \mathscr{E}') \times \begin{cases} N_0 & \text{absorption} \\ (N_0+1) & \text{emission} \end{cases} \tag{A.5}$$

where

$$\mathscr{E}' = \begin{cases} \mathscr{E} + \hbar\omega_{LO} & \text{absorption} \\ \mathscr{E} - \hbar\omega_{LO} & \text{emission} \end{cases} \tag{A.6}$$

Here, $\hbar\omega_{LO}$ is the longitudinal optical (LO) phonon energy, ε_0 is freee space permittivity, ε_∞ is the optical dielectric constant, and ε_s is the static dielectric constant. The function $F_0(\mathscr{E}, \mathscr{E}')$ takes p-state mixing into consideration.[12]

D. Piezoelectric scattering

The piezoelectric scattering rate is given by

$$\frac{1}{\tau_{PZ}} = \frac{(m^*)^{1/2} k_B T}{4\sqrt{2\pi} C_1 \hbar^2} \left(\frac{q P_Z}{\varepsilon_s \varepsilon_0}\right)^2 \frac{d\gamma(\mathscr{E})/d\mathscr{E}}{[\gamma(\mathscr{E})]^{1/2}} \ln\left(1 + \frac{8m^*}{\hbar^2} a^2 \gamma(\mathscr{E})\right) \tag{A.7}$$

where $C_l = \rho v_s^2$ is the longitudinal elastic constant, $\mathbf{P_Z}$ is the piezoelectric constant, and a is a screening length given by

$$a^2 = \frac{\varepsilon_s \varepsilon_0 k_B T}{q^2 n} + 4\pi^2 \frac{\varepsilon_s \varepsilon_0 \hbar^2}{3m^* q^2} \left(\frac{9\pi}{64}\right)^{1/3} \frac{1}{(N_I)^{1/3}} \tag{A.8}$$

In equation (A.7), n is the free-electron density and N_I is the ionized-impurity density.

E. Ionized-Impurity scattering

The scattering rate for ionized-impurity scattering is given by

$$\frac{1}{\tau_{IM}} = \left(\frac{2\pi^2}{m^*}\right)^{1/2} N_I R^2 \gamma(\mathscr{E}) \left(\frac{d\gamma(\mathscr{E})}{d\mathscr{E}}\right)^{-1} \tag{A.9}$$

Here, R is the radius of the effective sphere of influence which takes into account the fact that we must restrict the maximum impact parameter for any ionized-impurity scattering event to be less than one-half the average spacing between impurities. To ensure this we use the *ad hoc* equation

$$R^2 = \frac{r_m^2 b_m^2}{r_m^2 + b_m^2} \tag{A.10}$$

where

$$r_m = \frac{1}{2N_I^{1/3}} \tag{A.11}$$

and

$$b_m = \frac{2r/s}{(1+s^2)^{1/2}} \tag{A.12}$$

In equation (A.12), the quantities r and s can be expressed at

$$r = \frac{Zq^2}{16\pi\varepsilon_s\varepsilon_0 \gamma(\mathscr{E})} \frac{d\gamma(\mathscr{E})}{d\mathscr{E}} \tag{A.13}$$

and

$$s^2 = \frac{\hbar^2}{8m^* a^2 \gamma(\mathscr{E})} \tag{A.14}$$

Here, Z is the effective charge state of the impurity, N_I is ionized-impurity density, and a is the screening length defined by equation (A.8).

F. Alloy Scattering

The alloy scattering rate is given by

$$\frac{1}{\tau_{AL}} = \frac{3\pi\sqrt{2}(m^*)^{3/2}}{16\hbar^4} \Omega |\Delta u|^2 [c(\mathscr{E})]^{1/2} \frac{d\gamma(\mathscr{E})}{d(\mathscr{E})} \tag{A.15}$$

where Ω is the volume of the primitive cell and $|\Delta u|$ is the scattering potential.[19]

G. Non-equivalent Inter-valley Scattering

The total scattering rate between two non-equivalent energy bands, i to $\mathbf{j}$, is given by

$$\frac{1}{\tau_{ij}} = \frac{Z_j D_{ij}^2 (m_j^*)^{3/2}}{\sqrt{2}\pi\rho\hbar^2(\hbar\omega_{ih})} [\gamma_j(\mathscr{E})]^{1/2} \frac{d\gamma_j(\mathscr{E}')}{d\mathscr{E}'} \times \begin{cases} N_{ij} & \text{absorption} \\ (N_{ij}+1) & \text{emission} \end{cases} \tag{A.16}$$

where $\hbar\omega_{ij}$ is the inter-valley phonon energy, Z_j is the number of equivalent valleys, D_{ij} is the inter-valley deformation potential, and N_{ij} is the inter-valley phonon occupation number. In this case, the final state energy $\mathscr{E}'$ is given by

$$\mathscr{E}' = \begin{cases} \mathscr{E}_i - \Delta_j + \Delta_i + \hbar\omega_{ij} & \text{absorption} \\ \mathscr{E}_i - \text{v}\Delta_j + \Delta_i - \hbar\omega_{ij} & \text{emission} \end{cases} \tag{A.17}$$

where Δ_j and Δ_i are the energy gaps of the jth and ith valley, respectively, measured with respect to the valence band edge.

H. Equivalent Inter-valley Scattering

The scattering rate forinter-valley scattering between two equivalent engergy bands is given by

$$\frac{1}{\tau_{ii}} = \frac{(Z_i - 1) D_i^1 (m_i^*)^{3/2}}{\sqrt{2}\pi\hbar^2\rho(\hbar\omega_i)} [\gamma_i(\mathscr{E}')]^{1/2} \frac{d\gamma_i(\mathscr{E}')}{d\mathscr{E}'} \times \begin{cases} N_i & \text{absorption} \\ (N_i+1) & \text{emission} \end{cases} \tag{A.18}$$

where Z_i is the number of equivalent valleys, D_i is the deformation potential, $\hbar\omega_i$ is the inter-valley phonon energy, and N_i is the inter-valley phonon occupation number. The final state genergy $\mathscr{E}'$ is given by

$$\mathscr{E}' = \begin{cases} \mathscr{E} + \hbar\omega_i & \text{absorption} \\ \mathscr{E} - \hbar\omega_i & \text{emission} \end{cases} \tag{A.19}$$

GaInAsP Alloy Semiconductors
Edited by T. P. Pearsall

Chapter 11

High-field Transport Measurements

R. F. LEHENY

Bell Laboratories Inc., Holmdel, NJ 07733, USA

11.1 INTRODUCTION

Interest in alloy semiconductors arises in part from a practical hope that the correct combination of material properties can be found to optimize particular device performance and in part from a scientific interest in the relationship between specific material composition and such properties as band structure or carrier transport. In this chapter we review experimental investigations of the alloy system $Ga_xIn_{1-x}As_yP_{1-y}$, grown lattice-matched ($y \simeq 2.2x$) on InP, which relate to the properties of high-field electron transport. This subject is important for such device applications as field-effect transistors and transferred electron devices, topics covered in Chapter 17. The $Ga_xIn_{1-x}As_yP_{1-y}$ alloy system is particularly interesting from the point of view of transport devices since two of the constituent binary materials, GaAs and InP, have both demonstrated considerable importance for these device applications. Further, for some time there has been interest in the ternary alloy $Ga_xIn_{1-x}As$ grown on GaAs to take advantage of the increased electron mobility in these small-bandgap materials. Theoretical investigations (reviewed in Chapter 10), and low-field mobility measurements (reviewed in Chapter 8) have suggested that the quaternary alloys, lattice-matched to InP, can offer substantial improvements over the binary or non-lattice-matched alloys for these applications. Our discussion begins with a brief review of high-field electron transport in polar semiconductors, followed by measurements of physical properties important for transport, and finishes with specific measurements relating to high-field transport.

11.2 ELECTRON TRANSPORT IN POLAR SEMICONDUCTORS

It is convenient to view electron transport in semiconductors as the response of an average electron to an applied electric field and a series of independent momentum and energy scattering events. The motion of this average electron can be described by

$$\frac{\mathrm{d}(mv)}{\mathrm{d}t} = -eF - \frac{mv}{\tau_\mathrm{m}} \tag{11.1}$$

$$\frac{dE}{dt} = -evF - \frac{(E - E_0)}{\tau_e} \tag{11.2}$$

where τ_m (τ_e) is a momentum (energy) relaxation time, F is the applied field, and E_0 is the equilibrium average energy in the absence of the field. Electron interactions with the lattice, ionized impurities, and other electrons, all of which depend on electron energy, are taken to be contained in properly averaged momentum and energy scattering times. Also the variation of electron effective mass m with electron energy, as well as the details of a realistic band structure, must be taken account of (see, for example, Conwell and Vassell[1]). This subject has been of interest since the earliest investigations of semiconductor electrical properties and is discussed extensively in the literature (see, for example, Conwell,[2] Ridley,[3] and Ferry *et al.*[4]).

For a non-degenerate n-type semiconductor having the typical band structure illustrated in Figure 11.1, there is a continuous exchange of energy between the lattice at temperature T_0 ($E_0 = 3/2\,kT_0$) and the electrons with temperature T_e ($E = 3/2\,kT_e$) to maintain $T_e = T_0$ ($E = E_0$) and the electrons are distributed among the conduction band states with a Maxwell–Boltzman (MB) distribution. With a field applied the electrons acquire a drift in the direction of the field described by

$$v = \mu F \text{ cm s}^{-1} \tag{11.3}$$

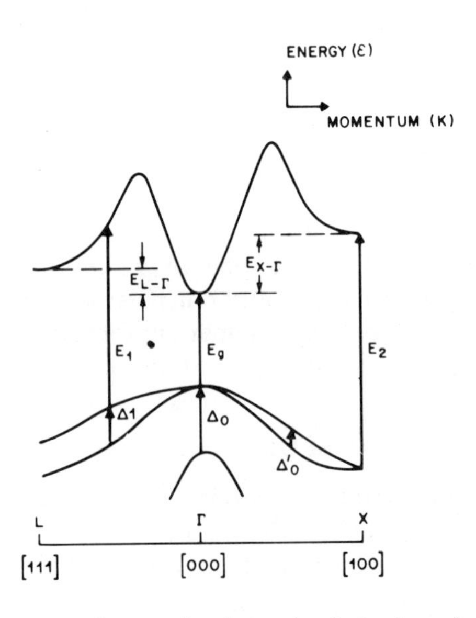

Figure 11.1 Typical direct-bandgap semiconductor band structure showing the variation in electron energy with wavevector along two of the symmetry axes in the Brillouin zone [100] and [111]. Also indicated are the transitions corresponding to critical points in the optical spectra of these materials

where the factor μ (electron mobility) can be related to the momentum relaxation time defined in equation (11.1) as

$$\mu = \frac{e}{m}\tau_m \, \mathrm{cm^2\,V^{-1}\,s^{-1}} \tag{11.4}$$

For GaAs-like materials $\tau_m \sim 0.3 \times 10^{-12}$ s at room temperature and varies significantly with temperature. The energy gained from the field is given by

$$\frac{E - E_0}{\tau_E} = evF \tag{11.5}$$

where τ_E characterizes the rate of energy exchange between electrons and the lattice.

Typically for III–V materials, of the various scattering processes contributing to τ_m, only electron–phonon scattering is effective in transferring energy to the lattice, and among the various phonons, LO phonons are the most effective at room temperature. Energy transfer is by phonon emission so that only electrons with energy greater than the LO phonon energy $\hbar\omega_{LO}$ are effective. Consequently the resultant electron distribution is not generally simply related to a MB distribution. Takenaka *et al.*[5] have discussed the influence of intercarrier scattering on the hot electron distribution function in GaAs and these workers[6] have also reported measurements of carrier distribution in the presence of applied field for this material. Deviation from an MB distribution complicates the problem of calculating average relaxation rates, and determining an analytic relationship describing the distribution for all values of electric field has proven so intractable a problem that most workers favour using numerical methods based on Monte Carlo calculations (see, for example, Fawcett *et al.*,[7] Littlejohn *et al.*,[8] and the discussion in Chapter 10).

To maintain some perspective on the physical origins of the phenomena important to high-field transport, we ignore this difficulty and discuss the problem with the assumption that electron–electron scattering is fast enough to maintain an MB distribution. With this assumption the effect of the field is to cause a drift of electrons, with the average electron energy $(3/2\,kT_e)$ increasing until the energy transferred to the lattice balances that gained from the field. This 'warm electron' approximation can be useful for obtaining an estimate of the actual electron dynamics.[3,9–11]

For an MB distribution, the energy loss per unit time from the electrons with temperature T_e to the lattice at temperature T_0 is given by[12]

$$\frac{E - E_0}{\tau_E} = -\left|\frac{2\hbar\omega_{LO}}{\pi m^*}\right|^{1/2} eF_0 \left|\frac{\exp(X_0 - X_e) - 1}{\exp(X_0) - 1}\right| X_e^{1/2} \exp(X_e/2)\, K_0(X_e/2) \tag{11.6}$$

In this expression,

$$X_0 = \hbar\omega_{LO}/kT_0$$
$$X_e = \hbar\omega_{LO}/kT_e$$

K_0 is a modified Bessel function

and F_0 is an effective field characterizing the electron–phonon coupling given by

$$F_0 = \frac{m_e^* \hbar\omega_{LO}}{\hbar^2}\left|\frac{1}{\varepsilon_\infty} - \frac{1}{\varepsilon_0}\right| \tag{11.7}$$

and ε_∞ (ε_0) is the high- (low-) frequency dielectric constant. At low temperatures, $T_e < 50$ K, the loss rate due to LO-phonon emission drops below that calculated for other electron–phonon interactions and these become dominant.[13]

Hilsum[10] and Ridley[3] have discussed how the physical properties such as effective mass, longitudinal optical phonon energy and the electron–phonon coupling combine to determine the high-field transport properties of polar semiconductors. In effect, there is a critical applied field F_c for which the rate of energy gained from the field exceeds the rate at which the electrons can dissipate this energy to the lattice. For higher fields there is a rapid increase in average electron energy. The variation of average energy with electric field calculated with these relationships, expressed as an effective electron temperature, for $Ga_{0.47}In_{0.53}As$[14] and GaAs,[9] are shown in Figure 11.2. Also shown, for comparison, is the result of a Monte Carlo calculation of electron temperature for GaAs and InP.[15] Variation of the critical field F_c with material composition for the alloy system $Ga_xIn_{1-x}As_yP_{1-y}$ is shown in Figure 11.6.[16]

With the dramatic increase in average electron energy for fields greater than F_c, electrons begin to occupy states high in the conduction band, and the simple parabolic band structure useful for consideration of low-field transport is no longer adequate. Under these conditions, there exists the possibility of direct relaxation to the bottom of the conduction band by ionization of electrons across

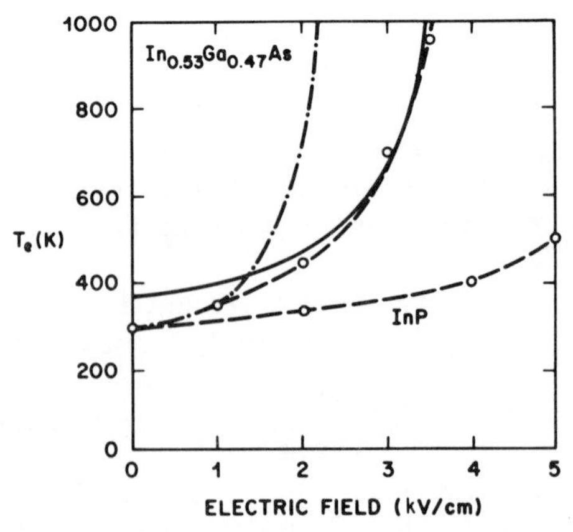

Figure 11.2 Variation of effective electron temperature with applied electric field for the materials $Ga_{0.47}In_{0.53}As$[14] and GaAs calculated according to the method described by Hilsum[10] (—·—·— Nahory *et al.*; ——— Hilsum) and for GaAs and InP calculated using Monte Carlo numerical techniques[15] (o——o Maloney and Frey)

the forbidden gap (impact ionization) or of scattering into indirect, satellite conduction band states.

For wide-bandgap materials ($E_g > E_{X\text{-}\Gamma}$ or $E_{L\text{-}\Gamma}$ in Fig. 11.1) the inter-conduction-band scattering from the Γ valley to the satellite valleys dominates. Of course, only electrons with sufficient energy can be scattered, but the scattering rate is very fast ($\tau \sim 10^{-14}\,s^{-1}$ in GaAs; see, for example, Conwell[2] or Pötzl[17]). Further, the large density of states associated with the satellite valleys ensures that most of the electrons with sufficient energy to be scattered remain in these states. Satellite-valley transport is limited by inter-valley scattering and the heavy satellite electron mass, with satellite electrons exhibiting low mobility ($\mu \simeq 150\,cm^2\,V^{-1}\,s^{-1}$). With the onset of inter-valley scattering, two distinct carrier types contribute to the net electron drift which is given by a weighted average as

$$v_D = \frac{n(\Gamma)}{n_0} v_\Gamma + \frac{n(s)}{n_0} v_s \tag{11.8}$$

with n_0 the total electron density. Increasing the applied field above F_c results in a decrease in drift velocity giving rise to a negative differential mobility that can result in current instabilities. Hauser *et al.*[18] have discussed how variation in the central-valley electron mass and electron–phonon coupling for hot electrons also contribute to negative differential mobility. These investigators suggest that these effects can be more significant than transfer to satellite valleys, particularly for materials with energy separation between the Γ valley minima and the lowest-lying satellite valley exceeding 0.5 eV.

For GaAs and InP, inter-valley transfer sets in at fields of about $4\,kV\,cm^{-1}$ and $10\,kV\,cm^{-1}$ respectively. For still higher fields these materials exhibit avalanche breakdown, and for this case the carrier dynamics become somewhat more complicated than we have described here (see, for example, Pearsall *et al.*[19]). Avalanche is typically observed in the depleted region of reverse-biased p–n junctions where the number of carriers is low and the high fields required can be achieved relatively easily.

11.3 PHYSICAL PROPERTIES INFLUENCING HIGH-FIELD TRANSPORT

Detailed discussion of experimental investigations for some of the material properties important for high-field electron transport are covered in specific chapters on measurements of material properties such as electron mass and phonon energies (Chapter 12), low-field electron mobility (Chapter 8), and the fundamental energy gaps E_0 (Chapter 12). In this section we focus on the measurement of phonon spectra and the question of electron–phonon coupling, and the experimental data which provides information on the higher-energy satellite valleys.

11.3.1 Phonon spectra

Considering the significance of optical phonon scattering on the transport properties of electrons in polar crystals, phonon spectra represent an important database for comparisons between materials. In addition to the explicit dependence on $\hbar\omega_{LO}$, the electron energy loss rate depends directly on the effective field F_0 associated with a phonon mode and this field is proportional to $(1/\varepsilon_\infty - 1/\varepsilon_0)$. While this electric field is not directly accessible to experimental measurement, the ratio of high-frequency (ε_∞) to low-frequency (ε_0) dielectric constants can be related to the phonon spectra by the Lyddane–Sachs–Teller relationship

$$\frac{\varepsilon_0}{\varepsilon_\infty} = \left|\frac{\omega_{LO}}{\omega_{TO}}\right|^2 \tag{11.9}$$

As a result the width of the *reststrahlen* band ($\Delta\omega = \omega_{LO} - \omega_{TO}$) associated with a specific lattice mode provides information on the strength of the electron coupling to that mode.

The possible vibration frequencies of an alloy material such as $Ga_xIn_{1-x}As_yP_{1-y}$ offer many possibilities for investigation of lattice dynamics. One would like to relate these dynamics to the constituent binary materials. Raman scattering and infrared reflectivity have been used[20] to determine the phonon modes in the ternary system $Ga_xIn_{1-x}As$ with the results shown in Figure 11.3(a). More recent investigations for the quaternary system, making contact with this work for the composition $x = 0.47$, $y = 1.0$, are shown in Figure 11.3(b). Brodsky and Luckovsky[20] characterized the infrared reflectivity spectra of $Ga_xIn_{1-x}As$ in terms of two reststrahlen bands. A 'strong' band which shifts down in energy and bandwidth monotonically from GaAs to the LO energy of InAs and a lower-frequency, 'weak' band, which they identify with InAs. Each of these bands shrinks to zero width consistent with a local 'impurity' mode at the opposite binary composition.

Brodsky and Lucovsky[20] point out that this behaviour is contrary to what had had previously been found for reststrahlen bands in mixed crystals. For one class of materials a single band was identified, with frequencies intermediate between the constituent binary materials. For a second class, two bands each lying within the range of the constituent compounds with relative strength varying with composition was identified. But, for the case of $Ga_xIn_{1-x}As$, the phonon spectra have characteristics of both these possibilities, exhibiting one-mode behaviour for the GaAs-rich compositions, two-mode behaviour for the InAs-rich compositions, and a smooth transition between the two cases for intermediate compositions. One result of this behaviour is that electron–phonon coupling can be expected to vary in a complex way for these mixed crystals. The evidence in the case of $Ga_{0.47}In_{0.53}As$ is that the 'GaAs-like' mode dominates this interaction as discussed below.

Similar variation of phonon frequency with composition has been found in the

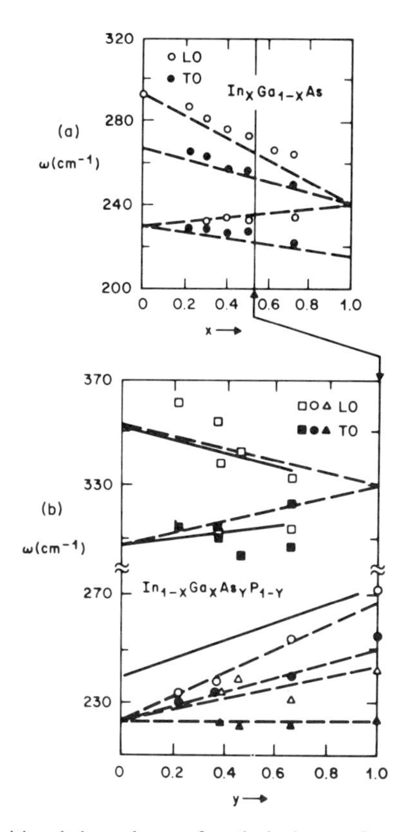

Figure 11.3 (a) Compositional dependence of optical phonon frequencies in $Ga_xIn_{1-x}As$ determined from infrared reflectivity.[20] Reproduced by permission of the Institute of Physics. (b) Compositional dependence of optical phonon frequencies in $Ga_xIn_{1-x}As_yP_{1-y}$ determined from infrared reflectivity.[23] Results of Raman scattering experiments[21, 22] are indicated schematically by the broken lines. Reproduced by permission of the authors, The Institute of Physics, and the American Physical Society

quaternary system (Figure 11.3(b)). Pinczuk *et al.*[21] have reported Raman scattering spectra spanning the entire compositional range $y = 0$ to 1 and identified two sets of modes, one 'InP-like' that extrapolated to a local mode at $y = 1$, and a second band identified as 'GaInAs-like'. Similar Raman results for $y = 0.23$ and $y = 1$ were found by Portal *et al.*[22] but these authors identified independent GaAs- and InAs-like modes in the $y = 1$ case, while assigning the two modes observed in the $y = 0.23$ material to InP- and InAs-like modes. Infrared reflectivity measurements spanning the entire composition range have been reported by Amirtharaj *et al.*[23] These investigators found the lower-energy modes identified in the Raman work could be resolved into two separate GaAs- and InAs-like modes which extrapolate to the similar binary-like modes in the $Ga_xIn_{1-x}As$ ternary system.

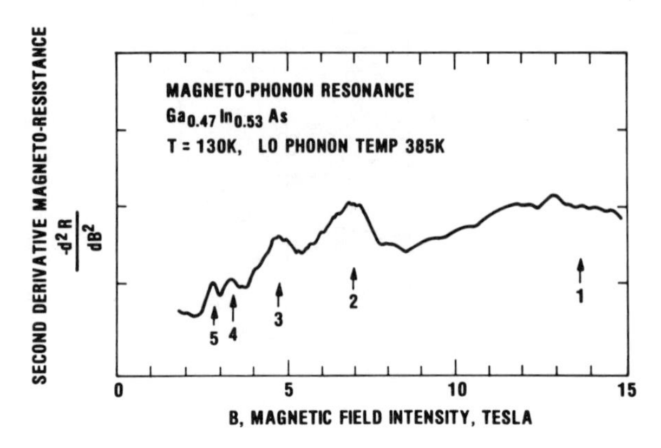

Figure 11.4 Experimental recording of the second derivative of the magnetoresistance for a $Ga_{0.47}In_{0.53}As$ sample at 130 K.[24] The upward arrows indicate the positions of the magnetophonon resonances for the 'GaAs'-like phonon mode. Reproduced by permission of the authors and the American Institute of Physics

Nicholas *et al.*[24] also reported the observation of magnetophonon oscillations in the transverse magnetoresistance for materials with $y = 0.23$ and $y = 1$. For the quaternary sample ($y = 0.23$) two distinct phonon modes could be resolved, while in the ternary sample only a single 'GaAs-like' phonon was identified (see Figure 11.4). Similar results, suggesting that electrons couple strongly to only the 'GaAs-like' mode in the ternary material have been reported by Shah *et al.*[25] who investigated the hot electron contribution to the photoluminescence spectrum of $Ga_{0.47}In_{0.53}As$. These luminescence measurements also permitted an indirect determination of the energy loss rate for the hot electrons and we discuss them further here.

The hot electron contribution to the total luminescence corresponds to the high-energy ($\hbar\omega > E_g$) portion of the luminescence spectrum. This luminescence results from recombination of electrons occupying states above the conduction band minima. For laser excitation a fraction of the incident photon energy $\Delta E = \hbar\omega_0 - E_g$ is available to heat carriers from the incident photons and the variation of luminescence intensity provides a measure of the resulting carrier temperature, T_c.[26] Spectra obtained for $Ga_{0.47}In_{0.53}As$ pumped by the 5145 Å line of an argon laser are shown in Figure 11.5(a).

These spectra can be analysed in the following straightforward way.[26] The excess energy of the photoexcited carrier is taken to be dissipated in one of two ways, either directly to the lattice by LO-phonon emission or by heating the steady-state distribution through carrier–carrier scattering. For low carrier density (low excitation intensity), the first process dominates and the smaller rate of energy exchange with the steady-state carrier population is taken to increase linearly with the number of carriers, i.e.$P_{in} \propto I$ where P_{in} is the power heating the carriers and I is the excitation intensity. The carrier temperature is then

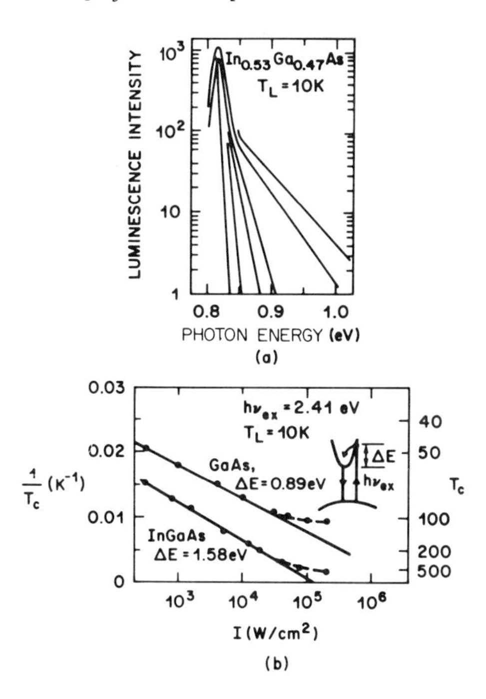

Figure 11.5 (a) Hot electron photoluminescence spectra for $Ga_{0.47}In_{0.53}As$. The slope of the variation of the high-energy tail of the spectrum plotted on semilog scale provides a measure of the carrier temperature. (b) Variation of the inverse carrier temperature with log of laser intensity for $Ga_{0.47}In_{0.53}As$ and GaAs. The experiment is illustrated schematically in the inset. The slope of the linear portion of these curves provides a measure of the LO phonon energy[25]

determined by setting P_{in} equal to the power flowing out of the steady-state population which is given for electrons by equation (11.6). With increasing laser excitation, T_c increases to maintain the balance between energy flow into and out of the carrier system. At the highest excitation intensities, the probability of photoexcited carrier relaxation by LO phonon emission becomes negligible compared to direct interaction with the steady-state carrier population and the rate of carrier heating, per carrier generated, saturates. Further increases in excitation intensity result in little further increase in T_c as indicated in Figure 11.5(b).

The energy of the phonon responsible for carrier energy relaxation can be determined from the slope of the variation of $1/T_c$ with log I[26] (see Figure 11.5). The ternary carrier temperature variation is fitted by a single phonon with energy 34 meV. This phonon energy is consistent with the 'GaAs-like' mode in the Raman and infrared data of Figure 11.3 and the oscillatory magnetophonon data of Figure 11.4. Comparison between data obtained for $Ga_{0.47}In_{0.53}As$ and GaAs, for comparable excitation conditions and accounting for the differences in E_g and electron mass, yields an energy loss rate three times smaller for the ternary than for the binary. This result is in agreement with a calculation of the carrier–phonon coupling strength for the 'GaAs-like' mode and the reststrahlen data.[25]

While not directly influencing the electron energy loss rate, elastic momentum scattering processes can still influence high-field transport. By reducing low-field mobility, momentum scattering increases the field required to achieve a given drift velocity and the energy gained from the field to maintain this velocity. As a result, the rate of energy transfer to the lattice must increase to maintain a steady state, resulting in a hotter electron distribution for a given net drift. In alloy materials the presence of an additional scattering mechanism due to the random distribution of the constituent atoms can have a significant effect on the transport of electrons, substantially reducing the peak drift velocity.[8]

This topic is more completely discussed in Chapter 8. However, for completeness, we point out that in discussing their results Pinczuk *et al.*[21] and Amirtharaj *et al.*[23] have commented on observations of additional structures which could not be assigned to specific modes. Pinczuk[21] suggested that the presence of a low-energy mode in the quaternary samples containing high concentrations of P, which was not observed in the ternary, might result from large wavevector acoustic modes. Not normally Raman-active, these modes may become active by random disorder. Low-field mobility data also suggest the presence of excessive scattering for the quaternary alloys (see, for example, Leheny *et al.*[27]) that is not present to the same extent in the ternary.

Once the critical field F_c is exceeded, electron temperature rises rapidly. A significant fraction of the electron population begins to occupy states high in the band. Scattering into satellite conduction band valleys begins to dominate electron transport. Zone edge LO and LA phonons are responsible for Γ–L,X inter-conduction-band scattering. Electron coupling to these short-wavelength phonons is principally via the deformation potential associated with each particular phonon mode. Mobility for satellite electrons is limited by their large effective mass ($m^*/m_o \sim 0.3$–0.4) and phonon scattering between equivalent valleys. At present there is little experimental data on the zone edge phonons for quaternary alloy semiconductors. Inter-conduction-band scattering rates for GaAs are in the range of $10^{14}\,s^{-1}$.

11.3.2 Band structure

The energy separation of the satellite conduction band minima and the Γ valley minima is critical to determining hot electron dynamics but the direct measurements of the satellite valley energy minima is difficult. Aspnes[28] has discussed some of the various experimental techniques used to establish the satellite conduction band minima energies and ordering in GaAs. Techniques such as electroreflectance in the visible and near-UV are straightforward to apply and can provide information on critical points in the joint density of states for transition from the valance to conduction band labelled E_g, E'_1, E'_0, etc., in Figure 11.1.

While not providing direct information on the satellite minima of the conduction bands, these techniques provide insight into the relative variation of the band structure with composition and these data can be related to variation in the relative positions of the conduction band minima by invoking the invariance of the *valence* bandwidth for germanium-type semiconductors. This invariance of the valence band has been cited by Auvergne *et al.*[29] for using the variation of critical point energies to determine the variation of the indirect Γ, L, and X minima with temperature or composition for III–V compounds. Measurements for the quarternary system have been reported by Perea *et al.*,[30] ad Laufer *et al.*[31] (see also Casper and Wieder,[32] and Yamzoe *et al.*[33]). The results obtained by Laufer *et al.*[31] are shown in Figure 11.6. The electroreflectance data indicate that there is little variation in the relative positions of the various conduction band minima with material composition.

Information on conduction band structure can also be determined using vacuum-UV reflectance techniques. These investigations probe dipole transitions ($h\nu \sim 17$ to 22 eV) originating with the cation core electron states which are nearly dispersionless across the Brillouin zone. Using vacuum-UV Schottky-barrier electroreflectance to investigate the Ga 3d core level transition to the conduction band, Aspnes[28] established that the L valley is the lowest satellite valley in GaAs. While comparable data are not available on the quaternaries, Kelso *et al.*[34] have reported on direct reflectance spectra in the same energy range. Probing

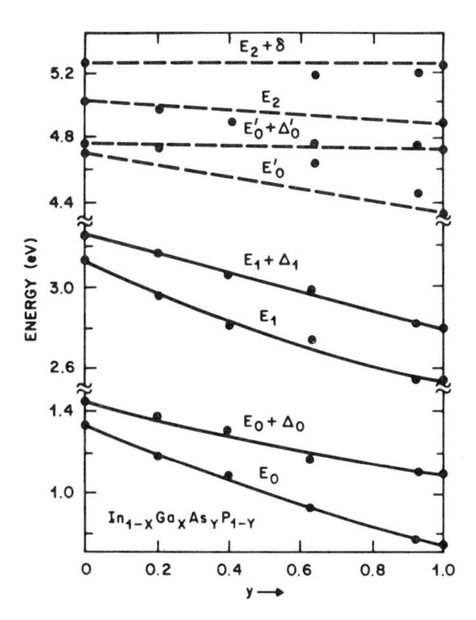

Figure 11.6 Variation of the critical point energies indicated in Figure 11.1 with material composition for $Ga_xIn_{1-x}As_yP_{1-y}$.[31] Reproduced by permission of the authors and Pergamon Press. Copyright 1980, Pergamon Press, Ltd

transitions from both the Ga and In core electron states, these investigators found spectral features associated with transition from both Ga and In core levels to minima (or maxima) in the conduction band. However, because of symmetry, transitions to the Γ minimum are forbidden and do not contribute to the spectra. The measurement does provide information on the relative energy spacing of the X and L minima and thus on the ordering of these higher-lying bands. Kelso *et al.*[34] conclude from their data that the ordering of these states is Γ–L–X as indicated in Figure 11.1 with the X–L spacing being 0.42 eV for $Ga_{0.47}In_{0.53}As$. This energy lies approximately half way between the measured spacing for GaAs (0.15 eV) and the calculated spacing for InAs (0.75 eV) and is only 0.03 eV less than for InP (0.45 eV).

11.4 HIGH-FIELD TRANSPORT MEASUREMENTS

Because of the very fast timescale for energy exchange with the lattice, and the relatively short lifetime of non-equilibrium carrier populations in direct-gap materials, the direct measurement of the dynamics of hot electrons and high-field transport properties is quite difficult. Nevertheless a variety of optical and electrical measurement techniques have been developed which provide information on electron dynamics in semiconductors and a recent review by Bauer[35] describes some of these techniques. Because of its technological importance, GaAs has been widely studied, and many of the techniques used to investigate this material can, in principle, be applied with some modification to the quaternary alloys.

One of the more direct high-field transport measurement techniques is the measurement of the transit time (T) required for electrons to travel a fixed distance (d) in a known field. For this case the drift velocity is given by $v_D = d/T$. Typically for such a measurement, electrons are injected into the depleted region of a back-biased Schottky diode by pulsed electron-beam excitation. The use of relatively low-energy electrons (1–3 keV) ensures that carriers are generated in a thin layer at the sample surface where the holes generated by the beam can be rapidly collected. Ruch and Kino[36] were the first to use this technique to measure the velocity–field relationship for electrons in GaAs. Evans and Robson[37] described an extension of the method that permits measurements in thin films. Recently this technique has been used to measure electron drift velocity for fields up to 220 kV cm^{-1} in GaAs.[38] However, the preparation of Schottky diode structure to perform this type of measurement with the quaternary materials presents some technological difficulties resulting from the low Schottky barrier potential for the narrow-bandgap quaternaries.

An alternative measurement technique is to deduce the electron velocity–field characteristic from the measurement of current vs voltage for a properly prepared sample. This technique provides information on the velocity–field relationship only up to the critical field. At higher fields current instabilities make the interpretation of the *I–V* relationship in terms of electron drift in a uniform field impossible. Drift velocity is deduced from the current density through a region of the sample where the field can be related to the applied voltage. Current injection at the contacts or field ionization is to be avoided.

Marsh *et al.*[16] have used this method to determine the compositional dependence of the peak velocity and threshold field for samples spanning the entire $Ga_xIn_{1-x}As_yP_{1-y}$ alloy system. Their results are summarized in Figure 11.7. Additional data points from other earlier investigations (Majerfeld *et al.*,[39] Houston *et al.*,[40] and Sasaki *et al.*,[41]) are included in the figure. The sample geometry is also illustrated. The large area of the sample at the contacts minimizes the influence of contact resistance and ensures that most of the applied voltage is actually across the narrow bar having a well defined length. For such a sample the resistance of the contact region can be accurately calibrated at low currents and is assumed to remain linear over the range of currents investigated. Current through the sample is monitored as a function of applied voltage up to the critical field F_c. At this point instabilities due to transfer of electrons to the satellite valley result in current pulsations. The variations of the critical field, calculated by Marsh *et al.*[16] using Hilsum's[10] method, is shown as a broken line in Figure 11.7. These investigators discuss their results in terms of separate measurements of the low-field mobility and the influence of alloy scattering.

The results of Marsh *et al.*[16] provide a basis for estimating the significance of the electron–phonon interaction on high-field transport for the quaternaries. Consider the product of the peak drift velocity and the critical field. This product provides a measure of the rate of energy transferred from the field to the electrons at the critical field. Clearly electrons in InP, with $F_c \geqslant 10\ \mathrm{kV\,cm^{-1}}$ and $v_p \sim 2.5 \times 10^7\ \mathrm{cm\,s^{-1}}$, have the highest capacity for transferring energy to the lattice, while the material with composition near $y = 0.85$ has the lowest ($F_c \simeq 2\ \mathrm{kV\,cm^{-1}}$ and $v_p \sim 1.5 \times 10^7\ \mathrm{cm\,s^{-1}}$). Phonon energy variation and the decrease in electron mass with increasing y influences this result. However a

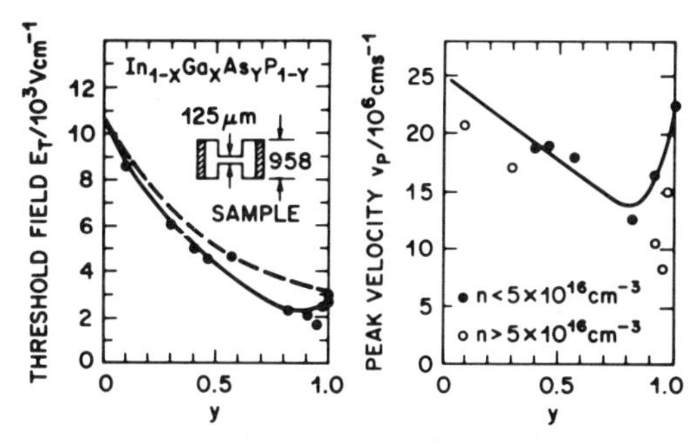

Figure 11.7 Compositional dependence of threshold field and peak drift velocity for $Ga_xIn_{1-x}As_yP_{1-y}$ deduced from *I–V* characteristics for samples having the shape indicated in the figure.[16] Reproduced by permission of the authors and the Institute of Physics

significant factor accounting for the difference between $y = 0.85$ and $y = 1$ must be the variation in electron–phonon coupling strength as suggested by the increase in F_c and v_p for $Ga_{0.47}In_{0.53}As$ which has the smallest electron mass and phonon energy for the alloy system.

The use of lasers to probe electron dynamics, particularly picosecond lasers, has been shown to offer unique opportunities to investigate hot electron phenomenon. Leheny *et al.*[42] have investigated hot electron relaxation in GaAs using an excite-and-probe technique with picosecond time resolution. These investigators found that for electron densities less than about $10^{17}\,cm^{-3}$, the relaxation of hot electrons to the lattice temperature, after photoexcitation, was well described by the LO-phonon emission process discussed in section 11.1 with results shown in Figure 11.8. However, at higher excitation densities, screening of the electron–phonon interaction significantly reduces the rate at which energy can be exchanged between the electron system and the lattice. The effects of screening can be seen in the higher effective temperature measured 10 ps after excitation in the presence of high electron densities. Since screening becomes important for electron densities in the range 10^{16}–$10^{17}\,cm^{-3}$, a range spanning the densities of interest for transport devices, the influence of screening on electron transport deserves more detailed investigation. An extension of the excite-and-probe technique has recently been used to investigate the transient velocity overshoot[15] phenomenon for electrons photoexcited in GaAs by Shank *et al.*[43]

One difficulty associated with measurements that rely on laser excitation is the

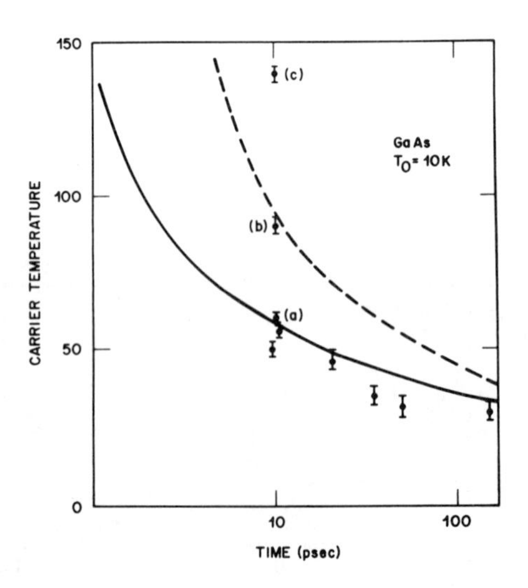

Figure 11.8 Variation of carrier temperature with time delay after excitation for GaAs. Points (a), (b), and (c) represent data for increasing pump excitation intensities, I_0(a), $4I_0$(b), and $15I_0$(c) with $I_0 = 2 \times 10^7\,W\,cm^{-2}$. The full curve is calculated for the known energy loss due to LO phonon scattering. The broken curve is calculated assuming this loss rate is decreased by a factor of five[42]

simultaneous generation of equal numbers of electrons and holes. Typically photoabsorption occurs on a scale of 0.2–1.0 μm and this often represents a significant fraction of the sample volume of interest. As a result, the dynamics of the photoinduced holes cannot be ignored. In fact, Gammel and Ballantyne[44, 45] have discussed the advantage that photoexcited holes offer for photoconductive gain in opto-FET structures used as photodetectors. Degani *et al.*[46] have described transport measurements in a similar photoconductive structure which avoids this problem by using p-type $Ga_{0.47}In_{0.53}As$ material. Nukermans and Kino[47] had demonstrated that p-type material could be used to investigate minority-carrier electron drift velocity in their investigation of the

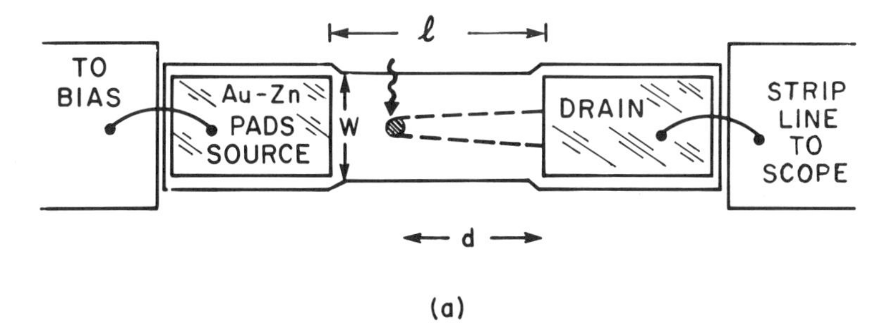

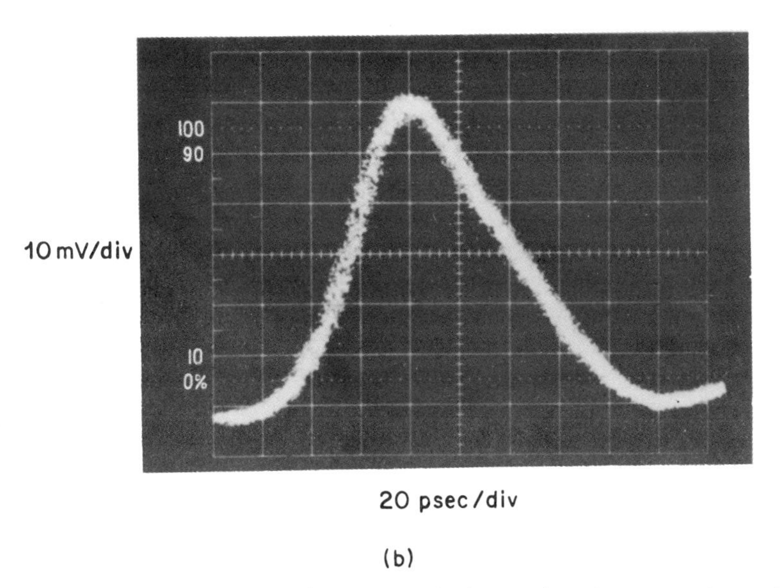

Figure 11.9 (a) Schematic diagram of the photoconductive sample used to measure the transit time of minority electrons in p-$Ga_{0.47}In_{0.53}As$. (b) The photograph illustrates the measured response of this device for the case where $\lambda = 13\ \mu m$[46]

velocity–field characteristics of InSb. For these measurements, if the density of photoexcited carriers is kept smaller than the background hole concentration, the high-mobility electrons dominate the induced current transient. The minority electrons cause an increase in current through the sample that persists until the electrons are collected at the contact.

Degani *et al.*[48] demonstrated that a sensitive, fast photoconductive detector taking advantage of the high electron drift velocity could be fabricated in this way. Figure 11.9 shows a typical photoresponse obtained for the structure shown ($l = 13\,\mu m$) with a 20 ps, 1.3 μm laser excitation source. By monitoring the variation of the transient current pulsewidth for a sample with $l = 100\,\mu m$ as a function of the distance between drain and the excitation spot (*d* in Figure 11.9), the drift-velocity-limited electron transit time was determined. Results obtained for applied fields up 7.5 kV cm^{-1} are shown in Figure 11.10. These results for minority electrons in p-type $Ga_{0.47}In_{0.53}As$, with $p_0 = 5 \times 10^{16}\,cm^{-3}$ are compared to the calculated velocity–field characteristic reported by Littlejohn *et al.*[8] for $N_D = 10^{17}\,cm^{-3}$ in n-type material. Degani *et al.*[46] point out that while the low-field drift velocity is in good agreement with measurements on n-type material having comparable doping,there is no evidence of a transferred electron effect in this p-type material. With a photoexcited electron density of about $10^{15}\,cm^{-3}$ these investigators attribute this result to the enhanced hot electron energy relaxation by collisions with holes, a direct result of the relatively large number of free holes in the sample investigated.

The importance of the background holes in cooling hot carriers has been discussed by Shah *et al.*,[49] who have investigated theoretically and experimentally the photoexcitation of hot carriers for the case of intentionally doped material. Their calculations show that, assuming all the steady-state carriers are at the same

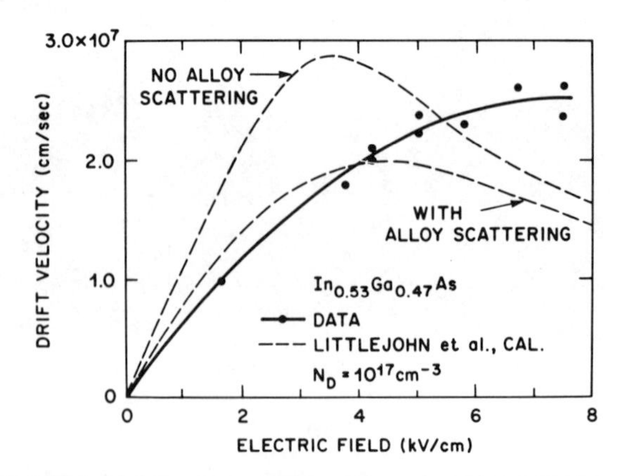

Figure 11.10 Measured minority electron drift velocity vs applied field for p-$Ga_{0.47}In_{0.53}As$.[46] Also shown for comparison are the results of the Monte Carlo calculations of Littlejohn *et al.*[8] for n-type material with $N_D = 1 \times 10^{17}\,cm^{-3}$

temperature, a background carrier population can influence the dynamics of carrier cooling in three ways. First, by increasing the probability of carrier–carrier scattering, secondly, by increasing the number of carriers which share the energy input, and, thirdly, by increasing the effective coupling to the lattice. However, because the electrons and holes have significantly different energy loss rates, the influence of these different factors depends critically on the majority carrier type. For n-type material these investigators find that the first factor dominates and they predict a hotter carrier distribution, for a given excitation intensity, than is observed for undoped material. For the case of p-type material, the third factor dominates and the carriers remain relatively cooler. These calculated results are supported by hot carrier photoluminescence measurements for p-type $Ga_{0.47}In_{0.53}As$.[49]

Further work is required to relate these photoluminescence results to the problem of high-field minority-carrier transport. However, if the effect of a background hole population in p-type $Ga_{0.47}In_{0.53}As$ is effectively to cool the minority electron temperature, then the transfer of electrons to satellite valleys can be significantly reduced, or entirely eliminated. Under these conditions electron drift velocity is determined by the dynamics of electron motion in the Γ valley. The average electron energy is still limited by LO-phonon emission to be approximately $\hbar\omega_{LO} = 34$ meV and the correponding electron drift velocity is found to be $v_d \sim 2.8 \times 10^7 \text{ cm s}^{-1}$, in reasonable agreement with the measurement.

ACKNOWLEDGEMENTS

The author would like to acknowledge the contributions to this work of Jagdeep Shah, R. E. Nahory, and Joshua Degani, and the continuing interaction with workers interested in quaternary alloy semiconductors at Bell Laboratories including M. A. Pollack, A. A. Ballman, J. C. DeWinter, R. J. Martin, and E. D. Beebe.

NOTES

Recent measurements by[50], using ultra violet photo emission techniques, have established that the Γ–L valley separation ($E_{L\text{-}\Gamma}$ in Figure 1) for $Ga_{0.47}In_{0.53}As$ is 0.55 ± 0.05 eV at 300 K. This value is $\sim 60\%$ higher than for GaAs. These investigators point out that electrons need to be heated to significantly higher temperatures in order for inter-valley scattering to occur.

T. H. Windhorn, *et al.*[51,52] have reported studies of electron velocity-field characteristics for $Ga_{0.47}In_{0.53}As$ using the method of microwave modulated electron beam excitation. They made use of a grown p–n junction for which the p layer was thinned to ~ 1000 Å to insure that the majority of electrons were generated in the depleted n^- ($n \sim 10^{14}\text{ cm}^{-3}$) layer. Their investigations were limited to applied fields above ~ 10 kV/cm and precluded measurement of the peak drift velocity. However, they established that at 300 K, and for fields below 18 kV/cm, the average electron velocity in $Ga_{0.47}In_{0.53}As$ exceeds that for GaAs. At higher fields they find the electron velocity drops below that for

GaAs. This result suggests that the fraction of carriers scattered into satellite valley excedes that for GaAs at these fields even though the Γ–L valley separation is ~ 200 meV greater. The value of peak drift velocity obtained by extrapolation to lower fields is in reasonable agreement with the measurement of Marsh, *et al.*[16] Further, these investigators report a negative differential mobility for the alloy that is ~1.7 times greater than for GaAs at 18 kV/cm.

J. Shah, *et al.*[53] have shown that the temperature dependence for hot electron energy loss in these alloy materials can be approximated by a single effective phonon (see Eq. 11.6) over a limited temperature range by taking a weighted average of the contribution from each of the constituent binary compounds LO-phonon.

For the case of $Ga_{0.28}In_{0.72}As_{0.6}P_{0.4}$, the energy of this effective phonon is ~ 31.6 meV in the temperature range between 35 and 160 K. This result is shown to be in reasonable agreement with hot carrier luminescence measurements.

REFERENCES

1. E. M. Conwell and M. O. Vassell, 'High field transport in n-type GaAs', *Phys. Rev.*, **166**, 797–821, 1968.
2. E. M. Conwell, 'High field transport in semiconductors', supplement to *Solid State Physics*, eds F. Seitz, D. Turnbull, and H. Ehrenreich, Academic Press, New York, 1967.
3. B. K. Ridley, 'Anatomy of the transferred-electron effect in III–V semiconductors', *J. Appl. Phys.*, **48**, 754–764, 1977.
4. D. K. Ferry, J. R. Barker, and C. Jacoboni (eds), *Physics of Nonlinear Transport in Semiconductors, NATO Advance Studies Institute Series*, Plenum Press, New York, 1979.
5. N. Takenaka, M. Inoue, and Y. Inuishi, 'Influence of intercarrier scattering on hot electron distribution functions in GaAs', *J. Phys. Soc. Jpn.*, **47**, 861–868, 1979.
6. M. Inoue, N. Takenaka, J. Shirafuji, and Y. Inuishi, 'Experimental and theoretical determination of hot electron distribution in GaAs and related mixed crystals', *Solid-State Electron.*, **21**, 29–34, 1977.
7. W. Fawcett, A. D. Boardman, and S. Swain, 'Monte Carlo determination of electron transport properties in gallium arsenide', *J. Phys. Chem. Solids*, **31**, 1963–1990, 1970.
8. M. A. Littlejohn, J. R. Hauser, T. H. Glisson, D. K. Ferry, and J. W. Harrison, 'Alloy scattering and high field transport in ternary and quaternary III–V semiconductors', *Solid-State Electron.*, **21**, 107–114, 1978.
9. C. Hilsum, 'Transferred electron amplifiers and oscillators', *proc. IRE*, **50**, 185–189, 1962.
10. C. Hilsum, 'A simple method for estimating peak electron velocities in polar semiconductors', *J. Phys. C: Solid State Phys.*, **9**, L629–L631, 1976.
11. E. M. Conwell and M. O. Vassell, 'High field distribution function in GaAs', *IEEE Trans. Electron. Devices*, **ED-13**, 22–27, 1966.
12. R. Stratton, 'The influence of interelectronic collisions on conduction and breakdown in polar crystals', *Proc. R. Soc.*, A, **246**, 406–422, 1958.
13. K. Seeger, 'Relaxation times in semiconductors', *Electronic Materials*, eds N. B. Hannay and U. Colombo, Plenum Press, New York, 1972.
14. R. E. Nahory and J. Shah, private communication, 1981.
15. T. J. Maloney and J. Frey, 'Transient and steady-state electron transport properties of GaAs and InP', *J. Appl. Phys.*, **48**, 781–787, 1977.
16. J. H. Marsh, P. A. Houston, and P. N. Robson, 'Compositional dependence of

mobility, peak velocity, and threshold field in $In_{1-x}Ga_xAs_yP_{1-y}$', *Gallium Arsenide and Related Compounds 1980*, Conf. Ser. no. 56, Institute of Physics, Bristol, 1981.
17. H. Pötzl, 'Hot electrons in III–V semiconductors', *Electronic Materials*, eds N. B. Hannay and U. Colombo, Plenum Press, New York, 1972.
18. J. R. Hauser, T. H. Glisson, and M. A. Littlejohn, 'Negative resistance and peak velocity in the central (000) valley of III–V semiconductors', *Solid-State Electron.*, **22**, 487–493, 1979.
19. T. P. Pearsall, R. E. Nahory, and J. R. Chelikowsky, 'Threshold energies for impact ionization by electrons and holes in GaAs–GaSb system', *Gallium Arsenide and Related Compounds* (*St Louis 1976*, Conf. Ser. no. 33b, Institute of Physics, Bristol, 1977.
20. M. H. Brodsky and G. Lucovsky, 'Infrared reflection spectra of $Ga_{1-x}In_xAs$: a new type of mixed-crystal behaviour', *Phys. Rev. Lett.*, **21**, 990–993, 1968; T. P. Pearsall, R. Bisaro, P. Merenda, G. Laurencin, R. Ansel, J. C. Portal, C. Houlbert, and M. Quillec, 'The characterization of $Ga_{0.47}In_{0.53}As$ grown lattice matched on InP substrates', *GaAs and Related Compounds, St Louis, 1978*, Inst. of Physics Conference Series 45, Institute of Physics, Bristol, 1979, pp. 94–102.
21. A Pinczuk, J. M. Worlock, R. E. Nahory, and M. A. Pollack, 'Lattice vibrations of $In_{1-x}Ga_xAs_yP_{1-y}$ quaternary compounds', *Appl. Phys. Lett.*, **33**, 461–463, 1978.
22. J. C. Portal, P. Perrier, M. A. Renucci, S. Askenazy, R. J. Nicholas, and T. Pearsall, 'A study of the condition band phonons of $Ga_xIn_{1-x}As_yP_{1-y}$ by the Shubnikov–deHaas effect, magnetophonon resonance and Raman Scattering', *Physics of Semiconductors 1978*, Conf. Ser. no. 43, Institute of Physics, Bristol, 1978, pp. 824–832.
23. P. M. Amirtharaj, G. D. Holah, and S. Perkowitz, 'Far-infrared spectroscopic studies of $In_{1-x}Ga_xAs_yP_{1-y}$', *Phys. Rev.*, B, **21**, 5656–5661, 1980.
24. R. J. Nicholas, J. C. Portal, C. Houlbert, P. Perrier, and T. P. Pearsall, 'An experimental determination of the effective masses of $Ga_xIn_{1-x}As_yP_{1-y}$ alloys grown on InP', *Appl. Phys. Lett.*, **34**, 492–494, 1979.
25. J. Shah, R. F. Leheny, R. E. Nahory, and M. A. Pollack, 'Hot-carrier relaxation in photoexcited $In_{0.53}Ga_{0.47}As$', *Appl. Phys. Lett.*, **35**, 475–477, 1980.
26. J. Shah, 'Hot electrons and phonons under high intensity photoexcitation of semiconductors', *Solid-State Electron.*, **21**, 45–50, 1978.
27. R. F. Leheny, A. A. Ballman, J. C. DeWinter, R. W. Nahory, and M. A. Pollack, 'Compositional dependence of the electron mobility in $In_{1-x}Ga_xAs_yP_{1-y}$', *J. Electron. Mater.*, **9**, 561–568, 1980.
28. D. E. Aspnes, 'GaAs lower conduction-band minima: ordering and properties', *Phys. Rev.*, B, **14**, 5331–5343, 1976.
29. D. Auvergne, J. Camassel, H. Mathieu, and M. Cardona, 'Temperature dependence of the band structure of germanium and zinc-blende semiconductors', *Phys. Rev.*, **B**, **9**, 5768–5177, 1974.
30. E. H. Perea, E. E. Mendez, and C. G. Fonstad, 'Electroreflectance of InGaAsP lattice match to InP', *Appl. Phys. Lett.*, **26**, 978–980, 1980.
31. P. M. Laufer, F. H. Pollack, R. E. Nahory, and M. A. Pollack, 'Electroreflectance investigation of $In_{1-x}Ga_xAs_yP_{1-y}$ lattice matched to InP', *Solid State Commun.*, **36**, 419–422, 1980.
32. H. H. Casper and H. H. Wieder, 'Optical studies on the band structure of $Ga_{0.08}In_{0.92}As_{0.18}P_{0.82}$', *Solid State Commun.*, **29**, 403–405, 1979.
33. Y. Yamazoe, T. Nishino, and Y. Harmakawa, 'Electroreflectance study of InGaAsP quaternary alloys lattice matched to InP', *IEEE J. Quantum Electron.*, **QE-17**, 139–144, 1980.
34. S. M. Kelso, D. E. Aspnes, C. G. Olson, D. W. Lynch, R. E. Nahory, and M. A. Pollack,

'Band structure and optical properties of $In_{1-x}Ga_xAs_yP_{1-y}$', *Japan. J. Appl. Phys.* **19**, Supp. 19–3, 327–331, 1980.

35. G. Bauer, 'Experimental studies of nonlinear transport in semiconductors', *physics of Nonlinear Transport in Semiconductors*, eds D. K. Ferry, J. R. Barker and C. Jacoboni, *NATO Advanced Studies Institute Series*, Plenum Press, New York, 1979.
36. J. G. Ruch and G. S. Kino, 'Measurement of the velocity–field characteristics of gallium arsenide', *Appl. Phys. Lett.*, **10**, 40–42, 1967.
37. A. G. R. Evans and P. N. Robson, 'Drift mobility measurements in epitaxial semiconductor layers using time-of-flight techniques', *Solid-State Electron.*, **17**, 805–812, 1974.
38. P. M. Smith, M. Inoue, and J. Frey, 'Electron velocity in Si and GaAs at very high electric fields', *Appl. Phys. Lett.*, **37**, 797–798, 1980.
39. A. Majerfeld, K. E. Potter, and P. N. Robson, 'Subthreshold velocity–field characteristics for bulk and epitaxial InP', *J. Appl. Phys.*, **45**, 3681–3682, 1974.
40. B. Houston, J. B. Restorff, R. S. Allgaier, J. R. Burke, D. K. Ferry, and G. A. Antypas, 'Hot electron and magneto-transport properties of $In_{1-x}Ga_xP_{1-y}As_y$ liquid phase epitaxial films', *Solid-State Electron.*, **21**, 91–94, 1978.
41. A. Sasaki, Y. Takeda, N. Shikagawa, and T. Takagi, 'Liquid phase epitaxial growth, electron mobility and maximum drift velocity of $In_{1-x}Ga_xAs$ ($x \simeq 0.5$) for microwave devices', *Japan. J. Appl. Phys.*, **16**, *Suppl.* 16-1, 239–243, 1977.
42. R. F. Leheny, J. Shah, R. L. Fork, C. V. Shank, and A. Migus, 'Dynamics of hot carrier cooling in photoexcited GaAs', *Solid-State Commun.*, **31**, 809–813, 1979.
43. C. V. Shank, R. L. Fork, B. I. Greene, F. R. Reinhart, and R. A. Logan, 'Picosecond nonequilibrium carrier transport in GaAs', *Appl. Phys. Lett.*, **38**, 104–105, 1981.
44. J. C. Gammel and J. M. Ballantyne, 'An integrated photoconductor and waveguide structure', *Appl. Phys. Lett.*, **36**, 149–151, 1980.
45. J. C. Gammel, H. Ohno and J. M. Ballantyne, *IEEE J. Quantum. Electron.*, **QE-17, 269**–272, 1981.
46. J. Degani, R. F. Leheny, R. E. Nahory, and J. P. Heritage, 'Velocity field characteristic of minority carriers (electrons) in p-$In_{0.53}Ga_{0.47}As$', *Appl. Phys. Lett.*, **39**, 569–572, 1981.
47. A. Nukermans, and G. S. Kino, 'Absolute measurement of the electron velocity–field characteristic of InSb'. *Phys. Rev.*, **B, 7**, 2703–2709, 1973.
48. J. Degani, R. F. Leheny, R. E. Nahory, M. A. Pollack, J. P. Heritage, and J. C. DeWinter, 'Fast photoconductive detector using p-$In_{0.53}Ga_{0.47}As$ with response to 1.7 μm', *Appl. Phys. Lett.*, **38**, 27–29, 1981.
49. J. Shah, R. E. Nahory, R. F. Leheny, and J. Degani, 'Influence of electron–hole interactions on hot carrier relaxation in p-$In_{0.53}Ga_{0.47}As$', *Appl. Phys. Lett.*, **40**, 505–507, 1982.
50. K. Y. Cheng, A. Y. Cho, S. B. Christman, T. P. Pearsall and J. E. Rowe, "Measurements of the Γ-L separation in $Ga_{0.47}In_{0.53}As$ by ultra violet photo emission", Applied Physics Letters, **40**, pgs. 429–431, 1982.
51. T. H. Windhorn, L. W. Cook and G. E. Stillman, "The electron velocity-field characteristics for n-$In_{0.53}Ga_{0.47}As$ at 300 K", Electron Device Letter, **EDL-3**, pgs. 18–20, 1982.
52. T. H. Windhorn, I. W. Cook and G. E. Stillman, "Electron drift velocities in $In_{0.53}Ga_{0.47}As$ at very high electric fields", *1981 International Electron Device Meeting Digest*, IEEE Electron Device Society, N. Y., pgs. 641–644, 1981.
53. J. Shah, B. Etienne, R. F. Leheny, R. E. Nahory and A. E. DiGiovanni, "Carrier-phonon interactions in 1.3 μm quaternary $In_{0.72}Ga_{0.28}As_{0.5}P_{0.4}$", accepted for publication–Journal of Applied Physics, 1982.

GaInAsP Alloy Semiconductors
Edited by T. P. Pearsall

Chapter 12

Electronic Structure of $Ga_xIn_{1-x}As_yP_{1-y}$ Alloys Lattice-matched to InP

T. P. PEARSALL

Bell Laboratories,
Murray Hill, NJ 07974, USA

12.1 INTRODUCION

The electronic band structure contains information on many parameters essential to the understanding of semiconductor device operation as well as the physics of the materials themselves. The energies of the conduction and valence band minima and the effective masses of electrons and holes are basic to the description of carrier mobility, impact ionization, radiative recombination, and non-radiative transitions such as Auger recombination. The physical properties contained in the electronic structure of the $Ga_xIn_{1-x}As_yP_{1-y}$ alloys lattice-matched to InP ($y \simeq 2.2x$) are of additional importance because this materials system provides a unique testing ground for the theoretical understanding of how these fundamental electronic properties vary with the chemical composition of a semiconductor alloy.

In initial studies of the fundamental energy gap of GaInAsP alloys lattice-matched to InP,[1-4] the experimentally measured direct gap was compared to a model of the compositional variation of the quaternary energy gap based on a linear interpolation of the measured ternary III–V alloy boundaries: $Ga_xIn_{1-x}As$, $InAs_yP_{1-y}$, $Ga_xIn_{1-x}P$, and $GaAs_yP_{1-y}$. It has long been recognized that this procedure always overestimates the bowing in the lattice-matched quaternary alloy. In a general semiconductor alloy, the crystal potential which produces the bandgap depends both on the average anion–cation bond length (lattice constant) and on non-periodic variations in the atomic potentials (alloy disorder).[5] The distinguishing feature in the GaInAsP/InP alloy system from ternary III–V alloys is that there is no change in bond length over the entire alloy system from InP to $Ga_{0.47}In_{0.53}As$ because the alloys are lattice-matched. Hence, an important variable affecting the crystal potential is held constant. We will show in this chapter that the compositional dependence of all the measured valence-to-conduction band transitions is accurately given by considering only the contribution of alloy disorder to the crystal potential using the Van Vechten–Bergstresser dielectric model.[5]

The curvature of the electronic band structure at the band edge ($\mathbf{k} = 0$) is inversely proportional to the effective mass. In section 12.3 we will present the experimental data for the electron and light hole mass as a function of alloy composition. Both masses show a smooth monotonic decrease from their values for InP ($m_e^* = 0.08$, $m_{lh}^* = 0.12$) to those for $Ga_{0.47}In_{0.53}As$ ($m_e^* = 0.041$, $m_{lh}^* = 0.051$). The measured results are in good agreement with hole and electron effective masses calculated using the $\mathbf{k}\cdot\mathbf{p}$ perturbation theory[6] to model the electronic structure at the band edges. As in the case of the energy gap data presented in section 12.2, some evidence exists in the measurements of effective mass for the presence of alloy disorder.

The measurements of various optical and electronic properties reviewed in this chapter of GaInAsP lattice-matched to InP indicate that these alloys have a well-defined energy band structure analogous to that for common III–V semiconductors such as GaAs or InP. This important result shows that any variations in the nominal alloy composition or crystal structure occur on a dimension much smaller than a free-electron radius, or about 100 Å. On the other hand, Raman scattering results[7,8] show the clear presence of interactions between optical phonons and binary semiconductor vibrational modes, thus indicating that these alloys do not behave like homogeneous semiconductors on the scale of the unit cell, which is about 5 Å. The effect of alloy disorder is therefore not to be seen in a 'smearing' in energy of optical transitions, but in the crystal potential resulting in a bowing in energy of the band structure in good agreement with the predictions of the Van Vechten–Bergstresser model.

12.2 OPTICAL TRANSITIONS IN GaInAsP ALLOYS

12.2.1 Bowing of the band structure and alloy disorder

The band structure[9] of GaAs along the three major symmetry axes is shown in Figure 12.1. The electronic structure of this material is typical of that for a III–V semiconductor. Shown in this figure are the principal optical transitions between the valence and conduction bands following the notation of Cardona.[10] In brief summary, the cross-section for optical transitions ($\Delta\hbar\mathbf{k} = 0$) is enhanced between the valence and conduction bands at points where the slopes $[\mathrm{d}E/\mathrm{d}\mathbf{k}]$ of the two energy bands are equal. There are only a few such regions, designated as critical points, in the band structure as can be seen in Figure 12.1. The transition energies associated with these critical points give fundamental, if somewhat limited, information about the band structure. The most important of these transitions for device applications is the direct bandgap energy E_0. The variation in the

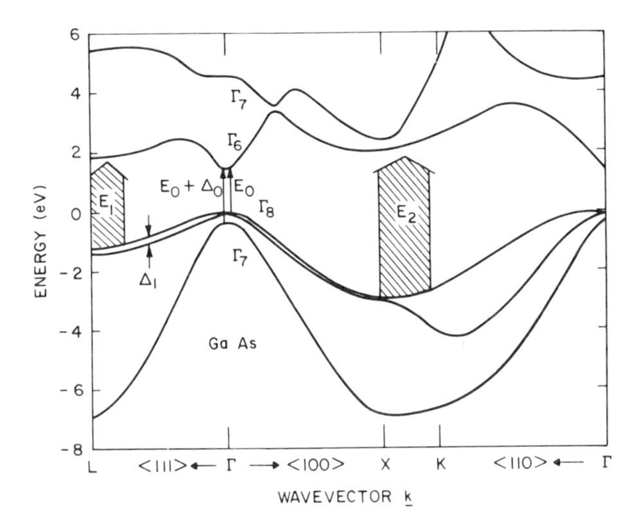

Figure 12.1 The electronic band structure of GaAs along the major symmetry axes, after Chelikowsky and Cohen.[9] Regions of the Brillouin zone where optical transitions are allowed between the valence and conduction bands are indicated by vertical arrows. Reproduced by permission of the authors and the American Physical Society

bandgap energy with composition of a ternary alloy MF_xG_{1-x} is expressed

$$E_0(x) = a + bx + cx(1 - x) \tag{12.1}$$

where a is the measured badgap at $x = 0$, and $(a + b)$ is the measured bandgap at $x = 1$. The last term depends on the bowing parameter c. Van Vechten and Bergstresser[5] have shown that the bowing is a measure of the change in crystal potential across the alloy field. This change is composed of a periodic component, which is related to the change in bond length (i.e. lattice constant) and an aperiodic component which is the root-mean-square fluctuation in the potential from its periodic amplitude. Since the lattice parameter in the GaInAsP alloy system may be kept fixed as the composition is varied, the periodic contribution to the bowing is, in principal, strictly zero for alloys lattice-matched to InP. The remaining aperiodic component is a direct measure of the effect of alloy disorder on the crystal potential.

Variations from perfect order in this quaternary alloy are subject to certain important restrictions so that not all spatial permutations of the constituent elements which give the nominal macroscopic composition are possible.[11,12] Because of the polar nature of the crystal, there is only mixing of Ga and In on group III sites and of As and P only on group V sites. This restriction is significant because it means that the bowing from the anion sublattice disorder is statistically independent of the disorder in the cation sublattice. (Here we ignore the 'lattice-pulling' effect where fluctuations in the group III sublattice influence the local composition on the group V sublattice in order to minimize local lattice-mismatch strain. Such a correlation effect would further reduce the bowing.) Hence, the total bowing in the quaternary is the simple sum of the contribution from each sublattice:

$$E_0(x, y) = a + by + c_{\rm III}x(1-x) + c_{\rm V}y(1-y). \tag{12.2}$$

The bowing parameter from alloy disorder is expressed according to standard perturbation theory arguments as the mean square fluctuation in the crystal potential divided by an appropriate bandwidth

$$c = \frac{C_{\rm FG}^2}{A} \tag{12.3}$$

where $C_{\rm FG}$ is the electronegativity difference between the elements F and G. The bowing term in equation (12.2) is therefore expressed as

$$D(y) = \text{bowing} = \frac{1}{A}[C_{\text{Ga–In}}^2 x(1-x) + C_{\text{As–P}}^2 y(1-y)]. \tag{12.4}$$

Remembering that for the alloy lattice-matched to InP $y = 2.2x$, equation (12.4) reduces to

$$D(y) = 0.219y - 0.149y^2 \text{ eV}. \tag{12.5}$$

The values for the electronegativity as well as the bandwidth $A = 1\,\text{eV}$ are taken

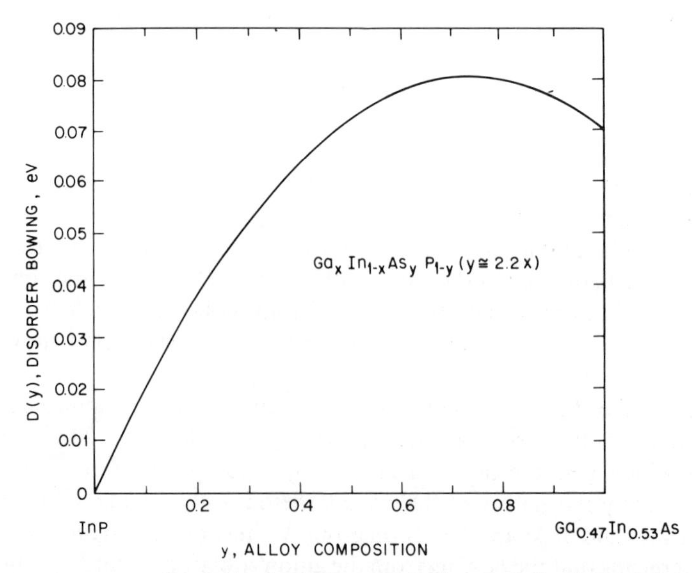

Figure 12.2 Disorder bowing in the GaInAsP system lattice-matched to InP calculated using the Van Vechten–Bergstresser dielectric model. The total bowing is the sum of the effects of disorder in the cation sublattice plus the disorder in the anion sublattice. Additional intrinsic bowing is strictly zero in these alloys because the lattice parameter does not change with composition

from reference 5. The disorder bowing is shown in Figure 12.2 as a function of arsenic alloy concentration.This expression gives the bowing for all transitions involving the valence and conduction bands: $E_0, E_0+\Delta_0, E_1, E_1+\Delta_1, E_2$, and $E_2+\delta$.

12.2.2 Experimental determination of optical transition energies

GaInAsP has a direct gap over the entire range of alloy compositions which are lattice-matched to InP. The E_0 gap has been measured extensively at room temperature by photoluminescence. The photoluminescence spectra of an n-type sample with $x = 0.15$ and $y = 0.34$ are shown in Figure 12.3 at 295 K, 77 K, and 2 K. The energy peak of the photoluminescence emission gives E_0 at 295 K. Some experimental results are shown in Figure 12.4 for E_0 determined from photoluminescence[3, 4] and electroreflectance[13, 14, 34] measurements. It can be seen that there is very little scatter in these data for E_0. We note that, however, the bandgap results obtained by Perea *et al.*[13] are consistently low. It appears that in the reduction of their electroreflectance data, the bandgap was identified as the lowest-energy peak. If the generally accepted '3-point method'[15] of Aspnes had been used instead, their data would probably be in good agreement with the other results shown in Figure 12.4. The full curve is calculated using equations (12.2) and (12.5)

$$T = 295\,\text{K}, \qquad E_0(y) = 1.35 - 0.775y + 0.149y^2\,\text{eV}. \qquad (12.6)$$

Below 10 K, the photoluminescence peak may no longer be a reliable measure of

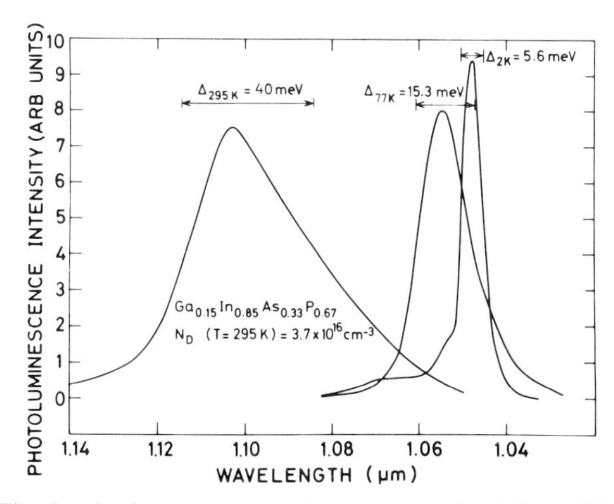

Figure 12.3 The photoluminescence spectra of an n-type sample of GaInAsP lattice-matched to InP with a room-temperature bandgap of 1.1 eV. The net carrier concentration is 2.8×10^{16} cm^{-3} and the Hall mobility at 295 K is 3850 cm^2 V^{-1} s^{-1}. The photoluminescence linewidths are typical of the total impurity concentration in the sample, and are similar to those measured in GaAs of the same impurity concentration

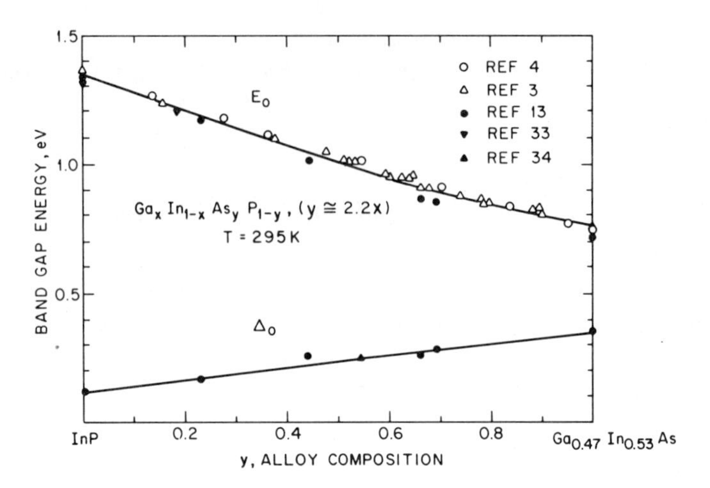

Figure 12.4 The energy gap (E_0) and spin–orbit splitting at the zone centre (Δ_0) in GaInAsP as a function of alloy composition at 295 K. The full curves are calculated using the expressions given in the text and the disorder bowing shown in Figure 12.2. The measured data show very little scatter and are in excellent agreement with the theoretical curves

the bandgap energy because the luminescence is dominated by donor–acceptor recombination whose characteristic luminescence energy may be shifted from about 10 to 40 meV below the band edge. For example, the 2 K luminescence in Figure 12.3 peaks at 1.18 eV while the actual bandgap is at 1.19 eV. The E_0 gap can be determined more accurately at low temperatures from the absorption edge, and these measurements[16–18] are shown in Figure 12.5. The full curve is calculated using the same expression for bowing:

$$T = 4\,\text{K}, \qquad E_0(y) = 1.425 - 0.7668y + 0.149y^2\ \text{eV}. \tag{12.7}$$

Electroreflectance and ellipsometric measurements have been used to resolve the transitions at energies above the fundamental gap. The data for $E_1(y)$[13, 14, 19] are in general agreement, but show more scatter than the $E_0(y)$ data. This difficulty may come from uncertainties in the actual alloy composition and errors introduced in taking the higher derivatives of the reflectance spectra in order to resolve the transition energies, as well as differences in the reduction of the data as noted above. However, the E_2 and $E_2 + \delta$ transitions are not well-resolved because of an overlap with the structure associated with E_0' (valence band to the next highest conduction band at $\mathbf{k} = 0$) over much of the alloy range.[14] The room-temperature data for E_1 are shown in Figure 12.6. The full curve is calculated using the bowing given in equation (12.5)

$$T = 295\,\text{K}, \qquad E_1(y) = 3.14 - 0.739y + 0.149y^2\ \text{eV}. \tag{12.8}$$

It is expected that the variation of E_2 with composition would be given by an

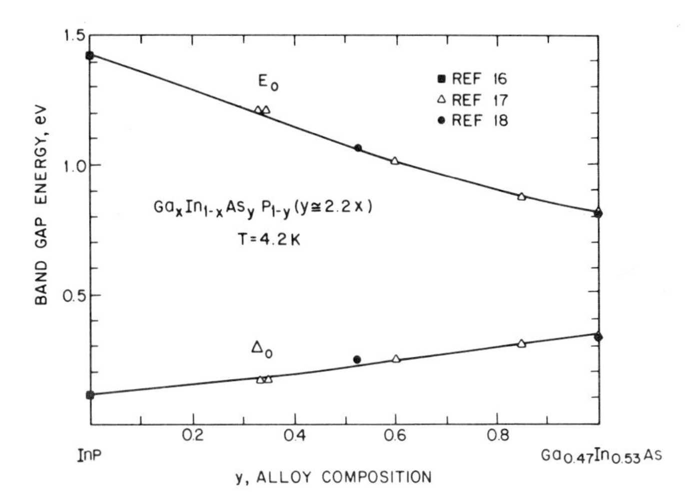

Figure 12.5 The energy gap (E_0) and spin–orbit splitting at the zone centre (Δ_0) in GaInAsP lattice-matched alloys at 4 K. The experimental data were obtained from absorption and magneto-oscillatory optical transmission. The theoretical variations of E_0 and Δ_0 with alloy composition are calculated using equations (12.7), and (12.10)–(12.12)

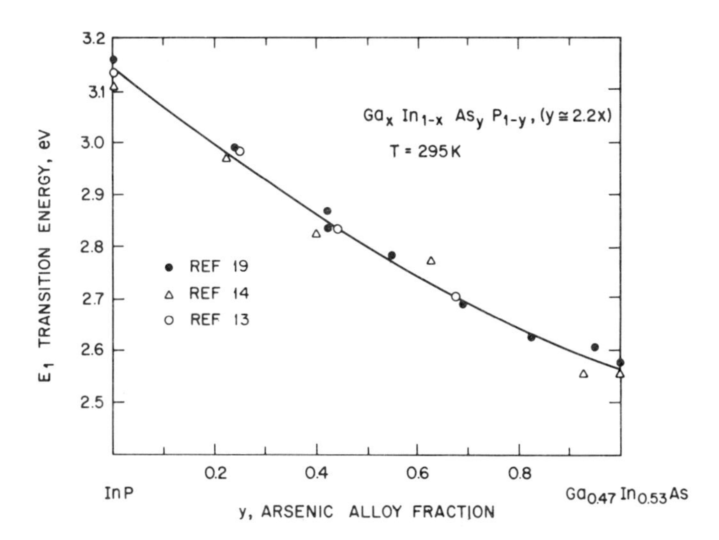

Figure 12.6 Electroreflectance and ellipsometric measurements of the E_1 transition energy in GaInAsP alloys lattice-matched to InP at 295 K. While the general agreement in the experimental data is generally good, there is more scatter than in the data for E_0. The full curve is the theoretical variation of the E_1 transition energy with alloy composition calculated using equation (12.8) with the same bowing as that for the E_0 gap. As in the case of Figure 12.4 and 12.5, the agreement between theory and experiment is extremely good

expression similar to equation (12.8), i.e.

$$T = 295\,\text{K}, \qquad E_2(y) = 5.04 - 0.309y + 0.149y^2\,\text{eV}. \tag{12.9}$$

Unfortunately, insufficient data are available at the present time for comparison with equation (12.9).

The Van Vechten–Bergstresser model for disorder-induced bowing has been extended to cover the behaviour of the spin–orbit splitting in the valence band with composition.[20, 21] At $\mathbf{k} = 0$ this splitting (Δ_0) is an effect primarily of the disorder-induced mixing of the conduction and valence band wavefunctions

$$\Delta_0(y) = A + By - f(y)(A + By). \tag{12.10}$$

In equation (12.10), A is the spin–orbit splitting of InP ($y = 0$) and B is the difference between the spin–orbit splitting for $Ga_{0.47}In_{0.53}As$ and InP. The function $f(y)$ is the fraction of conduction-band states mixed into the valence band by alloy disorder:

$$f(y) = \frac{D(y)}{\overline{E}_0(y)} \tag{12.11}$$

where $\overline{E}_0(y)$ is a weighted energy gap

$$\frac{3}{\overline{E}_0(y)} = \frac{2}{\langle E_0(y)\rangle} + \frac{1}{[\langle E_0(y)\rangle + \langle \Delta_0(y)\rangle]}. \tag{12.12}$$

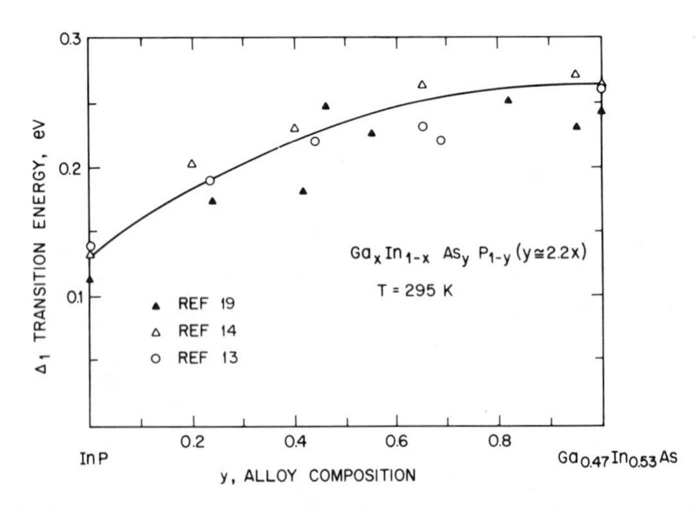

Figure 12.7 Measurements of the spin–orbit splitting of the valence band at the $\langle 111 \rangle$ critical point Δ_1. These data are determined by the difference in the energies of the E_1 and $E_1 + \Delta_1$ transition. Since $E_1 \gg \Delta_1$, the arithmetic difference of the measured transitions necessarily has a rather large percentage error, hence the scatter in the data for Δ_1. The theoretical variation of Δ_1 with alloy composition is calculated using the second-order perturbation results given in equation (12.13) with the bowing shown in Figure 12.4

In equation (12.12) $\langle E_0(y) \rangle$ and $\langle \Delta_0(y) \rangle$ are the compositionally linearly averaged energy gap and spin–orbit splitting. In Figure 12.4 we have also shown the measured spin–orbit splitting Δ_0 at 295 K. The full curve is calculated using equations (12.10)–(12.12) with the 295 K energy gaps, where appropriate. In Figure 12.5, we show the data and calculation at 4 K. The calculation in both cases shows that the bowing of the spin–orbit splitting is extremely small and very slightly positive.

The splitting at L, Δ_1, is calculated from second-order perturbation theory. Since the conduction and valence bands are separated in energy by more than 2.5 eV, interband mixing is negligible. The variation of the splitting Δ_1 is therefore not the same as that for Δ_0

$$\Delta_1(y) = \langle \Delta_1(y) \rangle + \frac{KD(y)}{\langle \Delta_1(y) \rangle} \tag{12.13}$$

In equation (12.13), K is a constant for all III–V materials. The value $K = 0.14$ is taken from reference 20. In Figure 12.7 we show the measured Δ_1 splitting for GaInAsP. As in the case of E_1, there is considerable scatter in these data. The Δ_1 splitting is calculated using equation (12.13) and equation (12.5) for the bowing.

12.2.3 Comparison between theory and experiment

It may be seen in Figures 12.4–12.7 that the Van Vechten–Bergstresser dielectric model can be used to describe the compositional variation of the critical points of the band structure of GaInAsP alloys lattice-matched to InP. The agreement between experiment and this theory is very good, showing that GaInAsP alloys have well-behaved electronic structure. The calculated bowing of the band structure at these critical points depends only on a single disorder parameter $D(y)$, because the intrinsic bowing, due to compositional variations in the lattice constant, is strictly zero in these lattice-matched compounds. This unusual feature of the GaInAsP/InP system leads to a reduced bowing of the bandgap E_0 compared to that estimated by earlier schemes[1-3] which have effectively ignored this very important point, and assumed implicitly both a disorder bowing and intrinsic bowing of approximately the same magnitude. The excellent agreement between theory and all available experimental optical data for this system would indicate that the experimental bowing is indeed a direct measure of the mean square fluctuations in the crystal potential caused by alloy disorder.

12.3 ELECTRON AND HOLE EFFECTIVE MASSES IN GaInAsP ALLOYS

12.3.1 The k·p theory

The energy band structure near the center of the Brillouin zone can be calculated in the **k·p** approximation

$$E(\mathbf{k}) = E_j + \frac{\hbar^2 k^2}{2m_0} + \sum_{i \neq j} \frac{|\langle u_j|P|u_i\rangle|^2}{E_j - E_i} \tag{12.14}$$

The momentum matrix elements $\langle u_j|P|u_i\rangle$ connect only states of differing symmetry such as the s-type conduction band with the p-type valence band. Because the energy denominator in (12.14) is large except for states whose energies are close to E_j, the **k·p** calculation is usually limited to include interactions between the valence bands (Γ_7^v and Γ_8^v) and conduction bands (Γ_6^c, Γ_7^c, and Γ_8^c) nearest the fundamental bandgap. These considerations simplify the evaluation of equation (12.14). The effective mass is inversely proportional to the energy band curvature,

$$\frac{1}{m_{ij}^*} = \frac{1}{\hbar^2}\frac{d^2}{dk_i dk_j}E(\mathbf{k}) \tag{12.15}$$

Near $\mathbf{k} = 0$, one obtains in a straightforward manner:[22]

$$\frac{1}{m_{s\text{-}o}} = \frac{1}{m_0} - \frac{P^2}{3m_0^2}\left(\frac{1}{E_0 + \Delta_0}\right) \tag{12.16}$$

$$\frac{1}{m_{hh}} = \frac{1}{m_0} \tag{12.17}$$

$$\frac{1}{m_{lh}} = \frac{1}{m_0} - \frac{P^2}{3m_0^2}\left(\frac{2}{E_0}\right) \tag{12.18}$$

and

$$\frac{1}{m_c} = \frac{1}{m_0} + \frac{P^2}{3m_0^2}\left(\frac{2}{E_0} + \frac{1}{E_0 + \Delta_0}\right) - \frac{P'^2}{3m_0^2}\left(\frac{2}{E(\Gamma_8^c) - E_0} + \frac{1}{E(\Gamma_7^c) - E_0}\right) + C \tag{12.19}$$

for the hole and electron band edge masses respectively. In these equations P is the momentum matrix element connecting the p-type valence band with the s-type conduction band, and P' is the momentum matrix element connecting the s-type conduction band with next higher-lying p-type conduction bands. C is a small correction for all higher-lying bands.

Because the conduction and valence band wavefunctions differ from one III–V compound to another, as Hermann and Weisbuch have shown explicitly,[23] it must be expected that P and P' will also be a function of composition. For the quaternary alloy, the assumption that the wavefunctions evolve monotonically with alloy composition leads to zero-order expressions for P^2 and P'^2 which are linear interpolations between measured values for InP, GaAs, and InAs

$$\frac{P_0^2(y)}{m_0} = 20.7 + 4.6y \, \text{eV} \tag{12.20}$$

$$\frac{P'^2_0(y)}{m_0} = 2.1 + 0.8y \, \text{eV} \tag{12.21}$$

$$C = \frac{-2}{m_0} \tag{12.22}$$

for GaInAsP alloys lattice-matched to InP.

12.3.2 The effect of alloy disorder on the interband matrix elements

As discussed in section 12.2, the effect of alloy disorder is to introduce a non-periodic fluctuation in the crystal potential. The bowing $D(y)$ gives the mean square magnitude of this variation. The effect of these fluctuations on the band structure is to mix the conduction and valence band wavefunctions. The fraction of conduction band s-character mixed into the p-type valence band is calculated directly from perturbation theory as the bowing energy divided by the bandgap and is given by equation (12.11). Since the momentum matrix element between s-

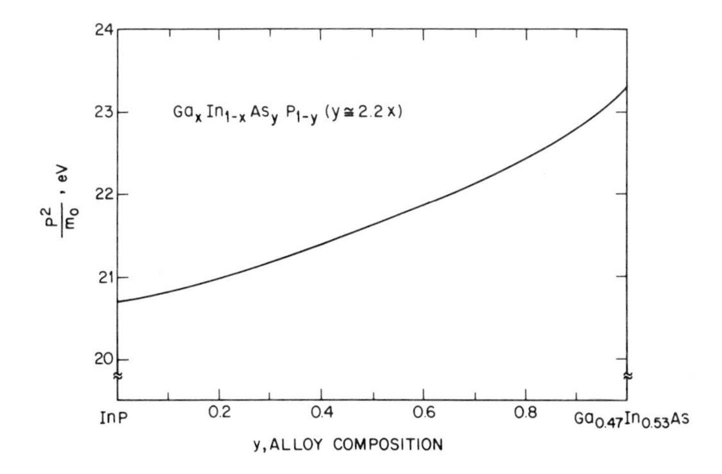

Figure 12.8 P^2/m_0, (equation (12.20)) as a function of alloy composition for GaInAsP. P^2/m_0 is maximum when the valence band and conduction band wave functions are orthogonal. Alloy disorder mixes these wavefunctions at the zone centre and so reduces P^2/m_0. The amount of mixing is given by equation (12.11)

states is zero, it is clear that the presence of alloy disorder will reduce $P_0^2(y)$ by the fraction of mixing $f(y)$

$$P^2(y) = [1 - f(y)]P_0^2(y) \tag{12.23}$$

where $P_0^2(y)$ is the matrix element estimated using (12.20). The variation of $P^2(y)$ with alloy composition corrected for alloy disorder is shown in Figure 12.8. The effect of disorder-induced mixing of $P'^2(y)$ is much smaller because of the larger energy denominator. For the GaInAsP system, the effects of alloy disorder increase the effective mass by about 10% above that which would be obtained using (12.20) in the $\mathbf{k} \cdot \mathbf{p}$ calculation.

12.3.3 Experimental determination of the effective masses in GaInAsP alloys

The experimental determination of effective masses in semiconductors most often involves the measurement of the magneto-transport properties of the appropriate carrier in the ballistic regime, where the carrier motion is largely independent of scattering. The experimental techniques used can be roughly divided into two areas—resonant methods such as cyclotron resonance or magnetophonon resonant scattering, and non-resonant methods such as optical pumping or the Shubnikov–deHaas effect. Resonant measurements give more precise resolution of the effective mass, and data from non-resonant techniques cannot be considered on the same footing as that from the more appropriate resonance methods.

Cyclotron resonance is the preferred technique for the measurement of electron and hole masses since the pioneering work of Dresselhaus, Kip, and Kittel[24] to determine these parameters in Ge and Si. Simply, a carrier of mass m^* has a well-defined orbital frequency in a magnetic field of intensity B given by

$$\omega_c = \frac{eB}{m^*} \tag{12.24}$$

providing that the measurement is made in a temperature range low enough (typically $2\,\mathrm{K} < T < 50\,\mathrm{K}$) and a high enough magnetic field (typically $B > 30\,\mathrm{kG}$) that the cyclotron orbit time is much less than the carrier scattering time. The resonance is usually detected by the absorption of the electromagnetic radiation whose frequency coincides with the cyclotron frequency. While in principle the interpretation of cyclotron resonance data is unambiguous, care must be taken to ensure that the absorption of the incident radiation is actually generated by cyclotron resonance. It is well-known that shallow donor levels in semiconductors are split by strong magnetic fields by an amount which can be quite similar to the typical photon energy of the incident electromagnetic field, i.e. 4 to 8 meV.[25] Since the splitting is proportional to the magnetic field, sharp absorption lines, corresponding to the $1s \rightarrow 2p$ impurity transitions or intra-p-band transitions may be taken erroneously for cyclotron resonance.[26] This difficulty is avoided either by making supplementary measurements at higher temperatures where the donors are ionized, or by using a different technique to confirm the mass values obtained by cyclotron resonance. Magnetophonon resonance has been used to complement cyclotron resonance data, and in the GaInAsP system these two techniques give excellent agreement.[8, 27] Magnetophonon resonance is seen in the sample conductivity as the magnetic field is varied. Increased carrier scattering by LO phonons occurs when the carrier cyclotron frequency is equal to the optical phonon frequency. Successful interpretation of this measurement requires the knowledge of an additional parameter, the optical phonon energies, which are usually determined by Raman scattering.[7, 8] This measurement technique avoids the problem of impurity transitions encountered in cyclotron resonance, because data are taken at temperatures ($\sim$ 100 K) where phonon scattering is stronger and all shallow donors are ionized.

Magneto-oscillatory absorption has been used to determine band structure at the zone centre.[16, 18] Measurements are taken at low temperatures in the quantum regime of magnetic field so that carriers are confined to Landau levels in the conduction and valence bands. The optical magnetoabsorption spectra show periodic oscillations related to optical transitions between these Landau levels. This method does not give as direct a measure of effective mass as cyclotron resonance. The data must be analysed in terms of the theory of Pidgeon and Brown[28] for coupled band magneto-optical effects. They are, however, extremely rich in information, and while this single technique can be used in principle to determine all the effective masses, the bandgap, and the spin–orbit splitting in a single sample, the interpretation of these data in InSb[28] still leads to controversies.

The Shubnikov–deHaas effect is a non-resonant technique which has been used to deduce the electron effective mass in GaInAsP lattice-matched alloys.[29, 30] This measurement is also taken in the quantum regime of magnetic fields where electrons are confined to Landau levels. The sample conductivity goes through a maximum where a Landau level energy, which is dependent on the magnetic field, is equal to the Fermi energy. The only direct information which is given by this effect is the Fermi surface cross-section and hence the carrier concentration. The effective mass can, in principle, be deduced from the variation of the envelope of conductivity oscillations with temperature. In order to obtain enough oscillation cycles to estimate the envelope accurately, the sample must be heavily doped (10^{18} cm^{-3}). In this case the effective mass is no longer characteristic of the band edge mass because of non-parabolicity of the band. At a carrier concentration of 6×10^{18} cm^{-3}, the effective mass in $Ga_{0.47}In_{0.53}As$ is nearly twice the band-edge value.[8] Effective masses deduced from Shubnikov–deHass measurements must be corrected for non-parabolicity effects. Such a model has been developed for GaAs.[31] While this procedure gives reasonable results for GaAs, it is unlikely that it can be used without modification for GaInAsP alloys. Thus until further work is done, Shubnikov–deHaas measurements in this alloy system can be regarded only as giving an approximate indication of the effective mass.

In Figure 12.9 we show the band edge electron effective mass as a function of alloy composition for GaInAsP lattice-matched to InP. The full curve is calculated from the **k·p** theory (equation (12.19)) using the 4 K measured values for E_0 and Δ_0 which were shown in Figure 12.5. The interband momentum matrix elements are corrected for alloy disorder and P^2/m_0 is shown in Figure 12.8. The theoretical variation of effective mass with alloy composition is seen to be in excellent agreement with experimental measurements by cyclotron resonance, magnetophonon resonance, and magneto-oscillatory absorption, while the Shubnikov–deHaas results give results which are consistently too low. The excellent agreement between theory and experiment can only be obtained if the effect of disorder is included in the calculation, showing that alloy disorder has an important effect on the band structure even for the ternary $Ga_{0.47}In_{0.53}As$.

The valence band structure in GaInAsP alloys has received considerably less attention than the conduction band. The valence bands are more complex, making interpretation of experiments subject to uncertainties not present in conduction band studies. The normally lower mobility of p-type material and the relatively high impurity levels (typically greater than $4 \times 10^{17}\,cm^{-3}$) have made conventional resonance detection of valence band hole masses impossible.

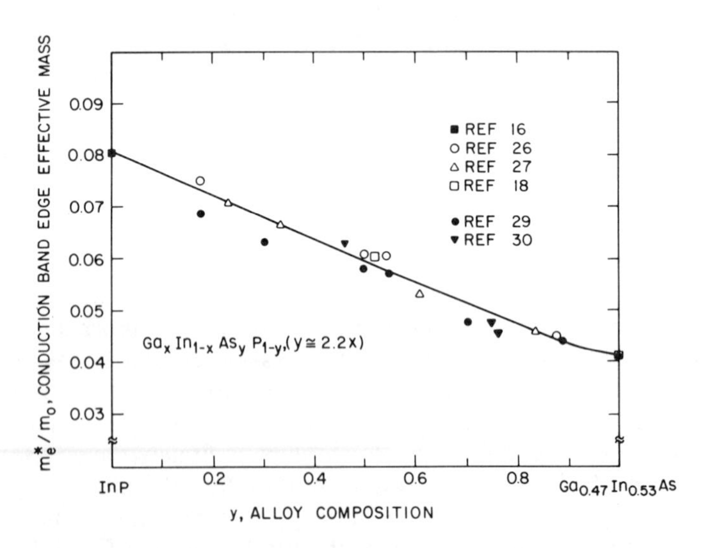

Figure 12.9 The conduction band edge electron effective mass in GaInAsP alloys as a function of composition for alloys lattice-matched to InP. The full curve is the calculated effective mass using the five-band **k · p** theory of Hermann and Weisbuch with $P^2(y)$ (Figure 12.8) corrected for alloy disorder. The agreement between this theory and the higher precision, $\pm 2\%$, measurements (cyclotron resonance, magneto-oscillatory absorption, and the magnetophonon effect) is excellent. The results obtained by the Shubnikov–deHaas effect are necessarily subject to a much larger error ($\pm 20\%$) and can be seen to give values of the effective mass which are consistently too low

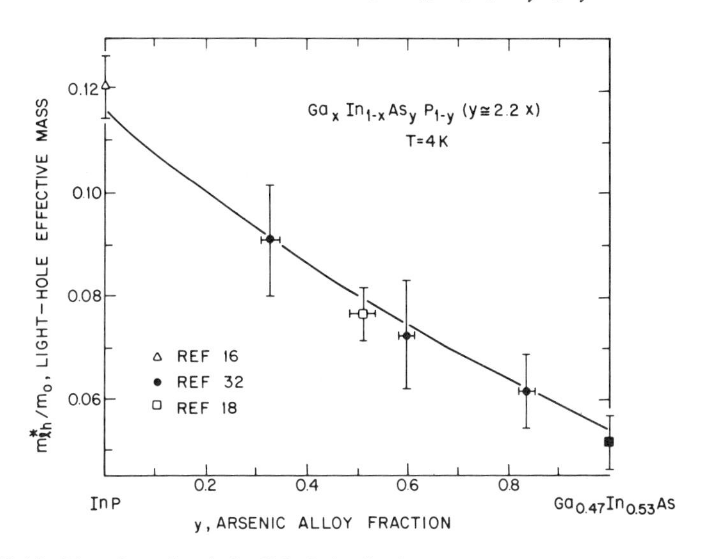

Figure 12.10 The valence-band edge light-hole effective mass in GaInAsP alloys lattice-matched to InP. Experimental data were obtained by magneto-oscillatory absorption and optical pumping. The relative precision of these two techniques is indicated by the error bars on the data. The light-hole mass can be calculated by the **k** · **p** theory and depends only on the momentum matrix element $P^2(y)$ and the bandgap. This theoretical result is shown as the full curve and is calculated using equation (12.16) and $P^2(y)$ corrected for alloy disorder

Optical pumping, which is the excitation of an electron distribution with a preferential spin population, has been used to deduce the light-hole effective mass from the dependence of the spin-polarized recombination radiation on the joint density of states.[17, 32] The reduction of the optical pumping data depends on knowledge of the electron and heavy-hole masses. This additional uncertainty limits the optical pumping technique to a precision of $\pm 10\%$. Magnetoabsorption measurements of valence band masses are more accurate to about $\pm 5\%$, but unfortunately these data exist for only two compositions. The light-hole effective masses obtained by these two techniques are in excellent agreement and are shown in Figure 12.10. The full curve is calculated from the **k** · **p** theory using equation (12.18). As is the case with Figure 12.9, excellent agreement is seen in Figure 12.10 between the data and the **k** · **p** theory. However the data lack both the precision and coverage of the alloy compositions which would permit the kind of conclusive comparison which can be made between theory and experiment of the effect of disorder in the case of the electron effective masses.

12.4 CONCLUSIONS

The band structure of GaInAsP alloys has been investigated by many different experimental methods to determine the energy separation of the conduction and valence bands and their shape at different regions of the Brillouin zone. In this chapter we have reviewed these data in terms of the Van Vechten and Bergstresser dielectric model of the band structure of alloys. We have shown how this model can be extended to quaternary alloys such as GaInAsP by noting that the disorder in the cation sublattice is independent of the disorder in the anion sublattice. Because there is no variation in lattice constant with the composition between InP and $Ga_{0.47}In_{0.53}As$, the intrinsic bowing in this alloy is, in principle, zero. A somewhat similar feature has been noted for the $Al_xGa_{1-x}As$ system. Hence the total measured bowing of the band structure is the sole result of alloy disorder. The simple expression for the bowing (equation (12.5)) is valid for all interactions between the valence and conduction bands. We have shown that this simple expression can be used to give the correct compositional variation of all known band structure data for this alloy system. The excellent agreement between theory and experiment has several important consequences. Most important of these is that the GaInAsP alloy band structure is similar to that of other better known III–V compounds. The effect of alloy disorder on the band structure does not therefore lead to a smearing of energy levels but rather a reduction in energy levels because of disorder-induced mixing of valence and conduction band states. This bowing of the band structure from alloy disorder is the mean square fluctuation in the periodic potential of the alloy lattice. Since these fluctuations would act to scatter charge carriers, one should expect to use this bowing in the calculation of alloy scattering mobility in these compounds, an important fundamental problem which is discussed in Chapters 8 and 9 of this book.

ACKNOWLEDGMENTS

The author wishes to thank S. M. Kelso and D. E. Aspnes for permission to use their data prior to publication. He also expresses his gratitude to C. Hermann, S. M. Kelso, J. C. Phillips, and M. D. Sturge for valuable comments and criticisms of this manuscript.

REFERENCES

1. R. L. Moon, G. A. Antypas, and L. W. James, 'Bandgap and lattice constant of GaInAsP as a function of alloy composition', *J. Electron. Mater.*, **3**, 635–644, 1974.
2. T. H. Glisson, J. R. Hauser, M. A. Littlejohn, and C. K. Williams, 'Energy bandgap and lattice constant contours of III–V quaternary alloys', *J. Electron. Mater.*, **7**, 1–17, 1978.
3. R. E. Nahory, M. A. Pollack, W. D. Johnston, Jr, and R. L. Barns, 'Bandgap versus composition and demonstration of Vegard's law for $In_{1-x}Ga_xAs_yP_{1-y}$ lattice-matched to InP', *Appl. Phys. Lett.*, **33**, 659–661, 1978.
4. K. Nakajima, A. Yamaguchi, K. Akita, and T. Kotani, 'Composition dependence of the band gaps of $In_{1-x}Ga_xAs_{1-y}P_y$ quaternary solids lattice-matched on InP substrates', *J. Appl. Phys.*, **49**, 5944–5950, 1978.
5. J. A. Van Vechten and T. K. Bergstresser, 'Electronic structures of semiconductor alloys', *Phys. Rev.*, **B**, **1**, 3351–3358, 1970.

6. T. P. Pearsall, '$Ga_{0.47}In_{0.53}As$: a ternary semiconductor for photodetector applications', *IEEE J. Quantum Electron.*, **QE-16**, 709–720, 1980.
7. A. Pinczuk, J. M. Worlock, R. E. Nahory, and M. A. Pollack, 'Lattice vibrations of $In_{1-x}Ga_xAs_yP_{1-y}$ quaternary compounds', *Appl. Phys. Lett.*, **33**, 461–463, 1978.
8. T. P. Pearsall, R. Bisaro, P. Merenda, G. Laurencin, R. Ansel, J. C. Portal, C. Houlbert, and M. Quillec, 'The characterization of $Ga_{0.47}In_{0.53}As$ grown lattice-matched on InP substrates', *Gallium Arsenide and Related Compounds 1978*, Conf. Ser. no. 45, Institute of Physics, Bristol, 1979, pp. 94–102.
9. J. R. Chelikowsky and M. L. Cohen, 'Non-local pseudopotential calculations for the electronic structure of eleven diamond and zinc-blende semiconductors', *Phys. Rev.*, B, **14**, 556–582, 1976.
10. M. Cardona, 'Optical absorption above the fundamental edge', *Semiconductors and Semimetals*, vol. 3, eds R. K. Willardson and A. C. Beer, Academic Press, New York, 1967, pp. 125–153.
11. J. C. Phillips, 'Topology of covalent non-crystalline solids I: short-range order in chalcogenide alloys', *J. Non-Cryst. Solids*, **34**, 153–181, 1979.
12. J. E. Fisher, M. Glicksman, and J. A. Van Vechten, 'Disorder effects in non-isovalent semiconductor alloys', *Proc. 13th Int. Conf. on the Physics of Semiconductors*, Rome, 1976, ed. F. G. Furmi, North-Holland, Amsterdam, 1976, pp. 541–544.
13. E. Perea, E. Mendez, and C. G. Fonstad, 'Electroreflectance of indium gallium arsenide phosphide lattice-matched to indium phosphide', *Appl. Phys. Lett.*, **36**, 978–980, 1980.
14. P. M. Laufer, F. H. Pollak, R. E. Nahory, and M. A. Pollack, 'Electroreflectance investigation of $In_{1-x}Ga_xAs_yP_{1-y}$ lattice-matched to InP', *Solid State Commun.*, **36**, 419–422, 1980.
15. D. E. Aspnes and J. E. Rowe, 'High resolution interband energy measurements from electroflectance spectra', *Phys. Rev. Lett.*, **27**, 188–190, 1971.
16. P. Rochon and E. Fortin, 'Photovoltaic effect and interband magneto-optical transitions in InP', *Phys. Rev.*, B, **12**, 5803–5810, 1975.
17. C. Hermann, G. Lampel, and T. P. Pearsall, 'Optical pumping measurement of valence-band spin–orbit splitting in $Ga_xIn_{1-x}As_yP_{1-y}$', *Proc. 15th Int. Conf. on the Physics of Semiconductors*, Kyoto, 1980; *J. Phys. Soc. Jpn.*, **49**, *Suppl.* A, 631–634, 1980.
18. K. Alavi, R. L. Aggarwal, and S. H. Groves, 'Oscillatory magneto-transmission of $In_{1-x}Ga_xAs_yP_{1-y}$ alloys', *J. Magn. Magn. Mater.*, **11**, 136–138, 1979, K. Alavi, R. L. Aggarwal, and S. H. Groves, 'Interband magneto-absorption of $In_{0.53}Ga_{0.47}As$', *Phys. Rev.*, B, **21**, 1311–1315, 1980.
19. S. M. Kelso, D. E. Aspnes, C. G. Olson, D. W. Lynch, R. E. Nahory, and M. A. Pollack, 'Band structure and Optical properties of $In_{1-x}Ga_xAs_yP_{1-y}$', *Japan J. Appl. Phys.* **19**, Supp. 19–3, 327–331, 1980.
20. J. A. Van Vechten, O. Berolo, and J. C. Woolley, 'Spin–orbit splitting in compositionally disordered semiconductors', *Phys. Rev. Lett.*, **29**, 1400–1403, 1972.
21. O. Berolo, J. C. Woolley, and J. A. Van Vechten, 'Effect of disorder on the conduction-band effective mass, valence-band spin–orbit splitting, and the direct band gap in III–V alloys', *Phys. Rev.*, B, **8**, 3794–3798, 1973.
22. D. Long and J. L. Schmit, 'Mercury-cadmium telluride and closely related alloys', *Semiconductors and Semimetals*, vol. 5, eds R. K. Willardson and A. E. Beer, Academic Press, New York, 1970, pp. 189–194.
23. C. Hermann and C. Weisbuch, '$\mathbf{K}\cdot\mathbf{p}$ perturbation theory in III–V compounds and alloys: a re-examination', *Phys. Rev.*, B, **15**, 823–833, 1977.
24. G. Dresselhaus, A. F. Kip, and C. Kittel, 'Cyclotron resonance of electrons and holes in silicon and germanium crystals', *Phys. Rev.*, **98**, 368–384, 1955.
25. R. J. Nicholas, A. M. Davidson, S. J. Sessions, and R. A. Stradling, 'Shallow donor

spectroscopy in $Ga_xIn_{1-x}As_yP_{1-y}$', *IEEE J. Quantum Electron.*, **QE-17**, 145–149, 1981.
26. R. J. Nicholas, S. J. Sessions, and J. C. Portal, 'Cyclotron resonance and the magnetophonon effect in $Ga_xIn_{1-x}As_yP_{1-y}$', *Appl. Phys. Lett.*, **37**, 178–180, 1980.
27. R. J. Nicholas, J. C. Portal, C. Houlbert, P. Perrier, and T. P. Pearsall, 'An experimental determination of the effective masses of $Ga_xIn_{1-x}As_yP_{1-y}$ alloys grown on InP', *Appl. Phys. Lett.*, **34**, 492–494, 1979.
28. C. R. Pidgeon and R. N. Brown, 'Interband magneto-absorption and Faraday rotation in InSb', *Phys. Rev.*, **146**, 575–583, 1966.
29. E. H. Perea, E. Mendez, and C. G. Fonstad, 'Electron effective masses in $In_xGa_{1-x}As_{1-x}P_y$ for $0 \leq y \leq 1$', *J. Electron. Mater.*, **9**, 459–466, 1980.
30. J. B. Restorff, B. Houston, R. S. Allgaier, M. A. Littlejohn, and S. B. Phatak, 'The electron effective mass in $In_{1-x}Ga_xAs_yP_{1-y}$ from Shubnikov–de Hass measurements', *J. Appl. Phys.*, **51**, 2277–2278, 1980.
31. A. Raymond, J. L. Robert, and G. Pistoulet, 'A new method for measuring the compensation and spatial fluctutations of impurities in n-type III–V compounds: applications to bulk and epitaxial GaAs', *Gallium Arsenide and Related Compounds (Edinburgh) 1976*, Conf. Ser. no. 33a, Institute of Physics, Bristol, 1977, pp. 105–112.
32. C. Hermann and T. P. Pearsall, 'Optical pumping and the valence-band light-hole effective mass in $Ga_xIn_{1-x}As_yP_{1-y}$ ($y \simeq 2.2x$)', *Appl. Phys. Lett.*, **38**, 450–452, 1980.
33. H. H. Caspers and H. H. Weider, 'Optical studies on the band structure of $Ga_{0.08}In_{0.92}As_{0.18}P_{0.82}$', *Solid State Commun.*, **29**, 403–406, 1979.
34. T. Nishino, Y. Yamazoe, and Y. Hamakawa, 'Electroreflectance of $In_{0.79}Ga_{0.21}As_{0.54}P_{0.46}$', *Appl. Phys. Lett.*, **33**, 861–862, 1978.

GaInAsP Alloy Semiconductors
Edited by T. P. Pearsall

Chapter 13

Photoluminescence and Optical Gain of GaInAsP

ERNST O. GÖBEL

Max-Planck-Institut Für Festkörperforschung, Heisenbergstrasse 1, D-7000 Stuttgart 80, FRG

13.1 INTRODUCTION

Photoluminescence has been developed as an important and powerful tool for the investigation and characterization of semiconductor materials. Direct-gap semiconductors (such as GaAs), which are important for optoelectronic applications, have been particularly intensively studied by photoluminescence. Most of the understanding of the various intrinsic and extrinsic excitations, which give rise to spontaneous and stimulated emission of radiation, is based on photoluminescence, electroluminescence, and optical gain studies (see e.g. references 1–4). This strong fundamental scientific background has elevated photoluminescence to a standard technique for the characterization of semiconductor materials.

Photoluminescence can be used to study spontaneous emission processes, which are essential for the understanding of semiconductor light sources. In contrast to electroluminescence, photoluminescence does not require further processing of the samples such as the formation of p–n junctions, heterojunctions or Schottky barriers, and thus provides more direct information on the intrinsic properties of the material. On the other hand, the relevance of photoluminescence results to device applications must be considered. This is especially important for semiconductor lasers. Generally, stimulated emission processes cannot be studied correctly by photoluminescence, because the occurrence of stimulated recombination rather complicates the interpretation of photoluminescence spectra.[3]

Thus, especially for the characterization of semiconductor laser material, it is more appropriate to measure the optical gain spectra directly. Three different experimental techniques have been developed for this purpose, namely:[4] (1) optical gain spectroscopy, which utilizes the amplification of the spontaneous emission in an optically pumped active region with variable length[5-7]; (2) measurements of the modulation depth of the Fabry–Perot resonances in a laser cavity[8-11]; and (3) transmission spectroscopy in which the amplification of an external laser beam of variable frequency in an optically pumped active volume is measured.[12,13] The potential of these techniques for the study of the origin, magnitude, spectral shape, and saturation behaviour of the optical gain has been widely demonstrated in the case of GaAs and GaAsGaAlAs-double heterostructure lasers.[3-17]

This article will be structured as follows. Some experimental aspects will be discussed in section 13.2. Data on bandgap energy vs composition of the GaInAsP as determined by photoluminescence experiments will be presented in section 13.3. Section 13.4 summarizes recent data on the photoluminescence of single-heterostructure GaInAsP epitaxial layers grown on InP as well as on GaAs substrates by liquid-phase and vapour-phase epitaxy. These results are compared to the photoluminescence data of GaInAsP/InP double-heterostructures in section 13.5. Optical gain spectra of free GaInAsP layers as well as double heterostructures will be discussed in section 13.6. Special attention will be paid to the very recent results on the temperature dependence of optical gain spectra of GaInAsP/InP double heterostructures. Finally in section 13.7 some summarizig and concluding remarks will be made.

13.2 EXPERIMENTAL CONSIDERATIONS

Photoluminescence today can be considered as a standard technique, so its experimental details will thus not be described in detail. In the case of GaInAsP epitaxial layers, any excitation source with photon energies higher than the fundamental bandgap energy such as Ar^+ or Kr^+ ion lasers, high-pressure mercury lamps, etc., can be used. For the study of GaInAsP/InP heterostructures, the wavelength of the excitation light must be longer than the absorption edge wavelength of the InP cladding layers, too. At room temperature this requires emission wavelengths longer than about 0.93 μm. For the technologically important GaInAsP/InP double heterostructures emitting light near 1.3 μm at room temperature, Nd:YAG lasers or dye lasers can be used as an excitation source. Depending on the composition of the GaInAsP, various photo-detectors have been used. As long as the emission wavelength is smaller than about 1.2 μm, photomultipliers provide convenient detection. Longer emission wavelengths require PbS cells or Ge-detectors.

The experimental set-up for the optical gain spectroscopy is depicted schematically in Figure 13.1. A pulsed N_2-pumped dye laser is used for optical excitation. The emission wavelength is chosen to about 0.95 μm.[18] The laser beam is focused by a cylindrical lens onto a slit with variable width. The slit then is imaged onto the sample, such that a small stripe of the InGaAsP active layer is excited. The emitted light is focused on to a spectrometer and detected by a PbS detector. A box car integrator is used for averaging.

By measuring the amplified spontaneous emission spectra for different excitation lengths L, unsaturated as well as saturated optical gain spectra can be

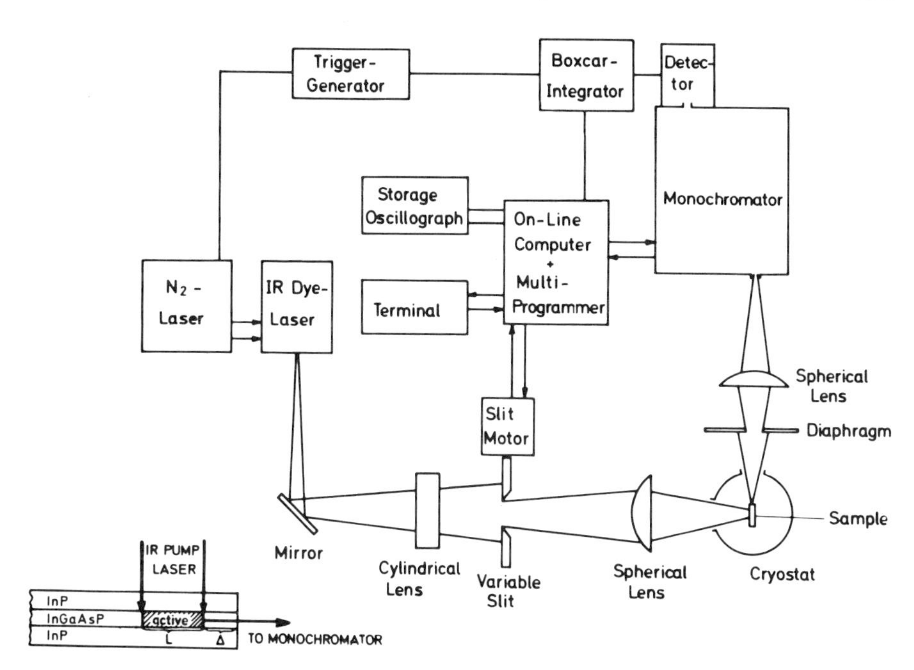

Figure 13.1 Experimental set-up for optical gain spectroscopy. The insert gives a schematic graph of the pumping configuration: the GaInAsP layer is excited directly using an IR dye laser. An unpumped region of width is introduced to minimize the influence of the light reflected from the surface

determined. It is particularly important in these experiments to introduce an unpumped region of length Δ between the excited stripe and the output surface of the crystal to minimize the intensity of light reflected back into the active volume.[7]

In the unsaturated region, the dependence of the amplified intensity I on the length of the excited stripe L simplifies to

$$I(L, \lambda) = (1-R)e^{-\Delta\alpha} \frac{I_{\text{spon}}(\lambda)}{g(\lambda)} (e^{g(\lambda)L} - 1) \qquad (13.1)$$

where R is the reflectivity of the front surface, Δ is the width and α the absorption coefficient of the unpumped region, and $I_{\text{spon}}(\lambda)$ and $g(\lambda)$ are the spontaneous emission intensity and gain factor, respectively. Two emission spectra $I(L, \lambda)$ thus determine the gain curve $g(\lambda)$.

13.3 DETERMINATION OF THE BANDGAP ENERGY BY PHOTOLUMINESCENCE

One purpose of photoluminescence studies is the determination of the bandgap energy versus composition of $Ga_xIn_{1-x}As_yP_{1-y}$ quaternary layers. As a result of these experiments a detailed knowledge of the dependence of the fundamental bandgap energy on composition x and y is expected. The earliest data by Moon *et al.*[19] on the bandgap variation of GaInAsP were based on experimental results of ternary materials. A list of the experimental data up to the end of 1976 is included in reference 20.

The bandgap energy of liquid-phase epitaxy $Ga_xIn_{1-x}As_yP_{1-y}$ lattice-matched on InP in the entire range $0 < x < 0.47$ and $0 < y < 1$ has been determined by photoluminescence by Nahory *et al.*[21] and Nakajima *et al.*[20] Figure 13.2 shows the experimental data reported in reference 21. The experimental points correspond to the energetic position of the maximum in the photoluminescence spectra. The full curve is a calculated curve via:

$$E_g(y) = 1.35 - 0.72y + 0.12y^2 \text{ eV} \tag{13.2}$$

A simple cubic relation to represent the bandgap surfaces $E_g(x, y)$, based on the values for the four ternary boundaries of GaInAsP (borken curve), does not describe the experimental data correctly,[20,21] (see Chapter 12 of this volume).

In the case of lattice-matched material, it is sufficient to represent the bandgap energy as a function of x or y alone, because the composition is additionally determined by the condition for equal lattice constants of the substrate and the

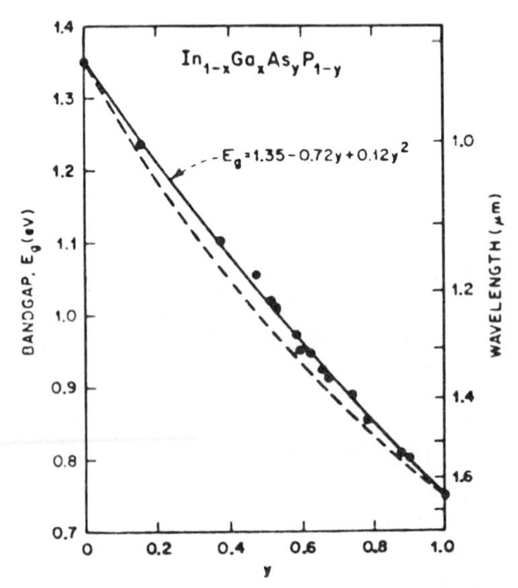

Figure 13.2 Bandgap energy E_g versus composition y for $Ga_xIn_{1-x}As_yP_{1-y}$ nearly lattice-matched to InP. The full curve is a fit to the data points (cf. equation (13.2)) and the broken curve is calculated from the constituent ternaries. From Nahory *et al.*[21] Reproduced by permission of the authors, Bell Laboratories, and the American Institute of Physics

quaternary material: $a_0(x, y) = a_0$ (substrate). The experimentally determined dependence of the lattice constant on the composition x and y in agreement with Vegard's law[20,21] is given by

$$a_0(x, y) = xya_0(\mathrm{GaAs}) + x(1-y)a_0(\mathrm{GaP}) + (1-x)ya_0(\mathrm{InAs}) + (1-x)(1-y)a_0(\mathrm{InP}) \quad (13.3)$$

The position of the room-temperature bandgap of GaInAsP versus the liquid atomic fraction of Ga in the melt has been measured by photoluminescence by Pollack *et al.*[22] These results together with the data for the dependence of liquid composition (As and P) as well as solid composition (x and y) on the liquid atomic fraction X_{Ga} [20,22] provide the recipe for growing lattice-matched GaInAsP with different bandgap energies.

Corresponding data in the bandgap range 0.75–1.35 eV for vapour-phase epitaxial GaInAsP on InP[23] have been reported by Hyder *et al.*[24] and on GaAs substrates by Sugiyama *et al.*[25] For liquid-phase epitaxial GaInAsP on GaAs substrates, the bandgap variation in the energy range 1.569–1.893 eV with composition has been measured by photoluminescence by Mukai *et al.*[26]

Though photoluminescence provides a convenient method to determine the relative position of the bandgap energy, the absolute value for E_g is more difficult to obtain. This is due to the fact that the origin of the transitions giving rise to luminescence in many cases is not identified in detail. In GaInAsP the near-band-edge emission observed so far does not show up sharp excitonic structures at low temperatures and thus generally is attributed to band-to-band transitions. However, the results cannot be explained on the basis of simple parabolic bands, as will be discussed in more detail in section 13.4. Thus the absolute position of the bandgap energy cannot be determined clearly without a careful analysis of the emission lineshape. A more accurate determination of the bandgap energy can be obtained by electroreflectance experiments under low-field conditions.[27] Comparison of photoluminescence and electroreflectance data[28,29] indicates that at room temperature the position of the bandgap determined by electroreflectance coincides with the maximum of the photoluminescence peak. For temperatures below 200 K, the photoluminescence peak is at even lower energies than the electroreflectance bandgap energy, because at the lower temperatures photoluminescence is more characteristic of the impurities. Nevertheless, photoluminescence remains a convenient method to obtain quickly information on the relative bandgap energies at room temperature. Furthermore, for applications in light-emitting devices the spectral position and shape of the photoluminescence might be even more relevant than the exact position of the bandgap energy.

Optical absorption (see e.g. references 30, 32) as well as photocurrent experiments (see e.g. references 31, 33) have been also used to determine the bandgap energy of GaInAsP.

13.4 PHOTOLUMINESCENCE OF GaInAsP EPITAXIAL LAYERS

Photoluminescence studies on GaInAsP epitaxial layers have been further stimulated by the observation that the electroluminescence spectra of GaInAsP heterostructures exhibit a rather large linewidth as compared to GaAs/AlGaAs LEDs.[34-42] (Note, however, that in some publications the halfwidth in angstroms is compared, where actually the energy width should be compared.) Photoluminescence experiments should help to clarify whether the broad emission band observed in electroluminescence is related to the fact that in a LED recombination takes place close to the heteroboundary where composition as well as impurity gradients occur.

Since in GaInAsP no well-resolved sharp emission lines resulting from, for example, bound exciton recombination as seen in GaAs[1] or InP[43] have so far been observed, the photoluminescence results should be discussed in terms of band-to-band transitions including impurity and tail states. A discussion of the spectral shape of these various transitions can be found in the literature.[1] As far as a band-to-band transition between simple parabolic band is concerned, Boltzmann statistics describe the spontaneous emission lineshape by

$$I(\hbar\omega) \sim (\hbar\omega - E_g)^{1/2} \exp\left[-(\hbar\omega - E_g)/kT\right] \tag{13.4}$$

This lineshape function $I(\hbar\omega)$ can be characterized as follows:

(i) the low-energy side of the spectrum increases as $(\hbar\omega - E_g)^{1/2}$,

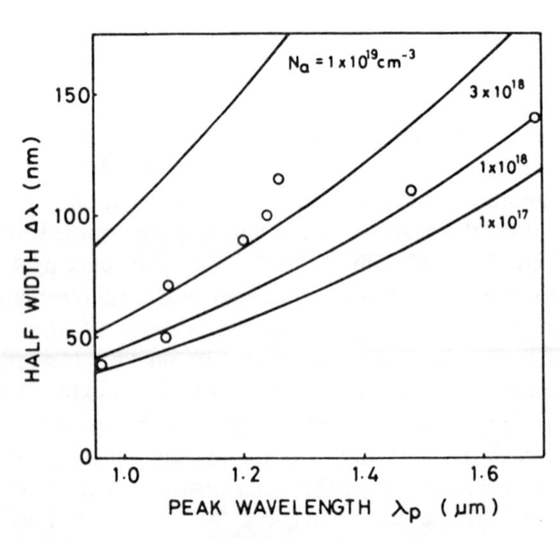

Figure 13.3 Dependence of the spectral halfwidth of the spontaneous emission on peak wavelength calculated for various acceptor concentrations. The circles refer to experimental data points. From Takagi.[50] Reproduced by permission of the author and the Publications Board, Japanese Journal of Applied Physics

(ii) the high-energy side decreases exponentially according to $\exp[-(\hbar\omega - E_g)/kT]$,
(iii) the maximum occurs at $\hbar\omega_{max} = E_g + kT/2$, and
(iv) the linewidth is given by $\Delta_{1/2} = 1.8kT$.

High doping[44] as well as high excitation of semiconductors[45-47] complicates the lineshape function. Emission below the actual bandgap results from impurity or carrier interaction. The most recent discussion on the photoluminescence lineshape at high doping has been given in references 48 and 49. Based on the model presented in reference 48, Takagi[50] and Dutta[51] have calculated emission spectra of GaInAsP at various doping levels. Figure 13.3 shows the calculated halfwidth of the spontaneous emission vs peak wavelength for different acceptor concentrations, together with some experimental data. As can be seen, acceptor concentrations up to about $5 \times 10^{18}\,cm^{-3}$ have to be required to explain for the observed spontaneous linewidth.

Low-temperature ($T = 2\,K$) photoluminescence spectra of epitaxial GaInAsP with different impurity concentration are depicted in Figure 13.4.[52] The samples were grown by LPE lattice-matched on ⟨100⟩ (P152, P166, P178) and ⟨111⟩B (P165) InP substrates.[53] The thickness of the epitaxial layers was 3 to 5 μm. In these experiments, excitation was provided by an Ar^+ ion laser (typically a few milliwatts excitation power) and thus only a region near the surface was excited directly. Figure 13.4 shows the influence of the impurity content on the emission lineshape and width. Whereas for the highest doped sample P152, a halfwidth of almost 20 meV is found, in the purest samples the linewidth is between 6 and 9 meV (see e.g. sample P166). These are the smallest linewidths for GaInAsP on InP reported so far. From a lineshape analysis, an effective carrier temperature of

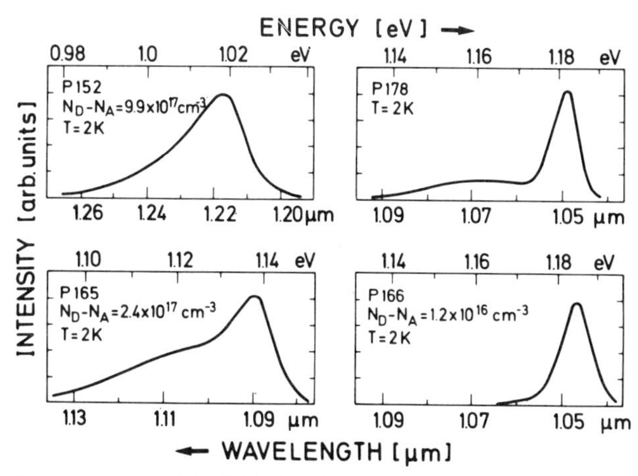

Figure 13.4 Low-temperature ($T = 2$ K) photoluminescence spectrum of GaInAsP epitaxial layers with different impurity content

3 K can be determined. Thus, for an 'intrinsic' band-to-band transition a linewidth of about 5 meV would be expected. However, even for the purest sample (P166) the low-energy side of the emission spectrum does not show a parabolic increase with photon energy, but rather is exponential. According to the assumption that the low-energy side of the spectrum increases like exp $(\hbar\omega - E_g)/\epsilon_t$, state parameter ϵ_t of about 3 to 5 meV is obtained. Because of the exponential increase of the low-energy side of the emission spectrum, it seems likely that localized states still participate in the recombination. These localized states could be due to shallow donors[54] as well as to compositional gradings within the epitaxial GaInAsP layers.[60] The spectra of samples P165 and P178 clearly exhibit two transitions shifted by about 20 meV. The low-energy band is attributed to a band-to-acceptor transition. The binding energy of about 20 meV obtained from the energy separation is lower than expected for a simple effective mass acceptor[32] and roughly corresponds to the value obtained for the Be acceptor by Feng *et al.*[55] For the highest doped sample, the band-to-band and the band-to-acceptor transitions cannot be resolved spectrally because of the formation of tail states at the high acceptor concentrations. More complicated low-temperature photoluminescence spectra are reported in reference 29. Figure 13.5 depicts these emission spectra which show some indication of fine structure

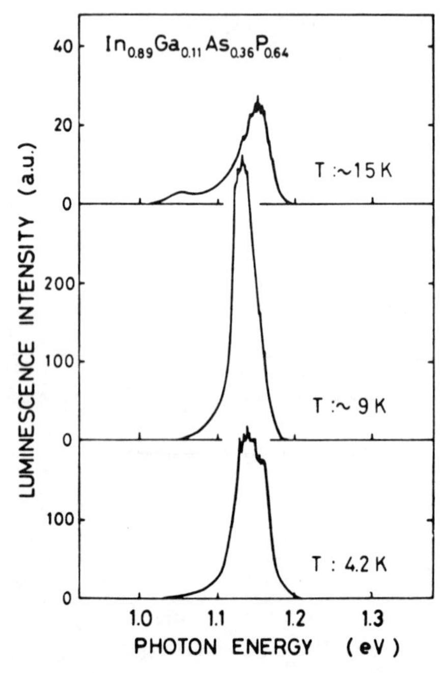

Figure 13.5 Low-temperature photoluminescence spectrum of $Ga_{0.11}In_{0.89}As_{0.36}P_{0.64}$. From Yamazoe *et al.*[29] Reproduced by permission of the authors and North-Holland Publishing Company

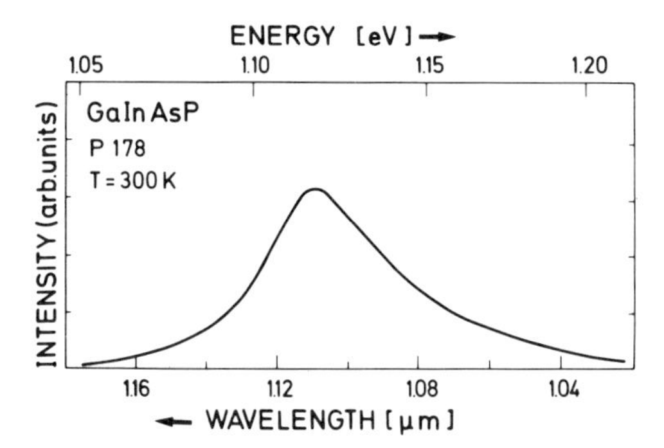

Figure 13.6 Room-temperature photoluminescence spectrum of an epitaxial GaInAsP layer

in the near-band-edge emission. So far it has not been decided whether bound exciton recombination is responsible for the occurrence of the sharp structures.

An example of room-temperature photoluminescence of free epitaxial GaInAsP layers (single heterostructure) is shown in Figure 13.6. Again at low excitation intensities, the high-energy side can be fitted by a Boltzmann distribution function using a carrier temperature of 300 K. The halfwidth of about 45 meV agrees exactly with the value of $1.8kT$ expected for a 'simple' band-to-band transition. However, the low-energy side again shows an exponential increase characterized by a tail state parameter $\epsilon_t = 20$ meV. This value agrees with the energy separation of the band-to-band and the band-to-acceptor transitions as observed for the same sample at low temperatures (cf. Figure 13.4). We therefore conclude that even though the halfwidth of the emission corresponds to that of a simple band-to-band transition, impurity states still influence the lineshape. Further support for this assumption is given by the photoluminescence and electroreflectance results reported in reference 28. Figure 13.7 compares room-temperature electroreflectance and photoluminescence results.[28] The photoluminescence clearly extends to energies below the bandgap (marked by an arrow in the electroreflectance curve) and again shows an exponential increase with a tail state parameter of 20 meV.

For GaInAsP layers with impurity content higher than $2 \times 10^{17}\,\text{cm}^{-3}$, the room-temperature linewidth of the spontaneous emission is larger than $1.8kT$. For sample P152 with $N_D - N_A = 9.9 \times 10^{17}\,\text{cm}^{-3}$ a halfwidth of 65 meV is measured. Transitions to deep levels have also been observed in photoluminescence.[52,56] In Figure 13.8 two transitions at 113 and 226 meV below the band-to-band transition can be resolved. The origin of the impurity states involved in the corresponding recombination process is not yet identified. In reference 56, deep centre photoluminescence of GaInAsP layers with various compositions corresponding to bandgaps ranging from 1.1 μm to 1.6 μm is reported. As a striking feature, they find a broad emission band located at about 1.6 μm, independent of quaternary composition. Therefore they conclude that in GaInAsP an energy

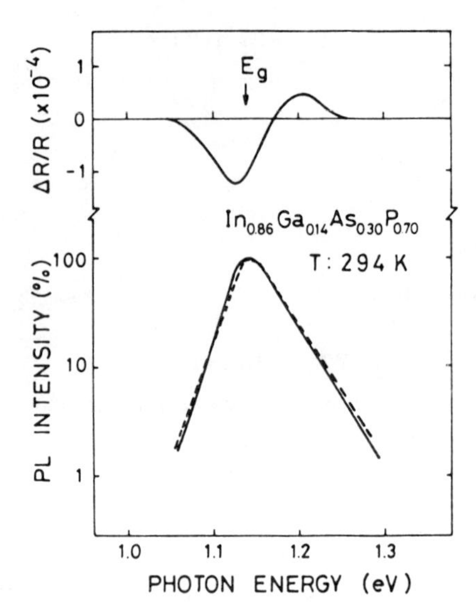

Figure 13.7 Comparison of electroreflectance (*upper part*) and photoluminescence spectrum (*lower part*) of a $Ga_{0.14}In_{0.86}As_{0.3}P_{0.7}$ LPE layer. E_g in the electroreflectance spectrum marks the position of the bandgap obtained by a three-point adjusting fit. From Yamazoe *et al.*[28] Reproduced by permission of the authors and *Japanese Journal of Applied Physics*

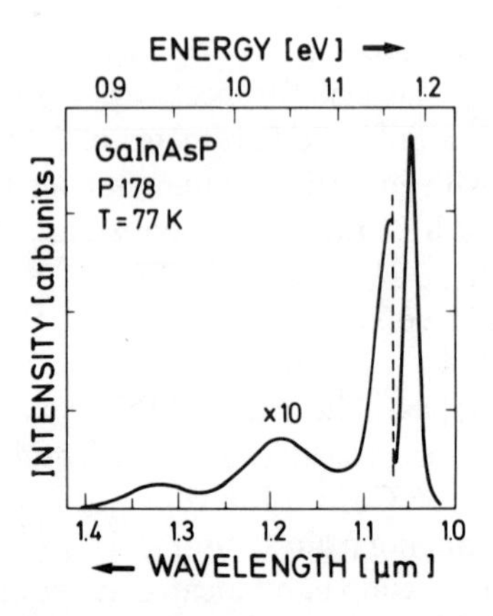

Figure 13.8 Deep-centre photoluminescence of GaInAsP at 77 K

level of about 0.75 eV exists, whose position is fixed to either the valence band or conduction band. This 0.75 eV level is probably related to a lattice defect in the GaInAsP epilayers. Deep levels in GaInAs have also been studied by deep level transient spectroscopy (DLTS).[57,58] Electron traps with activation energies of 0.27, 0.30, 0.51, 0.6, and 0.67 eV have been observed. Furthermore it was found that the concentration of the 0.67 eV trap depends markedly on the lattice mismatch.[57]

The influence of lattice mismatch on the absorption and photoluminescence spectra of GaInAsP epitaxial layers on InP substrates has been studied in reference 57. For GaInAsP grown on (111)B InP substrates the authors report an increase in Urbach tail extent and halfwidth of the absorption and photoluminescence spectra, as the lattice mismatch increases. This behaviour is explained in terms of local strain introduced by the lattice mismatch. However, surprisingly no

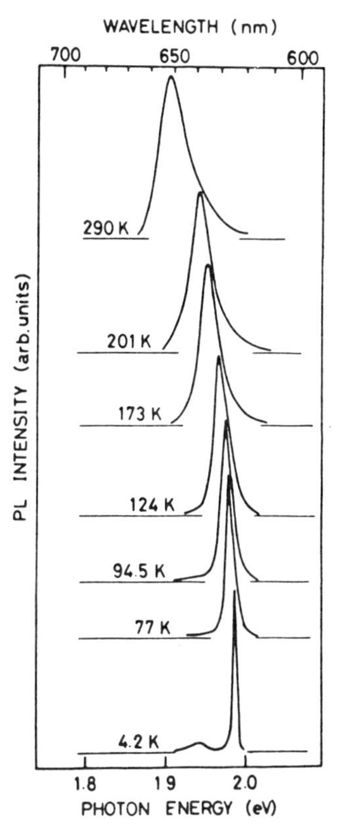

Figure 13.9 Photoluminescence spectra of an undoped n-GaInAsP epitaxial layer lattice-matched on GaAs at 4.2–290 K. From Kyuragi *et al.*[59] Reproduced by permission of the authors and the American Institute of Physics

dependence of the Urbach tail extent on lattice mismatch has been observed for GaInAsP on (100) InP substrates.

Photoluminescence of LPE GaInAsP on GaAs substrates has been studied by Mukai *et al.*[26] and Kyuragi *et al.*[59] Figure 13.9 shows emission spectra for temperatures between 4.2 K and 290 K. In the entire temperature range the halfwidth of the spontaneous emission corresponds to about $1.8kT$, indicating the good quality of the GaInAsP layers. However, as in the case of GaInAsP on InP, no sharp bound exciton transitions could be resolved at low temperatures.

To sum up the results of this section briefly, we can state that narrow near-band-edge emission of epitaxial layers of GaInAsP on InP as well as on GaAs substrates has been observed for material with impurity content smaller than $10^{17}\,cm^{-3}$. However, from the fact that in the low-temperature photoluminescence spectra no free and bound excition transitions have been found, as well as from lineshape analysis, it must be concluded that even for the best material fabricated so far localized states still contribute to the emission spectra. Whether these localized states are due to residual impurities or are possibly due to some compositional grading[59] has not yet been decided definitely.

13.5 PHOTOLUMINESCENCE OF GaInAsP DOUBLE HETEROSTRUCTURES

In this section, we summarize the results on photoluminescence of GaInAsP/InP double heterostructures and interfaces. As these data are compared to the results of section 13.4, it will be seen that in general the photoluminescence spectra of double heterostructures are more complicated than those for relatively thick free epitaxial layers.

The photoluminescence of GaInAsP–InP double heterostructures for different distances from the heteroepitaxial interfaces has been investigated by Fukui and Horikoshi.[40] Figure 13.10 shows experimental results for different structures: (a) an undoped GaInAsP layer on InP, measured after the outermost Cd-doped InP layer was removed; (b) a structure as in case (a) but with the GaInAsP etched to a thickness of 0.6 μm; (c) an as-grown GaInAsP wafer; and (d) a double heterostructure with a thin GaInAsP layer. The striking result is that the intensity of the low-energy band labelled C depends strongly on the detailed character of the surface. In the photoluminescence spectra measured at the InP/GaInAsP interface or for a double heterostructure with thin active layers, band C is strongest. Further support by electroluminescence led the authors to conclude that some kind of defects are introduced at the InP/GaInAsP interface during growth, owing to meltback of the GaInAsP quaternary solid as brought in contact with the InP liquid.[37]

Photoluminescence spectra very similar to those shown in Figure 13.10(c) have also been observed for GaInAsP/InP double heterostructures with 0.5 μm thick quaternary layers by Luz.[52]

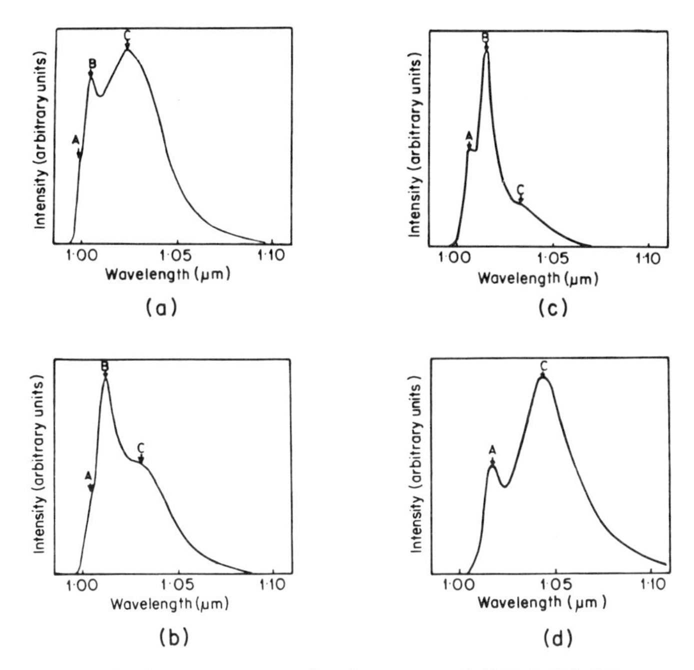

Figure 13.10 Photoluminesence spectra of various types of GaInAsP/InP heterostructures as described in the text. From Fukui and Horikoshi.[40] Reproduced by permission of *Japanese Journal of Applied Physics*

An elegant method of studying the effect of the GaInAsP/InP interface on the photoluminescence spectra has been employed by Rezek *et al.*[61] They compare emission spectra of photopumped cleaved rectangular double-heterostructure samples grown by two different procedures. In the first case a quaternary layer of 800 Å thickness is grown uninterrupted in about 1 s (1), whereas in the second case the active layer is grown in repetitive cycles that place the substrate in contact with the GaInAsP melt for about 18 ms followed by an approximately 400 ms waiting period (2). In both cases the composition of the liquidus solution was identical. As shown in Figure 13.11 the corresponding emission spectra are different. Curves (a) and (b) refer to the spontaneous and stimulated spectra, respectively, the latter ones clearly showing the Fabry–Perot resonances of the cleaved sample. Figure 13.11 (A) shows the result obtained for structure 1. Two set of modes corresponding to the width and length of the rectangular resonator structure are observed, respectively. These two sets of lasing modes are separated by about 15 meV. The energetic position of the low-energy modes coincides with the spontaneous and laser emission as measured for structure 2 (Figure 13.11 (B)). Based on the results shown in Figure 13.11, the authors conclude that the low energetic laser emission is related to the interface, because interface effects are expected to dominate in

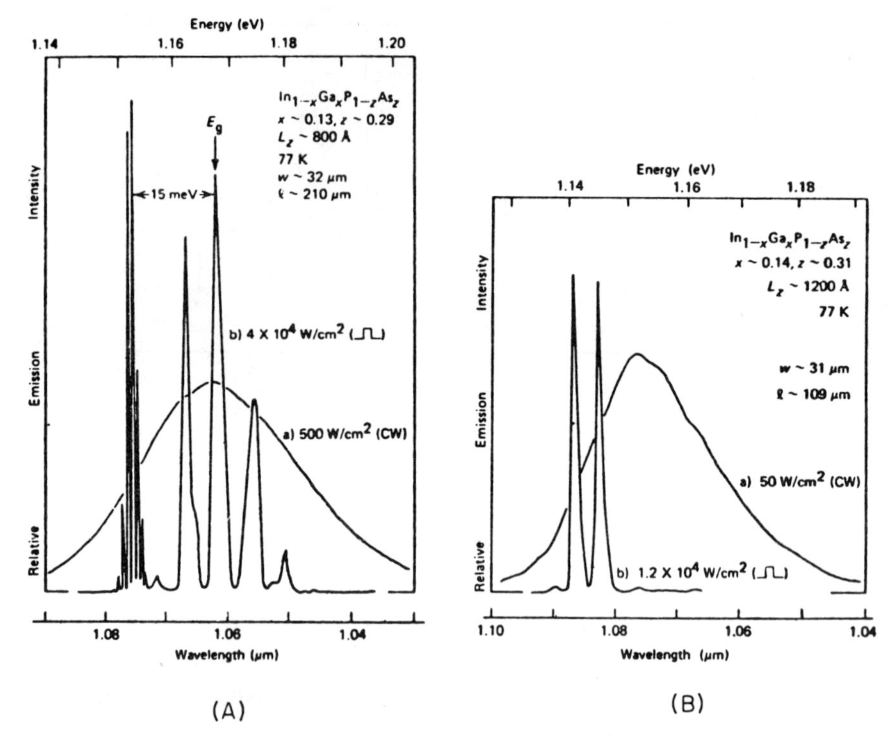

Figure 13.11 Photoluminescence spectra of rectangular cleaved samples of GaInAsP/InP double heterostructure. Figure 13.11A refers to a structure, where the 800 Å thick quaternary layer is grown uninterrupted in about 1 s. For the spectra shown in Figure 13.11B a structure is used where the active layer is grown in repetitive cycles as described in the text. Curves (a) and (b) correspond to the spontaneous and laser spectra, respectively. From Rezek *et al.*[61] Reproduced by permission of the authors and the American Institute of Physics

structure 2. The low-energy laser emission is attributed to the fact that in the initial stage of growth of the quaternary layer material with a composition corresponding to a 15 meV bandgap difference is deposited due to local depletion or accumulation of some or all the four liquidus components. Difficulties related to the initial growth kinetics have also been reported by other authors.[37,62,63]

Nijman *et al.*[64] have studied the temperature dependence of the photoluminescence of various GaInAsP single and double heterostructures in order to obtain a better understanding of the unusual strong increase of the laser threshold with temperature ('T_0 problem', cf. Chapter 15). The results for the relative integrated photoluminescence intensity versus temperature are summarized in Figure 13.12 and compared to AlGaAs as well as InP data. The photoluminescence intensity decreases exponentially with temperature ($I \sim \exp(-T/T_0)$ in all cases. However, for n-type AlGaAs only one T_0 value is needed to explain the results ($T_0 \sim 272$ K), while for the quaternary material a breakpoint temperature is observed, where T_0

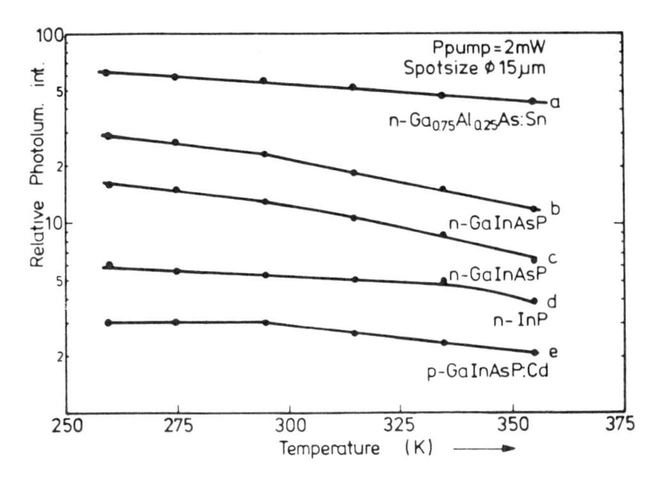

Figure 13.12 Relative photoluminescence intensity as a function of temperature for different GaInAsP/InP heterostructures as well as AlGaAs and InP epitaxial layers. From Nijman *et al.*[64] Reproduced by permission of the authors and The Institute of Physics

changes its value. Regardless of the doping, this breakpoint temperature is at about 295 to 300 K and thus is very close to the breakpoint temperature generally observed in the laser threshold data (cf. Chapter 15 and section 13.6). Furthermore, the 'breakpoint behaviour' is observed for free epitaxial layers as well as GaInAsP/InP double heterostructures excited indirectly via the InP. Thus the authors conclude that the T_0 problem is an intrinsic property of the GaInAsP. From an analysis of their data in terms of Auger recombination, the authors further conclude that the 'breakpoint behaviour' is not caused by Auger recombination. However, others have suggested Auger recombination as a possible mechanism to explain the breakpoint behaviour of the laser threshold data.[65–68] Additionally, Nijman *et al.* have determined the surface recombination velocity of n- and p-type GaInAsP from their photoluminescence experiments.[64] The values obtained are 5×10^5 cm s^{-1} for p-type and less than 10^4 cm s^{-1} for n-type free epitaxial layers, respectively.

The effect of intense optical excitation on the photoluminescence intensity of GaInAsP/InP double heterostructures has been investigated by Fukui and Horikoshi[69] and is compared to the behaviour of free GaInAsP epitaxial layers. The authors report that up to excitation intensities of 100 kW cm^{-2} (Nd: YAG, $\lambda = 1.06\,\mu$m), no dark regions are formed. This value exceeds by orders of magnitude those found for GaAs/AlGaAs double heterostructures. Since optically induced defects are identical to those found in degraded GaAs/AlGaAs double-heterostructure lasers,[70–72] the authors conclude that the GaInAsP/InP system is more stable against optical damage. In contrast to the double heterostructures, for the free GaInAsP single-heterostructure layers dark regions

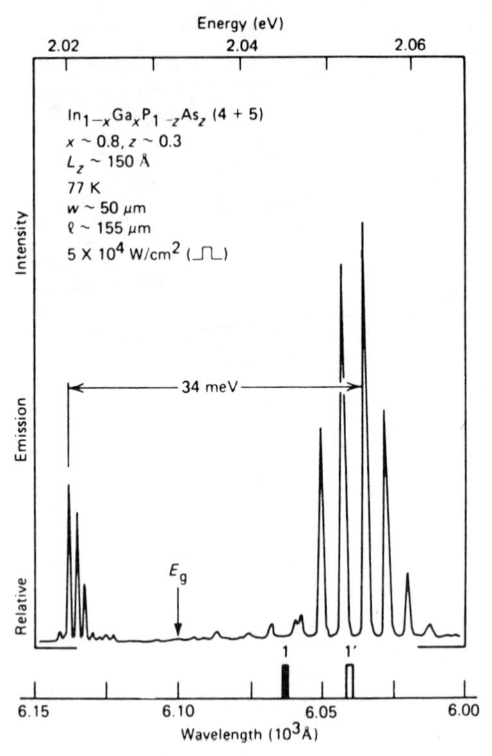

Figure 13.13 Pulsed laser spectrum of a cleaved, rectangular GaInAsP quantum well heterostructure. Stimulated emission is observed along the sample width as well as the sample length, the latter shifted by the energy of an optical phonon. From Chin *et al.*[74] Reproduced by permission of the authors and the American Institute of Physics

were formed by the optical irradiation. However, these dark regions are not caused by bulk degradation but are due to deterioration of the surface.[64,69]

Finally, in this section some of the work on GaInAsP quantum-well heterostructures should be mentioned.[73–77] Figure 13.13 shows a photopumped, pulsed laser emission spectrum of a visible $Ga_{x'}In_{1-x'}As_{z'}P_{1-z'}$/ $Ga_xIn_{1-x}As_zP_{1-z}$ ($x > x'$, $z > z'$) quantum-well heterostructure.[76,77] A rectangular sample (50 μm × 155 μm) cleaved at all four endfaces was used in these experiments. The high-energy set of laser modes is due to stimulated recombination between the $n = 1$ electron and $n = 1$ heavy hole subband as well as between the $n = 1$ electron and the $n' = 1'$ light hole subband.[74] As has been frequently observed in rectangular quantum-well lasers, the low-energy set of modes corresponds to laser oscillations perpendicular to the high-energy modes and is shifted by an optical phonon energy. For more details of the interesting properties of quantum-well lasers, the reader is referred to reference 77.

13.6 OPTICAL GAIN SPECTROSCOPY

Optical gain spectroscopy is particularly useful and important for characterizing and studying semiconductor laser material. Gain spectra of $Ga_{0.28}In_{0.72}As_{0.6}P_{0.4}$ have been calculated recently by Dutta.[51] Yano *et al.* have measured the rate of change of the gain with current in stripe geometry lasers.[78] Walpole *et al.*[79] reported on net gain spectra as obtained from the modulation depth of the Fabry–Perot resonances in the spontaneous emission of proton-bombarded stripe-geometry lasers. In this chapter some recent results on optical gain spectroscopy of GaInAsP epitaxial layers as well as GaInAsP/InP double heterostructures will be presented.

The principle of optical gain spectroscopy has been discussed in section 13.2 and is illustrated in Figure 13.14. In Figure 13.14(a) amplified emission spectra for three excitation lengths L of 100 μm, 187 μm, and 375 μm of a free GaInAsP epitaxial layer are depicted. Obviously the emission intensity increases super-linearly with length. The spectra for $L = 375\,\mu m$ and $L = 187\,\mu m$ show the typical cross-over at the high-energy side characteristic of the onset of gain saturation. This effect has been studied in detail for GaAs lasers.[80] The unsaturated optical gain spectrum as obtained from the two spectra with $L = 187$ and $L = 100 \mu m$ is shown in Figure 13.14(b). Maximum gain values of about $400\,cm^{-1}$ are observed for an optical pump intensity of $1\,MW\,cm^{-2}$. Similar values have been measured for high-purity as well as doped GaAs layers under comparable excitation conditions.[4,81] Spectrally the gain maximum agrees with the spontaneous luminescence maximum (cf. Figure 13.6). However, the contributions of tail states is less pronounced in the optical gain spectra. A lineshape fit of the optical gain spectrum obtained at 77 K is given in Figure 13.15. The lineshape can be described phenomenologically by a band-to-band transition with 'relaxed' k selection rule as is expected even for high-purity material.[45–47]

Gain spectra of GaInAsP/InP double heterostructures at temperatures of 2, 77,

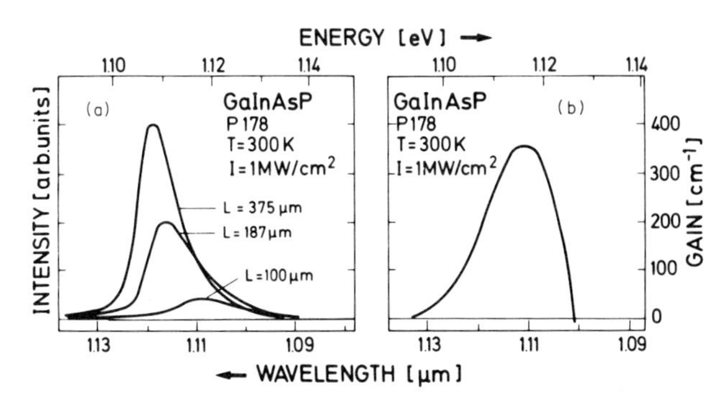

Figure 13.14 (a) Amplified spontaneous emission spectra for three different excitation lengths and (b) net gain spectrum as calculated from the lower two curves in (a) for a GaInAsP epitaxial layer at 300 K

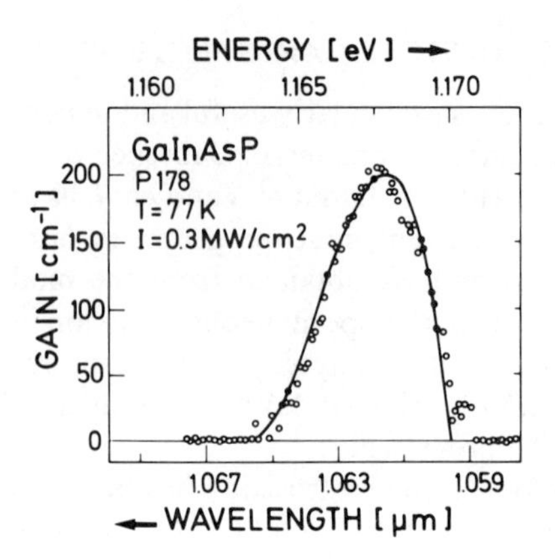

Figure 13.15 Net gain spectrum of a GaInAsP epitaxial layer at 77 K. The points are the experimental data, and the full curve represents a calculated gain spectrum (cf. text)

and 300 K are reported in reference 32. Figure 13.16 shows the room-temperature gain spectrum together with the low-intensity absorption curve. The absorption of this particular GaInAsP active layer is strongly determined by tail states. The optical gain is almost completely located in the tail state region. Again the maximum gain value ($g_{max} \sim 200\ cm^{-1}$) as well as the halfwidth of the gain spectrum ($\Delta \sim 40$ meV) are comparable to the results of GaAs/AlGaAs lasers.[7,81]

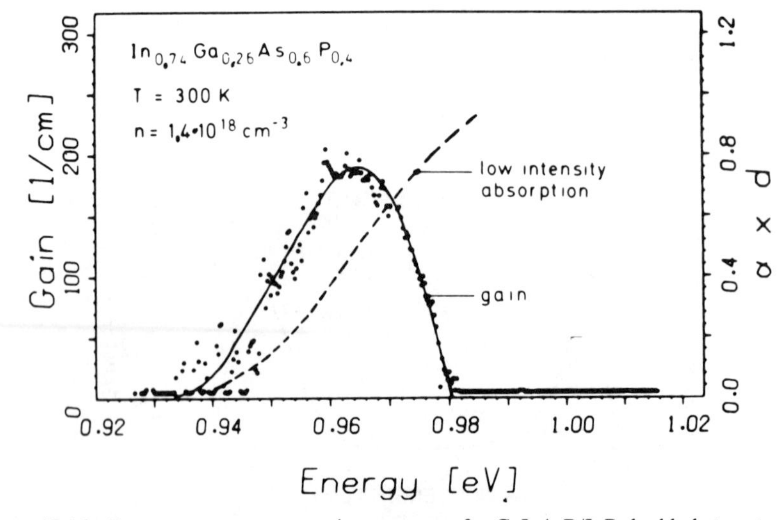

Figure 13.16 Room-temperature net gain spectrum of a GaInAsP/InP double heterostructure. The points correspond to the experimental data, and the full curve shows the gain spectrum calculated on the basis of the low-intensity absorption curve

The full curve through the measured gain data is calculated from the low-intensity absorption curve assuming an optically injected carrier density of about $1.4 \times 10^{18}\ cm^{-3}$.

More detailed studies on the temperature dependence of optical gain have been performed recently by Jung *et al.*[82,83] These investigations had been stimulated by the still open question on the origin of the so-called 'T_0 problem' (cf. Chapter 15). As an important extension of previous work on optical gain, the experimental results are compared directly with threshold data of GaInAsP/InP double-heterostructure injection lasers fabricated from the same material. Figure 13.17 shows the pulsed threshold current versus temperature as well as the results obtained from the optical gain experiments. The threshold data quantitatively agree with those observed by various authors for CW and pulsed GaInAsP/InP lasers (see Chapter 15). Especially, in the semilogarithmic plot based on the relation $J_{th} \sim \exp(T/T_0)$, a kink is observed at the breakpoint temperature $T_B \sim 255$ K, where T_0 changes its value from $T_0 = 96$ K at lower temperatures to $T_0 = 63$ K for higher temperatures. In the lower part of Figure 13.17 the dependence of the maximum optical gain on temperature is illustrated. For this purpose the optical pump power needed to maintain a constant maximum gain (nominal pump intensity) is plotted versus temperature. This type of plot allows a direct comparison of gain and threshold data.[84] As shown, identical results for the threshold current and the nominal pump intensity are obtained. In particular, the values for T_B and T_0 agree extremely well. This demonstrates that the T_0 problem of the quaternary lasers is caused by the gain process and is not due to other

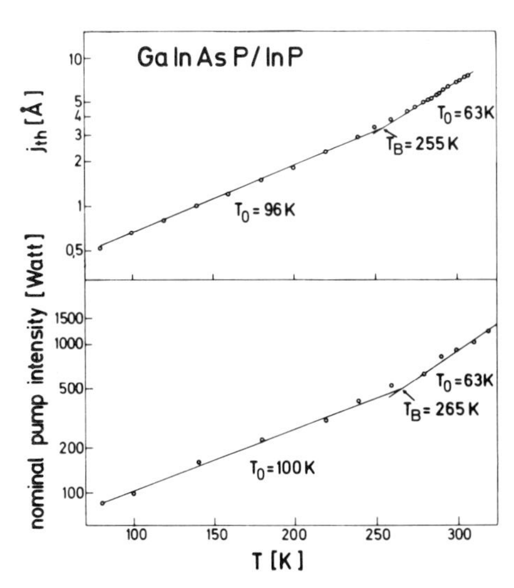

Figure 13.17 Pulsed threshold current J_{th} and nominal pump intensity of a GaInAsP/InP double heterostructure versus temperature. The T_0 values refer to the representation $J_{th} \sim \exp(T/T_0)$

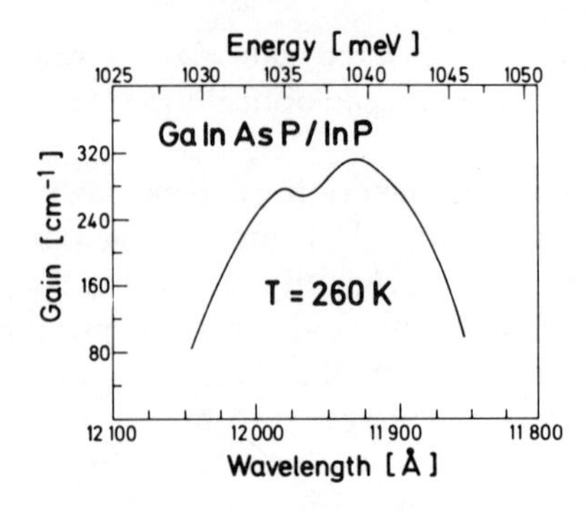

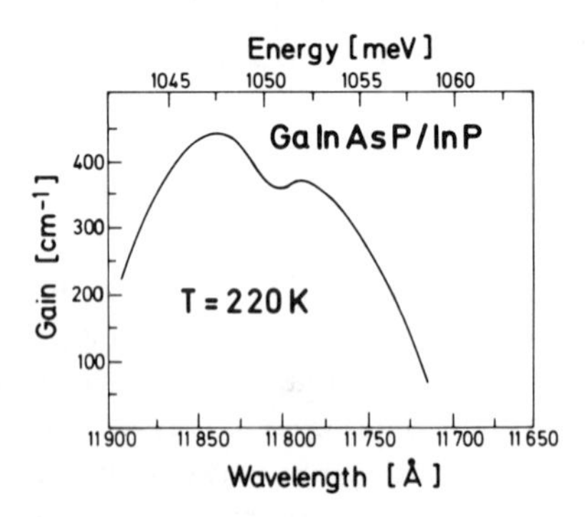

Figure 13.18 Optical gain spectra of a GaInAsP/InP double heterostructure at 260 K (*upper curve*) and 220 K (*lower curve*)

artifacts such as surface leakage currents. Besides a kink in the maximum gain values, spectral changes of the optical gain are observed close to the breakpoint temperature T_B. This is shown in Figure 13.18, where gain spectra at 260 K (*upper curve*) and 220 K (*lower curve*) are depicted, respectively. In both cases, two bands in the gain spectra separated by about 5 meV can be clearly resolved. The maximum gain corresponding to these two bands strongly depends on temperature. At low temperatures ($T < 220$ K) the low-energy gain band prevails, whereas at higher temperatures the high-energy band dominates. Thus at, for example 77 K and 300 K, essentially one gain band is observed, which corresponds to the low-energy or high-energy gain band, respectively. This behaviour is characteristic of different LPE double heterostructures; however, the energy separation of the two gain bands varied between 5 and 15 meV for different samples.[83]

The following conclusion must be drawn from these gain data. The gain spectra of the GaInAsP/InP double heterostructures investigated in references 82 and 83 cannot be explained by a simple band-to-band transition.[51] Rather it has to be assumed that contributions due to 'localized' or 'quasilocalized states' participate in the stimulated recombination and prevail at low temperatures. Whether these localized states originate from impurity (tail) states, compositional grading, or possibly from defects or composition inhomogeneities related to the heterointerface cannot be deduced. The fact that the gain spectra change at about the breakpoint temperature indicates that either the spectral shape of the intrinsic gain or of the losses changes. Indeed, the latter has been proposed by Adams *et al.* to explain the T_0 behaviour.[85] In any case, Auger recombination[65–68] and other non-radiative transitions[66,86,87] as well as leakage currents over the heterojunction barrier[88–92] cannot explain the change in the lineshape of the optical gain spectra, though they may contribute to an increase of the overall losses at higher temperatures.

13.7 SUMMARY

This chapter gives an overview on the diagnostic and scientific potential of photoluminescence and optical gain spectroscopy for the characterization and investigation of quaternary GaInAsP semiconductors. Photoluminescence has been widely used as a simple method to determine the bandgap energy for different compositions of the GaInAsP. However, in a quantitative analysis the details of the spontaneous emission lineshape have to be considered. Photoluminescence studies of free epitaxial layers lattice-matched on InP as well as GaAs have demonstrated that high-quality material can be grown by liquid-phase epitaxy as well as by vapour-phase epitaxy. The halfwidth of the emission spectrum of the best material is of the order of kT ($1.8 \times kT$) as expected for a band-to-band transition in relatively pure material. However, so far no single, sharp emission lines due to bound and free exciton luminescence have been observed in the low-temperature photoluminescence. Furthermore, detailed lineshape analysis leads to the conclusion that even in the best material localized states still influence the emission. Whether these localized states result from composition inhomogeneities or impurity and defect states has not been decided definitely. The photoluminescence of GaInAsP/InP heterostructures generally shows more complicated emission spectra than observed for free epitaxial layers. Besides possible contributions of impurities and compositional grading in the bulk quaternary layer, it has been shown by photoluminescence experiments that the heterobarrier interface may influence the spectra. This could be due to changes in composition as well as to formation of defect states during the initial stage of growth.

The results of the photoluminescence experiments have been further supported by optical gain spectroscopy. Generally, contributions from localized states must be assumed to explain the gain spectra. Furthermore, the optical gain spectra of GaInAsP/InP double heterostructures exhibit two bands similar to the spectra obtained in photoluminescence. An important additional result is that the optical gain experiments have shown that close to the breakpoint temperature in the threshold versus temperature data characteristic, changes in the gain spectra occur.

To summarize the present state of the art of the quaternary GaInAsP semiconductors on the basis of photoluminescence and optical gain experiments, we must state that the quality and properties of GaInAsP epitaxial layers and double heterostructures have not yet reached the high standard of the corresponding GaAs/AlGaAs binary–ternary system. The fundamental question, which has not been answered so far, is whether the complicated quaternary, small-band-gap compounds are inherently inferior to binary or elementary semiconductors. Further optical studies together with a high standard technology may contribute to answering this question.

ACKNOWLEDGMENTS

The author is indepted to G. Luz, H. Jung, A. Mozer, and K. M. Romanek for various contributions to the results reported in this article. Thanks are also due to M. H. Pilkuhn and H. J. Queisser for carefully reading the manuscript.

REFERENCES

1. E. W. Williams, and H. B. Bebb, 'Photoluminescence', *Semiconductors and Semimetals*, vol. 8, eds R. K. Willardson and A. C. Beer, Academic Press, New York, 1972, pp. 181–392.
2. P. J. Dean, 'III–V compound semiconductors', *Electroluminescence, Topics in Applied Physics*, vol. 17, ed. J. I. Pankove, Springer, Berlin, Heidelberg, New York, 1977, pp. 63–132.
3. E. Göbel, and M. H. Pilkuhn, 'Laser transitions in direct gap semiconductors', *J. Physique*, **35**, C3, 191–200, 1974.
4. M. H. Pilkuhn, 'Investigation on optical gain and its saturation behaviour in semiconductor lasers', *Proc. Optoelectronic School*, Cetniewo, Poland, 1978, ed. M. A. Herman, J. Wiley & Sons, 495pp.
5. K. L. Shaklee, and R. F. Leheny, 'Direct determination of optical gain in semiconductors', *Appl. Phys. Lett.*, **18**, 475–477, 1971.
K. L. Shaklee, R. F. Leheny, and R. E. Nahory, 'Optical gain in semiconductors', *J. Luminesc.*, **7**, 284–309, 1973.
6. P. S. Gross, and W. G. Oldham, 'Theory of optical gain measurements', *IEEE J. Quantum Electron.*, **QE–11**, 190–197, 1975.
7. O. Hildebrand, E. Göbel, and K. Löhnert, 'Investigations on unsaturated optical gain spectra of GaAs/GaAlAs DHS lasers', *Appl. Phys.*, **15**, 149–152, 1978.
8. B. W. Hakki, 'Carrier and gain spatial profiles in GaAs stripe geometry lasers', *J. Appl. Phys.*, **44**, 5021–5028, 1973.
9. B. W. Hakki, and T. L. Paoli, 'Gain spectra in GaAs double heterostructure injection lasers', *J. Appl. Phys.*, **46**, 1299–1306, 1975.
10. B. W. Hakki, 'Mode gain and junction current in GaAs under lasing conditions', *J. Appl. Phys.*, **45**, 288–294, 1974.
11. B. W. Hakki, 'GaAs double heterostructure lasing behaviour along the junction plane', *J. Appl. Phys.*, **46**, 292–302, 1975.
12. O. Hildebrand, B. Faltermeier, and M. H. Pilkuhn, 'Direct determination of reduced

band gap and chemical potential in an electron—hole plasma in high-purity GaAs', *Solid State Commun.*, **19**, 841–844, 1976.
13. R. F. Leheny, and J. Shah, 'Experimental evidence for the existence of an electron–hole liquid in II–VI compounds', *Phys. Rev. Lett.*, **37**, 871–874, 1976.
14. E. Göbel, H. Herzog, M. H. Pilkuhn, and K. H. Zschauer, 'Laser emission due to excitonic recombination processes in high purity GaAs', *Solid State Commun.*, **13**, 719–722, 1973.
15. E. Göbel, 'Recombination without *k*-selection rules in dense electron–hole plasmas in high purity GaAs lasers', *Appl. Phys. Lett.*, **24**, 492–494, 1974.
16. O. Hildebrand, E. O. Göbel, K. M. Romanek, H. Weber, and G. Mahler, 'Electron–hole plasma in direct gap semiconductors with low polar coupling: GaAs, InP, GaSb', *Phys. Rev.*, B, **17**, 4775–4787, 1978.
17. J. Bakker, and G. A. Acket, 'Single pass gain measurements on optically pumped AlGaAs–AlGaAs double heterojunction laser structures at room temperature', *IEEE J. Quantum Electron.*, **QE-13**, 567–573, 1977.
18. O. Hildebrand, 'Nitrogen laser excitation of polymethine dyes for emission wavelengths up to 9500Å', *Opt. Commun.*, **10**, 310–312, 1974.
19. R. L. Moon, G. A. Antypas, and L. W. James, 'Bandgap and lattice constant of GaInAsP as a function of alloy composition', *J. Electron. Mater.*, **3**, 635–637, 1974.
20. K. Nakajima, A. Yamaguchi, K. Akita, and T. Kotani, 'Composition dependence of the band gaps of InGaAsP quaternary solids lattice matched on InP substrates', *J. Appl. Phys.*, **49**, 5944–5950, 1978.
21. R. E. Nahory, M. A. Pollack, W. D. Johnston, Jr, and R. L. Barns, 'Band gap versus composition and demonstration of Vegard's law for InGaAsP lattice matched to Inp', *Appl. Phys. Lett.*, **33**, 659–661, 1978.
22. M. A. Pollack, R. E. Nahory, J. C. De Winter, and A. A. Ballmann, 'Liquid phase epitaxial InGaAsP lattice matched to (100) InP over the complete wavelength range $0.92 \leq \lambda \leq 1.65\,\mu m$', *Appl. Phys. Lett.*, **33**, 314–316, 1978.
23. G. H. Olsen, C. J. Nuese, and M. Ettenberg, 'Low threshold 1.25 μm vapor-grown InGaAsP cw lasers', *Appl. Phys. Lett.*, **34**, 262–264, 1979 and references therein.
24. S. B. Hyder, R. R. Saxena, and C. C. Hooper, 'Vapor phase epitaxial growth of quaternary InGaAsP in the 0.75–1.35 eV band-gap range', *Appl. Phys. Lett.*, **34**, 584–586, 1979.
25. K. Sugiyama, H. Kojima, H. Enda, and M. Shibata, 'Vapor phase epitaxial growth and characterization of GaInAsP quaternary alloys', *Jpn. J. Appl. Phys.*, **16**, 2197–2203, 1977.
26. S. Mukai, M. Matsuzaki, and J. Shmada, 'LPE growth and luminescence of InGaAsP on (100) GaAs with band gap energy in the region of $1.569\text{eV} \leq E_g \leq 1.893\text{eV}$', *Jpn. J. appl. Phys.*, **19**, L505–L508, 1980.
27. M. Cardona, *Modulation Spectroscopy*, Academic Press, New York, 1969.
28. Y. Yamazoe, T. Nishino, Y. Hamakawa, and T. Kariya, 'Bandgap energy of InGaAs quaternary alloy', *Jpn. J. Appl. Phys.*, **19**, 1473–1479, 1980.
29. Y. Yamazoe, H. Takakura, T. Nishino, and Y. Hamakawa, 'Characterization of the optical properties of LPE InGaAs thin layers grown on InP', *J. Cryst. Growth*, **45**, 454–458, 1978.
30. H. H. Wieder, A. R. Clawson, and G. E. McWilliams, 'InGaAsP/InP heterojunction photodiodes', *Appl. Phys. Lett.*, **31**, 468–470, 1977.
31. T. P. Lee, C. A. Burrus, Jr, and A. G. Dentai, 'InGaAsP/InP photodiodes: microplasma limited avalanche multiplication at 1–1.3 μm wavelength', *IEEE J. Quantum Electron.*, **QE-15**, 30–35, 1979.
32. E. O. Göbel, G. Luz, and E. Schlosser, 'Optical gain spectra of InGaAsP/InP double heterostructures', *IEEE J. Quantum Electron.*, **QE-15**, 697–700, 1979.

33. E. Göbel, H. Gottsmann, H. J. Herzog, P. Marschall, E. Schlosser, and E. A. Schurr, 'Preparation and characterization of GaInAsP/InP double heterostructure wafers and lasers for the 1.3 μm wavelength range', *Solid-State Electron. Devices*, **3**, 186–188, 1979.
34. J. J. Hsieh, J. A. Rossi, and J. P. Donnelly, 'Room temperature cw operation of GaInAsP/InP double heterostructure diode lasers emitting at 1.1 μm', *Appl. Phys. Lett.*, **28**, 709–710, 1976.
35. T. P. Pearsall, B. I. Miller, and R. J. Capik, 'Efficient lattice matched double heterostructure LED's at 1.1 μm from GaInAsP', *Appl. Phys. Lett.*, **28**, 499–501, 1976.
36. K. Oe, S. Ando, and K. Sugiyama, 'Surface emitting LED's for the 1.2–1.3 μm wavelength with GaInAsP/InP double heterostructures', *Jpn. J. Appl. Phys.*, **16**, 1693–1694, 1977.
37. H. Nagai, and Y. Nŏguchi, 'InP–GaInAsP double heterostructure for 1.5 μm wavelength', *Appl. Phys. Lett.*, **32**, 234–236, 1978.
38. I. Umebu, O. Hasegawa, and K. Akita, 'InGaAsP/InP D. H. LED's for fibre-optical communications', *Electron. Lett.*, **14**, 499–500, 1978.
39. S. Akiba, K. Sakai, and T. Yamamoto, 'InGaAs/InGaAsP double heterostructure lasers with emission wavelength of 1.67 μm at room temperature', *Jpn. J. Appl. Phys.*, **17**, 1899–1900, 1978.
40. T. Fukui, and Y. Horikoshi, 'Anomaleous luminescence near the InGaAsP/InP heterojunction interface', *Jpn. J. Appl. Phys.*, **18**, 961–965, 1979.
41. A. G. Dentai, T. P. Lee, C. A. Burrus, and E. Buehler, 'Small-area, high-radiance cw-InGaAsP LEDs emitting at 1.2 to 1.3 μm', *Electron. Lett.*, **13**, 484–485, 1979.
42. L. M. Dolginov, N. Ibrakhimov, M. G. Mil'vidskii, V. Yu. Rogulin, and E. G. Shevehenko, 'High efficiency electroluminescence of GaInAsP', *Sov. Phys.–Semicond.*, **9**, 871–873, 1975.
43. W. Ruehle, and W. Klingenstein, 'Excitons bound to neutral donors in InP' *Phys. Rev.*, B, **18**, 7011–7021, 1978 and references therein.
44. See, e.g., C. J. Hwang, 'Properties of spontaneous and stimulated emission in GaAs junction lasers, I. Density of states in the active region', *Phys. Rev.*, B, **2**, 4117–4125, 1970.
45. W. F. Brinkman, and P. A. Lee, 'Coulomb effects on the gain spectrum of semiconductors', *Phys. Rev. Lett.*, **31**, 237–240, 1973.
46. H. Haug, and D. B. Tran Thoai, 'Gain spectra of an e–h liquid in direct gap semiconductors', *Phys. Stat. Sol.*, 6, **98**, 581–589, 1980.
S. Schmitt-Rink, D. B. Tran Thoai, and H. Haug, 'On the gain spectrum of highly excited GaAs', *Solid State Commun.*, **36**, 381–382, 1980.
47. K. Arya, and W. Hanke, 'Coulomb effects on the gain spectra of the electron–hole plasma in GaAs', *Solid State Commun.*, **33**, 739–742, 1980.
48. H. C. Casey, Jr and F. Stern, 'Concentration-dependent absorption and spontaneous emission of heavily doped GaAs', *J. Appl. Phys.*, **47**, 631–643, 1976.
49. F. Stern, 'Calculated spectral dependence of gain in excited GaAs', *J. Appl. Phys.*, **47**, 5382–5386, 1976.
50. T. Takagi, 'Spectral half-width of spontaneous emission of GaInAsP lattice matched to InP', *Jpn. J. Appl. Phys.*, **18**, 2017–2018, 1979.
51. N. K. Dutta, 'Calculated absorption, emission, and gain in InGaAsP', *J. Appl. Phys.*, **51**, 6095–6100, 1980.
52. G. Luz, *Diploma Thesis*, Stuttgart, 1979, unpublished.
53. The samples were grown by T. Pearsall, Thomson-CSF, Paris.
54. R. J. Nicholas, A. M. Davidson, S. J. Sessions, and R. A. Stradling, 'Shallow donor spectroscopy in GaInAsP', *IEEE J. Quantum Electron.*, **QE-17**, 145–149, 1981.
55. M. Feng, J. D. Oberstar, T. H. Windhorn, L. W. Cook, G. E. Stillman, and B. G.

Streetman, 'Be implanted 1.3 μm InGaAsP avalanche photodetectors', *Appl. Phys. Lett.*, **34**, 591–593, 1979.
56. E. V. K. Rao, R. E. Nahory, B. Etienne, G. Chaminant, M. A. Pollack, and J. C. De Winter, 'Deep center photoluminescence in InGaAsP as a function of composition', *Gallium Arsenide and Related Compounds 1980*, Conf. Ser. no. 56, Institute of Physics, Bristol, 1981, p. 605–612.
57. A. Tamura, K. Oka, M. Inoue, J. Shirafugi, and Y. Inushi, 'Characterization of InGaAsP epitaxial layers and relation to lattice matching', *Proc. 11th Conf. on Solid State Devices; Jpn. J. Appl. Phys.*, **19–1**, 479–482, 1980.
58. Y. Sasai, Y. Yamazoe, M. Okuyama, T. Nishino, and Y. Hamakawa, 'Observation of deep impurity levels in InGaAsP', *Jpn. J. Appl. Phys.*, **18**, 1415–1416, 1979.
59. H. Kyuragi, A. Suzuki, S. Matsumura, and H. Matsunami, 'Photoluminescence of lattice matched InGaAsP layers on GaAs', *Appl. Phys. Lett.*, **37**, 723–724, 1980.
60. M. Quillec, T. Pearsall, J. Burgeat, G. Laurencin, J. L. Benchimol, P. Meranda, and C. Daguet, 'The compositional uniformity of GaInAsP grown by near-equilibrium cooling LPE', *Gallium Arsenide and Related Compounds 1980*, Conf. Ser. no. 56, Institute of Physics, Bristol, 1981, p. 105–113.
61. E. A.Rezek, B. A. Vojak, R. Chin, and N. Holonyak, Jr, 'Compositional inhomogeneity of liquid phase epitaxial InGaAsP layers observed directly in photoluminescence', *Appl. Phys. Lett.*, **36**, 744–746, 1980.
62. M. Feng, L. W. Cook, M. M. Tashima, G. E. Stillman, and R. J. Blattner, 'Auger profile study of the influence of lattice mismatch on the InGaAsP/InP heterojunction interface', *Appl. Phys. Lett.*, **34**, 697–699, 1979.
63. R. J. Nelson, 'Near equilibrium LPE growth of low threshold current density InGaAsP (λ = 1.3 μm) DH lasers', *Appl. Phys. Lett.*, **35**, 654–656, 1979.
64. W. Nijman, P. Post, and G. A. Acket, 'Temperature dependent photoluminescence of long wavelength GaInAsP/InP', *Gallium Arsenide and Related Compounds 1980*, Conf. Ser. no. 56, Institute of Physics, Bristol, 1981.
65. G. H. Thompson, and G. D. Henshall, 'Nonradiative carrier loss and temperature sensitivity of threshold in 1.27 μm GaInAsP/InP D. H. lasers', *Electron. Lett.*, **16**, 42–44, 1980.
66. Y. Horikishi, and Y. Furukawa, *Jpn. J. Appl. Phys.*, **18**, 809, 1977.
67. N. K. Dutta, and R. J. Nelson, 'Temperature dependence of threshold of InGaAsP/InP DH lasers and Auger recombination', *Gallium Arsenide and Related Compounds 1980*, Conf. Ser. no. 56, Institute of Physics, Bristol, 1981, p. 193–199.
68. T. Uji, K. Iwamoto, and R. Lang, 'Nonradiative recombination in InGaAsP/InP light sources causing LED output saturation and strong laser threshold–current temperature sensitivity', *Appl. Phys. Lett.*, **38**, 193, 1981., Erratum, *Appl. Phys. Lett.*, **39**, 182, 1981.
69. T. Fukui, and Y. Horikoshi, 'Optically induced low photoluminescence regions in InGaAsP', *Jpn. J. Appl. Phys.*, **18**, 127–132, 1979.
70. W. D. Johnston, and B. I. Miller, 'Degradation characteristics of cw optically pumped AlGaAs heterostructure lasers', *Appl. Phys. Lett.*, **23**, 192–194, 1973.
71. P. Petroff, W. D. Johnston, and R. L. Hartmann, 'Nature of optically induced defects in GaAlAs–GaAs double heterojunction laser structures', *Appl. Phys. Lett.*, **25**, 226–228, 1974.
72. R. Ito, H. Nakashima, S. Kishino, and O. Nakada, 'Degradation sources in GaAs–GaAlAs double heterostructure lasers', *IEEE J. Quantum Electron.*, **QE-11**, 551–556, 1975.
73. R. Chin, N. Holonyak, Jr, R. M. Kolbas, J. A. Rossi, D. L. Keune, and W. O. Groves, 'Single thin-active layer visible spectrum InGaAsP heterostructure lasers', *J. Appl. Phys.*, **49**, 2551–2556, 1978.

74. R. Chin, N. Holonyak, Jr, and B. A. Vojak, 'Visible spectrum multiple quantum well InGaAsP–InGaAsP heterostructure lasers', *J. Appl. Phys.*, **51**, 4017–4021, 1980.
75. E. A. Rezek, R. Chin, N. Holonyak, Jr, S. W. Kirchoefer, and R. M. Kolbas, 'Quantum well InP–InGaAsP heterostructure lasers grown by LPE?' *J. Electron. Mater.*, **9**, 1–27, 1980.
76. S. W. Kirchoefer, E. A. Rezek, B. A. Vojak, N. Holonyak, Jr, D. Finn, D. L. Keune, and J. A. Rossi, 'Continuous room-temperature photopumped lase operation of visible-spectrum LPE InGaAsP (λ = 6700Å)', *IEEE J. Quantum Electron.*, **QE-17**, 161–166, 1981.
77. N. Holonyak, Jr, R. M. Kolbas, R. D. Dupuis, and P. D. Dapkus, 'Quantum well heterostructure lasers', *IEEE J. Quantum Electron.*, **QE-16**, 170–186, 1980.
78. M. Yano, H. Nishi, and M. Takusagawa, 'Theoretical and experimental study of threshold characteristics in InGaAsP/InP DH lasers', *IEEE J. Quantum Electron.*, **QE-15**, 571–579, 1979.
79. J. N. Walpole, T. A. Lind, J. J. Hsieh, and J. P. Donelly, 'Gain spectra in GaInAsP/InP proton-bombarded stripe geometry DH lasers', *IEEE J. Quantum Electron.*, **QE-17**, 186–192, 1981.
80. E. O. Göbel, O. Hildebrand, and K. Löhnert, 'Wavelength dependence of gain saturation in GaAs lasers', *IEEE J. Quantum Electron.*, **QE-13**, 848–854, 1977.
81. E. Göbel, and M. H. Pilkuhn, 'Optische Verstärkung in Halbleiterlasern', *Research Report* T79-167, Federal Ministry for Science and Technology, 1979.
82. H. Jung, E. Göbel, K. M. Romanek, and M. H. Pilkuhn, 'Temperature dependence of optical gain spectra in GaInAsP/InP double heterostructure lasers', *Appl. Phys. Lett.*, **39**, 468–470, 1981.
83. H. Jung, *Diploma Thesis*, Stuttgart, 1980, unpublished.
84. G. Lasher, and F. Stern, 'Spontaneous and stimulated recombination radiation in semiconductors', *Phys. Rev.*, A, **133**, 553–563, 1964.
F. Stern, 'Effect of band tails on stimulated emission of light in semiconductors', *Phys. Rev.*, **148**, 186–194, 1966.
85. A. R. Adams, M. Asada, Y. Suematsu, and S. Arai, 'The temperature dependence of the efficiency and threshold current of InGaAsP lasers related to intervalence band absorption', *Jpn. J. Appl. Phys.*, **19**, L621–L624, 1980.
86. M. Ettenberg, and H. Kressel, 'Interfacial recombination at AlGaAs/GaAs heterojunction structures', *J. Appl. Phys.*, **47**, 1538–1544, 1976.
87. R. E. Nahory, M. A. Pollack, and J. C. De Winter, 'Temperature dependence of InGaAsP double heterostructure laser characteristics', *Electron. Lett.*, **15**, 695–696, 1971.
88. M. Yano, H. Nishi, and M. Takusagawa, 'Temperature characteristics of threshold current in InGaAsP/InP double heterostructure lasers', *J. Appl. Phys.*, **51**, 4022–4028, 1980.
89. D. L. Rode, 'How much Al in the GaAlAs–GaAs laser?', *J. Appl. Phys.*, **45**, 3887–3891, 1974.
90. A. R. Doodwin, J. R. Peters, M. Pion, G. H. B. Thompson, and J. E. A. Whiteway, 'Threshold temperature characteristics of double heterostructure GaAlAs lasers', *J. Appl. Phys.*, **46**, 3126–3131, 1975.
91. H. C. Casey, Jr, 'Room temperature threshold current dependence of GaAs–GaAlAs double heterostructure lasers on x and active layer thickness', *J. Appl. Phys.*, **49**, 3684–3698, 1978.
92. I. Ismailov, and I. M. Tsidulko, 'Influence of the potential barrier height in a heterojunction laser on the temperature dependence of the threshold current', *Sov. J. Quantum Electron.*, **9**, 1163–1166, 1979.

IV GaInAsP DEVICE TECHNOLOGY AND PERFORMANCE

GaInAsP Alloy Semiconductors
Edited by T. P. Pearsall

Chapter 14

Double-heterostructure Lasers

YASUHARU SUEMATSU*, KENICHI IGA†, AND KATSUMI KISHINO†

* *Tokyo Institute of Technology, Department of Physical Electronics, 2–12–1 O-okayama, Meguro-ku, Tokyo 152, Japan*
† *Research Laboratory of Precision Machinery and Electronics, 4259 Nagatsuta, Midori-ku, Yokohama 227, Japan*

14.1 INTRODUCTION

Injection lasers in the wavelength range of 1.1–1.67 μm can be made with the GaInAsP/InP lattice-matched system,[1–7] which is attractive as a light source for ultra-low-loss optical-fibre communications. In this chapter, the state of the art in GaInAsP/InP lasers will be described with strong emphasis on the technology of laser processing and some special structures such as mode-controlled lasers, lasers integrable to optical circuits, and distributed Bragg reflector (DBR) integrated lasers. Also a brief review will be made about laser characteristics such as the temperature dependence of the threshold and efficiency of the GaInAsP/InP lasers which has been one of the important questions as for the quaternary system.

14.2 LASER WAFERS

The GaInAsP/InP laser fundamentally consists of four or five layers of $Ga_xIn_{1-x}As_yP_{1-y}$ and InP epitaxially grown on a (100) InP substrate. An oxide-

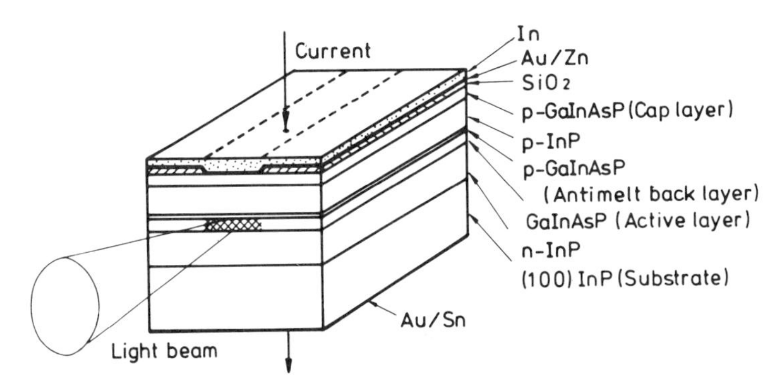

Figure 14.1 GaInAsP/InP laser with an oxide-stripe structure for 1.5–1.67 μm wavelengths

stripe laser, a typical example of a standard stripe laser, which emits in the 1.5–1.6 μm of wavelength range, is shown schematically in Figure 14.1. The double heterostructure (DH) confines injected carriers and the guided mode in the GaInAsP active region with the help of energy-step and refractive-index differences between the active layer and the cladding layer.

14.2.1 Wavelength dependence of threshold current density

The threshold current densities of prepared $Ga_xIn_{1-x}As_yP_{1-y}$/InP lasers are sufficiently low for emission wavelengths of 1.2 μm to 1.65 μm. The threshold current density J_{th} varies in proportion to the thickness d of an active layer when d is thicker than 0.2 μm.[6–8] Thus, the threshold current density normalized by active layer thickness, J_{th}/d, is a convenient parameter for the purpose of characterizing the quality of wafers. Experimental data for J_{th}/d are shown in Figure 14.2 against lasing wavelengths λ measured under pulsed conditions with broad contact lasers.[10]

Comparable values of J_{th}/d to those reported for GaAs/AlGaAs lasers,[9] i.e. 4–6 kA cm^{-2} μm^{-1}, are obtained for GaInAsP/InP DH lasers with emission wavelengths of 1.2–1.65 μm,[10–14] as shown in Figure 14.2. For lasers emitting at wavelengths longer than about 1.5 μm, the thin 'antimeltback' layer[15] mentioned in Chapter 2 is employed in order to prevent meltback of the active layer during the liquid-phase epitaxial (LPE) growth of the InP cladding layer, while the threshold current density is kept low. Figure 14.3 shows the light confinement

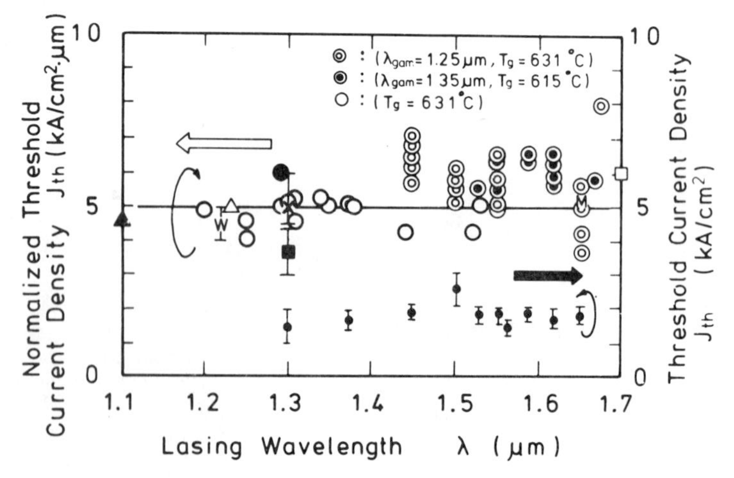

Figure 14.2 Wavelength dependence of normalized threshold current density: ○, Arai *et al.*[6, 10] (without AMB layer), ◎ ⊙, Arai *et al.*[10], and Itaya, *et al.*[25] (with AMB layer); ■, Yamamoto *et al.*[4]; ●, Oe *et al.*[11]; ▲, Hsieh *et al.*[1]; Δ, Nahory *et al.*[8]; □, Nuese *et al.*[13]; X, Coleman *et al.*[14]; M, Miller *et al.*[21]; W, Wakao *et al.*[12] Threshold currents: ⧫, Itaya, *et al.*[25]

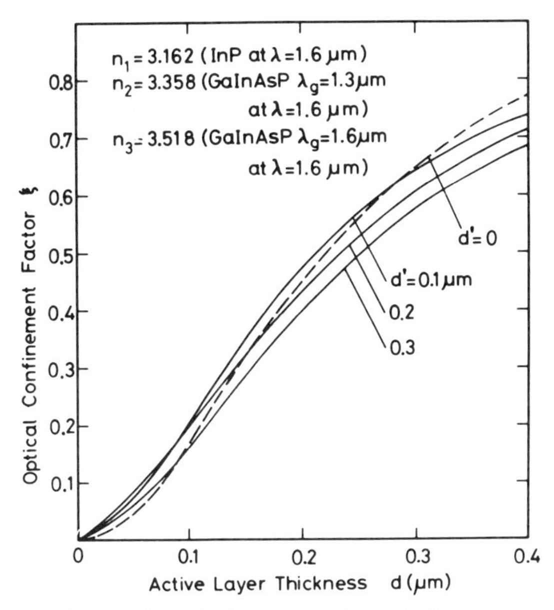

Figure 14.3 Light confinement factor in the active region of the laser structure with antimeltback layer as a function of active-layer thickness, where active and antimeltback layers are assumed to be 1.6 μm and 1.3 μm wavelength-composition, respectively and d' is the thickness of antimeltback layer

factor ξ associated with such a non-symmetric waveguide structure. The light confinement factor is not strongly dependent on the thickness d' of the antimeltback layer when d' is smaller than 0.3 μm. Furthermore, the GaInAsP active layer attached to the thin antimeltback layer is cladded by InP layers. Good carrier confinement can be also provided by the high heterobarrier at the heterojunction boundary with the InP layer. The feasibility of this structure has been proved because the laser with antimeltback layer has operated continuously at room temperature.[16, 17, 101]

The conventional DH wafers with emitting wavelength of 1.5–1.55 μm were also obtained by low-temperature and supercooling LPE techniques.[18, 102] However, for wavelengths longer than 1.6 μm, it is reported to be somewhat difficult to fabricate a DH structure without an antimeltback layer.[19] 1.5–1.7 μm wavelength GaInAsP/InP lasers have been fabricated by using LPE and molecular-beam epitaxy (MBE) techniques[20–24] without employing the anti-meltback layer and continuous-wave (CW) lasing operation has been achieved.[24]

14.2.2 Doping level and threshold current density

It is found that the concentration of p-type dopant in the active region significantly changes the threshold current density.[7] Usually Zn is used as a p-type dopant for the cladding layer. The threshold current density can be reduced by decreasing the acceptor concentration in the intentionally undoped active layer, which is doped by the diffusion of Zn from the cladding layer, down to a certain value where normal voltage–current (V–I) characteristics are maintained.[7]

The variation of the measured normalized threshold current J_{th}/d is plotted in Figure 14.4 against the Zn doping level of third layer, where the thickness d of the active layer is 0.1 μm, 0.2 μm, and 0.3 μm or larger and the lasing wavelength is typically 1.3 μm. The donor concentration N of the first InP layer is equally doped ($N = 7.5 \times 10^{18}\ \mathrm{cm}^{-3}$) for all wafers. The lowest value of J_{th} for a laser diode with the active layer thinner than 0.2 μm is obtained for $p = 1 \times 10^{17}\ \mathrm{cm}^{-3}$. For thicker active layers such as $d > 0.3\ \mu$m, J_{th} becomes a minimum even when $p > 1 \times 10^{17}\mathrm{cm}^{-3}$. In this case, however, the acceptor concentration in the active layer may be graded from the boundary of the third layer so that the mean value of acceptor concentration in the active layer would be less than that of the InP cladding layer. From this explanation, the threshold current density for thicker active layers is minimized at a relatively higher doping level compared with that for a thinner active layer. The lowest value of J_{th}/d for $d = 0.1\ \mu$m, on the other hand, is twice that for $d > 0.2\ \mu$m due to the poor light confinement, as discussed later. The threshold current density J_{th} increases again at Zn doping level lower than about $10^{17}\ \mathrm{cm}^{-3}$ as noted in Figure 14.4 (broken curves). This is due to another mechanism: since the residual doping level of the growth system used here is around $5 \times 10^{16}\mathrm{cm}^{-3}$, it is difficult to make the normal p–n junction

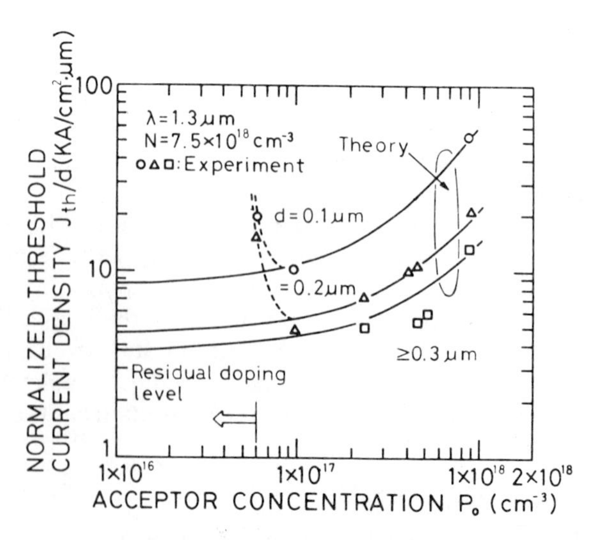

Figure 14.4 Variation of the normalized threshold current density for 1.3 μm wavelength lasers against the Zn doping level of the cladding layer[103]

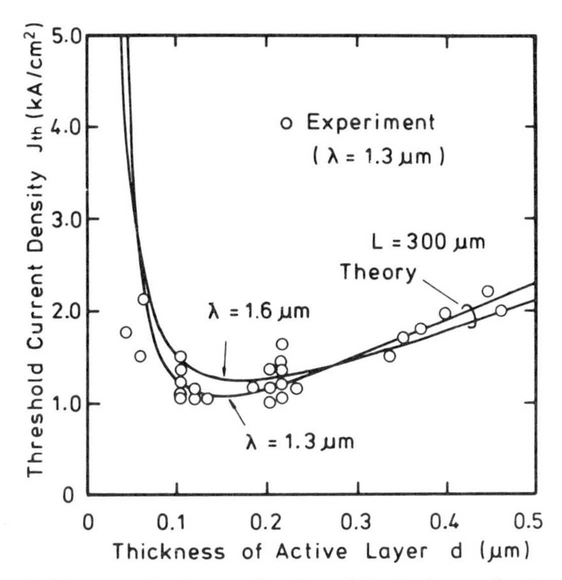

Figure 14.5 Threshold current density as a function of the thickness of active layer, where open circles are experimental values for 1.3 μm wavelength lasers, and full curves are theoretical results for 1.3 μm and 1.6 μm lasers. Data are from Itaya *et al* [7]

uniformly over the wafer at doping level less than $1 \times 10^{17}\,\mathrm{cm}^{-3}$. The full curves in Figure 14.4 indicate theoretical curves estimated by considering the absorption[26] between the acceptor level and the split-off band of the valence band for active and cladding layers. The experimental data are in good agreement with theoretical curves.

The dependence of threshold current density J_{th} on thickness d of the active layer of a laser diode at room temperature is shown in Figure 14.5 where the doping level into the cladding layer is optimized in the condition for the lowest threshold current density. The open circles are experimental results obtained for 1.3 μm quaternary lasers and full curves show calculated results of 1.3 μm and 1.6 μm wavelength DH lasers, respectively, based on laser parameters such as reported loss coefficients.[25,26,31] The threshold current density of broad contact lasers with 300 to 400 μm cavity length decreases linearly with decreasing thickness of the active layer down to $d = 0.2$ μm and is about $1.0\,\mathrm{kA\,cm}^{-2}$ for $d = 0.10$–0.22 μm. The threshold increases when the thickness is less than 0.07 μm. The lowest value of J_{th} which has been achieved is $0.77\,\mathrm{kA\,cm}^{-2}$ for a cavity length of 1500 μm.[27]

14.2.3 Beam divergence

The beam divergence angle $\theta_\perp$ is measured by scanning a small-area Ge avalanche photodiode perpendicular to the junction. Figure 14.6 shows the vertical beam

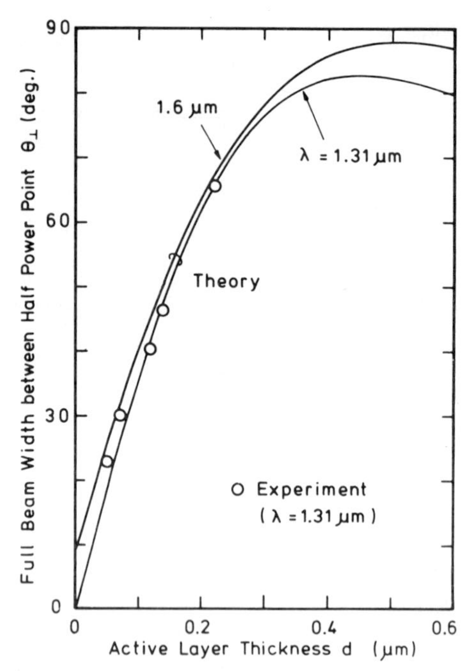

Figure 14.6 Full beamwidth at the half-power points as a function of active layer thickness. Data are from Itaya *et al.*[7]: ——, theory ($\lambda = 1.31$ and $1.6\ \mu m$); ○, experiment ($\lambda = 1.31\ \mu m$)

divergence angle $\theta_\perp$ as a function of active layer thickness d.[27] The open circle denotes the experimental value measured for $1.3\ \mu m$ DH lasers and full curves show theoretical results calculated by using the equations presented in reference 28 for $1.3\ \mu m$ and $1.6\ \mu m$ cases. First, we discuss the characteristic of $1.3\ \mu m$ DH lasers and then mention that of $1.6\ \mu m$ DH lasers. The following values of the refractive indices are used for $\lambda = 1.31\ \mu m$: $n_1 = 3.52$ for GaInAsP active layer (see reference 29), and $n_2 = 3.20$ for the InP layer (reference 30), that is, the refractive index difference $\Delta \simeq (n_1 - n_2)/n_1$ is 8.7%. In this case, the thickness of the active layer for which higher-order transverse modes in the plane perpendicular to the junction is cut off must be thinner than $0.45\ \mu m$. The theoretical curve of beam divergence angle is calculated only for the fundamental mode. The measured $\theta_\perp$ is 64° at $d = 0.22\ \mu m$, 40° at $d = 0.13\ \mu m$, and 23° at $d = 0.05\ \mu m$, respectively. The vertical beam angle $\theta_\perp$ can be narrowed by reducing the active-layer thickness in the single transverse-mode region (namely $d < 0.45\ \mu m$), because the intensity profile of the mode spreads out into the InP cladding layers. The experimental values agree well with the theory.

The beam divergence angle of a $1.6\ \mu m$ DH laser not decrease up to zero value with decreasing thickness of active layer. This is because of the non-symmetric waveguide structure of the laser mentioned previously. The thickness and wavelength of antimeltback layer assumed here are 0.05 μm and 1.30 μm, respectively, and the value of refractive index of GaInAsP is shown in Figure 14.23.

14.3 MODE-CONTROLLED LASERS

14.3.1 Various types of mode-controlled lasers

The transverse mode control which is obtained with the help of index-guiding structure is very important to improve lasing properties such as modulation linearity and longitudinal mode behaviour as confirmed experimentally and theoretically.[32,33] Various types of GaInAsP/InP lasers with index-guiding which operate with single transverse and longitudinal mode have been realized as shown in Figure 14.7.[34–47]

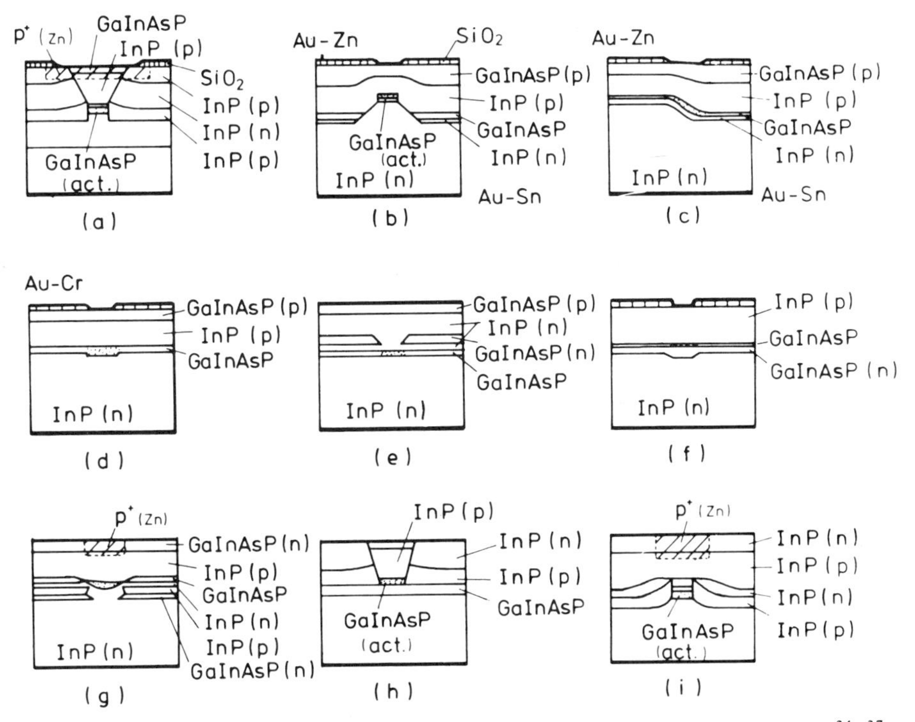

Figure 14.7 Mode-controlled quaternary lasers: (a) buried heterostructure (BH) lasers[34–37]; (b) mesa-substrate buried (MSB) heterostructure lasers[38,39,50]; (c) terraced-substrate (TS) lasers[40]; (d) rib-waveguide lasers[41]; (e) self-aligned structure (SAS) lasers[42]; (f) plano-convex waveguide (PCW) lasers[43,44]; (g) buried crescent (BC) lasers[45]; (h) strip-buried heterostructure lasers[46]; (i) planar buried heterostructure (PBH) lasers[47]

14.3.2 Buried-heterostructure (BH) lasers

The BH structure shown in Figure 14.7 (a), which was first demonstrated in the GaAs system,[48] is prepared from a two-step LPE growth process. We pick up, as an example, 1.5–1.6 μm wavelength BH lasers[35,49] here to explain the fabrication process of BH lasers.

First, the DH wafer is grown on a (100) InP substrate by the two-phase solution technique. After depositing an SiO_2 film on the DH wafer by a chemical vapour deposition (CVD) technique, narrow stripe masks are formed along the ⟨011⟩ crystal direction. Narrow strip mesas are etched and then InP blocking layers are grown on the mesa-etched wafer resulting in the BH structure. Usually, the etching is carried out down to below the active layer.[48] However, especially for the fabrication of GaInAsP/InP BH lasers, it is important to fabricate those without exposing the InP surface to a high-temperature atmosphere, because smooth regrowth of the blocking layers is quite sensitive to the surface condition. Thus, the DH wafer is etched first by a Br-CH_3OH solution to half-way of the p-type InP cladding layer and then etched by using a preferential etchant for InP, i.e. 4HCl + H_2O at about 1°C. This etch stops automatically just on the GaInAsP antimeltback layer.[103] After the etch, the DH wafer undergoes a regrowth process. In this method[35,49] the active layer is covered by the GaInAsP antimeltback layer during the saturation process of materials before the crystal growth of the blocking layers at high temperature. The surface damage of the quaternary crystal due to the dissociation of phosphorus is kept quite small compared with the InP surface.

In the regrowth cycle, p-InP, n-InP, and p-InP layers are successively grown on the etched wafer; the top of the mesa is covered with an SiO_2 mask. The exposed quaternary layers are melted back at 590°C, narrow strip active regions are formed by an unsaturated In–InP solution for 2 s, then three layers are

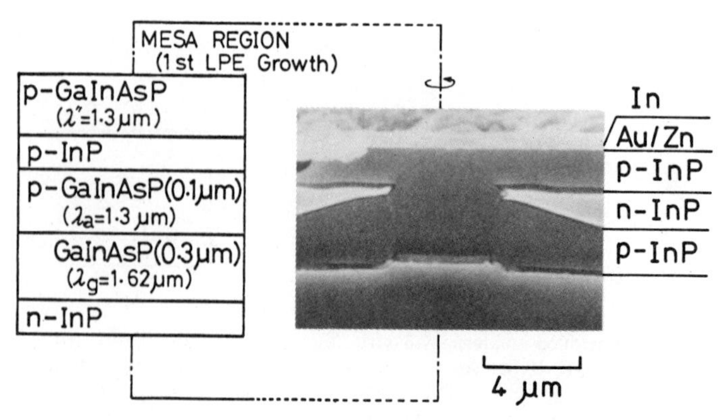

Figure 14.8 An SEM photograph of a cross-section of BH laser[49]

successively grown. For the purpose of achieving a small amount of meltback, 10 mg of InP source material and 5 g of In are used to make an unsaturated solution. Figure 14.8 shows a scanning electron (SEM) micrograph of the cross section of the obtained BH structure.

After the regrowth, the grown surface of the wafer is covered with an SiO_2 film by a CVD technique and 10 μm windows are opened to diffuse Zn for low-resistance contacts. The top p-type InP layer in the regrowth cycle is made to relax the alignment condition on the Zn diffusion region to the active region. After polishing the substrate side to about 100 μm, ohmic contacts are made by Au/Zn and Au/Sn for p- and n-type sides, respectively. After 2 min alloying at 420°C in purified hydrogen gas, indium metal is evaporated onto the p-type side for a soft contact.

Figure 14.9 shows (a) the etching characteristics on a DH wafer using 0.2 vol % Br-CH_3OH at about 20°C and (b) an example of the photograph of the etched cross section of an InP crystal. The mesa etched by this non-preferential etchant, which means that the etching speed is almost the same for both InP and GaInAsP crystals, has an undercut shape. The etching properties are complicated since width and depth of the mesa neck do not vary linearly with the etching time.

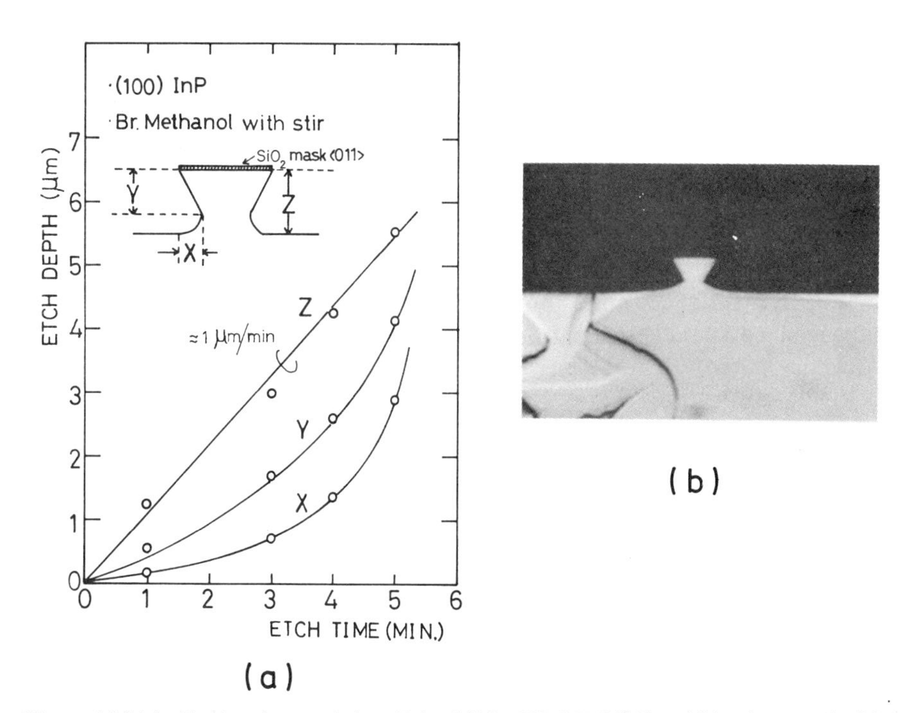

Figure 14.9 (a) Etching characteristics of 0.2 vol % Br-CH_3OH (20° C) and (b) a photograph of the etched cross-section of InP crystal [49]

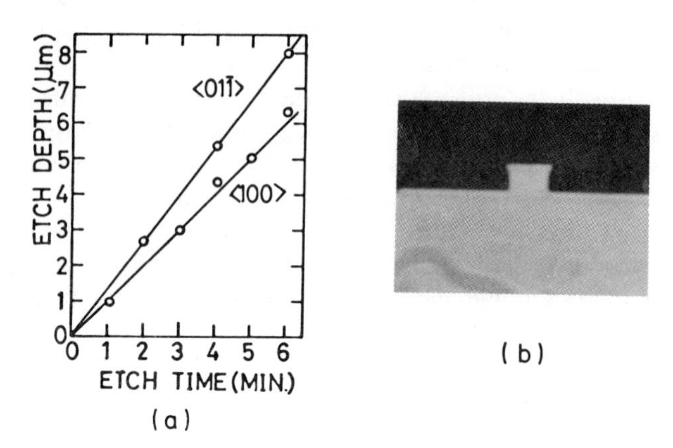

Figure 14.10 (a) Etch speed of $4HCl + H_2O$ for InP crystal at 1°C, and (b) an SEM photograph of the cross-section of the etched DH wafer by two-step etching process[49]

Therefore, if we want to make the mesa neck at the active layer, and to use the same stripe width of the SiO_2 mask for all wafers, the width of neck obtained will vary as much as 1 μm from wafer to wafer since the total thickness of p-type InP cladding layer and p-type GaInAsP cap layer grown by LPE fluctuates within 0.5 μm. To put it another way, the stripe width of SiO_2 mask should be set to an appropriate value for each DH wafer in order to get the required width of active region of the BH laser. However, by using the above-mentioned two-step etching process, the width of the mesa neck can be controlled because in the second step of the selective etch the mesa is reshaped to be rectangular owing to the existence of non-etched layers of GaInAsP above and below. In this etching process, the width of the mesa neck is determined by the etching time of the first step etching by Br-CH_3OH. Figure 14.10 shows (a) etch speed in InP using $4HCl + H_2O$ solution at about 1°C and (b) an SEM photograph of the cross-section of the DH wafer etched by the two-step etching process.

14.3.3 Mesa-substrate buried (MSB) heterostructure lasers

The schematic structure of the MSB laser[38, 50] is shown in Figure 14.7(b). In the MSB laser, the GaInAsP active region is grown separately on the top of the mesa-shaped (100) InP substrate and it is completely embedded in the subsequent InP crystal by employing single-step liquid-phase epitaxy. The buried active region, therefore, can be fabricated without exposing it to the air. The injection current is confined around the buried active region of width W and thickness d with an SiO_2 insulating window opened just above the buried active region. The current

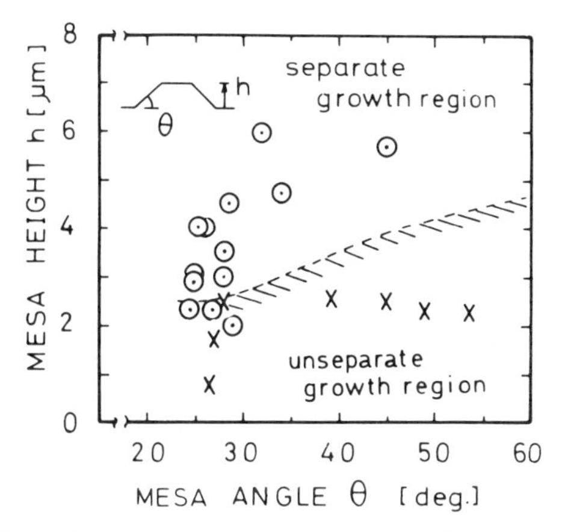

Figure 14.11 Conditions of mesa shape for separate growth of the buried active region on the mesa top of (100) InP substrate[50]

confinement of this structure by the oxide window should be appreciably improved compared with that of a conventional oxide stripe geometry because the p–n junction of InP is formed on the slope of the mesa with a wider energy gap than that of the active region.

As for its fabrication, the surface of a (100) oriented InP substrate is first chemically etched to form mesa structures along the $\langle 01\overline{1} \rangle$ crystal axis. Then, four layers with different compositions of $Ga_xIn_{1-x}As_yP_{1-y}$ are successively grown on the mesa-shaped substrate by conventional two-phase solution liquid-phase epitaxy.[7] If the mesa shape is chosen suitably, as will be mentioned later, the buried active region is grown separately on the mesa top, apart from that of the mesa bottom.[50] During the subsequent growth of InP, the active region is embedded in the InP crystal and the grown surface becomes flat when this layer is thick enough. Usually, an additional GaInAsP cap layer is grown in order to get good ohmic contact. Finally, the oxide-stripe electrode is formed on the MSB wafer.

The condition that the buried active layer be grown on the mesa top separately from that of the mesa bottom is influenced by the mesa height h and the side angle θ as shown in Figure 14.11. The circles and crosses represent experimental results for the separated and the unseparated cases of the active layer, respectively. When the mesa height is about 2.5–3 μm, separate growth of the buried active region is achieved if the side angle is about 28°. The side angle required for the separate growth increases with increasing mesa height as shown in Figure 14.11. Thus, separate growth is more easily obtained by increasing the mesa height. In order to avoid the deformation of the mesa shape during crystal growth, it is necessary that the thickness of the first grown InP buffer layer is thin enough. This is possible in the case of the GaInAsP/InP system because the substrate is transparent at the lasing wavelength.

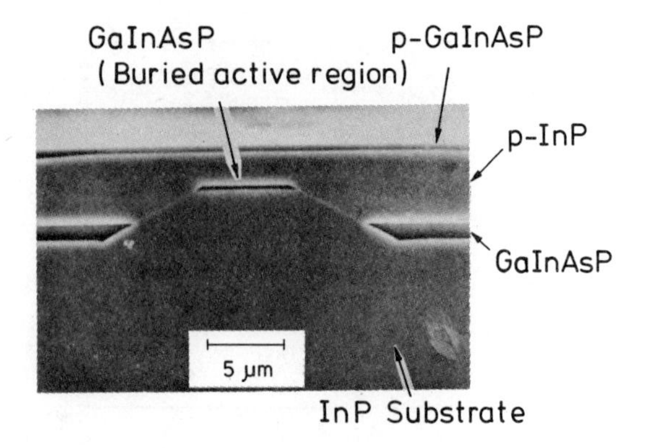

Figure 14.12 An SEM photograph of a cleaved cross-section of MSB wafer[50]

An SEM photograph of the cross-section of an MSB wafer is shown in Figure 14.12, where the width W of the active region is 5.5 μm, the thickness d is 0.3 μm, and the mesa height h is 3.5 μm. The as-grown surface of the wafer is flat.

14.3.4 Terraced-substrate (TS) lasers

The structure of a TS laser[40, 51] is shown in Figure 14.7(c). The waveguiding structure in the TS laser consists of a strip of non-uniform-thickness active layer which provides us with effective index guiding and is prepared by single-step LPE growth. The active layer is not exposed to the air, therefore, as in the fabrication of the buried heterostructure. This is one of the merits of the TS laser in addition to the simplicity of the process. The only TS laser which has been realized is from the GaAlAs/GaAs system.[52, 53]

It is important to control the shape of a built-in strip waveguide since lasing characteristics of the TS laser such as threshold current and transverse-mode behaviour depend strongly on the shape of the waveguide, i.e. the thickness and the width of the waveguide. The width and thickness of the strip waveguide necessary for guiding only a fundamental transverse mode has been studied theoretically.[40] Thus, well-controlled LPE growth is also required in order to have the expected shape of strip waveguide. As for the thickness, it is controlled by the amount of supercooling of the LPE solution for the active layer and by its growth time. As the growth on the terrace slope is remarkably faster than that on the flat regions outside the built-in waveguide, it is necessary to reduce the amount of supercooling when we want to grow a thin active layer. Therefore, the temperature difference between the saturation process and the growth of the active layer should be reduced, with the cooling rate being slowed. On the other hand, the width of the strip waveguide is controlled mainly by the terrace height of the substrate, i.e. the shallower the terrace is made, the narrower the strip width becomes. The shallow terrace is, therefore, required when we want to fabricate a narrow strip waveguide. The LPE growth process is carried out as follows. First, prior to crystal growth, terraces of 0.3–0.5 μm in height and about 120 μm in width are formed on a (100) InP substrate in the $\langle 011 \rangle$ direction by a chemical etch where a photoresist is used as an etch mask. The etchant for this process is a mixed solution of $HCl:CH_3COOH:H_2O_2$ (1:2:1 or 2:6:1) at 15° C.[54] These solutions provide us with mirror-like surfaces of the terrace and bottom. Next, four layers are grown on the terraced substrate by the two-phase solution method;[55] n-type InP buffer layer (Te-doped, 0.1–0.2 μm), $Ga_xIn_{1-x}As_yP_{1-y}$ active layer (undoped, 0.15–0.4 μm on the terrace slope, 0.05–0.25 μm on the flat regions, $x = 0.28$, $y = 0.62$ for $\lambda = 1.3$ μm), p-type InP cladding layer (Zn-doped, 2.5 μm), and p-type GaInAsP cap layer (Zn-doped). The saturation temperature and the growth temperature of the active layer are 645° C and 635° C, respectively. The terraced substrate is protected with an InP cover wafer during the saturation process and is melted back by a small amount (0.1–0.2 μm) in depth using an unsaturated In melt just before the first InP layer growth. The initial cooling rate is 0.8° C min^{-1}, and is slowed down to 0.2° C min^{-1} or 0.4° C min^{-1} about 15 min before the growth of the active layer in order to reduce the supercooling of the melt. By using these growth conditions, we have been able to obtain strip waveguides as thin as 0.2 μm on the terrace slope and as narrow as 2–3 μm in width, which are necessary for the reduction of the threshold current and for the stabilization of fundamental transverse mode. Figure 14.13 shows an SEM photograph of a cross-section of the TS laser. The thickness of the active layer is 0.3 μm on the terrace slope and 0.2 μm on the flat regions, and the width of the strip waveguide is narrower than 3 μm.

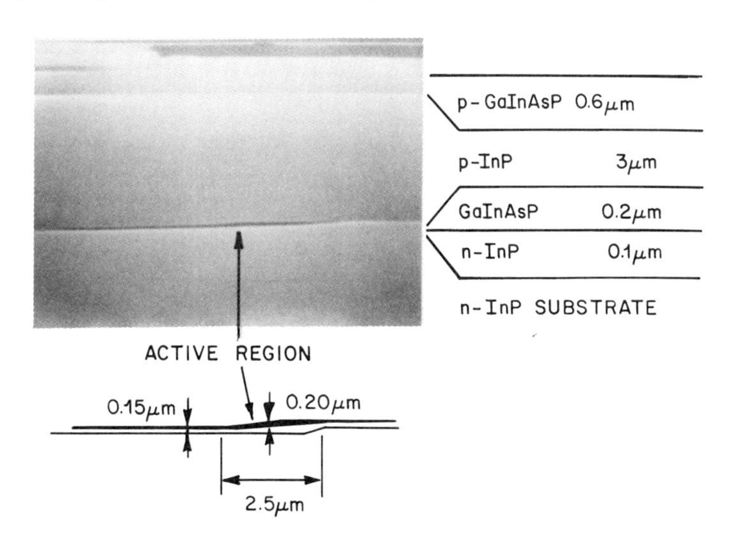

Figure 14.13 An SEM photograph of a cleaved cross section of TS wafer

In order to prepare striped electrodes, stripe windows (5–10 μm in width) are opened on an SiO_2 film formed on the as-grown surface of the DH wafer. Zn is diffused onto the cap layer of the wafer in order to obtain low series resistance necessary for CW operation. Ohmic contacts to the n-type and to the p-type surface are made by evaporating Au–Sn and Au–Zn metal, respectively, and alloying for 1 min at 420° C in vacuum. A laser chip is cleaved from the wafer and mounted on a copper heat sink.

14.3.5 Lasing characteristics of BH lasers

The BH wafer is cleaved into chips with cavity length of 250–350 μm and the laser chip is bonded on a Au–plated copper heat sink using an indium solder with p-side down. The threshold current is evaluated under pulsed excitation. Threshold currents as low as 23 mA with an average of 42 mA are obtained from 84 samples with cavity length of about 300 μm and width of the active region of 3–5 μm. Figure 14.14 shows *I–L* characteristics of selected samples with relatively low threshold. For most samples, light output increases linearly up to twice the threshold current. Transverse single-mode operation of the BH laser is obtained up to more than three times the threshold. Threshold current density of broad contact type lasers from this DH wafer is about 2kA cm^{-2}. Therefore, threshold current should be less than 30 mA for lasers with the above-mentioned size of the

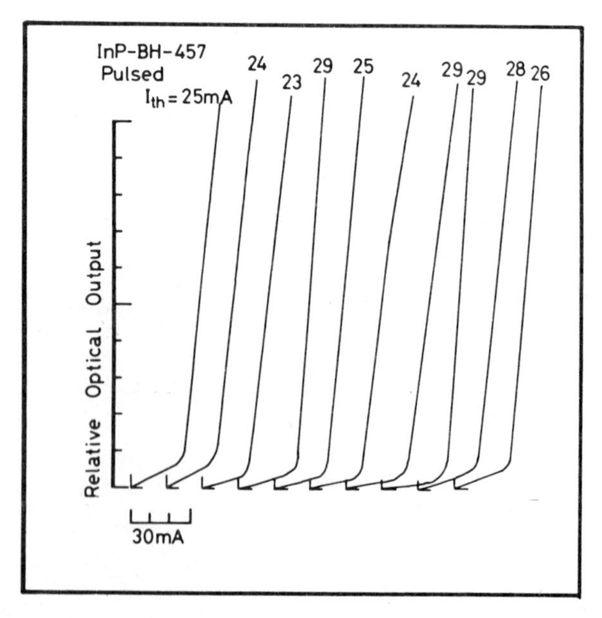

Figure 14.14 *I–L* characteristics of 10 samples obtained from the same wafer.[9] © 1981 IEEE

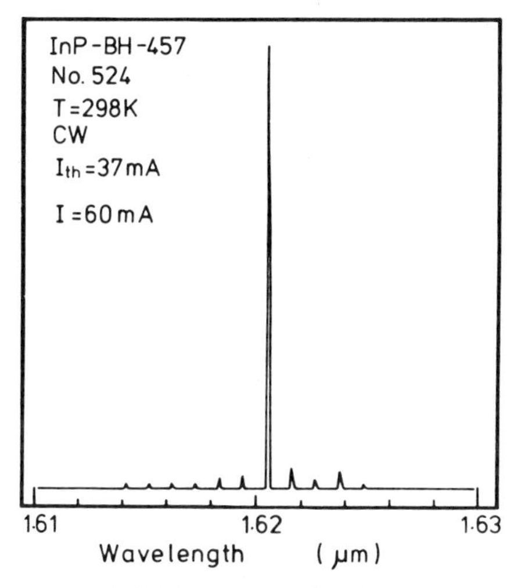

Figure 14.15 Lasing spectrum of a BH laser operated under room-temperature CW conditions [49]

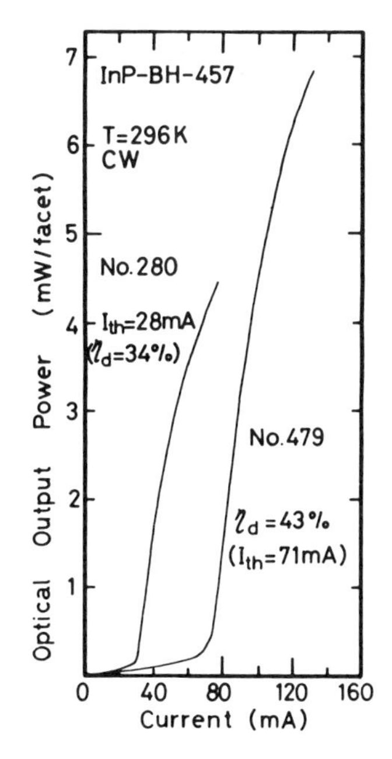

Figure 14.16 Light output vs DC current characteristics of BH lasers [49]

active region if there is neither leakage current nor additional loss due to the roughness of the boundary between the active region and regrown side-bounding InP layers. Figure 14.15 shows a lasing spectrum under the operation with the output power of 2 mW/facet. The lowest threshold current obtained is 28 mA and the highest differential quantum efficiency is 43 %, as shown in Figure 14.16. Maximum output power of about 10 mW/facet is obtained under room-temperature CW operation. Further improvement of the thermal resistance is still necessary to maintain high-power operation. Fundamental transverse-mode operation is observed up to more than three times the threshold, when the active region width is about 3 μm[35] or less.

14.3.6 Lasing characteristics of MSB lasers

A typical light output versus current (*L–I*) characteristic of the MSB laser with a 0.3 μm thick buried active region is shown in Figure 14.17. The width of the buried active region is $W = 5.5$ and 10 μm, and the corresponding stripe electrode width is 15 and 10 μm, respectively. It is seen that the MSB laser operates with a linear light output versus current characteristic over a range of more than three times the threshold current. The output power is about 60 mW at $3.4 \times I_{th}$ for W

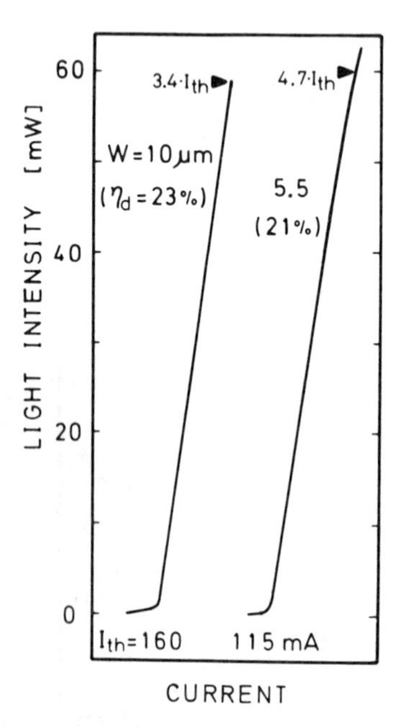

Figure 14.17 Light output vs current characteristics of MSB lasers with widths of buried active region of 5.5 and 10 μm[50]

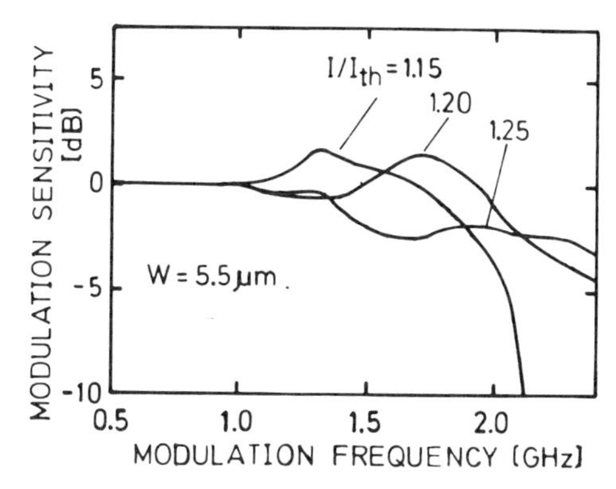

Figure 14.18 Modulation characteristics of an MSB laser [50]

$= 10\,\mu$m, and 60 mW at $4.7 \times I_{th}$ for $W = 5.5\,\mu$m, respectively. The minimum threshold current is 115 mA for a buried strip width of $W = 5.5\,\mu$m. The external quantum efficiency of about 12% is observed at a current level of three times the threshold for both strip widths. Recently, CW lasing operation of the device with the current-confinement structure has been reported by Nomura *et al.* with low threshold current of 20 mA for the emission wavelengths of 1.33 and 1.58 μm.[39] Stabilized fundamental-transverse-mode operation was also obtained with the stripe width of 2 μm.

A direct modulation characteristic of the MSB laser is measured by a sharp current pulse method.[56] The modulation efficiency is shown in Figure 14.18, where the buried strip width W is 5.5 μm. The resonance-like peak which often appears around the cut-off frequency is not significant and direct modulation up to about 2 GHz has been achieved.

14.3.7 Lasing characteristics of TS lasers

Threshold currents of most TS lasers are 50–150 mA and the lowest value so far obtained is 44 mA in a laser with cavity length 130 μm. Good linearity of the laser output power versus the injection current is achieved above the threshold . The differential quantum efficiency of 42% for both facet mirrors and laser output power of 170 mW are obtained in the best laser. CW operation has also been achieved at room temperature with higher threshold current (250 mA), which might be due to heating at the electrode contact which is not yet optimized.

An example of near-field patterns of a TS laser is shown in Figure 14.19. Stable fundamental-transverse-mode operation is achieved by reducing the width of the strip waveguide to about 3 μm. The spot size at half maximum intensity is about 2 μm and does not change much with increasing current level, indicating that the transverse mode is stabilized successfully by the built-in index waveguide. Single

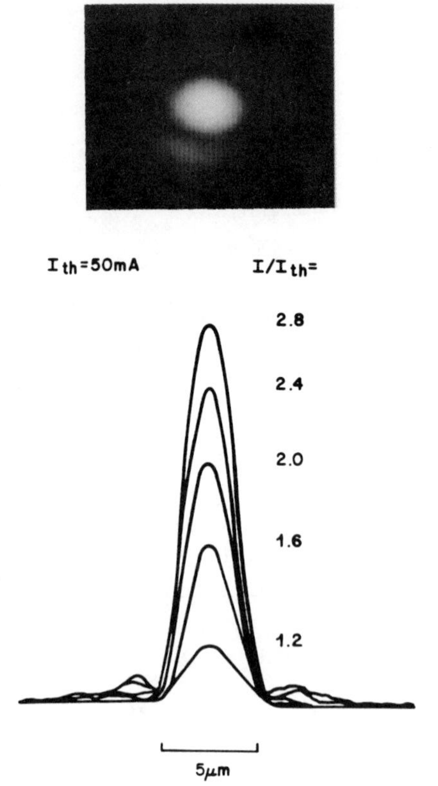

Figure 14.19 Near-field patterns observed in the junction plane. The spot size at half maximum is about 2 μm and does not change much with increasing current level. Dark bands seen in the photograph are from scanning lines of the TV monitor used and do not show the actual distribution of the light intensity of the lasing mode

wavelength operation is maintained up to 2.8 times the threshold in the better lasers.

Lasing wavelengths of TS lasers have been examined. There exists an almost linear relationship between lasing wavelengths and threshold current densities, i.e. lasing wavelengths decrease as threshold currents increase, which is considered to be due to the change of band filling. So far as lasing wavelengths of devices which have the same threshold current densities are concerned, the variation of the lasing wavelength within the unit wafer is 4.5 nm in variance, which is relatively small compared with the result obtained for broad contact lasers with uniform active layers.[55] Therefore, the compositional change originating in the selective epitaxial growth of GaInAsP quaternary layer on the terrace slope seems to be quite small.

14.4 TEMPERATURE DEPENDENCE OF THE EFFICIENCY AND THRESHOLD CURRENT

Assuming the temperature dependence of threshold current $I \propto \exp(T/T_0)$, the characteristic temperature T_0 of GaInAsP/InP DH lasers is 40–80 K at room temperature, where $T_0 \simeq$ 100–120 K at low temperature.[4, 10, 57, 58] The reason for such a small T_0 has been reported by a number of authors; (1) carrier leakage over the heterobarrier,[59] (2) Auger recombination,[60-62] and (3) inter-valence-band absorption of light.[26]

The differential quantum efficiency η_d of 1.6 μm GaInAsP/InP lasers decreases with increasing temperature.[26] The absorption loss coefficient in the active region α_{ac} estimated from the measured η_d is shown in Figure 14.20. The loss coefficient is fairly large compared with that of GaAs, and this cannot be explained by well-known loss processes such as free-carrier absorption. Therefore, light absorption due to electron transitions from the split-off valence band into holes injected into and thermally generated within the heavy-hole valence band has been introduced to explain the larger internal loss.[26]

The temperature dependence of the threshold current and the differential quantum efficiency are shown in Figure 14.21 in comparison with experiment and theory. The theoretical curve is derived by the above-mentioned split-off-band absorption process with measured carrier lifetime.[26] The measured carrier lifetime was inversely proportional to the carrier density which might be affected partly by the Auger effect and the carrier leakage, simultaneously with the other

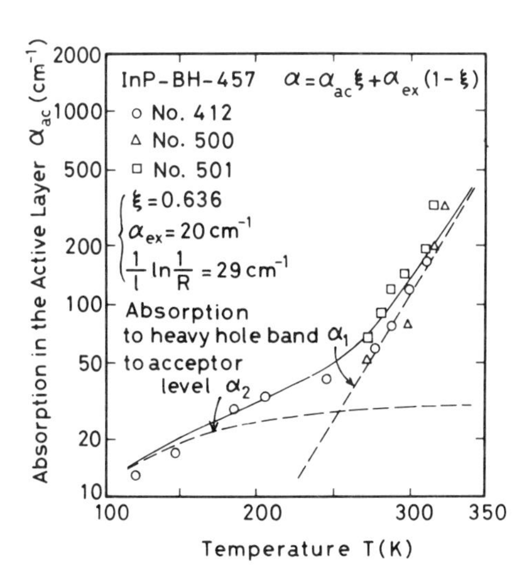

Figure 14.20 Temperature dependence of the absorption coefficient in the active region[26]

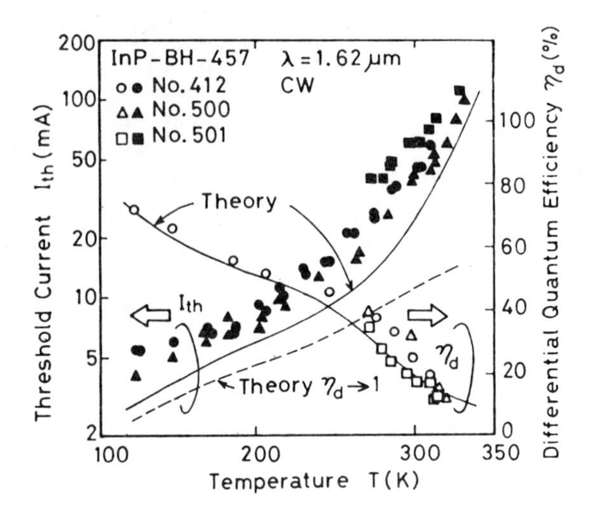

Figure 14.21 Temperature dependence of the threshold current I_{th} and differential quantum efficiency η_d. The broken curve indicates the predicted variation in the absence of inter-valence-band absorption

traps. According to the above-mentioned treatment, experiments can be explained fairly well. Thus, the absorption band is considered to be one of essential processes related with the temperature characteristics of GaInAsP/InP DH lasers.

14.5 RELIABILITY

GaInAsP/InP lasers have been shown to be much more reliable than GaAs lasers.[20, 63, 64] (a) The dark line which is often observed in GaAs lasers is observed only in unusual conditions.[65] It is believed that the effect of a dislocation centre is localized only around a limited region.[63, 66] (b) The maximum light power to degrade the mirror, 15 MW cm^{-2}, is about three times larger than for GaAs lasers.[67] The mirror facet is much more stable against water vapour.[68] (c) In solder increases thermal resistance and degrades the facet in a water vapour atmosphere.[63] However, Au–Sn solder improves these problems significantly.[69]

It was shown for 1.3 μm BH lasers that the laser lifetime at 70° C with 5 mW/facet is expected to be more than 9.5×10^3 h by an accelerated aging test.[70] From the result of these preliminary aging tests on new GaInAsP/InP lasers,[70, 71] it is believed that the operating lifetime is 10 times longer than that of GaAs lasers, even though the reliability problems are not yet fully understood.

14.6 INTEGRATED LASERS

14.6.1 Distributed Bragg reflector (DBR) integrated twin guide (ITG) lasers

Distributed Bragg reflector (DBR) GaInAsP injection lasers[72-76] are promising devices because they have properties of wavelength selectivity, wavelength stability against temperature and injection current, and wavelength control by the corrugation period of the DBR. The DBR and distributed feedback (DFB) laser with the corrugation, thus far, has been reported in GaAs and other materials and some of the above-mentioned properties have been also demonstrated.[77-82] Especially, stable single-longitudinal-mode operation of the laser even under high-speed direct modulation is essential for minimizing the delay distortion due to the fibre dispersion[87] at the ultimate low-loss wavelength region, that is 1.5 μm to 1.65 μm.[85,86] Moreover, in this device, TM mode can also be completely suppressed by loading a metal film[83] on the DBR region, whereas in DFB lasers[78,84] not only TE but also TM modes may oscillate.

In DBR lasers, the corrugation which forms the DBR is fabricated at a separate place from the active region, increasing the degree of freedom in the operational function. Thus, the following points are important: (1) efficient coupling between the active region and the DBR region and (2) the introduction of a low-loss waveguide on which the DBR is prepared. These two conditions can be satisfied in integrated twin guide (ITG) structures,[88] which consist of the active guide and the output guide coupled to each other in the manner of a directional coupler. In this structure, the waveguide parameters are chosen so that phase velocities of

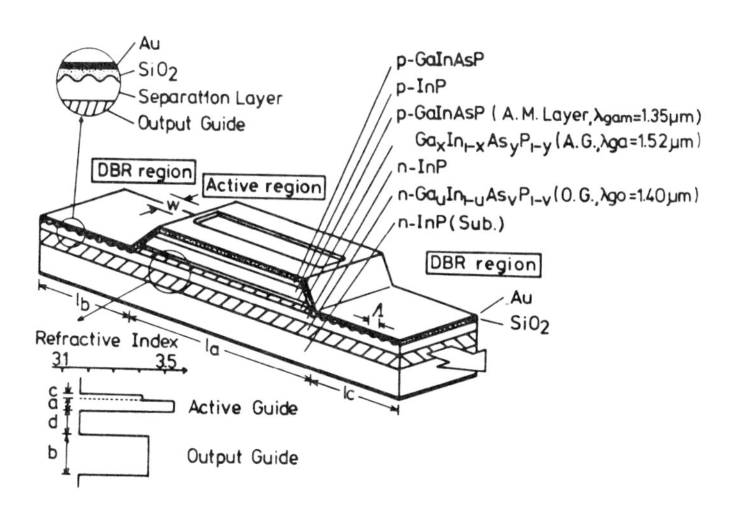

Figure 14.22 Structure of distributed Bragg reflector (DBR) integrated twin guide (ITG) GaInAsP/InP lasers.[74] Am = antimeltback, AG = active guide, OG = output guide

modes in both waveguides are equal.[89] DBR lasers with the integrated twin guide (ITG) structure are called as DBR-ITG lasers for short.

A GaInAsP/InP DBR-ITG laser for 1.5–1.6 μm wavelength is shown schematically in Figure 14.22. It consists of the twin guide (ITG) structure, that is the active guide with an antimeltback layer and the output guide coupled together. A thin antimeltback layer is introduced to prevent the meltback of GaInAsP active layer as described previously. The wafer is prepared by conventional liquid-phase epitaxy using the two-phase solution technique.[7]

14.6.2 Refractive index of GaInAsP crystals

In order to improve the coupling efficiency and to determine the corrugation period corresponding to the designed wavelength, exact knowledge of the refractive index of a quaternary GaInAsP crystal is required. The refractive index of the quaternary GaInAsP crystal can be estimated by using the concept of the single effective oscillator (SEO) method[90] confirmed for the ternary GaInP and GaInAs.[91,92] Using the modified single effective oscillator (SEO) method, the refractive index of the $Ga_xIn_{1-x}As_yP_{1-y}$ crystal at the photon energy $E(=h\nu)$ can be approximated by the characteristic parameters derived from the data on GaInP[91] and GaInAs[92] with an interpolation method as

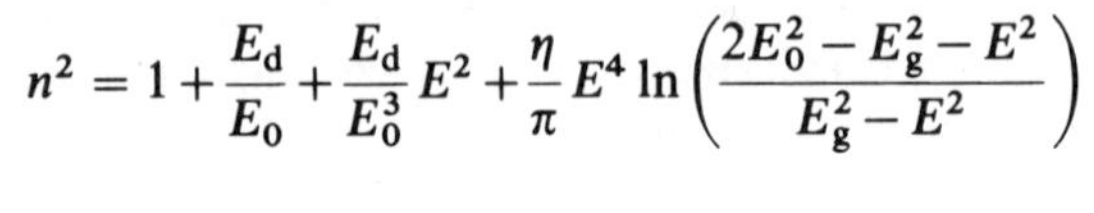

$$n^2 = 1 + \frac{E_d}{E_0} + \frac{E_d}{E_0^3}E^2 + \frac{\eta}{\pi}E^4 \ln\left(\frac{2E_0^2 - E_g^2 - E^2}{E_g^2 - E^2}\right)$$

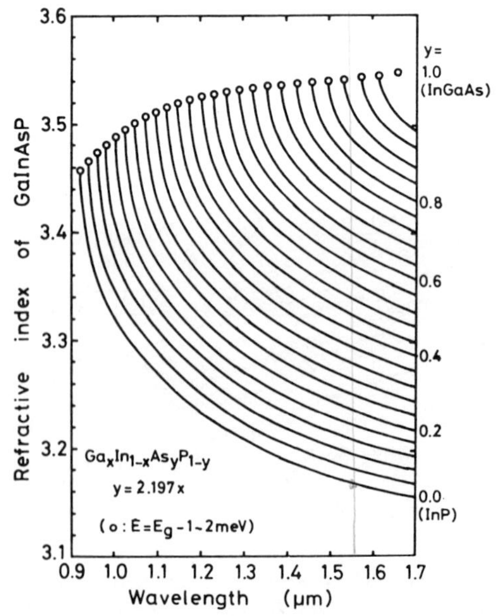

Figure 14.23 Refractive index of the lattice-matched GaInAsP crystal to InP crystal

where

$$\eta = \frac{\pi E_d}{2E_0^3(E_0^2 - E_g^2)}$$

and with

$$E_0 = 3.391 - 1.652y + 0.863y^2 - 0.123y^3$$
$$E_d = 28.91 - 9.278y + 5.626y^2$$
$$E_g = 1.35 - 0.72y + 0.12y^2$$

The bandgap energy E_g derived from the empirical fit to experimental results is used from reference 93. In the above equation, the lattice-matching condition, i.e. $y = 2.197x$,[93] has been used. The calculated curves of the refractive index of GaInAsP from the above expression are shown in Figure 14.23.

Using these theoretical refractive indices, the waveguide parameters for phase-matching between the active guide and the output guide is calculated. The design condition of a coupling system with the increased fabrication tolerance is given in reference 89.

14.6.3 Fabrication of DBR-ITG lasers

First, seven quaternary and InP lasers are successively grown on n-(100) InP substrate for $1.5 - 1.6\,\mu m$ wavelength DBR-ITG lasers, respectively. The saturation process temperature and the growth temperature of the active layer are 670°C and 615°C, respectively. The rather low cooling rate of $0.17°C\,min^{-1}$ is employed in order to obtain precise control of the growth thickness of each layer of the twin guide structure. In the case of 1.5–$1.6\,\mu m$ wavelength, an n-InP buffer layer (Te-doped, $2.5 \times 10^{18}\,cm^{-3}$; thickness $\sim 15\,\mu m$) is grown first, followed by the successive growth of an n-GaInAsP output guide layer (bandgap wavelength $\lambda_{g0} \simeq 1.40\,\mu m$; Te-doped, $5 \times 10^{18}\,cm^{-3}$; thickness $b \simeq 0.55\,\mu m$), n-InP separation layer (Te-doped, $5 \times 10^{18}\,cm^{-3}$; $d \simeq 0.4\,\mu m$), undoped GaInAsP active layer ($\lambda \simeq 1.57\,\mu m$; $a \simeq 0.2\,\mu m$), p-GaInAsP antimeltback layer ($\lambda_{gam} \simeq 1.35\,\mu m$; Zn-doped, $1 \times 10^{17}\,cm^{-3}$; thickness $\simeq 0.1\,\mu m$), p-InP cladding layer (Zn-doped,

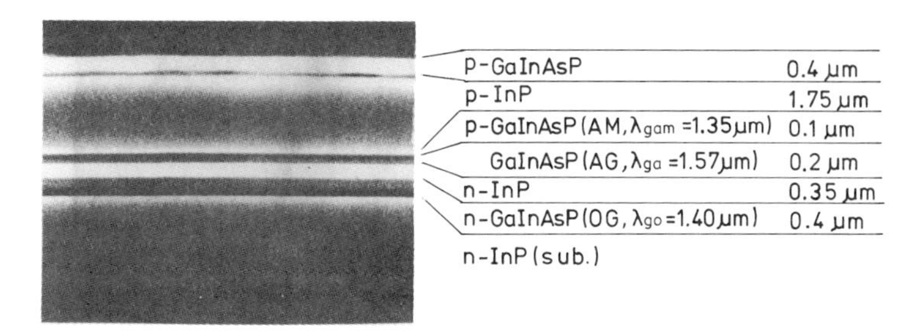

Figure 14.24 SEM photograph of ITG wafer, where AG = active guide, OG = output guide, and AM = antimeltback layer[76]

$1 \times 10^{17}\,\text{cm}^{-3}$; thickness $\sim 3\,\mu\text{m}$), and p-GaInAsP cap layer (Zn-doped, $3 \times 10^{18}\,\text{cm}^{-3}$; thickness $\sim 0.3\,\mu\text{m}$). An SEM view of the cross-section of an ITG wafer is shown in Figure 14.24. Those quaternary layers are lattice-matched to InP within 0.05%.

A part of the wafer thus prepared is chemically etched down to the separation layer to eliminate the active guide layer using the selective etchants $3H_2SO_4:H_2O:H_2O_2$ and $4HCl:H_2O$ for GaInAsP and InP, respectively. First-order corrugations for distributed Bragg reflectors (DBR) are then formed on the etched surface of the exposed InP separation layer by the interference method using the 3250 Å line of a He–Cd laser and chemical etching technique. The typical period of the first-order corrugation is 2363 Å for 1.525 μm wavelength DBR-ITG lasers. The SEM photographs of top view and cross-section of the corrugations thus fabricated are shown in Figure 14.25. The shape of the grooves is sine-like and the depth is about 450 Å. The coupling coefficient k between the forward and backward waves in the corrugation is estimated to be about $30\,\text{cm}^{-1}$ with the measured waveguide parameters of the output guide and the separation layer. In order to determine the corrugation period corresponding to a desired lasing wavelength, the refractive index of GaInAsP which is estimated from the SEO method is used to calculate the equivalent refractive index n_{eq} at the DBR region to obtain the corrugation period as $\Lambda = \lambda/2n_{eq}$.

After the fabrication of DBR regions, an SiO_2 film is deposited to get the oxide-

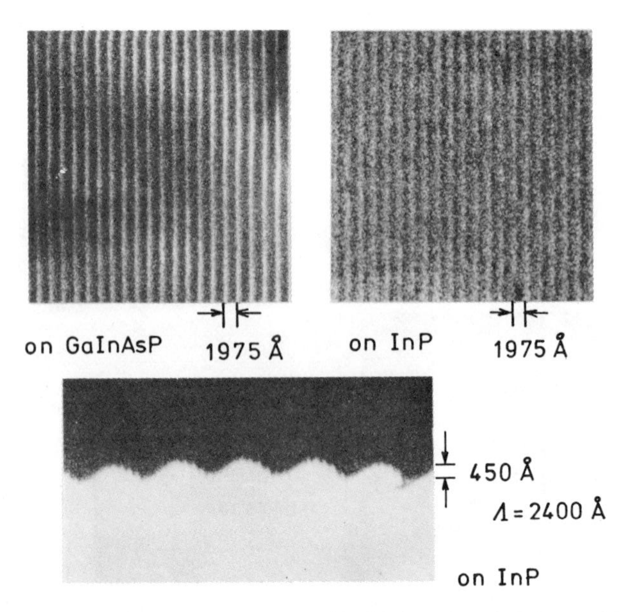

Figure 14.25 SEM photographs of grating fabricated on GaInAsP and InP[76]

stripe geometry for the confinement of current injection. Next the ohmic contacts are formed with Au/Zn and Au/Sn for the p- and n-sides, respectively. Then, a gold film is deposited on the entire area of the p-side to obtain a TE-mode filter[83] at the DBR region. Diodes are fabricated by cleaving at the active region and the DBR region, or at both the DBR regions. Consequently, two types of DBR-ITG lasers are fabricated, one consisting of a DBR and a cleaved facet, the other consisting of two DBRs. Typical lengths of the active region and the DBR region are $l = 200$–$400\ \mu m$ and $L_1, L_2 = 200$–$500\ \mu m$, respectively. The SiO_2 stripe width is $w \simeq 30\ \mu m$, as shown in Figure 14.22. The end of the DBR region, which is not used as the output facet, is roughened to avoid formation of an end mirror, which can cause spurious oscillation modes.

14.6.4 Lasing characteristics of DBR-ITG lasers

GaInAsP/InP DBR-ITG lasers thus fabricated operate under pulsed conditions at room temperature with a pulsewidth of about 200–300 ns. A typical example of a lasing spectrum is shown in Figure 14.26. Stable single-longitudinal-mode operations are observed in TE_0 mode at the emitting wavelength of $1.53\ \mu m$. Single-longitudinal-mode operation is maintained up to injection current levels of 1.39 times the threshold current. The upper limit of injection current for

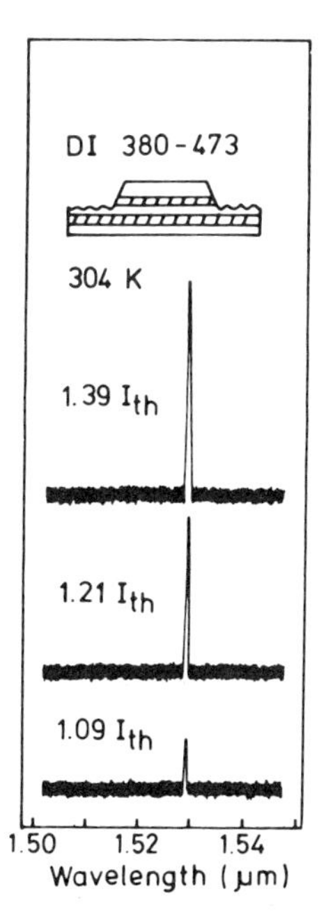

Figure 14.26 Lasting spectra of a DBR-ITG laser[76]

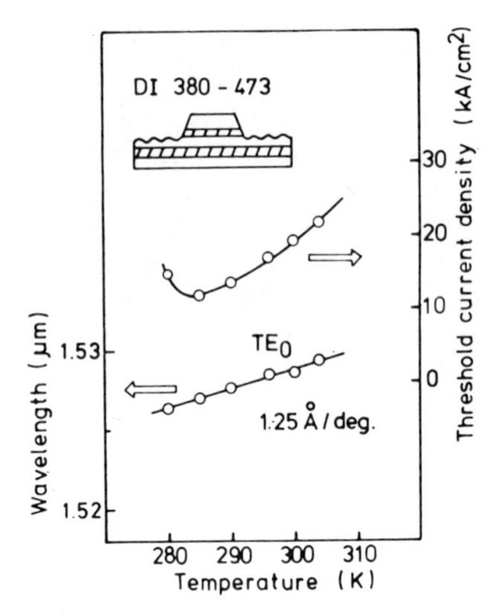

Figure 14.27 Temperature dependence of wavelength and threshold current density [76]

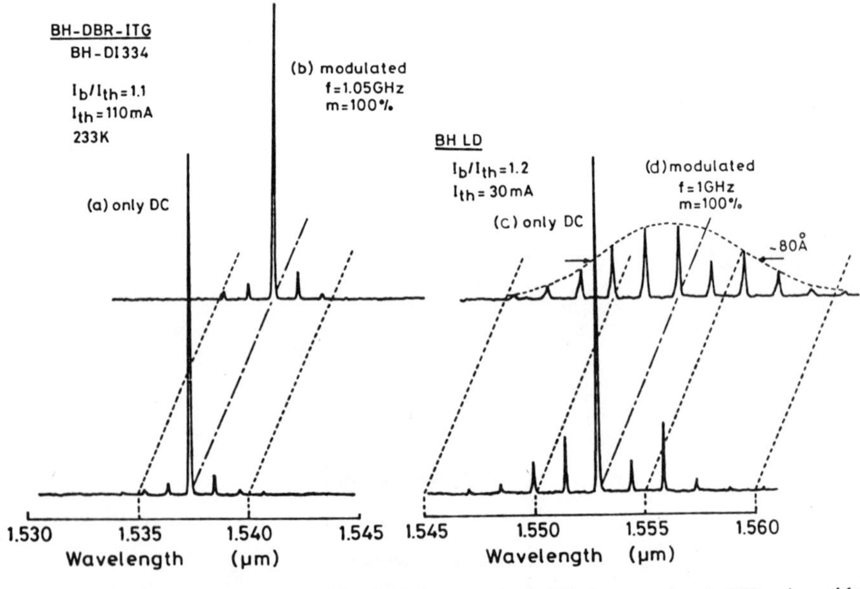

Figure 14.28 Lasing spectra of DBR-ITG laser and of BH laser under 1 GHz sinusoidal modulation. (a) DC operation and (b) 100% modulated condition, respectively, at DC bias of $I_b/I_{th} = 1.1$ for a DBR-ITG laser. (c) DC operation and (d) modulated condition, respectively, at DC bias of $I_b/I_{th} = 1.2$ for a BH laser[94]

obtaining a single-longitudinal-mode oscillation will be improved by optimizing the configuration parameters and reducing the loss of DBR-ITG lasers.

Temperature dependences of the lasing wavelength and the threshold current density are shown in Figure 14.27. The lasing wavelength varies with temperature by 1.25 Å deg^{-1} for this specific device and in general 0.8–1.25 Å deg^{-1}. The lasing wavelength is more stable against temperature compared with the conventional type of GaInAsP DH lasers. The threshold current density is about 10 kA cm^{-2} at room temperature, which is due to the lasing wavelength deviation from the peak wavelength of the spontaneous emission profile being shorter by about 500 Å. It follows from the theory that a lower threshold current density is realized if the Bragg wavelength is set to the gain peak wavelength. The differential quantum efficiency and the maximum output power are 11.7%/output facet and 25 mW, respectively.

High-speed direct modulation of the device with BH structure[94] is achieved by driving with a high-frequency sine-wave current. Lasing spectra of BH-DBR-ITG laser and of BH laser modulated by a 1 GHz sine wave with 100% modulation depth under a DC bias condition are shown in Figure 14.28.[94] In conventional DH lasers, it is known that rapid changes of the number of injected carriers under the high-speed modulation cause spectral broadening,[95] as shown in Figure 14.28(d). For a DBR-ITG laser, however, the single-longitudinal-mode oscillation is maintained with a fixed wavelength due to the wavelength selectivity of DBR, as shown in Figure 14.23(b), and it has also been reported for the modulation by a short current pulse with 1.5 ns pulsewidth.[75]

14.6.5 Lasers with etched mirrors

The monolithic fabrication of etched-mirror-facet lasers in the GaInAsP/InP system provides us with the capability not only of making very-short-cavity lasers which could possibly lase in a single longitudinal mode and operate at low current levels, but also of their fabrication on the same substrate with other optical or electronic components such as monitoring detectors or driving field-effect transistors and of easing the testing process.

We describe the fabrication of a monolithically etched-facet GaInAsP laser operating at 1.3 μm with one wet chemically etched (WCE) mirror and one cleaved mirror.

The structure of the etched-mirror laser is shown in Figure 14.29. The groove is etched in a batch process and one of the walls of the groove is used as a laser mirror. The other mirror is formed by cleaving and it is possible to get two etched-mirror lasers on one chip, as seen in Figure 14.29. This arrangement also eases the problem of mounting and handling of the laser chip.

In order to etch the facets into smooth vertical walls, we use an $HCl:CH_3COOH:H_2O_2$ (1:2:1) solution, the so-called KKI etchant.[96] This etchant works both on InP and GaInAsP, but acts differently in the $\langle 011 \rangle$ and

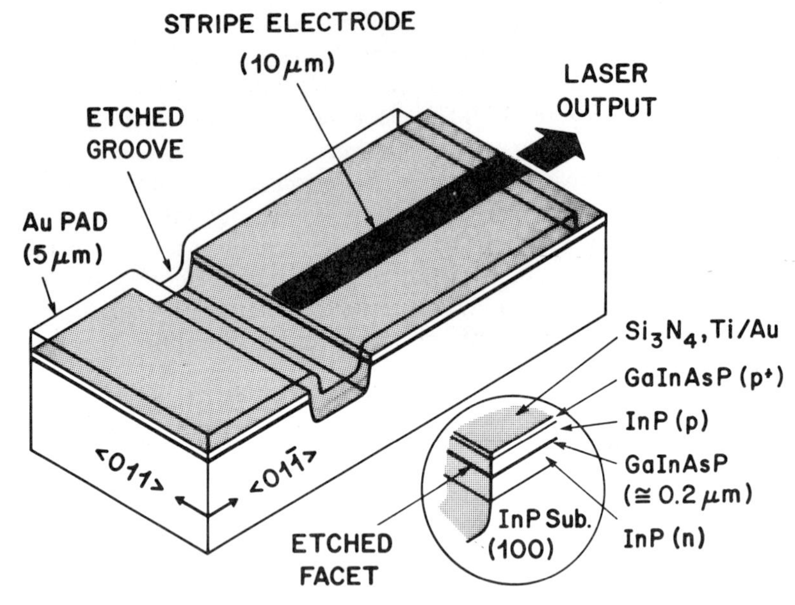

Figure 14.29 Structure of fabricated GaInAsP/InP laser with chemically etched mirror [99]

$\langle 01\bar{1} \rangle$ orientations associated with a (100) InP substrate with an HCl:H_2O (4:1) solution for 1 min at 20°C.[54] The angle of the face along the $\langle 01\bar{1} \rangle$ direction is about 23° while the wall along the $\langle 011 \rangle$ direction is −76° (reverse mesa).

We etch grooves at first and then make the stripe electrodes by using a segmented window stripe mask (width = 10 μm, length = 290 μm). On the stripe windows a gold–zinc–gold metal sandwich is evaporated or electroplated as a p-side electrode. We fire this plated wafer in a nitrogen atmosphere at 480°C. After thinning by mechanochemical polishing to 100 μm thick, plating the n-side electrode (Au–Ni/Sn–Au), and firing at 450°C, the gold metal is plated on both sides again. We evaporate 100 Å thick Ti layer and 800 Å thick gold on the wafer and thick gold pads (5 μm thick) are electroplated on. These metal coatings are effective for increasing the reflectivity at the etched facets. Finally, the central unetched parts of the mesas are cleaved and the other two sides are cut to form laser chips as shown in Figure 14.29. The cavity length is adjusted from 50 to 300 μm by controlling the cleaving position. The laser chip is bonded by using an In solder on a copper heat sink.

A typical light output versus injection current (*L–I*) characteristic of an etched laser with a 10 μm stripe is shown in Figure 14.30, for pulsed operation. The threshold current of this laser is 190 mA at 27°C, which is better than reported previously.[97,98] The cavity length *L* is about 260 μm. The differential quantum efficiency as measured from the cleaved front facet is 18% just above the

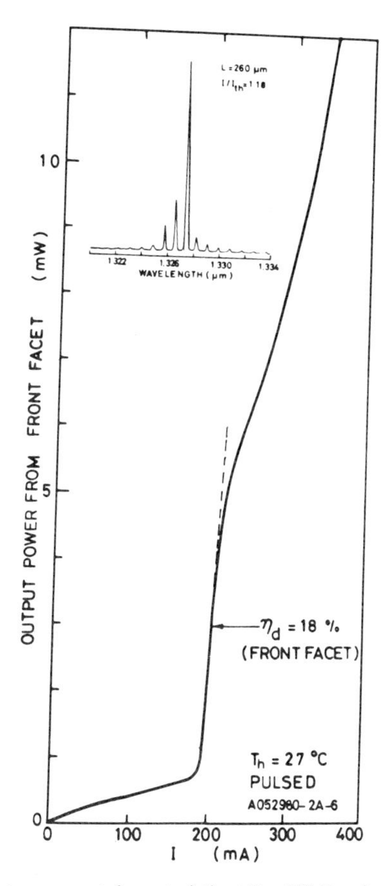

Figure 14.30 Light output vs current characteristic at $T = 27°C$ and spectrum for pulsed operation (*insert*) with $L = 260\ \mu m$ and $I/I_{th} = 1.18$[99]

threshold, but lower above the kink. We tested CW operation of this laser around room temperature.[99]

One of the merits of this structure is to integrate a monitoring detector, as shown in Figure 14.31.[98] The other half of the mesa is used as a monitoring detector to which the output light is coupled through the gap of the groove. The cavity length L_1 and the length of the detector L_2 are set by the cleaving position. The lengths L_1 and L_2 of the tested device are 268 μm and 100 μm, respectively.

The V–I characteristics of the laser and the detector in the same chip are nearly identical. The chip is mounted on a heat sink with epi-side up, and tested under pulsed conditions. The pulsewidth and repetition rate are 100 ns and 1 kHz, respectively. The reverse bias to the detector is -1.3 to 2.6 V. The output voltage at the load resistance (50 Ω) is shown by the broken line in Figure 14.32 against the

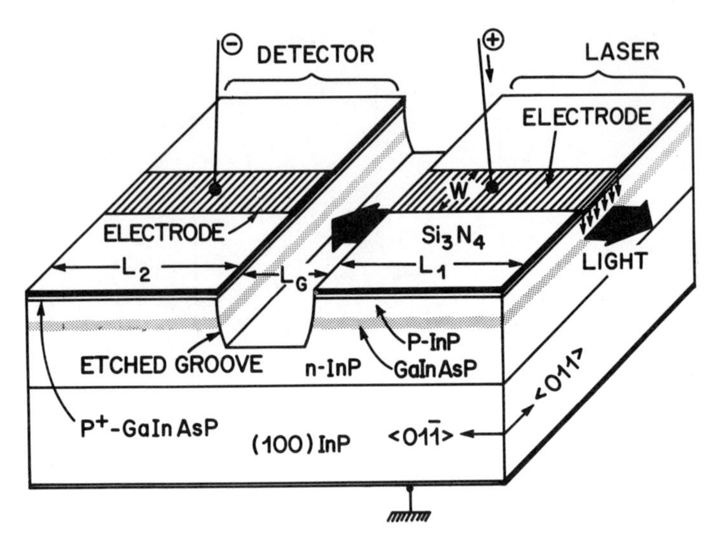

Figure 14.31 Structure of fabricated GaInAsP/InP laser with monitoring detector[98]

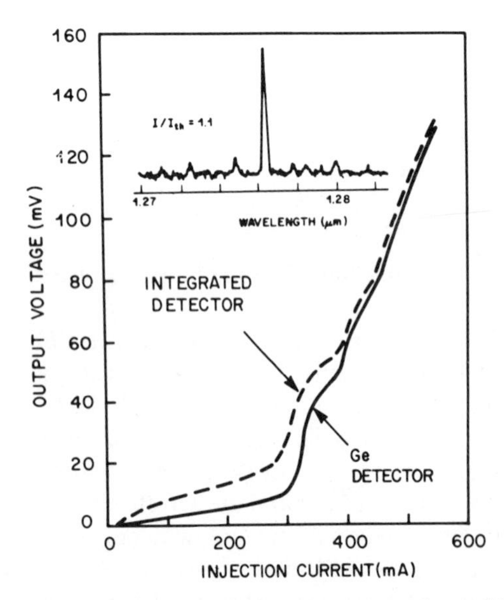

Figure 14.32 Light output plotted against the injection current detected by Ge photodiode and monitoring integrated detector, and (*insert*) a typical laser spectrum[98]

injection current to the laser, together with a typical spectrum. The output power P_c from the front end of the laser is monitored by a large-area Ge photodetector, and its output voltage at 50 Ω is shown by the full line in Figure 14.32. A voltage of 100 mA from the Ge detector corresponds to 6 mW. These two curves are correlated with each other, except below the lasing threshold. The output voltage of the integrated detector against the injection current shows non-linear behaviour below threshold. This is possibly due to the saturation of the spontaneous emission. The sensitivity of the detector does not change much with reverse bias.

The output power P_c from the etched end is measured for some laser diodes from which the detector has been removed. The ratio of the power from the etched end P_e to that from the front cleaved end P_c is nearly 0.5. If we accept also that the quantum efficiency of the Ge photodiode used is 60 % at 1.3 μm wavelength, the estimated effciency of the integrated detector appears to exceed 100 %. This apparent inconsistency may arise because the integrated detector receives light scattered at the etched facet and transmitted through the transparent InP substrate.

14.4.6 Surface emitting lasers

A conventional injection semiconductor laser consists of cleaved end mirrors perpendicular to the active layer. If an injection laser with light output perpendicular to the wafer surface could be obtained, it would be very easy to fabricate a laser cavity monolithically. The Fabry–Perot resonator of fabricated surface emitting injection lasers[100] consists of both surfaces of the wafer and its

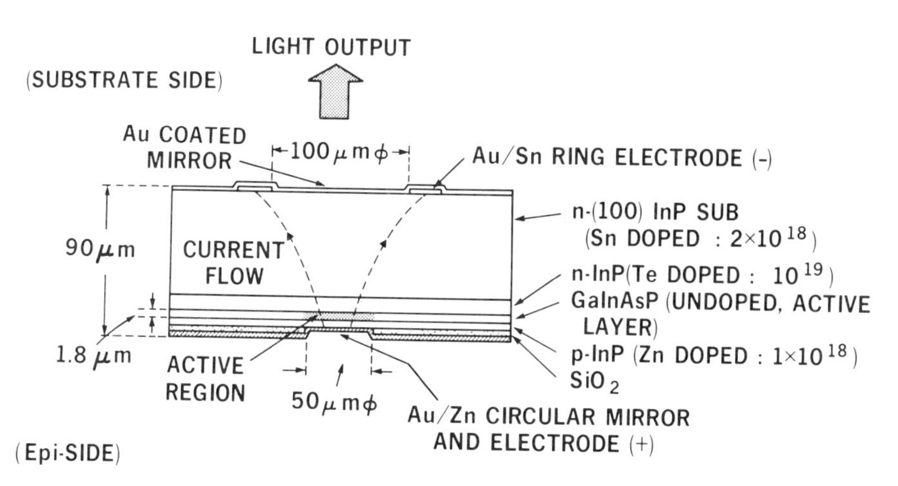

Figure 14.33 Schematic structure of GaInAsP/InP surface emitting injection laser[100]

axis being perpendicular to the active layer, as shown in Figure 14.33. The light output is obtained from one of the wafer surfaces.

Double-heterostructure (DH) GaInAsP wafers are prepared by the LPE technique. Three layers, an n-type InP layer (Te-doped), an undoped GaInAsP active layer, and a p-type InP layer (Zn-doped) are grown sequentially on a (100) n-type InP substrate in a rotary carbon boat at a cooling rate of 0.8°C min^{-1} and growth temperature of the active layer of 635°C. During the growth of DH wafers for surface emitting injection lasers, it is essential that the grown surface which acts as a resonator mirror should be mirror-like and the active layer should be thick enough to reduce the threshold current. As the active layer is close to the circular electrode, the current injection to the active layer is estimated to be localized near the circular electrode.

After sputtering an SiO_2 film onto the grown surface, circular holes of 50–100 μm diameter for p-side electrodes are opened by a conventional photolithographic technique. Next the n-side surface of the wafer is mechanically polished to form another reflector. In the polishing process, the polished surface is kept parallel to the grown surface. Following the reflector formation, n-side ring electrodes are provided by evaporating Au/Sn. Next, p-side circular electrodes of 50–100 μm diameter are formed by evaporating Au/Zn onto the circular holes. Alloying is done at 400°C for 5 min. Finally, a gold film a few tenths of a micrometre thick is coated on the n-side surface in order to increase the reflectivity of the resonator mirror. Then the wafer is cut into 500 μm × 500 μm chips and laser chips are mounted on Au-plated copper mounts with p-side down.

The device laser by 400 ns current pulses at a repetition rate of 1000 sec^{-1} at 77 K. The threshold current density J_{th} is 11 kA cm^{-2} circular electrode diameter of 100 μm) and the lasing wavelength is 1.18 μm. It is found that J_{th} of the laser with Au coating on the n-side mirror is less than a quarter of that with no Au coating. A light output of several milliwatts is obtained. From observation of the near-field pattern, the radiation area is located on the p-side circular electrode. It is estimated that the surface emitting laser could operate at room temperature by increasing the thickness of the active layer, reducing the diameter of the injection area, and increasing the reflectivity.

14.7 SUMMARY

We have illustrated the state of the art in $Ga_xIn_{1-x}As_yP_{1-y}$/InP double-heterostructure lasers which are attractive for light sources in long-wavelength optical-fibre communication. We tried to describe the technology of laser processing with regard to the preparation of DH wafers for 1.2–1.67 μm wavelength, doping techniques, mode-controlled lasers, and integrated lasers.

Firstly, we have presented the technology of the preparation of DH wafers for $Ga_xIn_{1-x}As_yP_{1-y}$ lasers. Low threshold current wafers have been obtained in the overall lasing wavelength range from 1.2 μm to 1.65 μm. Normalized threshold current density J_{th}/d which has been achieved is 4–6 kA cm^{-2} μm^{-1} and has almost no dependence on the lasing wavelength. The key technology was the control of the impurity to the p-InP cladding layer. Secondly, we have reviewed three types of mode-controlled lasers with narrow strip waveguides. Threshold currents as low as 25–40 mA emitting at 1.3 μm and 1.6 μm have been attained. As the result of the transverse mode control, the stable single-mode operation up to about three times the threshold has been obtained. At the present stage, it seems that the fundamental technology of the GaInAsP/InP laser is in our hands. However, it must be noted that the conventional DH laser usually operates in multi-longitudinal modes at the modulated condition. Dynamically single-mode lasers are necessary for the ultra-high-quality single-mode fibre system. Thirdly, the DBR-ITG integrated laser has been presented for this purpose and single-longitudinal-mode operation has been achieved even under high-speed direct modulation. Then, the technology of integrating lasers into optical circuits and fabricating laser resonators from a monolithic process has been illustrated.

It has been reported by a number of authors that the GaInAsP/InP laser is more reliable than the GaAlAs/GaAs laser. An operating lifetime longer than 10^6 h is estimated from the high-temperature acceleration test. The temperature dependence of the threshold and the efficiency have been also reviewed. It has been recognized that the temperature dependence of the GaInAsP/InP laser comes from the property of the material, not from the process technology and the structure of the devices.

We are still on the threshold of the development of a high-performance, dynamically stable single-wavelength laser. We have to develop high-precision epitaxies which provide us with the precise control of the composition and thickness, and related finer fabrication technologies. When these become possible, optical communication will find much wider applications.

REFERENCES

1. J. J. Hsieh, J. A. Rossi, and J. P. Donnelly, 'Room-temperature CW operation of GaInAsP/InP double-heterostructure diode lasers emitting at 1.1 μm', *Appl. Phys. Lett.*, **28**, 709–711, 1976.
2. K. Oe and K. Sugiyama, 'GaInAsP–InP double heterostructure lasers prepared by a new LPE apparatus', *Jpn. J. Appl. Phys.*, **15**, 2003–2004, 1976.
3. Y. Itaya, Y. Suematsu, and K. Iga, 'Carrier lifetime measurement of GaInAsP/InP double-heterostructure lasers', *Jpn. J. Appl. Phys.*, **16**, 1057–1058, 1977.
4. T. Yamamoto, K. Sakai, S. Akiba, and Y. Suematsu, '$In_xGa_{1-x}As_yP_{1-y}$/InP DH lasers fabricated on InP (100) substrates', *IEEE J. Quantum Electron.*, **QE-14**, 95–98, 1978.
5. M. A. Pollack, R. E. Nahory, J. C. DeWinter, and A. A. Ballman, 'Liquid phase epitaxial $In_{1-x}Ga_xAs_yP_{1-y}$, lattice matched to (100) InP over the complete wavelength range $0.92 \leqslant \lambda \leqslant 1.65$ μm', *Appl. Phys. Lett.*, **33**, 314–316, 1978.
6. S. Arai, Y. Itaya, Y. Suematsu, K. Kishino, and S. Katayama, 'Conditions of LPE growth for lattice matched GaInAsP/InP DH lasers with (100) substrate in the range of 1.2–1.5 μm', *Jpn. J. Appl. Phys.*, **17**, 2067–2068, 1978.

7. Y. Itaya, Y. Suematsu, S. Katayama, K. Kishino, and S. Arai, 'Low threshold current density (100) GaInAsP/InP double-heterostructure lasers for wavelength 1.3 μm', *Jpn. J. Appl. Phys.*, **18**, 1795–1805, 1979.
8. R. E. Nahory and M. A. Pollack, 'Threshold dependence on active layer thickness in InGaAsP/InP DH lasers', *Electron. Lett.*, **14**, 727–729, 1978.
9. H. Kressel and M. Ettenberg, 'Low-threshold double-heterojunction AlGaAs/GaAs laser diodes: theory and experiment', *J. Appl. Phys.*, **47**, 3533–3537, 1976.
10. S. Arai, Y. Suematsu, and Y. Itaya, '1.11–1.67 μm (100) GaInAsP/InP injection lasers prepared by liquid phase epitaxy', *IEEE J. Quantum Electron.*, **QE-16**, 197–205, 1980.
11. K. Oe, S. Ando, and K. Sugiyama, '1.3 μm CW operation of GaInAsP/InP DH diode lasers at room temperature', *Jpn. J. Appl. Phys.*, **16**, 1273–1274, 1977.
12. K. Wakao, K. Moriki, T. Kambayashi, and K. Iga, 'GaInAsP/InP DH laser grown by newly designed vertical LPE furnace', *Jpn. J. Appl. Phys.*, **16**, 2073–2076, 1977.
13. C. J. Nuese, R. E. Enstrom, and J. R. Appert, '1.7 μm heterojunction lasers and photodiodes of $In_{0.53}Ga_{0.47}As$/InP', *IEEE Trans. Electron Devices*, **ED-24**, 1213, 1977.
14. J. J. Coleman, P. W. Foy, R. B. Zetterstrom, S. Sumski, H. C. Casey, Jr, and G. A. Rozgonyi, 'Growth and characterization of InP–$Ga_xIn_{1-x}As_yP_{1-y}$ heterostructure lasers', *Gallium Arsenide and Related Compounds 1978*, Conf. Ser. no. 45, Institute of Physics, Bristol, 1979.
15. S. Arai, Y. Suematsu, and Y. Itaya, '1.67 μm $Ga_{0.47}In_{0.53}As$/InP DH lasers double cladded with InP by LPE technique', *Jpn. J. Appl. Phys.*, **18**, 709–710, 1979.
16. S. Akiba, K. Sakai, Y. Matsushima, and Y. Yamamoto, 'Room-temperature CW operation of InGaAsP/InP heterostructure lasers emitting at 1.56 μm', *Electron. Lett.*, **15**, 606–607, 1979.
17. S. Arai, M. Asada, Y. Suematsu, and Y. Itaya, 'Room-temperature CW operation of (100) GaInAsP/InP DH laser emitting at 1.5 μm', *Jpn. J. Appl. Phys.*, **18**, 2333–2334, 1979.
18. H. Kawaguchi, K. Takahei, Y. Toyoshima, H. Nagai, and G. Iwane, 'Room-temperature CW operation of InP/InGaAsP/InP double heterostructure diode lasers emitting at 1.55 μm', *Electron. Lett.*, **15**, 669–670, 1979.
19. Y. Noguchi, K. Takahei, Y. Suzuki, and H. Nagai, 'Low threshold current CW operation of InP/GaInAs buried heterostructure lasers', *Jpn. J. Appl. Phys.*, **19**, L759–L762, 1980.
20. G. H. Olsen, C. J. Nuese, and M. Ettenberg, 'Reliability of vapor-grown InGaAs and InGaAsP heterojunction laser structures', *IEEE J. Quantum Electron.*, **QE-15**, 688–693, 1979.
21. B. I. Miller, J. H. McFee, R. J. Martin, and P. K. Tien, 'Room-temperature operation of lattice-matched InP/$Ga_{0.47}In_{0.53}As$/InP double-heterostructure lasers grown by MBE', *Appl. Phys. Lett.*, **33**, 44–47, 1978.
22. J. P. Hirtz, J. P. Duchemin, P. Hirtz, B. de Cremoux, T. P. Pearsall, and M. Bonnet, '$Ga_xIn_{1-x}As_yP_{1-y}$/InP DH laser emitting at 1.15 μm grown by low-pressure metalorganic CVD', *Electron. Lett.*, **16**, 275–277, 1980.
23. W. D. Johnston, Jr, K. E. Strege, and A. A. Ballman, 'Uniformly low-threshold diode lasers at 1.5–1.55 μm from VPE $In_xGa_{1-x}As_yP_{1-y}$ material', *38th Annual Device Research Conference*, June 1980; *IEEE trans. Electron Dev.*, **ED-27**, 2191, 1980.
24. G. H. Olsen, T. J. Zamerowski, and N. J. DiGiuseppe, '1.5–1.7 μm V.P.E. InGaAsP/InP C.W. lasers', *Electron. Lett.*, **16**, 516–518, 1980.
25. Y. Itaya, S. Arai, K. Kishino, M. Asada, and Y. Suematsu, '1.6 μm wavelength GaInAsP/InP lasers prepared by two phase solution technique', *IEEE J.Q.E.*, **QE-17**, 635–640, 1981.
26. A. R. Adams, M. Asada, Y. Suematsu, and S. Arai, 'The temperature dependence of

the efficiency and threshold current of $In_{1-x}Ga_xAs_yP_{1-y}$ lasers related to intervalence band absorption', *Jpn. J. Appl. Phys.*, **19**, L621–L624, 1980.
27. Y. Itaya, S. Katayama, and Y. Suematsu, 'Narrow-beam divergence of the emission from low-threshold GaInAsP/InP double-heterostructure lasers', *Electron. Lett.*, **15**, 123–124, 1979.
28. H. C. Casey, Jr, M. B. Panish, and J. L. Merz, 'Beam divergence of the emission from double-heterostructure injection lasers', *J. Appl. Phys.*, **44**, 5470–5475, 1973.
29. H. Kawanishi, Y. Suematsu, Y. Itaya, and S. Arai, '$Ga_xIn_{1-x}As_yP_{1-y}$–InP injection laser partially loaded with distributed Bragg reflectors', *Jpn. J. Appl. Phys.*, **17**, 1439–1440, 1978.
30. G. D. Pettit and W. J. Turner, 'Refractive index of InP', *J. Appl. Phys.*, **36**, 2081, 1975.
31. M. Asada, A. R. Adams, K. E. Stubkjaer, Y. Suematsu, Y. Itaya, and S. Arai, 'The temperature dependence of the threshold current of GaInAsP/InP DH lasers', *IEEE J. Quantum Electron.*, **QE-17**, 611–619, 1981.
32. M. Yamada and Y. Suematsu, 'A condition of single longitudinal mode operation in injection lasers with index guiding structures', *IEEE J. Quantum Electron.*, **QE-15**, 743–749, 1979.
33. K. Aiki, M. Nakamura, T. Kuroda, and J. Umeda, 'Channelled-substrate-planar structure (AlGa)As injection lasers', *Appl. Phys. Lett.*, **48**, 649–651, 1977.
34. J. J. Hsieh and C. C. Shen, 'Room-temperature CW operation of buried-stripe double-heterostructure GaInAsP/InP diode lasers', *Appl. Phys. Lett.*, **30**, 429–481, 1977.
35. S. Arai, M. Asada, Y. Suematsu, Y. Itaya, T. Tanbun-ek, and K. Kishino, 'New 1.6 μm wavelength GaInAsP/InP buried heterostructure lasers', *Electron. Lett.*, **16**, 349–350, 1980.
36. H. Nagai, Y. Noguchi, K. Takahei, Y. Toyoshima, and G. Iwane, 'InP/GaInAsP buried heterostructure lasers of 1.5 μm region', *Jpn. J. Appl. Phys.*, **19**, L218–L220, 1980.
37. M. Hirao, S. Tsuji, K. Mizuishi, A. Doi, and M. Nakamura, 'Long wavelength InGaAsP/InP lasers for optical fiber communication systems', *J. Opt. Commun.*, **1**, 10–14, 1981.
38. K. Kishino, Y. Suematsu, and Y. Itaya, 'Mesa substrate buried heterostructure GaInAsP/InP lasers', *Electron. Lett.*, **15**, 134–136, 1979.
39. H. Nomura, M. Sugimoto, and A. Suzuki, 'InGaAsP current confinement mesa substrate buried heterostructure laser diodes fabricated by one LPE process', *Paper* OQE80-117, Technical Group of IECE of Japan, Optical and Quantum Electronics, 1981.
40. K. Moriki, K. Wakao, M. Kitamura, K. Iga, and Y. Suematsu, 'Single transverse mode operation of terraced substrate GaInAsP/InP lasers at 1.3 μm wavelength', *Jpn. J. Appl. Phys.*, **19**, 2191–2196, 1981.
41. A. Doi, N. Chinone, K. Aiki, and R. Ito, '$Ga_xIn_{1-x}As_yP_{1-y}$/InP rib-waveguide injection lasers made by one-step LPE', *Appl. Phys. Lett.*, **34**, 393–395, 1979.
42. H. Nishi, M. Yano, Y. Nishitani, Y. Akita, and M. Takusagawa, 'Self-aligned structure InGaAsP/InP DH lasers', *Appl. Phys. Lett.*, **35**, 232–234, 1979.
43. N. Matsumoto, K. Onabe, T. Yuasa, M. Ueno, Y. Matsumoto, T. Furuse, and I. Sakuma, '1.5 μm wavelength InGaAsP/InP plano-convex waveguide (PCW) lasers', *Japanese Society of Applied Physics, Digest of 41st Autumn National Convention*, 17a-Q-6, 1980.
44. K. Sakai, F. Tanaka, S. Akiba, Y. Matsushima, and T. Yamamoto, 'Characteristics of 1.5 μm wavelength PCW lasers with buffer layer', Ref. 43, *ibid.*, 17a-Q-8, 1980.
45. T. Murotani, E. Oomura, H. Higuchi, H. Namizaki, and W. Susaki, 'InGaAsP/InP

buried crescent laser emitting at 1.3 μm with very low threshold current', *Electron. Lett.*, **16**, 566–568, 1980.
46. R. J. Nelson, P. D. Wright, P. A. Barnes, R. L. Brown, T. Cella, and R. G. Sobers, 'High-output power InGaAsP ($\lambda = 1.3\,\mu$m) strip-buried heterostructure lasers', *Appl. Phys. Lett.*, **36**, 358–360, 1980.
47. I. Mito, K. Kaede, M. Kitamura, Y. Odagiri, M. Seki, and K. Kobayashi, 'InGaAsP planar buried heterostructure laser diode', *Paper* OQE80–116, Technical Group of IECE of Japan, Optical and Quantum Electronics, 1981.
48. T. Tsukada, 'GaAs–$Ga_{1-x}Al_xAs$ buried-heterostructure injection lasers', *J. Appl. Phys.*, **45**, 4899–4906, 1974.
49. S. Arai, M. Asada, T. Tanbun-ek, Y. Suematsu, Y. Itaya, and K. Kishino, '1.6 μm wavelength GaInAsP/InP BH lasers', *IEEE J. Quantum Electron.*, **QE-17**, 640–645, 1981.
50. K. Kishino, Y. Suematsu, Y. Takahashi, T. Tanbun-ek, and Y. Itaya, 'Fabrication and lasing properties of mesa substrate buried heterostructure GaInAsP/InP lasers at 1.3 μm wavelength', *IEEE J. Quantum Electron.*, **QE-16**, 160–164, 1980.
51. K. Wakao, K. Moriki, M. Kitamura, K. Iga, and Y. Suematsu, '$Ga_xIn_{1-x}As_yP_{1-y}$/InP terraced substrate single mode lasers', *IEEE J. Quantum Electron.*, **QE-17**, 1009–1013, 1981.
52. T. Sugino, K. Itoh, M. Wada, H. Shimizu, and I. Teramoto, 'Fundamental transverse and longitudinal mode oscillation in terraced substrate GaAs–(GaAl)As lasers', *IEEE J. Quantum Electron.*, **QE-15**, 714–718, 1979.
53. M. Wada, T. Sugino, H. Shimizu, K. Itoh, and I. Teramoto, 'Very low threshold visible and infrared TS lasers', *Proc. 7th IEEE Int. Semiconductor Laser Conf.*, September 1980.
54. T. Kambayashi, C. Kitahara, and K. Iga, 'Chemical etching of InP and GaInAsP for fabricating laser diodes and integrated optical circuits', *Jpn. J. Appl. Phys.*, **19**, 79–85, 1980.
55. K. Wakao, K. Moriki, T. Kambayashi, and K. Iga, 'GaInAsP/InP DH laser grown by newly designed vertical LPE furnace', *Jpn. J. Appl. Phys.*, **16**, 2073–2074, 1977.
56. Y. Sakakibara, K. Furuya, Y. Suematsu, and Y. Itaya, 'Direct modulation characteristics of GaInAsP/InP DH lasers with various stripe widths measured by sharp-pulse method', *Electron. Lett.*, **15**, 594–596, 1979.
57. Y. Horikoshi and Y. Furukawa, 'Temperature sensitive threshold current of InGaAsP–InP double heterostructure lasers', *Jpn. J. Appl. Phys.*, **18**, 809–815, 1979.
58. R. E. Nahory, M. A. Pollack, and J. C. DeWinter, 'Temperature dependence of InGaAsP double-heterostructure laser characteristics', *Electron. Lett.*, **15**, 695–696, 1979.
59. M. Yano, H. Nishi, and M. Tagusagawa, 'Temperature characteristics of threshold current in InGaAsP/InP heterostructure lasers', *J. Appl. Phys.*, **51**, 4022–4028, 1980.
60. G. H. B. Thompson and G. D. Henshall, 'Nonradiative carrier loss and temperature sensitivity of threshold in 1.27 μm (GaIn)(AsP)/InP DH lasers', *Electron. Lett.*, **16**, 42–44, 1980.
61. T. Uji, K. Iwamoto, and R. Lang, 'Nonradiative recombination in InGaAsP/InP light sources causing light emitting diode output saturation and strong laser-threshold-current temperature sensitivity', *Appl. Phys. Lett.*, **38**, 193–195, 1981.
62. A. Sugimura, 'Band to band Auger recombination effect on InGaAsP laser threshold', *IEEE J. Quantum Electron.*, **QE-17**, 627–635, 1981.
63. T. Yamamoto, K. Sakai, and S. Akiba, '10 000 h continuous CW operation of $In_{1-x}Ga_xAs_yP_{1-y}$/InP DH lasers at room temperature', *IEEE J. Quantum Electron.*, **QE-15**, 684–687, 1979.

64. I. Melngailis, 'Lasers at 1.0–1.3 μm for optical fiber communications', *Technical Digest of 1977 Int. Conf. on Integrated Optics and Optical Fiber Communication*, Tokyo, B2-1, 1977.
65. I. Hayashi, 'Recent progress in semiconductor laser research and development', *Proc. 11th Conf. on Solid State Devices; Jpn. J. Appl. Phys.*, **19**, *Suppl.* 19-1, 23–31, 1980.
66. S. Yamakoshi, M. Abe, O. Wada, S. Komiya, and T. Sakurai, 'Reliability of high radiance InGaAsP/InP LED's operating in the 1.2–1.3 μm wavelength', *IEEE J. Quantum Electron.*, **QE-17**, 167–173, 1981.
67. E. Oomura, T. Murotani, T. Ishii, and W. Susaki, 'Light-damage-limit of InP/InGaAsP laser', *Japanese Society of applied Physics, Digest of 26th Spring National Convention*, 27p-W-13, 1979.
68. M. Fukuda, K. Wakita, and S. Takahashi, 'Estimation of the facet of InGaAsP/Inp DH lasers (Part 2)', *Japanese Society of Applied Physics, Digest of 27th Spring national Convention*, 1p-G-6, 1980.
69. K. Fujiwara, T. Fujiwara, K. Hori, and M. Takusagawa, 'Aging characteristics of $Ga_{1-x}Al_xAs$ double-heterostructure lasers bonded with gold eutectic alloy solder', *Appl. Phys. Lett.*, **34**, 668–670, 1979.
70. K. Mizuishi, M. Hirao, S. Tsuji, H. Sato, and M. Nakamura, 'Accelerated aging characteristics of InGaAsP/InP buried heterostructure lasers emitting at 1.3 μm', *Jpn. J. Appl. Phys.*, **19**, L429–L432, 1980.
71. H. Imai, M. Morimoto, H. Ishikawa, K. Hori, and M. Takusagawa, 'Accelerated aging test of InGaAsP/InP double-heterostructure laser diodes with single transverse mode', *Appl. Phys. Lett.*, **38**, 16–17, 1981.
72. H. Kawanishi, Y. Suematsu, K. Utaka, Y. Itaya, and S. Arai, '$Ga_xIn_{1-y}As_yP_{1-y}$/InP injection laser partially loaded with first-order distributed Bragg reflector', *IEEE J. Quantum Electron.*, **QE-15**, 701–706, 1979.
73. K. Utaka, Y. Suematsu, K. Kobayashi, and H. Kawanishi, 'GaInAsP/InP integrated twin-guide lasers with first-order distributed Bragg reflectors at 1.3 μm wavelength', *Japan. J. Appl. Phys.*, **19**, L137–L140, 1980.
74. K. Utaka, K. Kobayashi, K. Kishino, and Y. Suematsu, '1.5–1.6 μm GaInAsP/InP integrated twin-guide lasers with first-order distributed Bragg reflectors', *Electron. Lett.*, **16**, 455–456, 1980.
75. Y. Sakakibara, K. Furuya, K. Utaka, and Y. Suematsu, 'Single-mode oscillation under high-speed direct modulation in GaInAsP/InP integrated twin-guide lasers with distributed Bragg reflectors', *Electron. Lett.*, **16**, 456–458, 1980.
76. K. Utaka, K. Kobayashi, and Y. Suematsu, 'Integrated twin-guide lasers with first-order distributed Bragg reflectors', *IEEE J. Quantum electron.*, **QE-17**, 651–658, 1981.
77. H. Kogelnik and C. V. Shank, 'Stimulated emission in a periodic structure', *Appl. Phys. Lett.*, **18**, 152–154, 1971.
78. M. Nakamura, K. Aiki, J. Umeda, and A. Yariv, 'CW operation of distributed feedback GaAs–GaAlAs diode lasers at temperature up to 300 K', *Appl. Phys. Lett.*, **27**, 403–405, 1975.
79. F. K. Reinhart, R. A. Logan, and C. V. Shank, 'GaAs–$Al_xGa_{1-x}As$ injection lasers with distributed Bragg reflectors', *Appl. Phys. Lett.*, **27**, 45–48, 1975.
80. W. T. Tsang and S. Wang, 'GaAs–$Ga_{1-x}Al_xAs$ double-heterostructure injection lasers with distributed Bragg reflectors', *Appl. Phys. Lett.*, **28**, 596–598, 1976.
81. H. Kawanishi, Y. Suematsu, and K. Kishino, 'GaAs–$Al_xGa_{1-x}As$ integrated twin-guide lasers with distributed Bragg reflectors', *IEEE J. Quantum Electron.*, **QE-12**, 64–65, 1977.
82. T. Kuroda, S. Yamashita, M. Nakamura, and J. Umeda, 'Channeled-substrate-

feedback semiconductor lasers', *Appl. Phys. Lett.*, **33**, 173–174, 1978.
83. Y. Suematsu, M. Hakuta, K. Furuya, K. Chiba, and R. Hasumi, 'Fundamental transverse electric field (TE_0) mode selection for thin-film asymmetric light guide', *Appl. Phys. Lett.*, **21**, 291–293, 1972.
84. A. Doi, T. Fukuzawa, M. Nakamura, R. Ito, and K. Aiki, 'InGaAsP/InP distributed-feedback injection lasers fabricated by one-step liquid phase epitaxy', *Appl. Phys. Lett.*, **35**, 441–443, 1979.
85. T. Miya, Y. Terunuma, T. Hosaka, and T. Miyashita, 'Ultimated low-loss single-mode fibre at 1.55 μm', *Electron. Lett.*, **15**, 106–108, 1979.
86. T. Moriyama, O. Fukuda, K. Sanada, K. Inada, T. Edahiro, and K. Chiba, 'Ultimately low OH content V.A.D. optical fibres', *Electron. Lett.*, **16,** 698–699, 1980.
87. D. N. Payne, and W. A. Gambling, 'Zero material dispersion in optical fibres', *Electron. Lett.*, **11**, 176–178, 1975.
88. Y. Suematsu, M. Yamada, and K. Hayashi, 'A multi-hetero-AlGaAs laser with integrated twin guide', *Proc. IEEE* (*Lett.*), **63**, 208, 1975.
89. Y. Suematsu and K. Kishino, 'Coupling coefficient in strongly coupled dielectric waveguides', *Radio Sci.*, **12**, 587–592, 1977.
90. S. H. Wemple and M. DiDomenico, Jr, 'Behavior of the electric dielectric constant in covalent and ionic materials, *Phys. Rev.*, **3**, 1338–1351, 1971.
91. M. A. Afromowitz, 'Refractive index of $Ga_{1-x}Al_xAs$', *Solid State Commun.*, **15**, 59–63, 1974.
92. T. Takagi, 'Refractive index of $Ga_{1-x}In_xAs$ prepared by vapor phase epitaxy', *Jpn. J. Appl. Phys.*, **17,** 1813–1817, 1978.
93. R. E. Nahory, M. A. Pollack, and W. D. Johnston, Jr, 'Band gap versus composition and demonstration of Vegard's law for $In_{1-x}Ga_xAs_yP_{1-y}$ lattice matched to InP', *Appl. Phys. Lett.*, **33**, 659–661, 1978.
94. K. Utaka, K. Kobayashi, F. Koyama, Y. Abe, and Y. Suematsu, 'Single-wavelength operation of 1.53 μm GaInasP/InP buried-heterostructure integrated twin-guide laser with distributed Bragg reflector under direct modulation up to 1 GHz', *Electron. Lett.*, submitted for publication.
95. T. Ikegami, 'Spectrum broadening and failing effect in directly modulated injection *lasers', 1st European Conf. on Optical Fibre Communication, Technical Digest*, 111, 1975.
96. K. Iga, T. Kambayashi, K. Wakao, and C. Kitahara, 'GaInAsP/InP DH lasers and related fabricating techniques for integration', *IEEE J. Quantum Electron.*, **QE-15**, 707–710, 1979.
97. B. I. Miller and K. Iga, 'GaInAsP/InP stripe laser with etched mirrors fabricated by a wet chemical etch', *Appl. Phys. Lett.*, **37**, 339, 1980.
98. K. Iga and B. I. Miller, 'GaInAsP/InP laser with monolithically integrated monitoring detectors', *Electron. Lett.*, **16**, 342–343, 1980.
99. K. Iga and B. I. Miller, 'CW operation of GaInAsP/InP laser with chemically etched mirror', *Electron. Lett.*, **16**, 830–832, 1980.
100. H. Soda, K. Iga, C. Kitahara, and Y. Suematsu, 'GaInAsP/InP surface emitting injection lasers', *Jpn. J. Appl. Phys.*, **18**, 2329–2330, 1979.
101. I. P. Kaminow, R. E. Nahory, M. A. Pollack, L. W. Stulz, and J. C. DeWinter, 'Single-mode CW ridge-waveguide laser emitting at 1.55 μm', *Electron. Lett.*, **15**, 763–765, 1979.
102. J. J. Hsieh, 'High-temperature CW operation of GaInAsP/InP lasers emitting at 1.5 μm', *Appl. Phys. Lett.*, **37**, 25–27, 1980.
103. T. P. Pearsall, R. Bisàro, R. Ansel, and P. Merenda, "The growth of $Ga_xIn_{1-x}As$ on (100) InP by liquid-phase epitaxy", *Appl. Phys. Lett.*, **32**, 497–499, 1978.

GaInAsP Alloy Semiconductors
Edited by T. P. Pearsall

Chapter 15

Temperature Dependence of Laser Threshold Current

YOSHIJI HORIKOSHI

Musashino Electrical Communication Laboratories, Nippon Telegraph and Telephone Public Corporation, Musashino-shi, Tokyo 180, Japan

15.1 INTRODUCTION

The usefulness of GaInAsP–InP double-heterostructure (DH) lasers for optical communication line sources is now well-established due to their wavelength and stable life.[1,2] These lasers, however, present difficulties when put to practical use because their lasing characteristics, especially threshold current, are very sensitive to the ambient temperature,[3-5] as compared with 0.8 μm region AlGaAs–GaAs DH lasers. Experimentally, the threshold current at temperature T, $I_{th}(T)$, is expressed by the value at temperature T', $I_{th}(T')$, as

$$I_{th}(T) = I_{th}(T')\exp[(T-T')/T_0] \tag{15.1}$$

where T_0 is the characteristic temperature which is often used to express the temperature sensitivity of threshold current. Around room temperature, $T_0 \geq 120\,K$ for AlGaAs–GaAs DH lasers,[6] while $T_0 = 50\text{–}70\,K$[3-5,7-12] for GaInAsP–InP DH lasers. If the threshold current is very sensitive to the ambient temperature, in other words if T_0 is very small, the laser-operation current increases steeply as the temperature increases. Laser diodes will then be heated by the increased operation current, which leads to a further increase in threshold current. This thermal runaway process limits the temperature range for stable continuous-wave (CW) operation.

The heat sink temperature of a laser diode set in a repeater of an optical communication line system often becomes much higher than room temperature, for example, 50°C. Therefore, it is necessary that the lasers can be operated stably even at such a high temperature with a sufficient external differential quantum efficiency. This is why a high-temperature sensitivity of threshold current is considered to be a serious problem. Therefore, since the CW operation was realized, this problem has been studied extensively by many workers. Two different approaches to solving this problem have been taken up to the present.

Reducing threshold current and thermal resistance so as to minimize the junction temperature increase during operation is one way to attain stable CW operation at fairly high temperatures. A remarkable result of this effort my be the buried-heterostructure lasers which have extremely low threshold currents of 10–30 mA and high CW operation temperature up to 90–100°C.[10,11,14] Thus, the most substantial difficulty for practical application was overcome, although the problem of the high-temperature sensitivity of threshold current is still left unsolved. The other approach is to clarify the origins of the temperature sensitivity and to remove them if possible. Up to the present, several mechanisms which explain the observed results have been proposed. They are (1) temperature dependence of the optical gain,[16,21] (2) leakage of the injected carriers to the confinement layers,[16-18,20,25] (3) band-to-band Auger effect,[4,5,21] (4) temperature dependence of the optical absorption loss,[26] and (5) non-radiative recombination centres.[4,5,22-24]

A rapid decrease in external differential quantum efficiency with increasing temperature is another difficulty. Fortunately, this does not pose a severe problem as long as the lasers are used in a system whose bandwidth does not exceed 400 Mb s^{-1}.[15] However, in principle, this problem may be more difficult to circumvent than the temperature sensitivity of threshold current, because the former cannot be compensated for by only changing the DC bias. Meanwhile, saturation of the output light intensity has been observed in surface emitting GaInAsP–InP DH LEDs as the driving current increases, and the same origins as those proposed for the low T_0 of lasers and superluminescence along the plane of the active layer have been suspected.

However, at the present stage, no mechanism proposed can completely explain the experimental results obtained from lasers and LEDs. Therefore, in this chapter, the proposed mechanisms as well as the experimental results are summarised and discrepancies between them are discussed, which may present basic ideas for a further discussions on this matter.

15.2 EXPERIMENTAL RESULTS

First we will review experimental results on the temperature dependence of threshold current and external differential quantum efficiency of GaInAsP–InP DH lasers, as well as the experimental results on the light output saturation in surface emitting GaInAsP–InP DH LEDs, and these experimental results will be compared to those on lasers and LEDs made from the other DH materials, such as AlGaAs–GaAs, AlGaAsSb–GaAsSb, and AlGaAsSb–GaInAsSb. Figure 15.1 shows threshold current–temperature ($I_{th} - T$) relations for GaInAsP–InP DH lasers, which suggests that the T_0 values at room temperature are between 50 and 70 K, and that some lasers show distinct breakpoints in the temperature range 250–280 K and above room temperature.[4, 5, 7-14, 26] On the contrary, 0.8 μm region AlGaAs–GaAs DH lasers exhibit much higher T_0 values of $T_0 \geq 120$ K

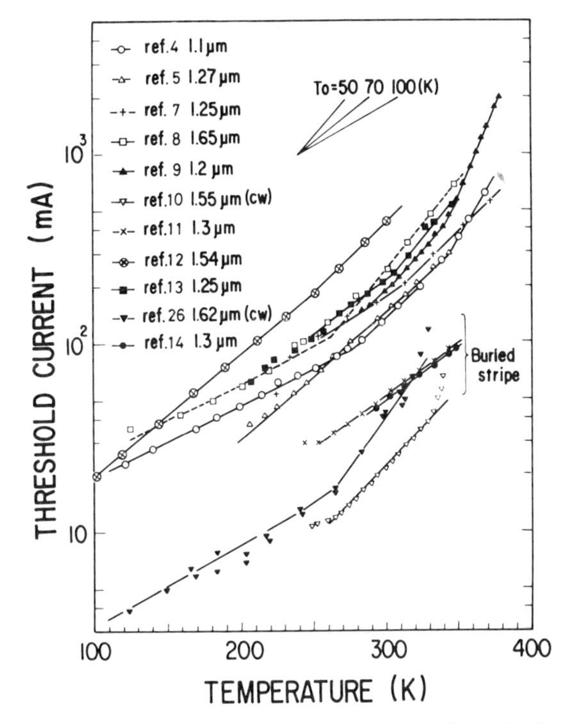

Figure 15.1 Temperature dependence of threshold current in GaInAsP–InP DH lasers with various emission wavelengths

around room temperature,[6,27] and good agreement with the theory based on the temperature dependence of the optical gain has been reported.[6,27] $I_{th} - T$ relations of 1.8 μm AlGaAsSb–GaInAsSb DH lasers[29] and 1.0 μm AlGaAsSb–GaAsSb lasers[30] are shown in Figure 15.2 for comparison. Although the emission wavelengths of these lasers almost correspond to the shortest and longest emission wavelengths of GaInAsP–InP DH materials, T_0 values around room temperature are much higher than those shown in Figure 15.1. Similar high T_0 values have also been reported for 1.1 μm AlGaAsSb–GaAsSb DH lasers.[31] Recently, Dolginov *et al.*[34] reported low T_0 values of 60–80 K, which are very similar to those for GaInAsP–InP DH lasers, for GaAlAsSb–GaSb DH lasers with 1.5–1.8 μm emission wavelengths, where variation of J_{th} with emission wavelength is strongly affected by the solid composition of the confinement layers. Similar low T_0 values have been observed in the 1.4–1.5 μm emission wavelength by Sasaki *et al.*[19] and Motosugi *et al.*[33] Therefore, available data of T_0 values for lasers with AlGaAsSb confinement layers scatter over a considerable range. However, it should be noted that a relatively high T_0 value has been obtained for the lasers with very low threshold current density of 2.5 kA cm^{-2}.

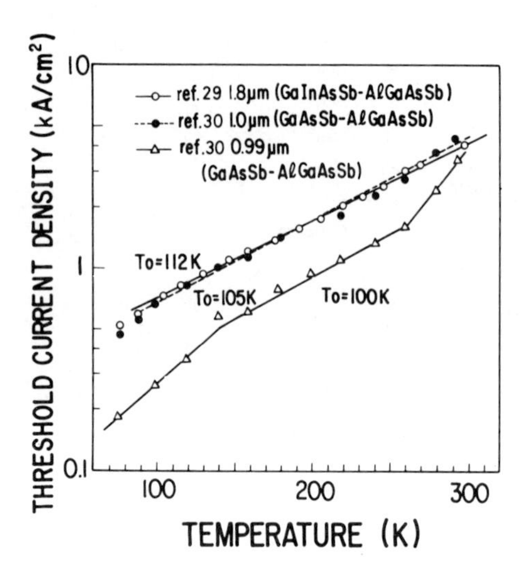

Figure 15.2 Temperature dependence of threshold current in GaInAsSb–AlGaAsSb, and GaAsSb–AlGaAsSb DH lasers

μm at room temperature.[29] Thus, the experimental results shown in Figure 15.1 and 15.2 may suggest that the high-temperature sensitivity of threshold current of GaInAsP–InP DH lasers is a phenomenon peculiar to this quaternary material.

Additional information which seems to be very useful for the discussion on the origins of the temperature characteristic is shown in the following. Figure 15.3 shows emission wavelength (λ) dependence of T_0 values at room temperature reported in the past. Although considerable scattering exists among the data, no distinct correlation can be seen between λ and T_0. On the other hand, when there is a shunt current path in the diode, excess current due to the shunt path affects the observed T_0 unless the excess current is proportional to the total injection current. Indeed, stripe buried-heterostructure lasers with poor current blocking exhibit slightly higher T_0 values and, moreover, a temperature-dependent shunt current path made T_0 for AlGaAs–GaAs DH lasers extremely high.[40] The latter phenomenon has been attributed to the temperature dependent current focusing in the active region. Figure 15.4 shows T_0 values for GaInAsP–InP DH lasers with various emission wavelengths as a function of the threshold current density normalized to the 1 μm active-layer thickness, J_{th}/d, where d represents the active-layer thickness in micrometres. The data in the figure include the results of the lasers with various active region thicknesses, various stripe widths, and those of buried-heterostructure lasers. Therefore, the contributions of interface recombination, the effect of shunt path as well as optical confinement factor may be different for each laser. Nevertheless, there seems to be no distinct correlation between T_0 and J_{th}/d, when $J_{th}/d \leq 25\,\text{cm}^{-2}\,\mu\text{m}^{-1}$. This implies that the temperature dependence of the above factors may not be responsible for observed

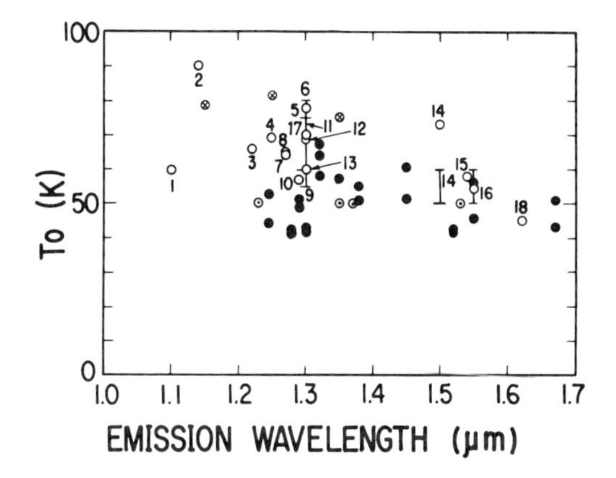

Figure 15.3 T_0 plotted as a function of emission wavelength: ●, ref. 18; ⊗, ref. 13; ⊙, ref. 23; 1, ref. 4; 2, ref. 32; 3, ref. 42; 4, ref. 7; 5, ref. 11; 6, ref. 14; 7, ref. 5; 8, ref. 3; 9, ref. 35; 10, ref. 36; 11, ref. 39; 12, ref. 37; 13, ref. 25; 14, ref. 20; 15, ref. 12; 16, ref. 10; 17, ref. 38; 18, ref. 26

low T_0 values. J_{th} values reported in the past scatter over a considerable range. However, determining the lowest values for each wavelength may be very useful in order to study the origin of the temperature sensitivity. Figure 15.5 shows relatively low J_{th}/d values which have been obtained for lasers with relatively thick active layers as a function of emission wavelength. Note that J_{th}/d values are between 4 and 5 kA cm^{-2} μm^{-1} regardless of λ, and therefore regardless of the bandgap difference at the heterointerfaces, ΔE_g. A similar observation has been reported recently.[47]

In Figure 15.6, reported experimental results on the external differential

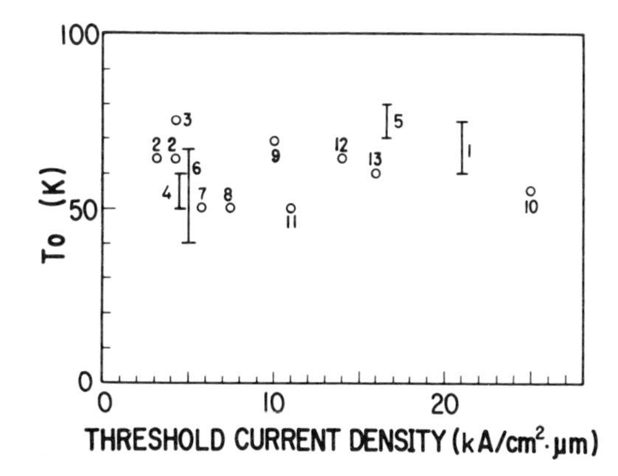

Figure 15.4 T_0 plotted as a function of threshold current density per 1 μm active layer thickness: 1, ref. 39; 2, ref. 3; 3, ref. 13; 4, ref. 20; 5, ref. 14; 6, ref. 18; 7, ref. 23; 8, ref. 12; 9, ref. 37; 10, ref. 10; 11, ref. 26; 12, ref. 5; 13, ref. 25

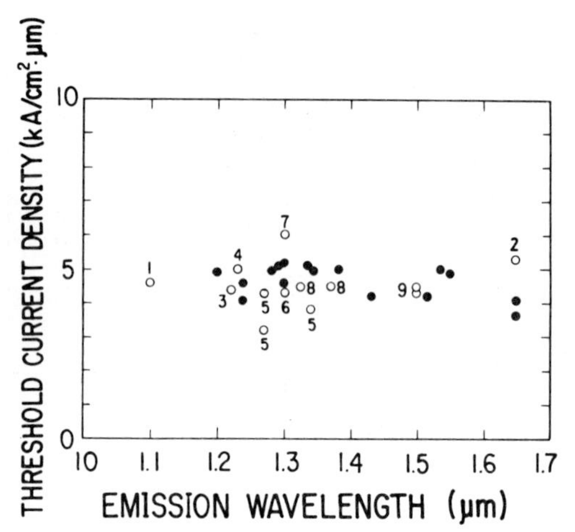

Figure 15.5 Emission wavelength dependence of relatively low threshold current density: ●, ref. 18; 1, ref. 41; 2, ref. 45; 3, ref. 42; 4, ref. 43; 5, ref. 3; 6, ref. 13; 7, ref. 44; 8, ref. 23; 9, ref. 20

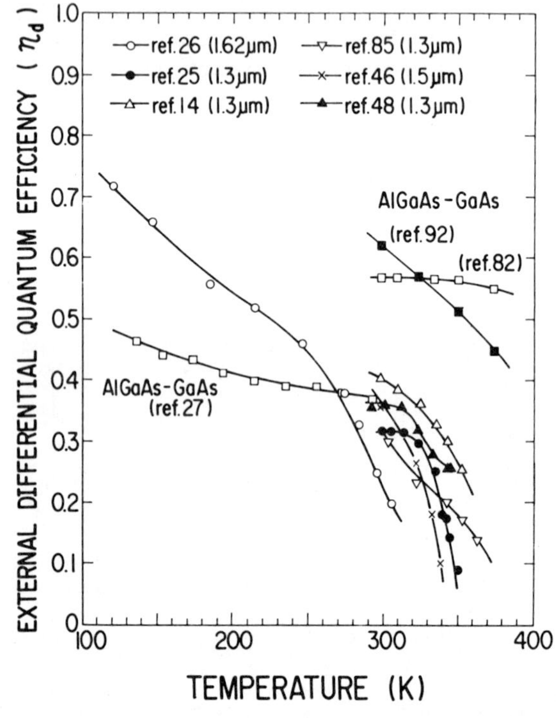

Figure 15.6 Temperature dependence of the external differential quantum efficiency

quantum efficiency η_d are shown as a function of temperature. Some papers show η_d explicitly, but others show the output–current $(L-I)$ relations measured at several temperatures. In the latter case, η_d was estimated from the linear part just above the threshold. Since η_d depends strongly upon the cavity length l, the absolute values of η_d in this figure should not simply be compared with each other, because the data are taken from lasers with various cavity lengths. Above room temperature, η_d values for GaInAsP–InP DH lasers decrease rapidly with increasing temperature, although some lasers keep fairly constant η_d up to 50–60°C.[14, 25, 48] However, the η_d–T relations for AlGaAs–GaAs DH lasers[82, 92] show much more gradual variation as shown in Figure 15.6.

Next, experimental results on the saturation characteristics in L–I relation of surface-emitting GaInAsP–InP DH LEDs are describd. A saturation phenomenon is often observed in small-area, high radiance surface emitting LEDs, even in AlGaAs–GaAs DH LEDs. This phenomenon should be discussed for LEDs with relatively high internal quantum efficiency, because otherwise measured L–I relations would be controlled by a dominant non-radiative recombination process, the effect of which may vary from diode to diode depending upon the amount and characteristics of the non-radiative centres involved. Figure 15.7 shows L–J relations measured at room temperature for LEDs with estimated external quantum efficiency of more than 1%.[49–53, 93] The result on a Burrus type AlGaAs–GaAs DH LED is also shown for comparison.[55] Even in pulsed operation, a distinct saturation is observed in both GaInAsP–InP and AlGaAs–GaAs DH LEDs, which suggests that light output saturation is not a phenomenon peculiar to GaInAsP–InP DH LEDs. However, according to the

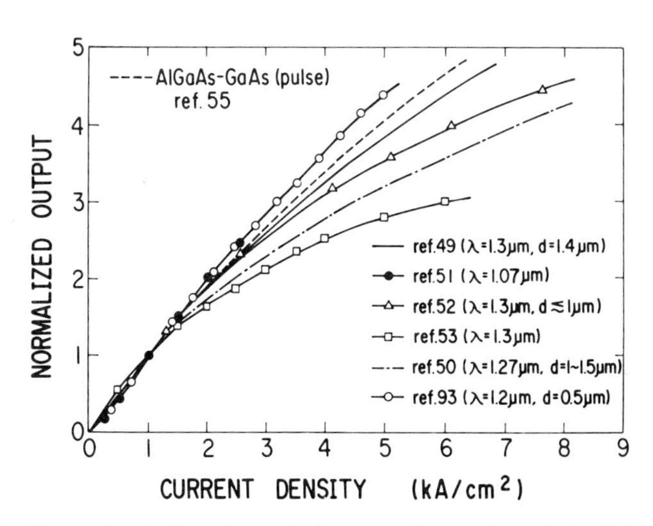

Figure 15.7 Current density dependence of light output from surface emitting GaInAsP–InP DH LEDs. Each output was normalized to the value at 1 kA cm^{-2}

experimental results reported so far, this phenomenon occurs more distinctly in GaInAsP–InP DH LEDs. Although the saturation behavior varies considerably from diode to diode as shown in Figure 15.7, it has been found that the saturation becomes more severe when the active-layer thickness or emitting area decreases.[49] In the case of Burrus type AlGaAs–GaAs DH LEDs, the saturation phenomenon has been attributed to the junction heating when they are operated continuously.[56] However, the saturation is apparently observed in pulsed operation as mentioned above, and therefore some additional loss process is to be considered at higher current densities. Since Burrus type AlGaAs–GaAs DH LEDs show a saturation at much lower current levels as compared with single-heterostructure and diffused homostructure counterparts,[56] carrier leakage beyond the heteroboundaries may be one possible candidate. Although some observations on light output saturation have been reported for GaInAsP–InP DH lasers and edge-emitting LEDs,[13, 54] usually such observations are very difficult because of the effect of dominant optical amplification along the active layer.

15.3 I_{th}–T RELATION DETERMINED BY TEMPERATURE DEPENDENCE OF THE OPTICAL GAIN

Although experimental results for GaInAsP–InP DH lasers shown in Figure 15.1 suggest the existence of additional loss processes, temperature dependence of the optical gain is still one of the most important factors which determine the I_{th}–T characteristics. The threshold condition is established when the maximum optical gain g_{max} for the gain spectrum reaches the total cavity loss. This can be expressed as

$$g_{max}^{th} = \frac{1}{\Gamma}\left[\alpha + \frac{1}{l}\ln\left(\frac{1}{R}\right)\right] \tag{15.2}$$

where g_{max}^{th} represents the optical gain at threshold, Γ is the optical confinement factor which is determined by the refractive index step at the heteroboundaries, emission wavelength and active layer thickness, α is the internal loss, l is the cavity length, and R is the mirror reflectivity. The threshold current is taken as one which yields a gain given in equation (15.2). For AlGaAs–GaAs DH lasers with complete carrier confinement, the I_{th}–T relation is considered to be almost completely explained by the temperature dependence of I–g_{max} relation, because the refractive index step, internal loss, and mirror reflectivity have a much smaller effect. However, the agreement between the experimental I_{th}–T relations and theory [6, 60] may be fortuitous, because I_{th}–T relations can be influenced by so many factors which are not well understood quantitatively. Moreover, theoretical I_{th}–T relations depend strongly upon the radiative recombination models and the shape of the band tails assumed. Therefore, the I_{th}–T relations for AlGaAs–GaAs DH lasers are still unexplained quantitatively. However, in the following, a theoretical method used for AlGaAs–GaAs DH lasers is followed to obtain the temperature dependence of the optical gain. For AlGaAs–GaAs DH lasers, many models have been used for the theoretical calculation of the gain spectrum. Depending upon the recombination mechanisms, there are band to band calculations with or without the *k*-selection rule, band to impurity states, band to band-tail, and band-tail to band-tail calculations. Among these calculations, experimental results on some AlGaAs–GaAs DH lasers have been well-reproduced by Stern's method which employs the Gaussian–Halperin–Lax band tail model.[6, 57] However, no theoretical work on the temperature dependence of J_{nom}–g_{max} relation, where J_{nom} represents the nominal current density, of GaInAsP–InP DH lasers has been reported probably because of the much lower accuracy of the quaternary material parameters.

Göbel *et al.* showed that the laser transition in GaInAsP–InP DH lasers occurs between the band tail states even in unintentionally doped active material from the optical gain measurement.[58] Therefore, as in the case of AlGaAs–GaAs DH lasers, the theoretical calculation based on the inter-band-tail transition model is expected to be appropriate for GaInAsP–InP DH lasers. Following Stern's method,[57] which employs a Kane function fit to the Halperin–Lax density of states, Sugimura calculated the gain spectrum for GaInAsP–InP DH lasers as a function of current and temperature.[21] Figure 15.8 shows the temperature dependence of the nominal threshold current density ($J_{nom,th}$–T relations) calculated by assuming the total cavity loss of 50 cm^{-1} and the perfect optical confinement. $J_{nom,th}$ increases gradually with temperature and T_0 values estimated around room temperature are between 210 and 225 K. Since the $J_{nom,th}$–T relations depend strongly upon the recombination mechanism used, the result shown in Figure 15.8 should be modified when the other models are

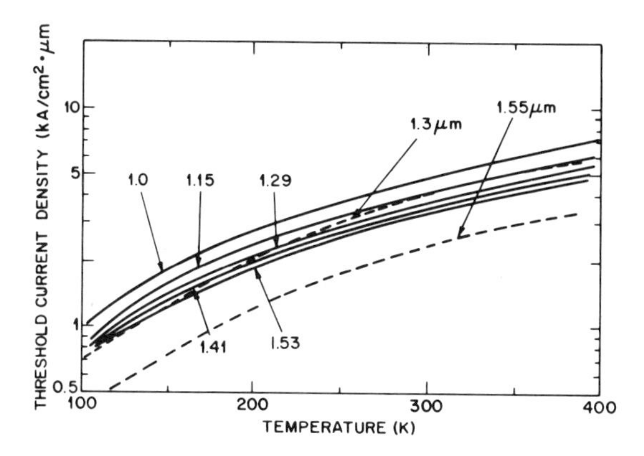

Figure 15.8 J_{th}–T relations for the GaInAsP–InP DH laser with various wavelengths; full curves show the results calculated by Sugimura[21] with the total cavity loss of 50 cm^{-1}, and broken curves show those calculated by Dutta *et al.*[61] Reproduced by permission of the authors and the American Institute of Physics

employed. However, it should be noted that T_0 values calculated by a model which explains the behaviour of some AlGaAs–GaAs DH lasers very well are much higher than the experimental results. A similar calculation has been performed recently by Dutta and Nelson[61] who applied a Gaussian fit to the Halperin–Lax density of states and Stern's matrix element, and obtained $J_{nom,th}$–T relations for 1.3 μm and 1.55 μm lasers both with a 0.2 μm thick active region by using the total cavity loss of 60 cm^{-1} at room temperature and Γ of 0.48. In order to compare their results with those by Sugimura in Figure 15.8, it is necessary to obtain $J_{nom,th}$ from the original $J_{th,r}$ data in reference 61, by considering both d and Γ used in the calculation. This has been done by applying the calculated temperature dependence of g_{max}–J_{nom} relationship[95] and the results are shown by the dashed curves in Figure 15.8. Note that the results are very similar to those shown by the full curves, except that the emission wavelength dependence of $J_{nom,th}$ is slightly different with each. Although this discrepancy may be serious when we evaluate the non-radiative recombination processes by comparing with the experimental result, the agreement between two independent calculations seems to be fairly good, and it should be emphasized that T_0 values predicted at 1.3 μm and 1.55 μm emission wavelength by the latter calculation are 202 K and 194 K, respectively. These values are very similar to those obtained from the full curves in Figure 15.8.

The T_0 values determined by the J–g_{max} relations have also been estimated on the basis of the experimental results. Yano *et al.* calculated the I_{th}–T relation by applying the I–g_{max} relations for GaAs laser calculated by Stern[64] and experimental temperature dependence of the radiative recombination constant, $B \propto T^{-0.8}$, and estimated the value of T_0 to be 140–150 K.[16,25] Adams *et al.* analysed their experimental results on 1.6 μm buried-heterostructure lasers by assuming a model in which the observed high-temperature sensitivity is considered to be caused by the increased optical absorption loss, and obtained $T_0 \simeq 145$ K for the absorption-loss-free case.[26] Experimentally, Thompson *et al.* obtained $T_0 \simeq 150$ K from the I_{th}–T relation corrected by the measured external quantum efficiency.[5] All these T_0 values are much higher than the observed results given in Figure 15.1.

15.4 DEVIATION FROM THE LOSS-FREE J_{TH}–T RELATION

As shown in the previous section, T_0 values expected from the temperature dependence of J_{nom}–g_{max} relation are much higher than the experimental results. This result suggests that some additional carrier or photon loss processes are operating in GaInAsP–InP DH lasers, although, at this stage, it is not known whether or not the theoretical model used for the gain calculation is appropriate to the GaInAsP quaternary system. In the case of AlGaAs–GaAs DH lasers, deviation from the theoretical J_{th}–T relation occurs at low temperatures below 100 K and above room temperature. The former is considered to be caused by the destruction of the carrier confinement due to the reduced diffusion length by the enhanced impurity scattering.[27] The latter is caused by the carrier leakage from the active region and is often very important because this process occurs in the temperature range where most of the devices are operated. However, this problem almost disappears if ΔE_g is greater than 0.4 eV.[6]

Several models have been proposed to explain the temperature characteristics of GaInAsP–InP DH lasers. They are described in the following.

15.4.1 Carrier leakage to the confinement layers

In semiconductor DH lasers, carriers injected into the active region partially escape beyond the heterobarriers without contributing to the optical gain in the active region, and are lost through wide bandgap confinement layers by the transport mechanism of either drift or diffusion. Since ΔE_g of 1.3 μm GaInAsP–InP lasers is very similar to that of $Al_{0.3}Ga_{0.7}As$–GaAs DH lasers, it is very natural to suspect the carrier leakage effect as an origin of the high-temperature sensitivity of I_{th}. Moreover, the conduction band discontinuity ΔE_c of 1.3 μm GaInAsP–InP DH lasers is much smaller than that of $Al_{0.3}Ga_{0.7}As$–GaAs DH lasers. The study on the quantum-well photoluminescence revealed that $\Delta E_c \simeq 0.67\Delta E_g$ for GaInAsP–InP heterointerface,[68] while $\Delta E_c \simeq 0.85\Delta E_g$ for AlGaAs–GaAs interface.[67] Therefore, the GaInAsP–InP heterobarrier seems to be less effective for electron confinement, although this problem may be alleviated by the diffusion potential.[59] In addition, electron effective mass m_c for GaInAsP is much smaller than that for GaAs. For example, m_c for GaInAsP with the composition corresponding to 1.3 μm emission is $0.052\,m_0$,[69] while $m_c = 0.067m_0$ for GaAs, where m_0 represents the free-electron mass. Therefore, at threshold, the effective barrier height at the GaInAsP–InP heterointerface would be further reduced because of a deep penetration of the electron quasi-Fermi level into the conduction band.

Meanwhile, Anthony *et al.* showed that the leakage loss depends not only upon ΔE_g but also upon the carrier concentration of p-type confinement layers.[70] The drift leakage is expected to be dominant and temperature-sensitive when the resistance of p-confinement layer is fairly high. However, when p-confinement layer with hole concentration greater than $(3–5) \times 10^{17}\,\text{cm}^{-3}$ is used, the drift leakage becomes less important[59] and leakage loss due to diffusion takes place.[71] Dutta *et al.* studied the differential resistance of GaInAsP–InP DH lasers and concluded that the drift leakage mentioned above is not responsible for the temperature sensitivity of I_{th}.[24] Indeed, since the p-confinement layers conventionally used have fairly high hole concentrations, the drift leakage should not be a dominant effect in a majority of GaInAsP–InP DH lasers. On the other hand, Yano *et al.* studied the contribution of the carrier leakage due to the diffusive current by using Casey's method[73] and assuming parabolic band structure.[16] The resulting T_0 at and below room temperature was approximately equal to 150 K,

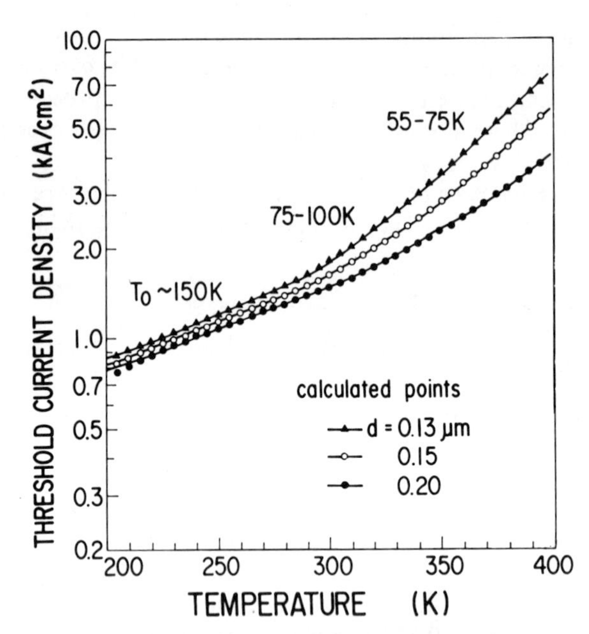

Figure 15.9 J_{th}–T relations calculated by considering the diffusion leakage in parallel with the radiative recombination process. After Yano *et al.*[16] Reproduced by permission of the authors and IEEE. Copyright 1981 IEEE

which was attributed to the temperature dependence of the optical gain. Above room temperature, I_{th} becomes more sensitive to temperature due to the diffusion leakage, and the corresponding T_0 is about 100 K for the laser with 0.2 μm thick active layer. T_0 values around room temperature decrease as d decreases as shown in Figure 15.9, and become 75 K when d is as thin as 0.13 μm. This is very close to the experimental result. However, experimentally, low T_0 values have been observed even in lasers with much thicker active layers, and, moreover, most of the lasers show low T_0 values even far below room temperature.

Figure 15.10 shows the energy distribution of electrons in the active region of 1.3 μm GaInAsP–InP DH lasers at 25° C and 70° C.[16] The full curves were calculated by considering the diffusion leakage, while the broken curves correspond to perfect confinement. Leakage loss represented by the hatched area in the figure is very small at 25° C, while it grows appreciably at 70° C. The above result suggests that the leakage loss is partially responsible for the observed J_{th}–T relations, at least above room temperature. In order to examine the effect of ΔE_g on the I_{th}–T relations, I_{th} (70° C)/I_{th} (22° C) was plotted as a function of ΔE_g or λ in Figure 15.11 from the available I_{th}–T relations. Except for the data around 1.1 μm, there is no distinct correlation between I_{th} (70° C)/I_{th} (22° C) and ΔE_g. At larger ΔE_g, therefore longer-wavelength region, the effect of larger ΔE_g may be compensated for appreciably by the smaller electron effective mass. However,

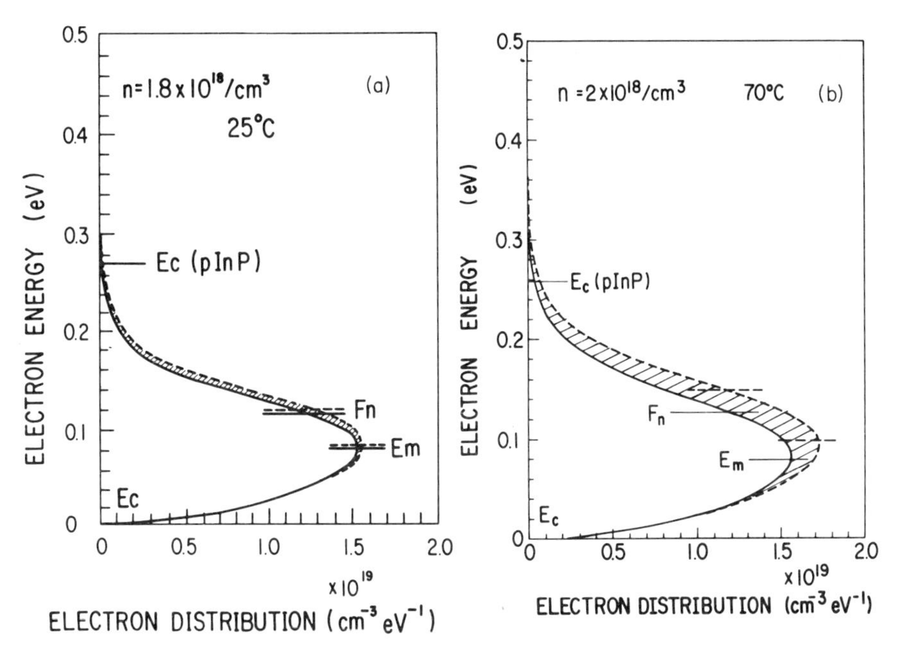

Figure 15.10 Energy distribution of electrons calculated at (a) 25°C and (b) 70°C for 1.3 μm GaInAsP–InP DH lasers at threshold in the active region. Hatched areas correspond to the electrons which may escape by dffusive leakage. After Yano *et al.*[16] Reproduced by permission of the authors and the American Institute of Physics

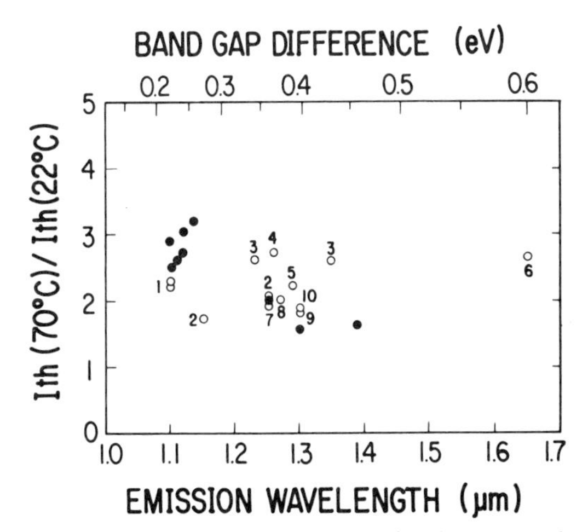

Figure 15.11 I_{th} (70°C)/I_{th} (22°C) plotted as a function of λ of ΔE_g: ●, ref. 17; 1, ref. 4; 2, ref. 13; 3, ref. 23; 4, ref. 72; 5, ref. 36; 6, ref. 8; 7, ref. 7; 8, ref. 5; 9, ref. 11; 10, ref. 25

Nahory *et al.* found that similar emission wavelength lasers with various ΔE_g, which were fabricated by changing the confinement layer material, showed similar temperature characteristics.[23] Moreover, electroluminescence from the confinement layers is very weak.[4] All these results suggest that the contribution of the leakage current to the total threshold current cannot be very large around room temperature.

Carrier leakage might be responsible for the saturation of LED output power since the carrier leakage current over the heterojunction barrier depends strongly upon the injection level. The observations in GaAs-based LEDs described in section 15.2 and the discussions on the laser characteristics mentioned above suggest that the leakage effect is partially responsible for the LED output saturation at room temperature and elevated temperatures. Indeed, Sanada *et al.*[62] found that the light output power and linearity of the *L–I* relation for 1.5 μm surface emitting DH LEDs are considerably improved when InP was used for the p-confinement layer material instead of GaInAsP with 0.95 eV bandgap energy. This result suggests that the carrier leakage effect is one of the most important factors responsible for the LED output saturation, although, in this case, the GaInAsP confinement layer provides very small bandgap difference ($\Delta E_g \simeq 0.12$ eV). However, it should be noted that the output saturation still remains even in the LEDs with p-InP confinement layer, where ΔE_g is expected to be as large as 0.5 eV.[62] Therefore, some other effects are to be considered in addition to the carrier leakage effect. Moreover, as reported in reference 49, LED output saturation occurs even at fairly low temperatures and low current levels where no appreciable leakage can be expected. Therefore, it is difficult to explain the LED output saturation by only the leakage effect. More recently, Goodfellow *et al.*[63] found that −3 dB output saturation of 1.06 and 1.3 μm GaInAsP–InP DH LEDs occurs at fairly high current densities of 30–300 kA cm^{-2}, and that the temperature dependence of LED output (*L–T* relation) does not

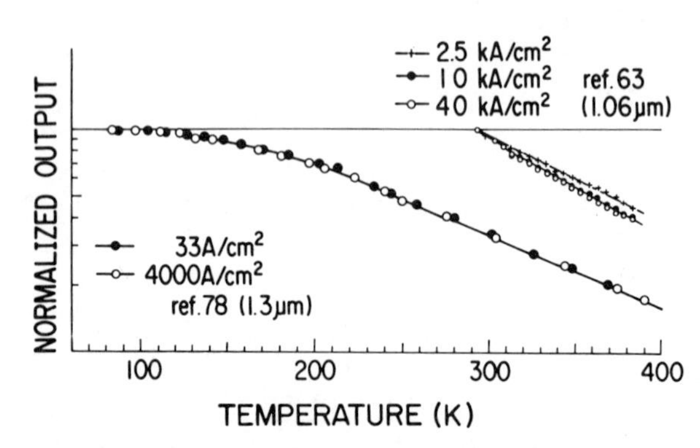

Figure 15.12 Temperature dependence of surface emitting GaInAsP–InP DH LEDs with various current densities from references 63 and 78

depend upon the current densities up to 40 kA cm^{-2} and 72 kA cm^{-2} for 1.06 and 1.3 μm LEDs, respectively. Similar observations have been made by Okuda *et al.*[78] who showed that L–T relations of 1.3 μm LEDs with 1 μm thick active layer do not depend upon the current density if $J \lesssim 4$ kA cm^{-2}. These results are shown in Figure 15.12. Because of these observations, the authors mentioned above concluded that LED output saturation cannot be predominantly caused by the carrier leakage effect. Goodfellow *et al.*[63] proposed superluminescence in the active layer plane for the dominant mechanism determining the LED output saturation characteristics. However, output saturation occurs more severely at higher temperatures,[49, 61] where superluminescence is expected to be less intense, and moreover, experimental results on edge emitting GaInAsP–InP DH LEDs showed fairly linear[77] or even saturating[13,54] L–I relations and no distinct superlinear increase in light output with increasing driving current.

15.4.2 Band-to-band Auger effect

Band-to-band Auger effect essentially includes strong temperature and carrier concentration dependence. Therefore, this effect has been suspected to be responsible for the high-temperature sensitivity of I_{th}.[4, 5, 21, 61] Recently, experimental and theoretical reports on the LED output saturation have been published, where the band-to-band Auger process was considered to be the most dominant effect.[49, 53, 74, 76] For GaInAsP–InP DH lasers, experimental J_{th}–T relations can be explained by the non-radiative recombination process, the rate of which is proportional to n^2, where n represents injected carrier density.[5] Similarly, LED saturation is explained[49] by the non-radiative process with the rate proportional to n.$^{2-4}$ These observations suggest a recombination process in which more than two carriers are expected to participate, such as band-to-band Auger effect.

Possible band-to-band Auger effect in III–V compound semiconductors includes following three direct processes:

(i) the CCHC process, in which two conduction band electrons interact with one hole leaving a high-energy electron in the conduction band;
(ii) the CHHL process, in which two holes interact with one conduction band electron leaving a high-energy hole in the light hole band; and
(iii) the CHHS process, in which two holes interact with one conduction band electron just as in the CHHL process, but leaving one hole in the split-off valence band.

Transition models for these Auger processes are shown in Figure 15.13, where C, H, L, and S designate conduction band, heavy-hole, light-hole, and spin split-off valence band, respectively. It is very important to evaluate the transition probabilities of these processes. Since the activation energies of the CCHC and CHHL processes are similar, they may have a similar importance for determining

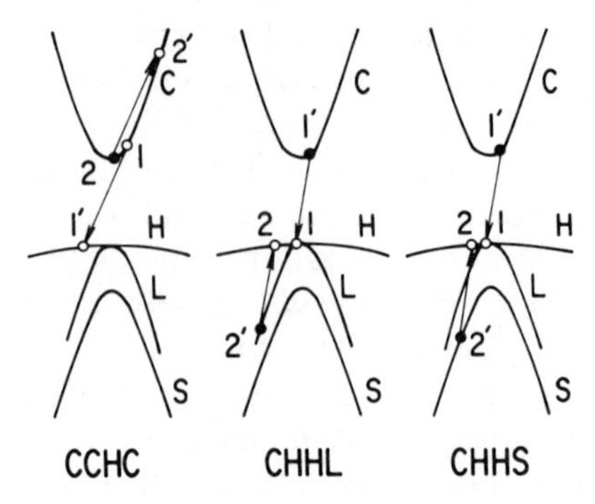

Figure 15.13 Transition models for three band-to-band Auger processes; CCHC, CHHL, and CHHS transitions. C, H, L, and S represent conduction band, heavy-hole, light-hole, and spin split-off valence band, respectively

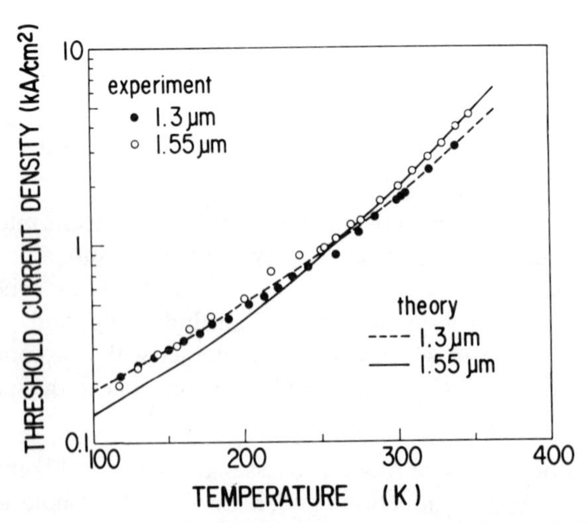

Figure 15.14 Comparison between the experimental J_{th}–T relations and those theoretically estimated by combining the Auger component due to the CCHC process with the radiative component given by broken curves in Figure 15.8. After Dutta *et al.*[61] Reproduced by permission of the authors and the American Institute of Physics

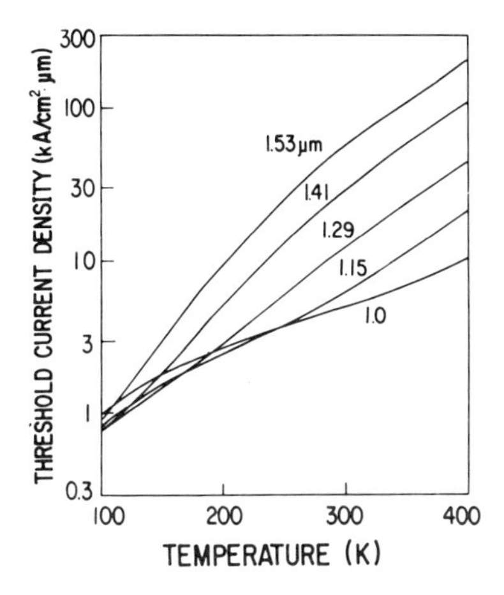

Figure 15.15 Theoretical J_{th}–T relations estimated by combining the Auger component due to the CHHS process with the radiative component given by full curves in Figure 15.8. After Sugimura.[21] Reproduced by permission of the author and IEEE. Copyright © 1981 IEEE

the total carrier lifetime.[83] On the other hand, the CHHS process has been found to be much more probable than the CHHL process in p-type III–V compound semiconductors.[75] However, transition probabilities of these processes are strongly affected by the material parameters, most of which are still uncertain at the present stage. Among these processes, the CCHC process has first been studied by many workers.[4,61,83,84] Dutta *et al.*[61] calculated J_{th}–T relations for 1.3 and 1.55 μm lasers by combining their theoretical result on $J_{nom,th}$–T relations shown by broken curves in Figure 15.8 with the Auger component due to the CCHC process. Their experimental J_{th}–T relations are well-reproduced by the theoretical calculation as shown in Figure 15.14. It is noteworthy that the theoretical J_{th} values at 300 K for 1.3 μm and 1.55 μm lasers are very similar to each other, which is consistent with the experimental J_{th}/d–λ relation given in Figure 15.5. However, the agreement mentioned above may be fortuitous, since in the calculation of the Auger component in J_{th}, $J_{th,a}$, parabolic band structure has been assumed, which may result in an overestimation of the Auger transition probability, and the CHHS process, which may be more important than the other processes, has not been considered. Meanwhile, Sugimura calculated Auger lifetime due to the CHHS process in GaInAsP–InP DH lasers by using Fermi functions for the carrier distribution.[21] The J_{th}–T relation and T_0 values were thus calculated by connecting this result with the $J_{nom,th}$–T relations shown by full curves in Figure 15.8. Results are shown in Figure 15.15, which shows much more rapid increase of J_{th} with temperature as compared with the ideal J_{th}–T relation. Figure 15.16 shows T_0 values at 300 K as a function of λ, together with those for the ideal case deduced from Figure 15.8. Calculated T_0 values are very close to the experimental result given in Figure 15.3 when $\lambda \gtrsim 1.2\,\mu$m. This may suggest that

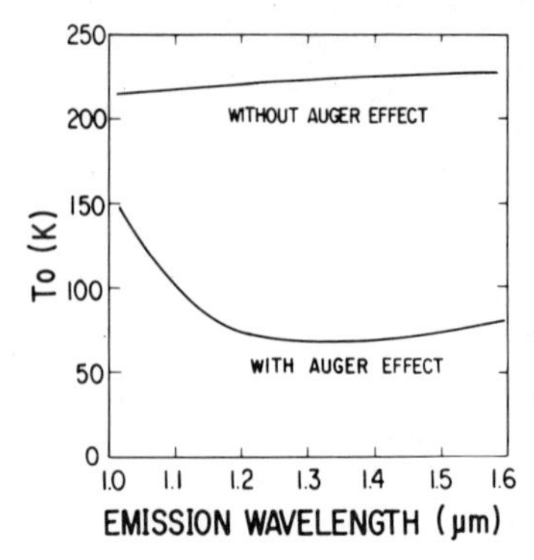

Figure 15.16 Room-temperature T_0 values with or without the band-to-band Auger effect due to CHHS process as a function of emission wavelength. The T_0–λ curve due to the Auger effect is determined from Figure 15.15, and that for the ideal case from Figure 15.8. After Sugimura.[21] Reproduced by permission of the author and IEEE. Copyright © 1981 IEEE

the Auger process is one of the dominant effects in determining J_{th}. However, J_{th} at room temperature increases rapidly as λ increases, as shown in Figure 15.15, while experimental J_{th} values at room temperature are almost independent of λ as shown in Figure 15.5.

Therefore, there is a marked discrepancy between two theoretical calculations; one which employs the CCHC process and the other the CHHS process. This discrepancy may arise from the difference between the Auger transition models employed, and between the theoretically estimated $J_{nom,th}$ values. The Auger lifetime essentially decreases; therefore, $J_{th,a}$ increases and the radiative component of J_{th}, $J_{th,r}$, decreases as λ increases, although their functional forms depend upon the transition mechanisms involved. In the case of the calculation by Dutta *et al.* an increase of $J_{th,a}$ for the longer-wavelength laser is almost completely compensated for by the decrease of $J_{th,r}$. Thus, the total J_{th} increases only slightly when λ increases from 1.3 μm to 1.55 μm. It should be noted that the CHHS process predicts much shorter lifetimes, and therefore much higher $J_{th,a}$ values, than the CCHC process. Nevertheless, the theoretical calculation based on the CCHC process reproduces the experimental result well as mentioned above. This contradiction may be due both to the incompleteness of the transition models employed and to the uncertainty in the available material parameters. Therefore, it may be true to say that, at present, it is difficult to compare quantitatively the experimental J_{th}–T relations with the theoretical result, since the accuracy of the latter is still rather poor.

The contribution of the band-to-band Auger effect to the LED output saturation has also been studied by considering the same Auger processes.[63, 76] Observation of light output from surface emitting LEDs seems to be useful to investigate the non-radiative process in GaInAsP–InP DH devices, because LED efficiency is determined much more simply than the laser threshold current.

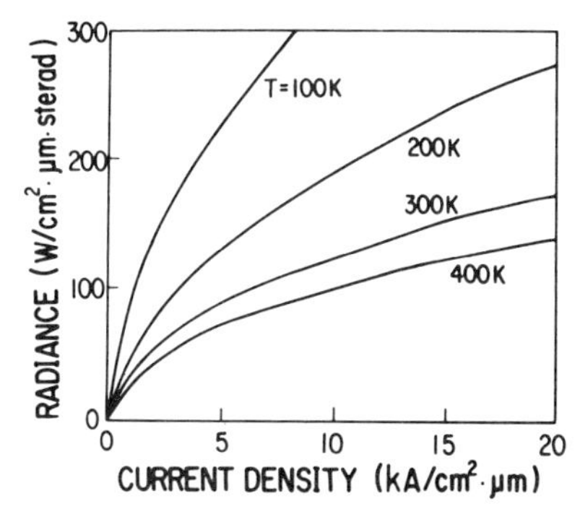

Figure 15.17 *L–I* relations of 1.3 μm surface emitting GaInAsP–InP DH LEDs calculated by considering the CHHS process. After Sugimura.[76] Reproduced by permission of the author and IEEE. Copyright © 1981 IEEE

Moreover, theoretical results on the LED efficiency can be compared with the experimental result at any injection level, while in the case of lasers, the comparison may be possible only at the threshold condition. Figure 15.17 shows calculated results on 1.3 μm GaInAsP–InP DH LEDs by applying the CHHS process,[76] which qualitatively explain the observed room-temperature characteristics shown in Figure 15.7. It should be noted that the Auger recombination coefficient obtained in this calculation without using any fitting parameters is very close to that estimated by Chik *et al.*[53] from adjusting a simple theory to the experimental saturation behaviour. Dutta *et al.*[61] compared their experimental results on the 1.3 and 1.55 μm emission wavelength LEDs with the theoretical calculation based on the CCHC process. Although qualitative agreement is reasonable, there are marked discrepancies, especially, in low-temperature characteristics. Besides, experimental *L–I* relations at room temperature show much more distinct saturation as compared with the theoretical results. Similar results have been obtained by Goodfellow *et al.*[63] Moreover, as described in section 15.4.1, *L–T* relations of some surface emitting LEDs are not very sensitive to the injection current. Therefore, it is still unknown whether or not the Auger effect is dominates the output characteristics of surface emitting GaInAsP–InP DH LEDs.

15.4.3 Effect of optical absorption loss

According to equation (15.2), optical gain required for laser oscillation increases as optical absorption loss α increases. Optical absorption also affects the external differential quantum efficiency η_d. Therefore, if there is strong optical absorption which increases rapidly with temperature, both J_{th}–T and η_d–T relations will be explained at the same time. Possible optical absorption mechanisms in GaInAsP–InP DH materials include those associated with impurity states and

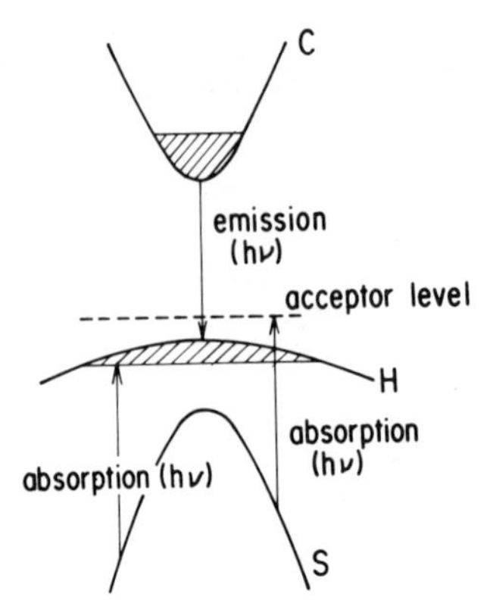

Figure 15.18 Electronic transition model for the inter-valence-band absorption. C, H, and S represent the conduction band, heavy-hole valence band, and spin split-off band, respectively. After Adams *et al.*[26] Reproduced by permission of the authors and Publication Manager, Japanese Journal of Applied Physics

free-carrier absorption. Since relatively pure material is used for the majority of lasers and since the room-temperature value of the free-carrier concentration is as high as $10^{18}\,cm^{-3}$ at threshold, optical absorption associated with impurity states may have a much smaller effect than free-carrier absorption. According to the theoretical consideration by Fan,[86] free-carrier absorption due to the scattering of carriers inside the band has been evaluated for GaInAsP–InP DH lasers at the emission wavelengths to be less than $10\,cm^{-1}$ at room temperature.[21,26] Therefore, free-carrier absorption does not seem to explain the observed temperature sensitivity, since this value is much smaller than the cavity loss, which may be 40–60 cm^{-1} for 200–300 μm long lasers. However, free carriers can absorb photons through inter-conduction-band and inter-valence-band electronic transitions. The former will be less important, because the inter-conduction-band transition in the energies corresponding to the emission wavelengths requires an indirect process. The latter, however, may have an optical absorption strong enough to determine the characteristics of GaInAsP–InP DH lasers. Adams *et al.*[26] proposed the inter-valence-band absorption shown in Figure 15.18 which involves an electronic transition between the split-off valence band and the heavy-hole band or acceptor levels, and showed good agreement between the theory and the experimental results on 1.62 μm buried-heterostructure lasers. This is shown in Figure 15.19, where full curves correspond to the theoretical J_{th}–T and η_d–T relations. According to this model, α as large as $200\,cm^{-1}$ is required to explain η_d and J_{th} at room temperature. Experimentally fairly high values of α have been measured in the past. For 1.1 μm and 1.3 μm GaInAsP–InP DH lasers, $\alpha = 68\,cm^{-1}$ and $\alpha = 98\,cm^{-1}$ have been obtained, respectively.[41,81] However, these values involve both the optical

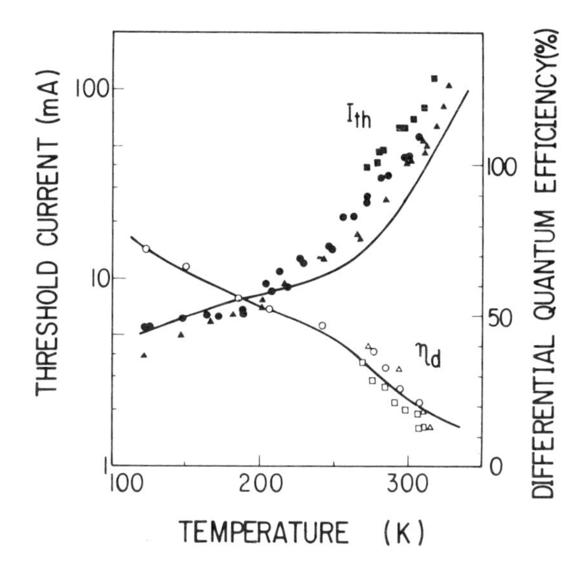

Figure 15.19 I_{th}–T and η_d–T relations for 1.62 μm buried-heterostructure lasers. Full curves represent theoretical results due to the inter-valence-band absorption, while each point shows an experimental result. After Adams *et al.*[26] Reproduced by permission of the authors and Publication Manager, Japanese Journal of Applied Physics

absorption loss and the background loss caused by the quantity α_0 in the gain–current relationship, $g = \beta J_{nom}$-α_0, where β is the gain constant. In recent well-characterized stripe-geometry lasers, $\alpha = 20$–$40\ \text{cm}^{-1}$, which is very similar to that of AlGaAs–GaAs DH lasers, have been estimated for 1.3 μm and 1.5 μm GaInAsP–InP DH lasers.[47,66,96] Moreover, Tokunaga *et al.*[65] reported α of 20–50 cm^{-1} for 1.55 μm GaInAsP–InP buried heterostructure lasers through the measurement of the cavity length dependence of the differential quantum efficiency. Therefore, optical absorption loss does not seem to be a major effect determining the temperature characteristics of GaInAsP–InP DH lasers, although it may be partially responsible for them. The inter-valence-band absorption considered occurs when $E_g \geq \Delta_0$, where Δ_0 is the spin–orbit splitting at the Γ point, and will be enhanced when $E_g - \Delta_0$ becomes smaller. Although the compositional dependence of Δ_0 for GaInAsP is not known, experimental results on the related ternary alloys[94] suggest that Δ_0 increases as E_g decreases, and therefore $E_g - \Delta_0$ decreases as emission wavelength increases. Therefore, the contribution of the inter-valence-band absorption will be clarified through the careful measurement of α as a function of emission wavelength. Optical absorption loss, in principle, does not explain the LED output saturation, because the output radiates normal to the junction plane and, therefore, cannot be absorbed appreciably in a very thin active region.

15.4.4 Non-radiative recombination centres

Since a relatively high-temperature stability is predicted by theoretical considerations, and since the carrier leakage, Auger effect, and optical absorption loss are not complete explanations for the observed temperature sensitivity, as described in the preceding sections, non-radiative recombination centres are to be considered. Indeed, non-radiative recombination centres control I_{th}–T relations especially at low temperatures. Figure 15.20 shows I_{th}–T and τ_s–T relations of GaInAsP–InP DH lasers reported in reference 4, together with the traces of the optical output response measured at temperatures below and above the breakpoint in the I_{th}–T relation, where τ_s represents the spontaneous carrier lifetime measured at threshold. At the same temperature, breakpoint also appears in the τ_s–T relation. Below the breakpoint, I_{th} becomes less temperature-sensitive and τ_s tends to saturate as temperature decreases. Moreover, the pulse response below the breakpoint exhibited a rather long risetime and a reduced relaxation oscillation compared with the room-temperature response, as shown in Figure 15.20(b). On the other hand, in lasers which do not show a distinct breakpoint nor very high T_0 values at lower temperatures, τ_s increases continuously as the temperature decreases and no anomaly is observed in the output response, as shown in Figure 15.21 for a 1.3 μm laser[91] and in Figure 15.22 for a 1.54 μm laser.[22] The contrast between these figures and Figure 15.20 can be understood by considering non-radiative recombination centres which are active below the breakpoint. A similar correlation has been observed in AlGaAs–GaAs DH lasers.[22] In the AlGaAs–GaAs DH laser which shows very high T_0, saturating τ_s, and anomalous optical output response, several traps have been observed in deep-

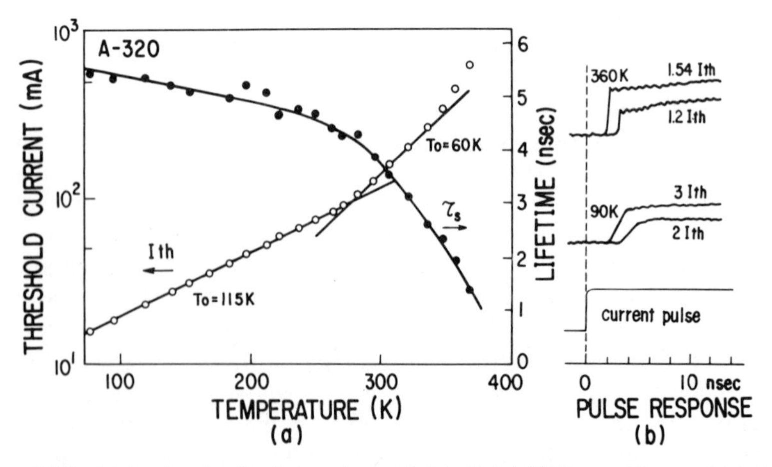

Figure 15.20 (a) I_{th}–T and τ_s–T relations for the GaInAsP–InP DH laser with a very high T_0 value at lower temperatures. (b) Pulse response of optical output observed at temperatures below and above the breakpoint

level transient spectroscopy (DLTS) spectra which are very similar to those observed by Uji.[79] Although no DLTS signal was observed for GaInAsP–InP DH lasers with characteristics shown in Figure 15.20, probably because of the relatively high background impurity concentration, it may be true to say that similar traps are responsible for the experimental results given in Figure 15.20. However, at room temperature, T_0 is not very high and no anomaly is observed in the optical output response. Therefore, traps mentioned above cannot be an origin of the temperature sensitivity around room temperature.

It is interesting to note that as shown in Figures 15.21 and 15.22, spontaneous carrier lifetimes at threshold for the lasers with relatively normal characteristics increase continuously without showing saturation as temperature decreases. Moreover, lifetimes as long as 10 ns have been observed for 1.3 μm laser at low temperatures.[91] Nevertheless, theoretical considerations[21,61] predict saturating τ_s with a saturating value of 5–6 ns, which is determined by the radiative recombination process. This fact again suggests the incompleteness of the theory for the $J_{nom,th}$ calculation. Experimentally, non-radiative recombination centres with activation energies of 0.1–0.3 eV[4,23] have been considered to explain room-temperature characteristics of GaInAsP–InP DH lasers. Thompson *et al.*[5] proposed a complex process composed of non-radiative recombination with rate proportional to n^2 and that due to the non-radiative recombination centres with 0.09 eV activation energy. The same activation energy has been obtained from the LED output characteristics.[78] More recently, however, Thompson[83] found that the activation energy of about 0.1 eV observed in the temperature dependence of the non-radiative current component is very close to the minimum total kinetic energies of interacting particles in the band-to-band Auger processes, and

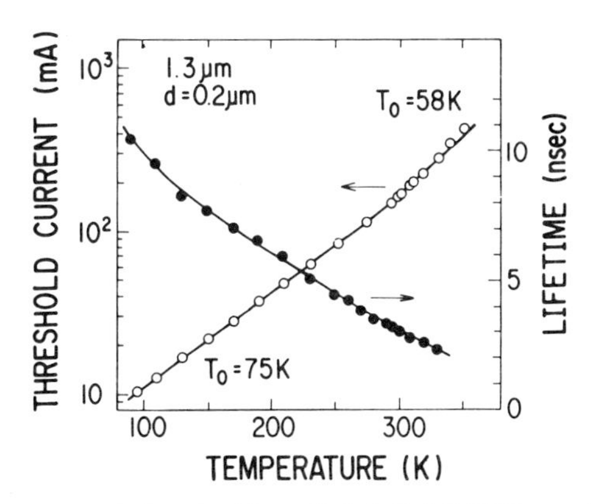

Figure 15.21 I_{th}–T and τ_s–T relations for 1.3 μm GaInAsP–InP DH laser which shows no anomaly in the optical output response at low temperatures. After Mukai.[91] Reproduced by permission of the author

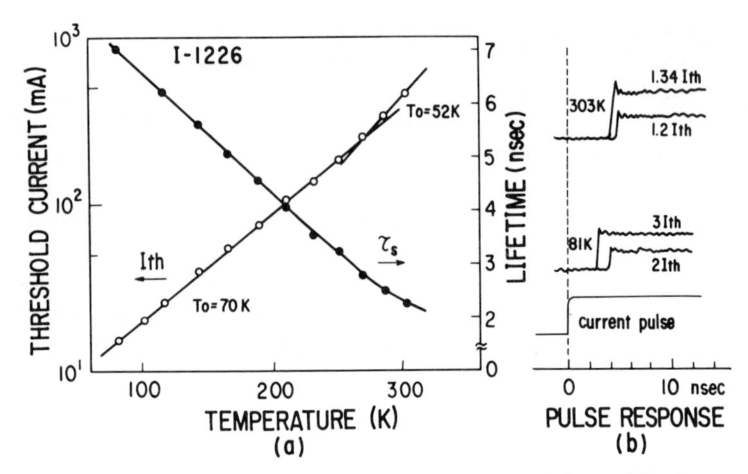

Figure 15.22 (a) I_{th}–T and τ_s–T relations for the GaInAsP–InP DH laser which does not show a distinct breakpoint nor a very high T_0 at lower temperatures. (b) Corresponding output waveforms observed at 81 K and 303 K

concluded that the observed result on 1.3 and 1.5 μm lasers is consistent with the presence of Auger processes on parallel with the radiative recombination.

Although no detailed discussions have been reported concerning the structure of non-radiative recombination centres, multivalent centres[4] such as phosphorus vacancies or non-radiative recombination centres related to the presence of indium in the active region[87] have been proposed so far. Non-radiative recombination at the heterointerfaces would also have an effect on the laser characteristics. As in the case of AlGaAs–GaAs DH lasers,[80] the recombination velocity at the heteroboundaries has been evaluated through the measurement of τ_s for DH lasers with various active-layer thicknesses. Although interface recombination seems to be partially responsible for the temperature sensitivity when the lattice mismatch is fairly large, a negligibly small contribution has been observed for closely lattice-matched lasers.[25] Therefore, this factor should not be blamed, because high-temperature sensitivity of laser threshold appears even in closely lattice-matched GaInAsP–InP DH lasers.

More recently, a hot carrier study on the GaInAsP–InP and GaInAs–GaInAsP lattice-matched n–n heterojunctions in the hetero-type avalanche photodiode structure revealed that there exist high-density hole traps with activation energies of 0.09 and 0.15 eV in the higher-bandgap material side of the heterojunction.[88] It should be noted that the activation energy of 0.09 eV mentioned above is very similar to the activation energies of 0.09–0.1 eV determined experimentally through the characteristics of lasers and LEDs as described before.

As mentioned in this section, non-radiative recombination centres and traps in the vicinity of the heterojunction seems to be very important to explain the observed temperature sensitivity. Further investigation is required to clarify the contribution of the non-radiative recombination centres and traps.

15.5 DISCUSSION ON η_d–T RELATION

As shown in Figure 15.6, external differential quantum efficiency of GaInAsP–InP DH lasers is very sensitive to the ambient temperature. η_d can be expressed by the internal quantum efficiency above threshold η_i, the carrier injection efficiency η_c,[40] and optical absorption loss α as

$$\eta_d = \eta_c \eta_i \frac{(1/l)\ln(1/R)}{\alpha + (1/l)\ln(1/R)} \tag{15.3}$$

Since the quantity $(1/l)\ln(1/R)$ is relatively temperature-insensitive, the temperature dependence of η_d may be caused by the temperature dependence of α and/or $\eta_c\eta_i$. Optical absorption was first suspected by Adams *et al.*[26] who considered several absorption mechanisms and concluded that optical absorption due to the inter-valence-band transition is responsible for the observed η_d–T relations. This model successfully explains the result on 1.62 μm lasers as shown in Figure 15.19. Although, as mentioned in section 15.4(c), experimentally estimated absorption coefficients for 1.3 and 1.5 μm lasers are much smaller than those required for the complete explanation, optical absorption due to this mechanism may be partially responsible for η_d–T relations. Next, we discuss the effect of η_c and η_i in equation (15.3). η_i can be simply related to the carrier lifetimes as

$$\eta_i = \frac{1}{1 + \tau_r^{st}/\tau_n} \tag{15.4}$$

where τ_r^{st} and τ_n represent the stimulated radiative recombination lifetime and non-radiative recombination lifetime, respectively. τ_r^{st} is limited by the intra-band relaxation time of injected carriers and has been shown to have a value of 10^{-12}–10^{-13} s.[89, 90] A non-radiative recombination process such as those associated with crystal imperfections and band-to-band Auger process may have lifetimes of 10^{-8}–10^{-9} s. Therefore, η_i cannot be affected by the non-radiative recombination process, and will be very close to unity[26] because $\tau_r^{st}/\tau_n \simeq 10^{-3}$–$10^{-5}$.

However, carrier leakage beyond the heterobarriers may degrade the carrier injection efficiency η_c which represents the ratio of the number of carriers recombining in the active region to that of the total number of carriers injected above threshold. When J_{th} includes appreciable leakage component, a portion of carriers injected above threshold may have energies larger than the heterobarriers. Most of these 'high-energy' carriers are expected to recombine with the stimulated radiative recombination lifetime τ_r^{st} in the active region. However,

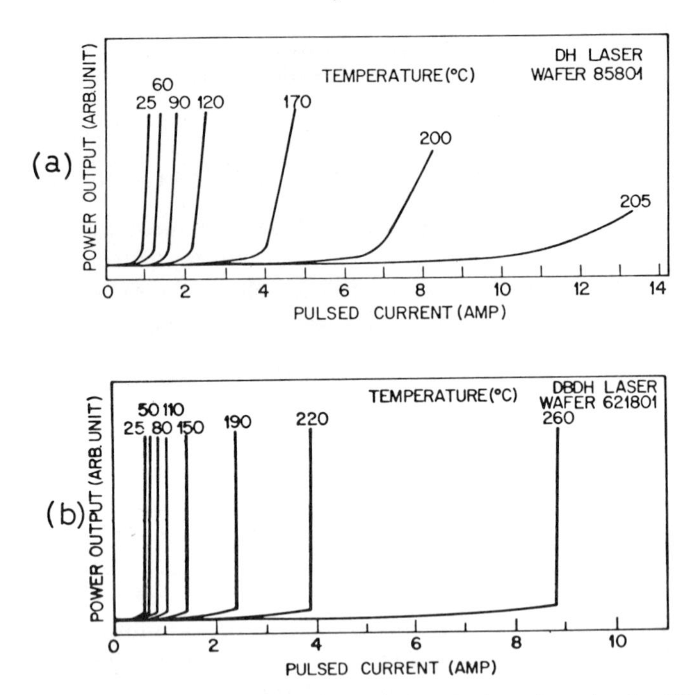

Figure 15.23 Temperature dependence of *L–I* relations for large heterobarrier AlGaAs–GaAs DBDH laser. After Tsang.[95] (a) *L–I* relations for AlGaAs–GaAs lasers with a conventional DH structure where $\Delta x = 0.25$. (b) *L–I* relations for DBDH laser. Reproduced by permission of the author and the American Institute of Physics

carrier leakage will compete with this process and some portion may overcome the heterobarriers. Although evaluating the leakage component above threshold requires the spatial distributions of optical gain and injected carriers, the contribution of this factor may be suggested by comparing v_t/d with $1/\tau_r^{st}$, where v_t represents the thermal velocity of injected carriers and d is the active-layer thickness. If we take $d = 0.1\ \mu m$, and $\tau_r^{st} = 10^{-13}$ s, the ratio $(v_t/d)/(1/\tau_r^{st})$ becomes ~ 0.1, since $v_t \simeq 10^7$ cm s^{-1}, which implies that an appreciable portion ($\sim 10\%$) of 'high-energy' carriers may escape beyond the heterobarriers. Thus, carrier leakage over the heterojunctions may be partially responsible for the observed η_d–T relations, because the number of 'high-energy' carriers increases as temperature increases. Indeed, a direct correlation between the carrier leakage and the reduction in η_d has been shown by Thompson *et al.*[97] for AlGaAs–GaAs heterostructure lasers. More recently, Tsang[98] showed that very high heterobarriers in the AlGaAs–GaAs double-barrier double-heterostructure (DBDH) laser considerably improve η_d–T relations at high temperatures as shown in Figure 15.23. Although carrier leakage beyond the stripe boundaries could affect the behaviour of η_d, this factor should be of secondary importance, because the η_d–T

relations of stripe buried-heterostructure lasers are very similar to those of conventional contact stripe lasers.

In this chapter, we implicitly assumed carrier temperature to be equal to lattice temperature. More recently, however, Shah *et al.*[99] found that the carrier temperature in forward biased 1.3 μm surface emitting GaInAsP LEDs is significantly greater than the lattice temperature. Although, the mechanism generating hot carriers is not identified yet, this effect should be taken into account in the theoretical models for loss processes discussed in this chapter.

15.6 SUMMARY

Temperature dependence of the threshold current of GaInAsP–InP DH lasers is much more sensitive to temperature than that predicted by the theoretical optical gain calculations. The I_{th}–T relations of recent well-characterized lasers are very similar to those of lasers in the very early stage. This implies the existence of some dominant intrinsic loss processes in parallel with radiative recombination. Therefore, intrinsic carrier or photon loss processes such as band-to-band Auger effect, carrier leakage beyond the heterobarriers, and optical loss due to the inter-valence-band absorption have been considered. However, as discussed in this chapter, these factors cannot completely explain the observed I_{th}–T relations. Non-radiative recombination centres and mixed processes of the above factors have been considered to compensate for the discrepancy between theoretical models and experimental results. However, explanations by them are still unsatisfactory. There are also considerable discrepancies among the experimental results reported in the past. The discrepancies mentioned above may be due to the incompleteness of the theoretical models including less reliable material parameters and to the difference in the device characteristics probably caused by the difference in the material quality and in the device structure. Therefore, improvement in the theoretical models for the calculations of the optical gain and band-to-band Auger effect, as well as in the material parameters, are required for further investigation. Also required are careful observations on the temperature characteristics of high-quality simple-structured GaInAsP–InP DH lasers. The same thing can be said for the output saturation phenomenon in surface emitting GaInAsP–InP DH LEDs. If the intrinsic carrier loss processes such as band-to-band Auger effect and carrier leakage beyond the heterobarriers are responsible for the temperature sensitivity of laser threshold current, they could be also responsible for the LED output saturation. Therefore, the contribution of these processes can be investigated through the LED output characteristics rather than laser characteristics, which seems to be very useful, because LED output is determined much more simply than the laser threshold current. In addition to the intrinsic loss processes, non-radiative recombination centres peculiar to the GaInAsP quaternary or to the GaInAsP–InP heterojunctions, and Auger effect associated with impurity centres are to be considered, since, at present, the intrinsic processes do not give complete explanations as described in this chapter.

REFERENCES

1. H. Osanai, T. Shioda, T. Moriyama, S. Arai, M. Horiguchi, T. Izawa, and H. Takata, 'Effect of dopants on transmission loss of low-OH-contents optical fibers', *Electron. Lett.*, **12**, 549–550, 1976.
2. Y. Horikoshi, T. Kobayashi, and Y. Furukawa, 'Lifetime of InGaAsP–InP and AlGaAs–GaAs DH lasers estimated by the point defect generation model', *Jpn. J. Appl. Phys.*, **18**, 2237–2244, 1979.
3. T. Yamamoto, K. Sakai, S. Akiba, and Y. Suematsu, '$In_{1-x}Ga_xAs_yP_{1-y}$/InP DH lasers fabricated on InP(100) substrates', *IEEE J. Quantum Electron.*, **QE-14**, 95–98, 1978.
4. Y. Horikoshi and Y. Furukawa, 'Temperature sensitive threshold current of InGaAsP–InP double heterostructure lasers', *Jpn. J. Appl. Phys.*, **18**, 809–815, 1979.
5. G. H. B. Thompson and G. D. Henshall, 'Nonradiative carrier loss and temperature sensitivity of threshold in 1.27 μm (GaIn)(AsP)/InP DH lasers', *Electron. Lett.*, **16**, 42–44, 1980.
6. H. C. Casey, Jr and M. B. Panish, *Heterostructure Lasers*, part B, Academic Press, New York, 1978, p. 187.
7. G. H. Olsen, C. J. Nuese, and M. Ettenberg, 'Low-threshold 1.25 μm vapor-grown InGaAsP CW lasers', *Appl. Phys. Lett.*, **34**, 262–264, 1979.
8. Y. Itaya, S. Arai, K. Kishino, M. Asada, and Y. Suematsu, '1.6 μm wavelength GaInAsP/InP lasers prepared by two-phase solution technique', *IEEE J. Quantum. Electron.*, **QE-17**, 635–640, 1981.
9. M. Yano, H. Nishi, and M. Takusagawa, 'Theoretical and experimental study of threshold characteristics in InGaAsP/InP DH lasers, *IEEE J. Quantum. Electron.*, **QE-15**, 571–579, 1979.
10. H. Nagai, Y. Noguchi, K. Thakahei, Y. Toyoshima, and G. Iwane, 'InP/GaInAsP buried heterostructure lasers of 1.5 μm region', *Jpn.J. Appl. Phys.*, **19**, L218–L220, 1980.
11. E. Oomura, T. Murotani, H. Namizaki, M. Ishii, W. Sasaki, and K. Shirahata, 'Low threshold InGaAsP/InP buried crescent laser with double-current-confinement structure', *7th IEEE Int. Semiconductor Laser Conf.* Abstract of Paper 29, September 1980.
12. N. Kobayashi and Y. Horikoshi, '1.5 μm InGaAsP/InP DH laser with optical waveguide structure', *Jpn.J. Appl. Phys.*, **18**, 1005–1006, 1979.
13. P. D. Greene, and G. D. Henshall, 'Growth and characteristics of GaInAsP/InP double heterostructure lasers', *Solid State Electron Devices*, **3**, 174–178, 1979.
14. M. Hirao, K. Mizuishi, S. Tsuji, and M. Nakamura, 'Accelerated aging of 1.3 μm InGaAsP/InP lasers', *7th IEEE Int. Semiconductor Laser Conf.*, Abstract of Paper 30, September 1980.
15. T. Ito, private communication (Yokosuka Electrical Communication Laboratory, Nippon Telegraph and Telephone Public Corporation, Yokosuka-shi, Kanagawa, Japan).
16. M. Yano, H. Imai, and M. Takusagawa, 'Analysis of threshold temperature characteristics for InGaAsP/InP double heterojunction lasers' *J. Appl. Phys.*, **52**, 3172–3175, 1981. M. Yano, H. Imai, and M. Takusagawa, 'Analysis of electrical, threshold and temperature characteristics of InGaAsP/InP double heterojunction lasers', *IEEE J. Quantum. Electron.*, **QE-17**, 1954–1963, 1981.

17. M. Ettenberg, C. J. Nuese, and H. Kressel, 'The temperature dependence of threshold current for double-heterojunction lasers', *J. Appl. Phys.*, **50**, 2949–2950, 1979.
18. S. Arai, Y. Suematsu, and Y. Itaya, '1.11–1.67 μm (100) GaInAsP/InP injection lasers prepared by liquid phase epitaxy', *IEEE J. Quantum. Electron.*, **QE-16**, 197–205, 1980.
19. A. Sasaki, Y. Takeda, A. Ooishi, and S. Mizugaki, 'Temperature dependence of threshold current of AlGaAsSb/GaSb DH lasers', *Trans. IECE of Japan*, **J64-C**, 335–336, 1981.
20. G. D. Henshall and P. D. Greene, 'InGaAsP/InP DH lasers for cw operation at 1.5 μm wavelength', *7th IEEE Int. Semiconductor Laser Conf.*, Abstract of Paper 32, September 1980.
21. A. Sugimura, private communication. A. Sugimura, 'Band to band Auger recombination effect on InGaAsP laser threshold', *IEEE J. Quantum. Electron.*, **QE-17,** 627–635, 1981.
22. Y. Horikoshi, H. Saito, M. Kawashima, and Y. Takanashi, 'Low-temperature behavior of the threshold current and carrier lifetime of InGaAsP–InP DH lasers', *Jpn. J. Appl. Phys.*, **18**, 1657–1658, 1979.
23. R. E. Nahory, M. A. Pollack, and J. C. DeWinter, 'Temperature dependence of InGaAsP double-heterostructure laser characteristics', *Electron. Lett.*, **15**, 695–696, 1979.
24. N. K. Dutta, R. J. Nelson, and P. A. Barnes, 'Temperature dependence of threshold and electrical characteristics of InGaAsP–InP DH lasers', *Electron. Lett.*, **16**, 653–654, 1980.
25. M. Yano, H. Nishi, and M. Takusagawa, 'Influences of interfacial recombination on oscillation characteristics of InGaAsP/InP DH lasers', *IEEE. J. Quantum. Electron.*, **QE-16**, 661–667, 1980. M. Yano, H. Nishi, and M. Takusagawa, 'Temperature characteristics of threshold current in InGaAsP/InP double-heterostructure lasers', *J. Appl. Phys.*, **51**, 4022-4028, 1980.
26. A. R. Adams, M. Asada, Y. Suematsu, and S. Arai, 'The temperature dependence of the efficiency and threshold current of $In_{1-x}Ga_xAs_yP_{1-y}$ lasers related to intervalence band absorption', *Jpn. J. Appl. Phys.*, **19**, L621–L624, 1980.
27. C. J. Hwang, N. B. Patel, M. A. Sacilotti, F. C. Prince, and D. J. Bull, 'Threshold behavior of (GaAl)As–GaAs lasers at low temperatures', *J. Appl. Phys.*, **49**, 29–34, 1978.
28. A. R. Goodwin, J. R. Peters, M. Pion, G. H. B. Thompson, and I. E. A. Whiteaway, 'Threshold temperature characteristics of double heterostructure $Ga_{1-x}Al_xAs$ lasers', *J. Appl. Phys.*, **46**, 3126–3131, 1975.
29. N. Kobayashi, Y. Horikoshi, and C. Uemura, 'Room temperature operation of the InGaAsSb/AlGaAsSb DH laser at 1.8 μm wavelength', *Jpn. J. Appl. Phys.* **19**, L30–L32, 1980.
30. R. E. Nahory and M. A. Pollack, 'Low-threshold room-temperature double-heterostructure $GaAs_{1-x}Sb_x/Al_yGa_{1-y}As_{1-x}Sb_x$ injection lasers at 1 μm wavelengths', *Appl. Phys. Lett.*, **27**, 562–564, 1975.
31. C. Chaminant, J. Charil, J.-C.Bouley, and E. V. K. Rao, 'Growth and properties of GaAsSb/GaAlAsSb double heterostructure lasers', *Solid-State Electron Devices*, **3**, 196–200, 1979.
32. J. A. Rossi, J. J. Hsieh, and J. P. Donnelly, '$In_{1-x}Ga_xAs_yP_{1-y}$–InP double-heterostructure lasers', *Gallium Arsenide and Related Compounds* (*St Louis*) 1976, Conf. Ser. no. 33b, Institute of Physics, Bristol, 1977, pp. 303–310.
33. G. Motosugi and T. Kagawa, 'Temperature dependence of the threshold current of AlGaAsSb/GaSb DH lasers', *Jpn. J. Appl. Phys.*, **19**, 2303–2304, 1980.
34. L. M. Dolginov, A. E. Drakin, L. V. Druzhinina, P. G. Eliseev, M. G. Milvidsky, V.

A. Skripken, and B. N. Sverdlov, 'Low threshold heterojunction AlGaAsSb/GaSb lasers in the wavelength range of 1.5–1.8 μm', *IEEE J. Quantum Electron.*, **QE-17**, 593–597, 1981.
35. H. Imai, M. Morimoto, H. Ishikawa, K. Hori, M. Takusagawa, K. Wakita, M. Fukuda, and G. Iwane, 'Accelerated aging test of InGaAsP/InP double-heterostructure laser diodes with single transverse mode', *Appl. Phys. Lett.*, **38**, 16–17, 1981.
36. J. J. Hsieh, 'Zn-diffused, stripe-geometry, double-heterostructure GaInAsP/InP diode lasers', *IEEE J. Quantum. Electron.*, **QE-15**, 694–697, 1979.
37. D. Renner and G. D. Henshall, 'CW and pulsed operation of GaInAsP/InP narrow stripe injection lasers', *7th IEEE Int. Semiconductor Laser Conf.*, Abstract of Paper 35, September 1980.
38. I. Sakuma, T. Furuse, Y. Matsumoto, S. Matsumoto, M. Ueno, H. Kawano, and F. Saito, 'Fundamental transverse mode InGaAsP PCW lasers' *7th IEEE Int. Semiconductor Laser Conf.*, Abstract of Paper 33, September 1980.
39. H. Ishikawa, H. Imai, K. Hori, Y. Nishitani, and M. Takusagawa, 'Separated multiclad layer stripe geometry InGaAsP/InP DH laser', *7th IEEE Int. Semiconductor Laser Conf.*, Abstract of Paper 31, September 1980.
40. D. Botez and J. C. Connolly, 'Low-threshold high-T_0 constricted double heterojunction AlGaAs lasers', *Tech. Digest Int. Electron Devices Meeting*, Washington DC, December 1980, pp. 357–361. D. Botez, J. C. Connolly, D. B. Gilbert, and M. Ettenberg, 'Very low threshold-current temperature sensitivity in constricted double heterojunction AlGaAs lasers', *J. Appl. Phys.*, **51**, 3840–3844, 1981.
41. J. J. Hsieh, 'Room-temperature operation of GaInAsP/InP double heterostructure diode lasers emitting at 1.1 μm', *Appl. Phys. Lett.*, **28**, 283–285, 1976.
42. K. Wakao, K. Moriki, T. Kambayashi, and K. Iga, 'GaInAsP/InP DH laser grown by newly designed vertical LPE furnace', *Jpn. J. Appl. Phys.*, **16**, 2073–2074, 1977. K. Wakao, T. Kambayashi, H. Morita, K. Moriki, and K. Iga, 'Characteristics of homo-isolation stripe GaInAsP/InP DH lasers', Papers Tech. Group Opt. Quantum Electron., IECE of Japan, *Paper OQE*-21, 1978 (in Japanese).
43. R. E. Nahory and M. A. Pollack, 'Threshold dependence on active-layer thickness in In GaAsP/InP DH lasers', *Electron. Lett.*, **14**, 727–729, 1978.
44. J. C. Bouley, G. Chaminant, J. Charil, P. Devoldere, and M. Gilleron, 'A simple stripe structure for GaInAsP–InP CW laser', *7th IEEE Int. Semiconductor Laser Conf.*, Abstract of Paper 28, September 1980.
45. B. I. Miller, J. H. McFee, R. J. Martin, and P. K. Tien, 'InP/$Ga_{0.47}In_{0.53}As$/InP double-heterostructure lasers grown by MBE', *Appl. Phys. Lett.*, **33**, 44–47, 1978.
46. H. Imai, H. Ishikawa, T. Tanahashi, and M. Takusagawa, 'InGaAsP/InP separated multiclad layer stripe geometry lasers emitting at 1.5 μm', *Electron. Lett.*, **17**, 17–19, 1981.
47. D. Botez, 'InGaAsP/InP double-heterostructure lasers: simple expressions for wave confinement, beam width, and threshold current over wide ranges in wavelength (1.1–1.65 μm)' *IEEE J. Quantum Electron.*, **QE-17**, 178–186, 1981.
48. R. J. Nelson, R. B. Wilson, P. D. Wright, P. A. Barnes, and N. K. Dutta, 'CW electrooptical properties of InGaAsP ($\lambda = 1.3\ \mu$m) buried-heterostructure lasers', *IEEE J. Quantum. Electron.*, **QE-17**, 202–207, 1981.
49. T. Uji, K. Iwamoto, and R. Lang, 'Nonradiative recombination in InGaAsP/InP light sources causing LED output saturation and strong laser-threshold-current temperature-sensitivity', *Appl. Phys. Lett.*, **38**, 193–195, 1981.
50. O. Wada, S. Yamakoshi, M. Abe, Y. Nishitani, and T. Sakurai, 'High radiance InGaAsP–InP lensed LEDs for optical communication systems at 1.2–1.3 μm', *IEEE J. Quantum Electron.*, **QE-17**, 1981.

51. P. D. Wright, Y. G. Chai, and G. A. Antypas, 'InGaPAs–InP double-heterojunction high-radiance LEDs', *IEEE Trans. Electron Devices*, **ED-26**, 1220–1227, 1979.
52. R. C. Goodfellow, A. C. Carter, I. Griffith, and R. R. Bradley, 'GaInAsP/InP fast high-radiance 1.05–1.3 μm wavelength LEDs with efficient lens coupling to small numerical aperture silica optical fibers', *IEEE Trans. Electron. Devices*, **ED-26**, 1215–1220, 1979.
53. K. D. Chik, A. J. Springthorpe, T. F. Devenyi, and B. A. Richardson, 'Effect of Auger recombination on double heterojunction 1.3 μm GaInAsP/InP light-emitting diodes', *Tech. Digest Int. Electron Devices Meeting*, Washington DC, December 1980, pp. 374–377.
54. P. Devoldere, G. Chaminant, J. Charil, and J. C. Bouley, 'Sub-linearity in GaInAsP/InP edge-emitting LED's,' *7th IEEE Int. Semiconductor Laser Conf.*, Abstract of Paper 49, September 1980.
55. F. D. King, A. J. Springthorpe, and O. I. Szentesi, 'High-power long-lived double heterostructure LEDs for optical communications', *Tech. Digest Int. Electron Devices Meeting*, Washinton DC, December 1975, pp. 480–483.
56. C. A. Burrus, 'Small-area high-radiance light-emitting diodes coupled to multimode optical fibers', *Topical Meeting on Integrated Optics*, Abstract of Paper WB3-1, 1972.
57. F. Stern, 'Calculated spectral dependence of gain in excited GaAs', *J. Appl. Phys.*, **47**, 5382–5386, 1976.
58. E. O. Göbell, G. Luz, and E. Schlosser, 'Optical gain spectra of InGaAsP/InP double heterostructures', *IEEE J. Quantum Electron.*, **QE-15**, 697–700, 1979.
59. W. W. Ng and P. D. Dapkus, 'Growth and characterization of 1.3 μm CW GaInAsP/InP lasers by liquid-phase epitaxy', *IEEE J. Quantum Electron.*, **QE-17**, 193–198, 1981.
60. H. Kressel and H. F. Lockwood, 'Lasing transitions in p^+–n–n^+ (AlGa)As–GaAs heterojunction lasers', *Appl. Phys. Lett.*, **20**, 175–177, 1972.
61. N. K. Dutta and R. J. Nelson, 'Temperature dependence of threshold of InGaAsP/InP double heterostructure lasers and Auger recombination', *Appl. Phys. Lett.*, **38**, 407–409, 1981.
62. T. Sanada, O. Wada, and Y. Nishitani, *28th Meeting of Appl. Phys. Japan*, March 1981, Abstracts, p. 191 (in Japanese).
63. R. C. Goodfellow, A. C. Carter, G. J. Rees, and R. Davis, 'Radiance saturation in small-area GaInAsP/InP and GaAlAs/GaAs LEDs', *IEEE Trans. Electron Devices*, **ED-28**, 365–371, 1981.
64. F. Stern, 'Gain–current relation for GaAs lasers with n-type and undoped active layers', *IEEE J. Quantum Electron.*, **QE-9**, 290–294, 1973.
65. M. Tokunaga, Y. Nakano, K. Takahei, Y. Noguchi, H. Nagai, and K. Nawata, 'Effect of cavity length on 1.55 μm buried-heterostructure DH laser characteristics', *Electron. Lett.*, **17**, 234–236, 1981.
66. J. N. Walpole, T. A. Lind, J. J. Hsieh, and J. P. Donnelly, 'Gain spectra in GaInAsP/InP proton-bombarded stripe-geometry DH lasers', *IEEE J. Quantum Electron.*, **QE-17**, 186–192, 1981.
67. R. Dingle, *Festkorper-Problems XV, Advances in Solid State Physics*, Pergamon-Vieweg, 1975, p. 21
68. R. Chin, N. Holonyak, Jr, S. W. Kirchoefer, R. M. Kolbas, and E. A. Rezek, 'Determination of the valence-band discontinuity of InP–$In_{1-x}Ga_xP_{1-z}As_z$ ($x = 0.13$, $z = 0.29$) by quantum-well luminescenece', *Appl. Phys. Lett.*, **34**, 862–864, 1979.
69. R. J. Nicholas, J. C. Portal, C. Houlbert, P. Perrier, and T. P. Pearsall, 'An experimental determination of the effective masses for $Ga_xIn_{1-x}As_yP_{1-y}$ alloys grown on InP', *Appl. Phys. Lett.*, **34**, 492–494, 1979.

70. P. J. Anthony and N. E. Schumaker, 'Ambipolar transport in double heterostructure injection lasers', *IEEE Electron Devices Lett.*, **EDL-1**, 58–59, 1980.
71. P. J. Anthony and N. E. Schumaker, 'Temperature dependence of the lasing threshold current of double heterostructure injection lasers due to drift current loss', *J. Appl. Phys.*, **51**, 5038–5040, 1980.
72. E. Oomura, K. Takahashi, T. Murotani, I. Ishii, and W. Susaki, 'InGaAsP/InP laser diode with single-mode operation', *Paper* ED79-77, Technical Group of IECE of Japan, Electron Devices, 1979 (in Japanese).
73. H. C. Casey, Jr, 'Room-temperature threshold-current dependence of GaAs–$Al_xGa_{1-x}As$ double-heterostructure lasers on x and active-layer thickness', *J. Appl. Phys.*, **49**, 3684–3692, 1978.
74. R. C. Goodfellow, A. C. Carter, I. Griffith, and S. Hersee, 'Radiance saturation in small-area GaInAsP/InP LEDs', *IEEE Trans. Electron Devices*, **ED-26**, 1843–1844, 1979.
75. M. Takeshima, 'Effect of electron–hole interaction on the Auger recombination process in a semiconductor', *J. Appl. Phys.*, **46**, 3082–3088, 1975.
76. A. Sugimura, 'Band to band Auger effect on the output power saturation in InGaAsP LEDs', *IEEE Quantum Electron. Lett.*, **QEL-17**, 441–444, 1981.
77. G. H. Olsen, F. Z. Hawrylo, D. J. Channin, D. Botez, and M. Ettenberg, 'High performance 1.3 μm InGaAsP edge-emitting LEDs', *Tech. Digest Int. Electron Devices Meeting*, Washington DC, December1980, pp. 530–533.
78. H. Okuda and Y. Sasatani, *41st Meeting of Appl. Phys. Japan*, 1980, Abstracts, p. 172 (in Japanese).
79. T. Uji, 'Deep levels in the n-$Al_{0.3}Ga_{0.7}As$ layer of (AlGa)As double-heterostructure lasers', *Jpn. J. Appl. Phys.*, **17**, 727–728, 1978.
80. M. Ettenberg and H. Kressel, 'Interfacial recombination at (AlGa)As/GaAs heterojunction structure', *J. Appl. Phys.*, **47**, 1538–1544, 1976.
81. Y. Itaya, S. Katayama, Y. Suematsu, K. Kishino, and S. Arai, 'Oscillation characteristics of low threshold (100) GaInAsP/InP DH lasers', *Paper* OQE78-114, Technical Group of IECE of Japan, Optical and Quantum Electronics, 1978 (in Japanese).
82. R. L. Hartman and R. W. Dixon, 'Reliability of DH GaAs laser at elevated temperature', *Appl. Phys. Lett.*, **26**, 239–245, 1975.
83. G. H. B. Thompson, 'Temperature dependence of threshold current in (GaIn)(AsP) DH lasers at 1.3 and 1.5 μm wavelength'), *IEE Proc. Part I, Solid State Electron Devices*, **128**, 37–43, 1981.
84. W. Nijman, P. Post, and G. A. Acket, 'Temperature-dependent photoluminescence of long wavelength GaInAsP/InP', *Gallium Arsenide and Related Compounds* 1980, Conf. Ser. no. 56, Institute of Physics, Bristol, 1981, pp. 679–688.
85. M. Hirao and M. Nakamura, 'Long wavelength semiconductor lasers for optical communications', *Oyo Butsuri*, **49**, 1207–1213, 1980 (in Japanese).
86. H. Y. Fan, *Semiconductors and Semimetals*, vol. 3, eds R. K. Willardson and C. A. Beer, Academic Press, New York, 1967, p. 409.
87. G. H. Olsen, 'Laser diodes for the 1.5 μm–2.0 μm wavelength range', *J. Opt. Commun.*, **2**, 11–19, 1981.
88. Y. Takanashi and Y. Horikoshi, 'Hot carrier study on heterostructure avalanche photodiodes', Proc. of 1981 Int. Symposium on GaAs and Related Compounds, Institute of Physics, Bristol, 1982.
89. H. Haug and H. Haken, 'Theory of noise in semiconductor laser emission', *Z. Phys.*, **204**, 262–275, 1967.
90. M. Yamada and Y. Suematsu, 'Theory of single mode injection lasers taking account

of electric intra-band relaxation', *Proc. 10th Conf. Solid State Devices*, Tokyo, 1978; *Jpn. J. Appl. Phys.*, **18**, *Suppl.* 18-1, 347–354, 1978.
91. T. Mukai, private communication (Musashino Electrical Communication Laboratories, Nippon Telegraph and Telephone Public Corporation, Musashino-shi, Tokyo, Japan).
92. P. A. Kirkby, 'Channelled-substrate narrow-stripe GaAs/GaAlAs injection lasers with extremely low threshold current', *Electron. Lett.*, **15**, 824–826, 1979.
93. K. Oe, S. Ando, and K. Sugiyama, 'Surface emitting LEDs for 1.2–1.3 μm wavelength with InGaAsP/InP double heterostructure', *Jpn. J. Appl. Phys.*, **16**, 1693–1694, 1977.
94. O. Berolo and J. C. Woolley, 'Spin–orbit splitting in III–V alloys', *Proc. 11th Int. Conf. on the Physics of Semiconductors*, July 1972, pp. 1420–1425.
95. N. K. Dutta, private communication (Bell Laboratories, Murray Hill New Jersey).
96. K. Wakita, private communication (Musashino Electrical Communication Laboratories, Nippon Telegraph and Telephone Public Corporation, Musashino-shi, Tokyo, Japan).
97. G. H. B. Thompson, G. D. Henshall, J. E. A. Whiteaway, and P. A. Kirkby, 'Narrow-beam five-layer (GaAl)As/GaAs heterostructure lasers with low threshold and high peak power', *J. Appl. Phys.*, **47**, 1501—1514, 1976.
98. W. T. Tsang, 'A new current-injection heterostructure laser: the double-barrier double-heterostructure laser', *Appl. Phys. Lett.*, **38**, 835–837, 1981.
99. J. Shah, R. F. Leheny and R. E. Nahory "Hot carrier effects in 1.3 μm $In_{1-x}Ga_xAs_yP_{1-y}$ light emitting diodes", *Appl. Phys. Lett.*, vol. **39**, pp. 618–620, Oct. 1981.

GaInAsP Alloy Semiconductors
Edited by T. P. Pearsall

Chapter 16

Photodetectors

Y. MATSUSHIMA AND K. SAKAI

KDD Research and Development Laboratories, 2-1-23 Nakameguro, Meguro-ku, Tokyo 153, Japan

16.1 INTRODUCTION

First-generation optical-fibre communication systems were composed of AlGaAs/GaAs luminescent sources at 0.8 μm wavelength and silicon detectors with well-characterized performance.[1-3] Advances in low-loss, low-dispersion silica fibres for the 1.0–1.7 μm wavelength region have stimulated the concurrent development of several light-emitting sources and photodetectors for these longer wavelengths. Germanium avalanche photodiodes (APDs) are the principal commercially available detectors in this wavelength region.[4-6] However, Ge APDs have at present high dark current and excess noise in comparison with Si APDs. A photodiode superior to the Ge APD is desirable especially for optical-fibre communication systems with long repeater spacing and large transmission capacity operating at the longer wavelengths.

Figure 16.1 shows the recent development of fibre loss[7] and the spectral response of several semiconductor photodiodes. There are two lattice-matched alloy systems which are of particular interest for optical communication in the 1.0 μm wavelength region: $Al_xGa_{1-x}As_ySb_{1-y}$ grown on GaSb, and $Ga_xIn_{1-x}As_yP_{1-y}$ grown on InP. Photodiodes made from the AlGaAsSb alloy have demonstrated high-speed response (< 100 ps) and high quantum efficiency (> 60 % at 1.3 μm) in the 1.0–1.4 μm wavelength range.[8-11] However, GaSb p–n junctions, whose wavelength spectral sensitivity (< 1.7 μm) is the longest in the AlGaAsSb alloy system, are known to have a prohibitively large dark current.[12,13] The other candidate, GaInAsP, is a quaternary alloy whose absorption edge can be varied between 0.9 μm (InP) and 1.65 μm ($Ga_{0.47}In_{0.53}As$). The general advantages of GaInAsP photodiodes over Ge photodiodes are as follows. (1) This quaternary alloy is a direct-gap semiconductor with a high absorption coefficient (~ 10^4 cm^{-1}) above the bandgap, and its dielectric constant is smaller than that of Ge. These features mean that the depletion layer in GaInAsP can be reduced to obtain the same quantum efficiency and capacitance as Ge, so that high efficiency and high-speed response can be expected. (2) The effects of surface

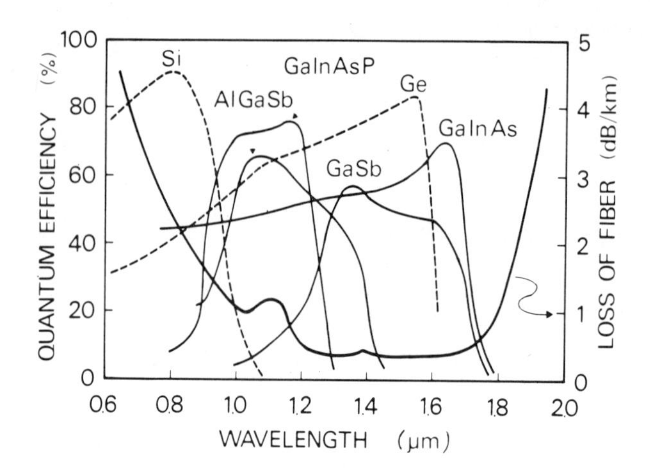

Figure 16.1 Spectral response of the following semiconductor photodiodes in the 0.6–2.0 μm wavelength region: Si, ref. 2; Ge, ref. 5; $Al_{0.42}Ga_{0.58}As_{0.02}Sb_{0.98}/Al_{0.16}Ga_{0.84}Sb$, ref. 9; $InP/Ga_{0.27}In_{0.73}As_{0.57}P_{0.43}$, ref. 26; $Ga_{0.47}In_{0.53}As$, ref. 47; $Al_{0.3}Ga_{0.7}Sb/GaSb$, ref. 13. Si and Ge photodiodes are antireflection coated. Diodes made from compound semiconductors are not coated. The recent development[7] of loss spectra of the silica fibre is also shown for reference

recombination on the quantum efficiency can be eliminated by growing a higher bandgap window layer over the absorption region, which can be accomplished easily by the epitaxial techniques of GaInAsP/InP heterostructure. (3) The ionization coefficient ratio β/α, which is an important factor especially for an APD, is not unity. It implies a lower noise performance in GaInAsP APDs. (4) High-quality GaInAsP epitaxial layers in which the dark current through the p–n junction can be significantly reduced are attainable by liquid-phase epitaxy (LPE) and vapour-phase epitaxy (VPE).

At the present time, the GaInAsP alloy seems to be a very promising material for use as a photodetector in 1.0–1.7 μm wavelength optical-fibre communication and information processing systems. In this article the current developments of the technology and performance of GaInAsP/InP photodetectors are briefly reviewed.

16.2 DESCRIPTION OF PHOTODETECTOR

16.2.1 Avalanche and p–i–n photodiodes

The avalanche photodiode (APD) and the p–i–n photodiode made from GaInAsP alloys are candidates for high-speed photodetectors used in the 1.0–1.7 μm wavelength range. The performance of the photodetector is defined by the minimum detectable power which is related to the signal-to-noise ratio. Comparison of the sensitivity in the two types of photodetectors is necessary for the system design.[14–16] The signal-to-noise power ratio (s.n.r.) at the detector output can be estimated simply as[17]

$$\text{s.n.r.} = \frac{\frac{1}{2}(\eta eP/hv)^2 M^2}{[(2\eta e^2 P/hv) + 2eI_d]M^2 FB + (4k_B TB/R_L)} \tag{16.1}$$

where P is the mean optical power, 100% modulated, η is the quantum efficiency, e is the electronic charge, h is Planck's constant, v is the optical frequency, I_d is the dark current, M is the multiplication gain, F is the excess noise factor, B is the modulation frequency bandwidth, k_B is Boltzmann's constant, T is the absolute temperature, and R_L is the load resistance. The excess noise factor is written as

$$F = M\left[1 - (1-k)\left(\frac{M-1}{M}\right)^2\right] \tag{16.2}$$

for the electron initiated multiplication, where $k(=\beta/\alpha)$ is the ratio of the electron and hole ionization coefficients, α and β.[18,19] If $\alpha = \beta$, then F is equal to M so that the noise power increases as M^3. When the ionization coefficients are very much different, $F = 2$, and the noise power increases as M^2.[20] Therefore, materials with significantly different ionization coefficients are desirable for low-noise APDs. In the case of a p–i–n photodiode, $F = M = 1$.

In the design of digital communication systems, the required s.n.r. is determined by the bit error rate of the system, and s.n.r. is about 12 for a bit error rate of 10^{-9}, which is the standard for telecommunications. Although equation (16.1) is slightly different for a digital signal system, we can take the input optical power P_o, putting s.n.r. = 12, as a figure of merit. The calculated P_o for the wavelength of 1.55 μm is shown in Figure 16.2 as a function of bandwidth. In the calculation, $R_L = 1/BC$ is assumed, where C is the total capacitance including the junction capacitance and the stray capacitance. The other parameters are chosen by taking into account the current experimental results for the GaInAsP alloy described below. Figure 16.2 illustrates that the P_o for p–i–n photodiodes is less sensitive to dark current than that of APDs when I_d is less than 1 μA, and mainly depends on the capacitance. Therefore the reduction of the capacitance is essential for p–i–n photodiodes, as well as an improvement of η. In APDs, many parameters determine P_o, but, for a frequency range larger than 10 MHz, the receiver sensitivity of an APD is superior to that of a p–i–n photodiode, as shown in Figure 16.2.

16.2.2 Ionization coefficients of the GaInAsP alloy

The ionization coefficient ratio k is the important factor which determines the excess noise factor of an APD as seen in equation (16.2). The carrier with the larger ionization coefficient should be injected into the avalanche region to achieve low noise gain, so that k data are indispensable for the design of the structure. Although the ionization coefficients of GaInAsP alloys have not been

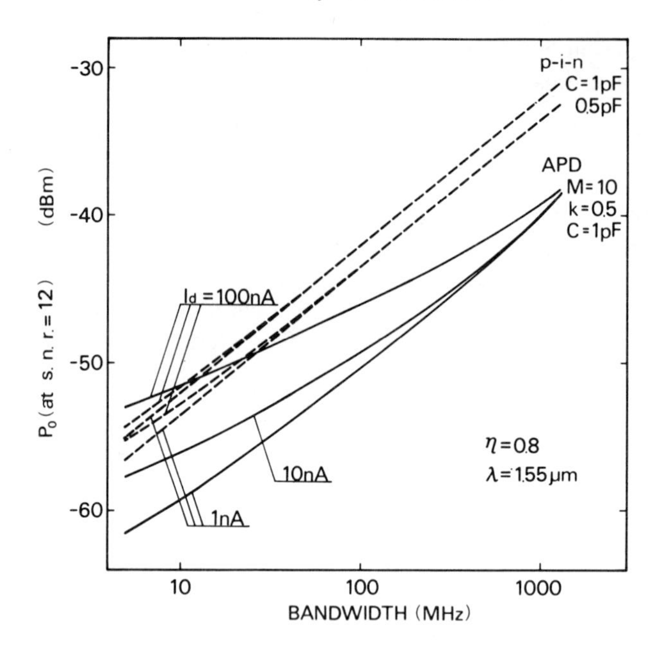

Figure 16.2 Input optical power P_0 at s.n.r. of 12, against frequency bandwidth B for two types of GaInAsP photodetector: p–i–n photodiode and APD. Supposed quantum efficiency in both diodes is 80% at 1.55 μm. Other parameters are shown in the figure

determined for the whole compositional range, some data have been published.[21–28] The ratio k at 300K for an electric field of $(2–5) \times 10^5$ V cm^{-1} in the $\langle 100 \rangle$ and $\langle 111 \rangle$ directions is plotted in Figure 16.3 as a function of arsenic composition y. From the figure the ionization coefficient ratio k of InP is greater than 2, and the values for GaInAsP are estimated to be less than 0.5 for the alloy compositions with $\lambda_g > 1.1$ μm. The ratio k is generally dependent on the electric field direction,[29,30] thus another direction may give rise to an even more desirable value of k. In the Ge APD, k is almost unity. The excess noise is expected to become smaller, therefore, in the GaInAsP APD.

16.2.3 Avalanche and Zener breakdown

The dark current not only generates a background noise but limits the maximum multiplication gain, so that low dark current is required to obtain a high-performance photodiode. The dark current comes from the surface leakage current, the generation–recombination current, the diffusion current, and the tunnelling current through the p–n junction.[31–33] In the GaInAsP system, where the bandgap energy varies from 1.35 eV (InP) to 0.75 eV ($Ga_{0.47}In_{0.53}As$), the tunnelling current cannot be disregarded, and at high donor concentration Zener

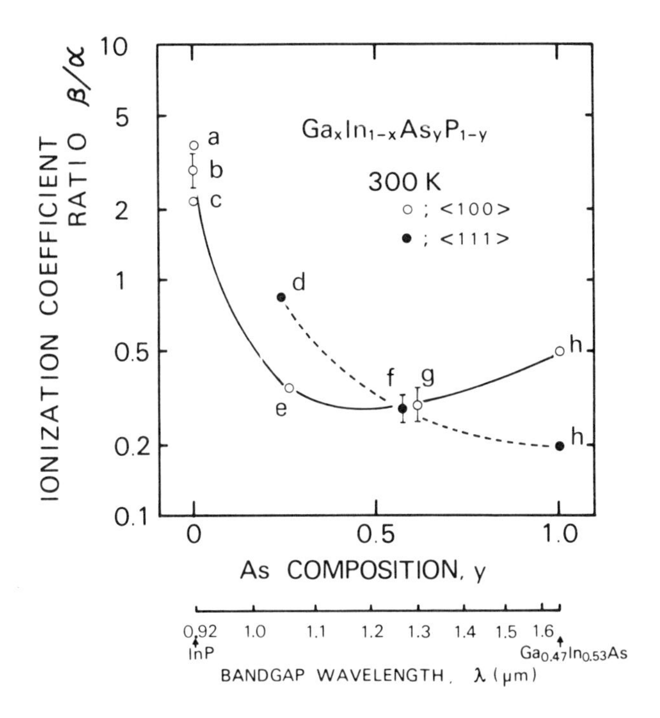

Figure 16.3 Experimentally determined ionization coefficient ratio β/α with the electric field oriented in the ⟨100⟩ and ⟨111⟩ directions as a function of arsenic composition y in $Ga_xIn_{1-x}As_yP_{1-y}$ alloy. The data indicated by a–h correspond to the following references: a, refs 21 and 28; b, ref. 22; c, ref. 23; d, ref. 24; e, ref. 25; f, ref. 26; g, estimated from low-temperature data in ref. 22; h, ref. 27

breakdown takes place before the avalanche breakdown occurs.[34–37] Takanashi *et al.* investigated Zener breakdown characteristic in GaInAsP alloy in detail.[34] They measured the reverse bias voltage V_R at the current density of 10^{-2} A cm^{-2} for various temperatures. Generally the temperature coefficient β_o is positive for avalanche breakdown, where β_o is defined by

$$V_R = V_R(T_o)[1 - \beta_o(T - T_o)], \tag{16.3}$$

whereas β_o is negative for Zener breakdown. The transition from Zener breakdown to avalanche breakdown with the decrease of the donor concentration has been clearly demonstrated by the measurement of β_o in $Ga_{0.11}In_{0.89}As_{0.26}P_{0.74}$. The approximate donor concentration for the avalanche–Zener transition is calculated and illustrated in Figure 16.4 for GaInAsP alloys.

As the obtained donor concentration of GaInAsP grown by liquid-phase epitaxy is usually of the order of 10^{16} cm^{-3}, it is very difficult to make a high-gain homojunction GaInAsP APD with a bandgap wavelength larger than 1.3 μm. The heterostructure avalanche photodiode (HAPD), in which light absorption

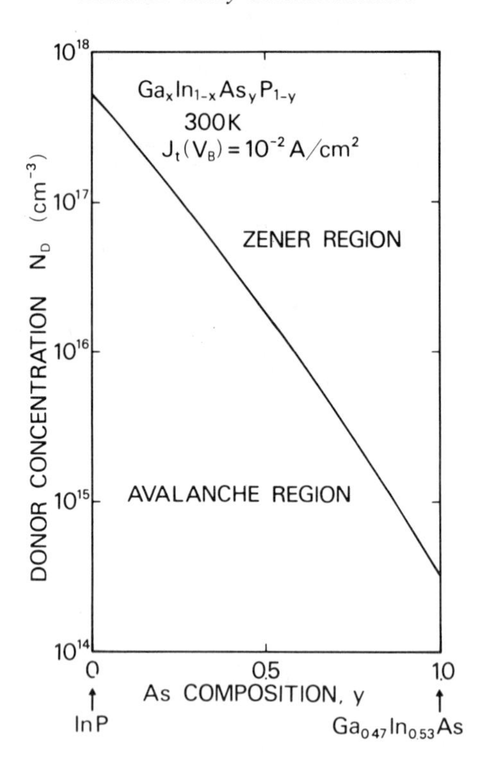

Figure 16.4 Calculated donor concentration for the avalanche–Zener transition as a function of arsenic composition *y*. The current density at breakdown is assumed to be 10^{-2} A cm^{-2}, and calculation process is the same as described by Takanashi *et al*[34]

region and avalanche region are composed of different materials, is a promising approach to a high-performance APD with high multiplication gain and low dark current, as described in section 16.4.[38–40] This heterostructure is also applicable to a p–i–n photodiode with low dark current. These heterostructure photodiodes are now under intensive development.

16.3. DEVICE FABRICATION AND TECHNOLOGY

16.3.1 Basic structure

The basic structures of photodiodes are roughly separated into two types; one is a mesa structure and the other is a planar structure. The majority of GaInAsP photodetectors previously reported have been mesa structures, since the fabrication technology of the mesa type was considered to be somewhat simpler than that of planar, and suitable for checking the principal characteristics.[41–43] A planar-type diode has advantages over a mesa type with respect to reliability and reproducibility, even though the fabrication techniques become complicated due to the need for surface passivation and a guard-ring structure.[44-46]

16.3.2 Junction formation

In the early stage of GaInAsP mesa-type photodiode development, p–n homojunctions were formed by several methods, such as successively grown p–n layers,[47] the diffusion of Zn from a heavily doped window layer or substrate,[35,48,49] Be ion implantation,[50,51] and Zn or Cd diffusion.[52,53] The schematic cross-sectional views of these mesa-type photodiodes are illustrated in Figure 16.5. Hurwitz *et al.* reported that GaInAsP APD fabricated from successively grown homojunction wafers and heterojunction wafers did not exhibit avalanche gain.[43] Takanashi *et al.* suggested that the high dark current of the GaInAsP APD with grown junction was due to the existence of carrier generation centres in the heterointerface.[48] From these points of view, the interface of the grown junction seems to be undesirable for an APD.

The Zn and Cd diffusion techniques now seem to be effective in forming well-controlled p–n junctions in GaInAsP photodiodes. Figure 16.6 shows the relationship between the junction depth and the square root of the diffusion time for (a) Zn diffusion, and (b) Cd diffusion into InP and GaInAs, respectively.[52,54,55] Although both Zn and Cd provide a steep diffusion front suitable for the formation of abrupt junctions, Cd diffusion is slower and thus easier to control. As for the characteristics of InP p–n junctions, Cd diffusion was reported

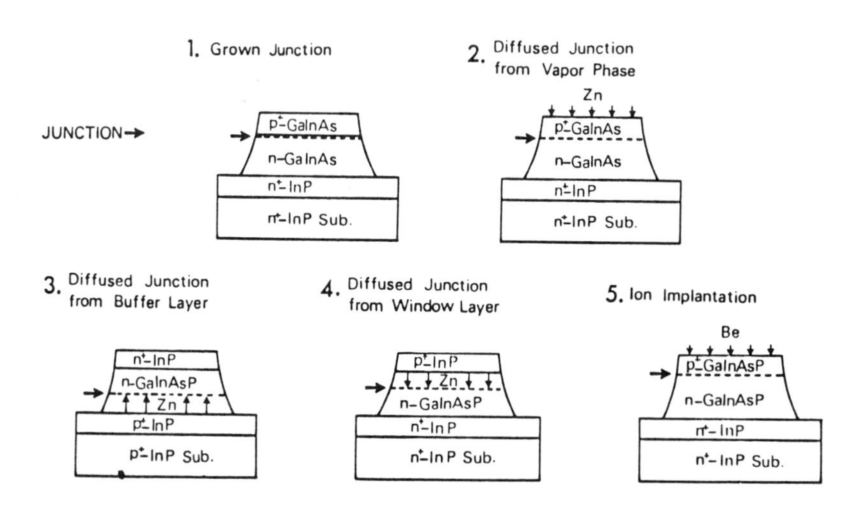

Figure 16.5 Schematic views of reported mesa-type photodiodes in which the p–n junction is formed by different methods

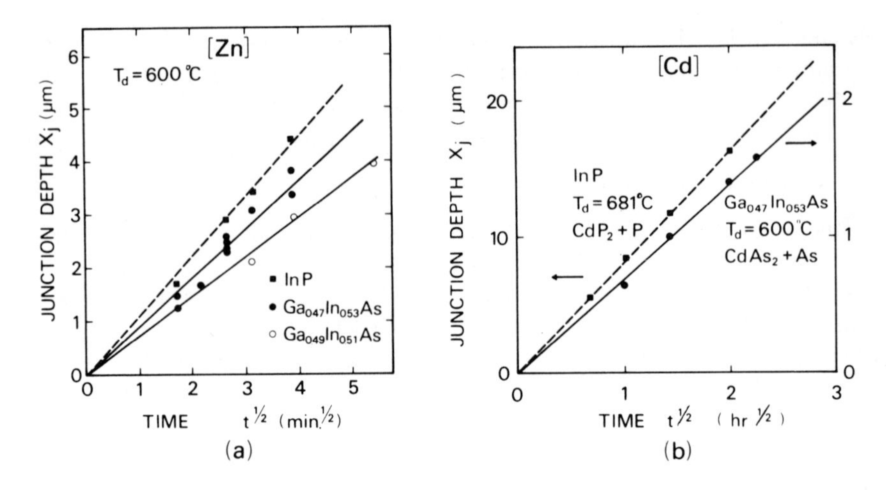

Figure 16.6 The relation between the junction depth x_j and the square root of diffusion time for (a) Zn diffusion and (b) Cd diffusion into InP and GaInAs. In (a) the plots of ■ and ● are from ref. 52 and those of ○ are from ref. 54. respectively. In (b) ● is from ref. 56 and ■ from ref. 55, respectively

to be superior to Zn diffusion.[53] The source materials used for Zn or Cd diffusion were $ZnAs_2$, ZnP_2 or $CdAs_2$, CdP_2, respectively. The diffusion temperature adopted was usually in the range 500–600° C, which was lower than the crystal growth temperature and higher than the electrode alloying temperature. In order to obtain a graded impurity profile or to reduce the hole concentration in the diffused layer, a drive-in diffusion technique is effective, especially in the formation of a guard-ring in a planar structure. However, investigations of the diffusion process and technology for GaInAsP alloys are not sufficiently advanced at present to permit the fabrication of high-performance photodiodes.

In Si and Ge planar-type photodiodes, guard-ring structures have usually been employed to eliminate the possibility of surface leakage current and local breakdown at the junction periphery. Some technologies for fabricating guard-ring structures have been found in the literature, such as a dual-ion implantation technique for InP APD,[57] two-step Zn diffusion for GaInAsP APD,[31] a selective diffusion of Zn and Cd for $Ga_{0.47}In_{0.53}As$ APD,[58] and a Cd diffusion into mesa structure for InP/GaInAsP APD.[58] Typical planar diodes with guard-ring structures are illustrated in Figure 16.7. Further investigations in the process and technology of guard-ring structures seem necessary for the future development of planar photodiodes.

16.3.3 Some aspects of wafer growth

Epitaxial wafers of GaInAsP, up to now, have been almost exclusively made by LPE due to its simple apparatus and the ease of obtaining high-quality crystals. In

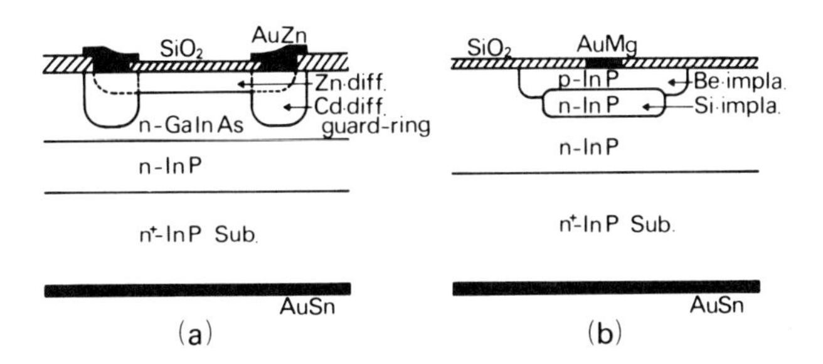

Figure 16.7 Typical planar diodes with guard-ring structures for (a) $Ga_{0.47}In_{0.53}As$[58] and (b) InP[57]

contrast, VPE has several advantages over LPE: the ease of making high-purity epitaxial layers, high controllability of layer thickness, and mass producibility.[59–61] Furthermore, in the VPE process there is no problem concerning a meltback phenomenon, which causes some trouble in the LPE overgrowth of InP on GaInAsP ($\lambda_g \gtrsim 1.4\ \mu m$). For example, to obtain a structure with InP (window layer) on $Ga_{0.47}In_{0.53}As$ is very difficult with the LPE process, and some kind of buffer layer between the two layers is necessary.[62] In contrast, it is easy to grow InP on $Ga_{0.47}In_{0.53}As$ by the VPE process, and preliminary examinations have reported an InP/$Ga_{0.47}In_{0.53}As$ photodiode by VPE.[63]

One of the serious problems in photodiodes is the presence of dislocations in the grown wafer, which cause an excess generation–recombination current and microplasmas at high electric fields. The microplasma not only gives an excess leakage current near the breakdown but also limits multiplication gain to lower values. The related phenomena with microplasma in InP and GaInAsP alloy have been investigated by use of a detailed measurement of I–V characteristics,[49,64] an observation of etch pits, and an electron-beam-induced current (EBIC) technique.[65] The microplasma density is approximately equal to the etch pit density (EPD) of the InP substrate. Supposing an EPD of $10^5\ cm^{-2}$ in the grown layer and a diode diameter of 150 μm, each diode contains about two dislocations. Therefore dislocation densities much less than $10^4\ cm^{-2}$ will probably be a prerequisite for practical long-wavelength APDs in the GaInAsP system.

The background carrier concentration N_B in the depletion layer is also an important factor in the design of a high-performance photodiode. In the GaInAsP photodiode, the p^+–n junction has usually been adopted. Thus the residual donor concentration determines the breakdown voltage, the depletion layer width, and the diode capacitance. Especially for p–i–n photodiodes, the diode must be designed with a very low capacitance and a wide depletion layer, so the donor concentration must be less than $1 \times 10^{16}\ cm^{-3}$. Such a low concentration $Ga_{0.47}In_{0.53}As$ layer has already been produced by LPE and VPE, and p–i–n photodiodes made by these wafers have shown a low capacitance.[66–68]

Figure 16.8 shows the relation between breakdown voltage and background

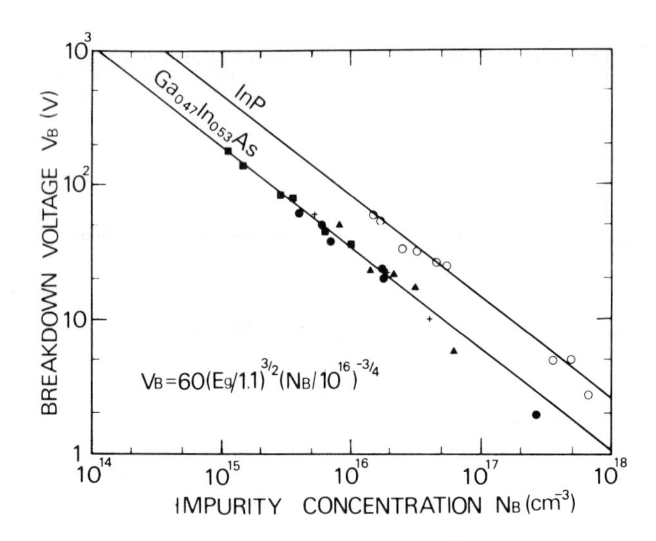

Figure 16.8 Calculated avalanche breakdown voltage (from equation (16.4)) for $Ga_{0.47}In_{0.53}As$ and InP p^+–n abrupt junctions: ○ ●, Matsushima *et al.*, unpublished; ■, Leheny *et al.*[67]; ▲, Takanashi *et al.*[34]; +, Pearsall[37]

donor concentration in InP and $Ga_{0.47}In_{0.53}As$. The full curves are calculated avalanche breakdown voltages from the semi-empirical expression of Sze,[69]

$$V_B = 60(E_g/1.1\,\text{eV})^{3/2}\,(N_B/10^{16})^{-3/4}. \tag{16.4}$$

The experimental values correspond to this expression in the lower concentration region, but in the higher region observed values fall below the theoretical curve due to the tunnelling mechanism.[37] The tunnelling current near breakdown becomes a serious problem, especially for longer-wavelength compositions such as $Ga_{0.47}In_{0.53}As$, because the tunnelling current increases with decreasing bandgap and carrier effective mass as discussed in section 16.2.3. To eliminate tunnelling current near breakdown voltage in GaInAsP APD, a heterostructure avalanche photodiode (HAPD) has been proposed, where an avalanche region is separated from a light absorption region using different materials.[70,71] An example of the HAPD is the $Ga_{0.47}In_{0.53}As$/InP structure shown in Figure 16.9(a), where a GaInAs layer is used for light absorption and InP is used in the avalanche region. Some of the HAPDs reported are also illustrated in Figure 16.9.

16.3.4 Technology of device fabrication

Chemical etching is an important process in making a mesa structure with smooth peripheral walls. The GaInAsP photodiodes are defined in the grown epitaxial wafer by photolithography, using a resin such as AZ-1350 or KPR resist,

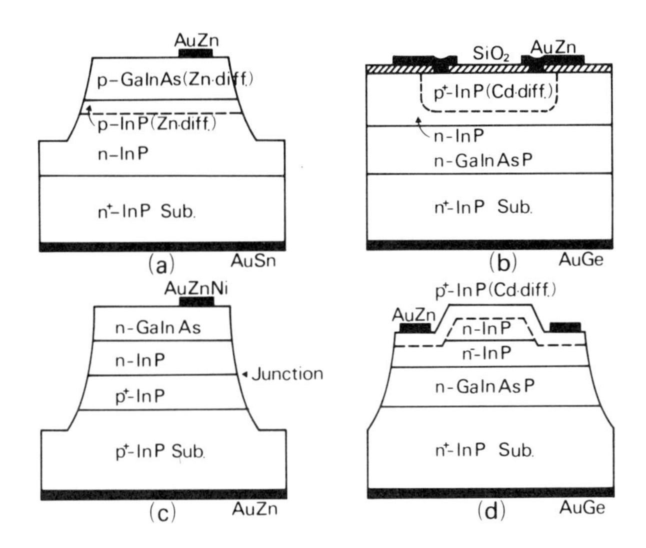

Figure 16.9 Some kinds of heterostructure avalanche photodiodes (HAPDs) in which the light absorption region is made with GaInAsP alloy and p–n junction is located in InP: (a) Matsushima *et al.*[70]; (b) Nishida *et al.*[39]; (c) Susa *et al.*[71]; (d) Osaka *et al*[58]

and chemical etching with dilute 1–3% Br-methanol or H_2O_2:H_2SO_4:H_2O to produce a mesa structure. Typical etchants used for GaInAsP/InP wafers are summarized in Table 16.1. Masks for the chemical etching process have used photoresist, SiO_2, and Si_3N_4 films. Another etching method reported is an ion milling using an argon ion beam.[72]

Surface passivations in both mesa- and planar-type diodes have been found to be effective and critical to the attainment of reproducibly low dark current. In particular, for the fabrication of planar photodiodes, the technology of surface passivation becomes of great interest. Susa *et al.* reported that the dark current of a $Ga_{0.47}In_{0.53}As$ planar diode with sputtered SiO_2 film attached, which was used

Table 16.1 Etching solution for mesa fabrication

Material	Solution	Temperature (°C)	Reference
InP	3% Br-CH_3OH:1 H_3PO_4	45	21, 74
	1–3% Br-CH_3OH	25	51
GaInAs	1 H_2O_2:8 H_2SO_4:1 H_2O	25	84, 52
	1 H_2O_2:3 H_2SO_4:1 H_2O	60	61
	6 H_2O_2:1 H_2SO_4:10 H_2O	50	47
InP/GaInAsP	1 HCl:2CH_3COOH:1 H_2O_2	15	88
	1 HCl:1 CH_3COOH:1 H_2O_2	50	91
	1–3% Br-CH_3OH	25	51, 33

for Zn-preferential diffusion, was as large as 10^{-5} A at 5 V bias. Thus they concluded that a sputtered SiO_2 film on GaInAs seemed to cause a dark current increase.[61] On the other hand, planar diodes with chemical vapour deposition (CVD) grown SiO_2 on $Ga_{0.47}In_{0.53}As$ and GaInAsP surface did not exhibit such a large increment of dark current.[44,56] As for the InP surface, the evaluation of the passivation films among sputtered SiO_2, CVD-grown SiO_2, and plasma-deposited Si_3N_4 indicated clearly that the diodes with sputtered SiO_2 caused a large dark current, whereas those with CVD SiO_2 and Si_3N_4 did not show a large leakage current.[73] From the observation of the cleaved surface after Zn diffusion using these films as a preferential diffusion mask, a diffusion wing at the periphery was found in the diode with sputtered SiO_2 film. The wing may have been induced by the Zn diffusion along the SiO_2–InP interface. A detailed characterization of surface passivation for InP and GaInAsP mesa-type p–n junctions was recently made by Diadiuk *et al.*[74] According to their results, devices consisting entirely of InP were passivated with plasma-deposited Si_3N_4, and those with a GaInAsP layer but with the p–n junction in InP were passivated with polyimide. Neither of these techniques successfully reduced dark currents in the devices with the p–n junction in the GaInAsP, but a film of photoresist sprayed with SF_6 as the propellant gave excellent results. Judging from these results, a plasma-deposited Si_3N_4 film is the most promising material for InP surface passivation. However, much work remains to be done on the GaInAsP passivation problem.

The process and performance of ohmic contacts and bonding on a photodiode do not seem to be as serious a problem as they are with light-emitting devices, because thermal conduction from laser to heat sink is an important factor for a reliable continuous-wave (CW) laser, but the photodiode is essentially a small-current device. Typical contacts onto InP or GaInAsP photodiode are usually Au/Zn, Au/Mg, Ti/Au, and Ag/Zn for the p-side, and Au/Sn, Au/Ge/Ni, and Au/Sn/Ti for the n-side. The contacts are made by alloying or plating. Diode chips are mounted on packages such as TO-18 for APD and DIL-14 lead package for p–i–n/field-effect transistor (FET) modules to reduce stray capacitance. The increase of receiver capacitance due to the packaging is small—as low as 0.4 pF.[68]

16.4 CURRENT STATE OF THE ART IN GaInAsP PHOTODETECTORS

16.4.1 Quantum efficiency

The quantum efficiency, which is the ratio of the number of electrons collected at the contacts to the number of photons in the incident light, is essentially determined by reflectivity, absorption coefficient, and carrier diffusion length. As the absorption coefficient is dependent on the incident light wavelength, the quantum efficiency has a spectral response, as shown in Figure 16.1. One would conclude from Figure 16.1 that Si and Ge both have higher quantum efficiencies than GaInAsP photodiodes, which is not the case. The Si and Ge diodes have been antireflection coated. Quantum efficiencies of $Ga_{0.47}In_{0.53}As$ structures greater than 60 % have been reported with no antireflection coatings, for example. This means that the internal quantum efficiency is nearly 100 % when the effects of surface reflection loss are taken into account. [75] Similar results, η of 50–70 %, are reported for GaInAsP alloys.[76] With an appropriate antireflection coating such as Si_3N_4 and a window layer to reduce the surface recombination, η of over 80 % is anticipated.

16.4.2 Dark current

Dark current in photodetectors limits the sensitivity, especially in APDs, as shown in Figure 16.2, and the reduction of dark current is important for the improvement of the minimum detectable power. Smaller energy bandgap causes a larger generation–recombination current, diffusion current, and also tunnelling current density. Figure 16.10 shows the dark current of $Ga_{0.47}In_{0.53}As$ APD for various temperatures.[32] The current increases suddenly above 30 V, and the voltage corresponding to the current of 10 μA decreases with increasing

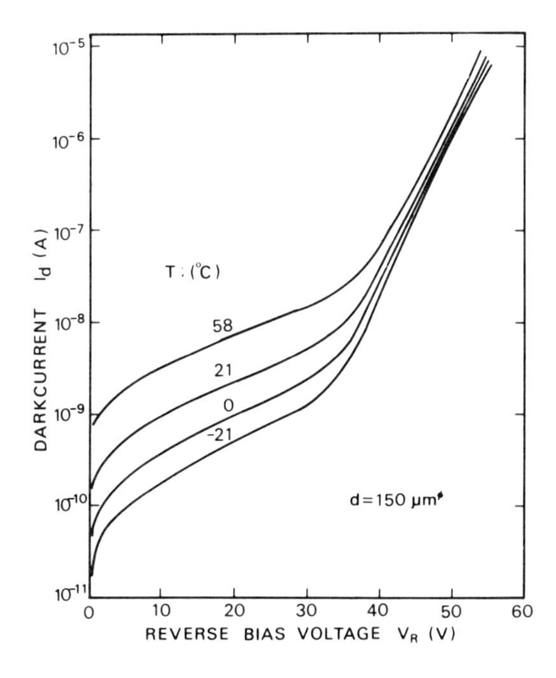

Figure 16.10 Dark current of $Ga_{0.47}In_{0.53}As$ homojunction APD with mesa diameter of 150 μm for various temperatures[32]

temperature, suggesting that tunnelling current is dominant in this voltage region, whereas the dominant dark current component at $0.5 V_B$ is the generation–recombination current.[77] The tunnelling current near breakdown cannot be reduced in small-bandgap GaInAsP, as mentioned in section 16.2.3. In the GaInAsP/InP HAPD, where the p–n junction is located in InP, the maximum electric field exists in InP, so that the tunnelling current is very small compared to the usual GaInAsP homojunction APD. The reported dark current density at $0.9 V_B$ is plotted in Figure 16.11. The dark current density of homojunction APD increases about four orders of magnitude with decreasing bandgap from 1.35 eV (InP) to 0.75 eV ($Ga_{0.47}In_{0.53}As$), whereas that of HAPD increases only one order of magnitude. Dark current densities as low as in InP photodiodes are expected in HAPDs with a suitable design.[96]

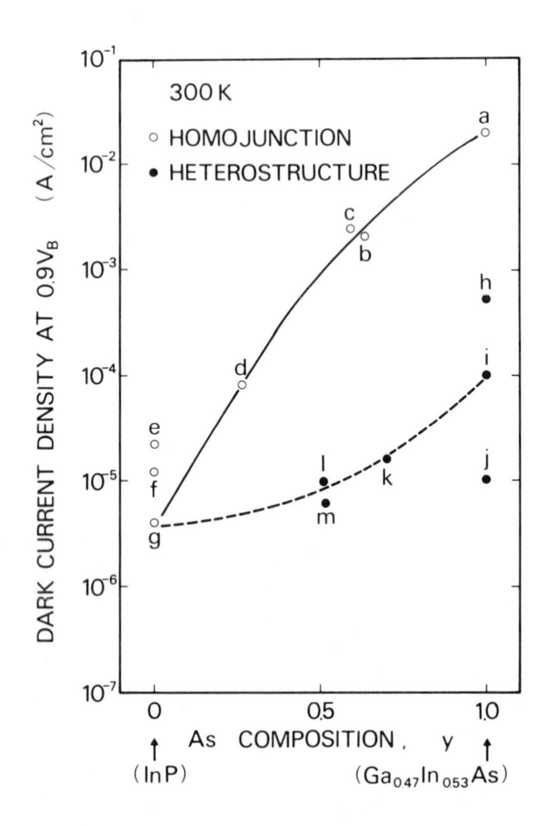

Figure 16.11 Reported dark current density at $0.9 V_B$ for homojunction and heterostructure $Ga_xIn_{1-x}As_yP_{1-y}$ APD. The data indicated by a–m correspond to the following references: a, ref. 32; b, ref. 78; c, ref. 59; d, ref. 79; e, ref. 64; f, ref. 80; g, ref. 21; h, ref. 71; i, ref. 62; j, ref. 70; k, ref. 58; 1, ref. 39; m, ref. 83

16.4.3 Response time

The high-speed response of photodiodes is characterized by a risetime and a full width at half maximum (FWHM) of the response when a narrow light pulse irradiates the photodiode. The photoresponse of GaInAsP photodiodes has been measured by many researchers, and the reported data are almost of the same magnitude, although early data have shown a relatively slow response.[81] The photoresponse of a $Ga_{0.47}In_{0.53}As$ planar APD illuminated by 1.3 μm light pulses from a GaInAsP/InP laser is shown in Figure 16.12.[46] The risetime is less than 100ps, and FWHM is 160 ps. In Table 16.2 the response times of GaInAsP APDs are summarized. For high-speed operation of APDs, an electric field punch-through structure is required. Delay time is limited by a *RC* response time, a transit time across a depletion layer, and an avalanche process. The data are affected by optical pulsewidth, diode structure, as well as material parameters, but we can anticipate faster response times than from Ge photodiodes, where the risetime and FWHM are 100 and 200 ps, respectively.[5]

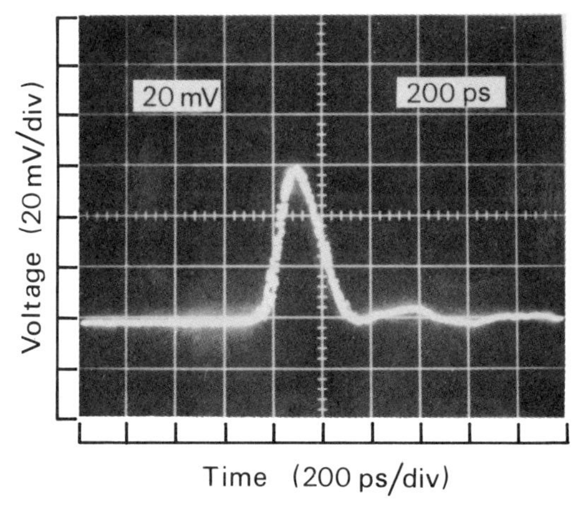

Figure 16.12 High-speed pulse response of $Ga_{0.47}In_{0.53}As$ planar APD illuminated by 1.3 μm light pulses from GaInAsP/InP laser.[46] The risetime of less than 100 ps and FWHM of 160 ps are shown

Table 16.2 Response time of $Ga_xIn_{1-x}As_yP_{1-y}$ photodiodes

As composition, y	Risetime (ps)	FWHM (ps)	Reference
0.34	60	120	50
0.52	160	–	83
0.55	700	–	81
0.58	150	–	43
1	160	200	47
1	82	126	60
1	100	160	46
1	120	225	71

16.4.4 Multiplication gain

The multiplication gain increases with the reverse voltage and shows a maximum near the breakdown voltage. The maximum value M_{max} is determined by

$$M_{max} = \left(\frac{V_B}{n R_s I_{p0}} \right)^{1/2}, \tag{16.5}$$

where R_s is the series resistance, I_{p0} is the initial photocurrent, and n is a parameter depending on the material, the doping profile, and the wavelength. To achieve large M_{max}, low dark current and small resistance, that is hard breakdown characteristics, are required. In GaInAsP alloys, the band-to-band tunnelling current becomes dominant especially when the bandgap is small. In GaInAsP homojunction APDs, it is very difficult to obtain a multiplication gain of over 50. On the other hand, a multiplication gain of 2×10^4 has been obtained in InP.[80] HAPDs, in which the p–n junction is located in InP and therefore the tunnelling current is small enough, are a way to realize a high-gain APD with a wide spectral response extending to longer wavelength. Although the HAPD structure is in a

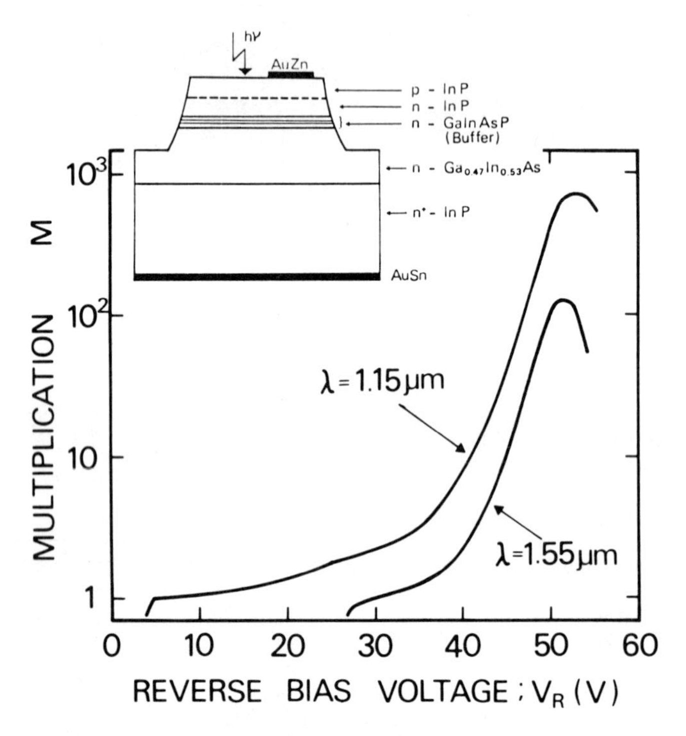

Figure 16.13 The multiplication gain versus reverse bias voltage of $Ga_{0.47}In_{0.53}As$/InP HAPD with GaInAsP buffer layers for light of wavelength 1.15 μm and 1.55 μm.[62] Schematic structure is shown in the inset

preliminary stage, M_{max} of 3000 and 16000 have been attained for $Ga_{0.23}In_{0.77}As_{0.51}P_{0.49}$[39] and $Ga_{0.47}In_{0.53}As$[70] light absorption regions, respectively. In Figure 16.13 the multiplication gain of a GaInAs/InP HAPD with GaInAsP buffer layers is illustrated, as an example.[62] In this type of HAPD, the voltage where photocurrent begins to flow depends on the wavelength, so that the multiplication gain is different for various wavelengths. This wavelength dependence on the part of M will vanish if the buffer layers are very thin or are removed. In Tables 16.3 and 16.4, the reported maximum multiplication gain is summarized.

16.4.5 Excess noise

Excess noise, inherent in APDs, is directly described by the ionization coefficient ratio k given in equation (16.2). Experiments concerning the excess noise factor F of GaInAsP APD at 300 K are very few, although low-temperature experiments have been reported.[82] In InP, $F = 23$ at a gain of 120 was reported when the electric field was applied in the $\langle 111 \rangle$ direction.[80] This value is much smaller

Table 16.3 Maximum multiplication gain of homojunction APDs

As composition, y	M_{max}	Reference
0	20000	80
0.26	200	34
0.26	86	79
0.34	12	50
0.52	2	51
0.55	42	81
0.56	20	59
0.57	5	26
0.58	12	43
0.63	100	78
1	12	47
1	5	60
1	32	52

Table 16.4 Maximum multiplication gain of heterostructure APD

As composition, y	M_{max}	Reference
0.51	3000	39
0.52	700	83
1	45	71
1	16000	70
1	900	62

than that of a Ge APD, $F = 90$, at a gain of 100.[5] For an HAPD with an InP avalanche region, $F = 3$–5 has recently been reported at $M = 10$,[83] and this value is about half that in a Ge APD at the same gain. These data are consistent with the excess noise factor and ionization coefficient ratio measured in an InP APD. A GaInAsP/InP HAPD is superior to a Ge APD also from the viewpoint of excess noise, for 1 μm wavelength region photodetectors.

16.5 ADVANCES IN INTEGRATED PHOTODETECTORS

16.5.1 Hybrid integration of p–i–n photodiodes and FETs

The combination of the photodiode and preamplifier is an important aspect in the realization of a practical optical receiver, especially for p–i–n photodiodes. Recently a $Ga_{0.47}In_{0.53}As$ p–i–n photodiode has been hybrid-integrated with a GaAs MESFET preamplifier having a high-impedance front-end design.[84] Two preamplifier designs have been investigated: and FET bipolar cascade with emitter follower buffer and an FET bipolar shunt feedback stage. In both designs the gain is sufficient to render negligible the noise from following amplifier stages. The total input capacitance achieved with the two designs is the same and consists

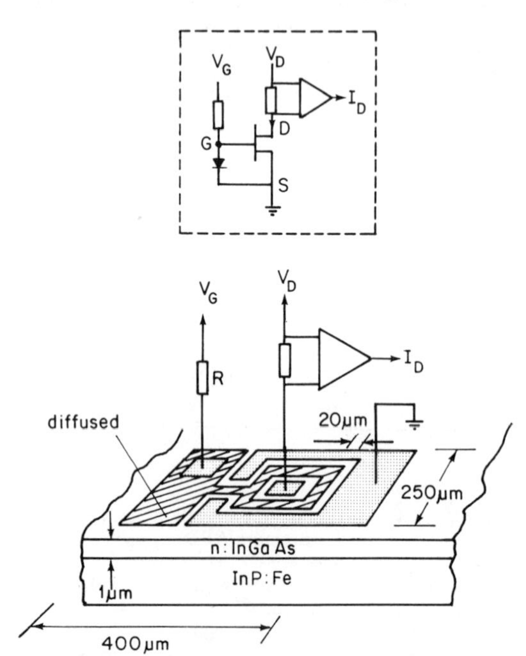

Figure 16.14 Diagram of p–i–n FET photoreceiver. A diffused gate electrode (cross-hatched area) is extended to form p–i–n diode region. Dotted region corresponds to the metallized contacts. Equivalent electrical circuit is shown in the inset above. From Leheny *et al.*[86] Reproduced by permission of the authors and IEE

of the p–i–n photodiode capacitance, FET gate-source capacitance, and stray capacitance. InGaAs/InP p–i–n photodiode-GaAs FET hybrid receivers achieved a receiver sensitivity of -38 dBm for a 280 Mbit s^{-1} NRZ system at 1.3 μm for a 10^{-9} error rate.[85]

16.5.2 Integrated photoreceiver

A preliminary study on an integrated p–i–n photodiode-FET amplifier on a single wafer of $Ga_{0.47}In_{0.53}As$ has recently been reported, as shown in Figure 16.14. The improvement of the fabrication process and diode design for an InGaAsP FET should stimulate interest in integrated optical receivers.[87]

Other designs using intergrated photodiodes include a monolithically integrated laser and monitoring detector,[88] a p–n–p–n optical detector and light-emitting diodes,[89] a multiwavelength-detection photodiode,[90] and phototransistors.[91,92] These new functional detectors will become important components in the future development of optical communication and information processing systems.

16.6 SUMMARY

One of the major motivations for the development of GaInAsP photodetectors is their use in optical-fibre telecommunications. Recent developments on the attenuation of silica fibres have shown that the wavelength which offers minimum loss extends from 1.3 μm to the 1.5–1.6 μm region, with $\alpha_{min} = 0.2$ dB km^{-1}.[93] As reviewed in this chapter, although GaInAsP alloys are very suitable materials for photodiodes in the 1.0–1.7 μm wavelength region, the choice of materials for a high-performance photodiode for 1.5–1.6 μm appears to be limited to $Ga_{0.47}In_{0.53}As/InP$. Therefore, many studies have been especially concentrated on $Ga_{0.47}In_{0.53}As$.[94] At 1.3 μm wavelength, the best overall performance is attained using $Ga_{0.27}In_{0.73}As_{0.60}P_{0.40}$ p–n photodetectors, because of their sufficiently low dark current due to their energy gap (0.9 eV versus 0.75 eV for $Ga_{0.47}In_{0.53}As$). The promising results obtained in GaInAsP photodetector performance indicate that the next appropriate step in the development of this material concerns the growth and processing of this quaternary alloy, such as a vapour-phase epitaxy and a diffusion and a surface passivation technique with high controllability. A serious drawback in GaInAsP homojunctions is the somewhat elevated dark current near breakdown, especially for the compounds with lower energy gap. This problem can be avoided by utilizing a combination of p–i–n photodiode and FET amplifier and can be overcome by the HAPD structure, where the p–n junction is formed in a higher-bandgap material such as InP. Although there are some unsolved problems, GaInAsP alloys now offer superior characteristics to Ge APDs. A GaInAs FET integrated with a GaInAsP photodiode to produce a monolithic optical receiver[95] is a particularly attractive example of the future possibilities offered by this alloy.

REFERENCES

1. J. K. Butler, *Semiconductor Injection Laser*, IEEE Press, New York, 1980.
2. R. C. Goodfellow, R. Davis, and T. P. Pearsall, 'Optical source devices' and 'Photodetectors for communication by optical fibers', *Optical Fiber Communications*, eds M. J. Howes and D. V. Morgan, John Wiley & Sons, Chichester, 1980, pp. 27–164.
3. H. Melchior, M. B. Fisher, and F. R. Arams, 'Photodetectors for optical communication systems', *Proc. IEEE*, **58**, 1466–1486, 1970.
4. H. Melchior and W. T. Lynch, 'Signal and noise response of high speed germanium avalanche photodiodes', *IEEE Trans. Electron Devices*, **ED-13**, 829–838. 1966.
5. H. Ando, H. Kanbe, T. Kimura, T. Yamaoka, and T. Kaneda, 'Characteristics of germanium avalanche photodiodes in the wavelength region of 1–1.6 μm', *IEEE J. Quantum Electron.*, **QE-14**, 804–809, 1978.
6. T. Kaneda, S. Kagawa, T. Mikawa, T. Toyama, and H. Ando, 'An n^+–n–p germanium avalanche photodiode', *Appl. Phys. Lett.*, **36**, 572–574, 1980.
7. T. Moriyama, O. Fukuda, K. Sanada, K. Inada, S. Tanaka, K. Chida, and T. Edahiro, 'Fabrication of ultra-low-OH content optical fibers with VAD method', *6th European Conf. on Optical Communication*, York, 1980, pp. 18–21.
8. H. D. Law, L. R. Tomasetta, K. Nakano, and J. S. Harris, '1.0–1.4 μm high-speed avalanche photodiodes', *Appl. Phys. Lett.*, **33**, 416–417, 1978.
9. L. R. Tomasetta, H. D. Law, R. C. Eden, I. Deyhimy, and K. Nakano, 'High sensitivity optical receivers for 1.0–1.4 μm fiber-optic systems', *IEEE J. Quantum Electron.*, **QE-14**, 800–804, 1978.
10. T. Kagawa and G. Motosugi, 'AlGaAsSb photodiodes lattice matched to GaSb', *Jpn. J. Appl. Phys.*, **18**, 1001–1002, 1979.
11. T. Kagawa and G. Motosugi, 'AlGaAsSb avalanche photodiodes for 1.0–1.3 μm wavelength region', *Jpn. J. Appl. Phys.*, **18**, 2317–2318, 1979.
12. T. Sukegawa, T. Hiraguchi, A. Tanaka, and M. Hagino, 'Highly efficient p-GaSb–n-$Ga_{1-x}Al_xSb$ photodiodes', *Appl. Phys. Lett.*, **32**, 376–378, 1978.
13. F. Capasso, M. B. Panish, S. Sumski, and P. W. Foy, 'Very high quantum efficiency GaSb mesa photodiodes between 1.3 and 1.6 μm', *Appl. Phys. Lett.*, **36**, 165–167, 1980.
14. S. D. Personik, 'Receiver design for digital fiber optic communication systems, I', *Bell. Syst. Tech. J.*, **52**, 843–874, 1973.
15. S. Hata, K. Kajiyama, and Y. Mizushima, 'Performance of p–i–n photodiode compared with avalanche photodiode in the longer-wavelength region of 1 to 2 μm', *Electron. Lett.*, **13**, 668–669, 1977.
16. D. R. Smith, R. C. Hooper, and I. Garrett, 'Receivers for optical communications: a comparison of avalanche photodiodes with PIN-FET hybrids', *Opt. Quantum Electron.*, **10**, 293–300, 1978.
17. L. K. Anderson and B. J. McMurty, 'High-speed photodetector', *Proc. IEEE*, **54**, 1339–1349, 1966.
18. R. J. McIntyre, 'Multiplication noise in uniform avalanche diodes', *IEEE Trans. Electron Devices*, **ED-13**, 164–168, 1966.
19. G. E. Stillmann and C. M. Wolfe, *Semiconductors and Semimetals*, vol. 12, ed. R. K. Willardson and A. C. Beer, Academic Press, New York, 1977, pp. 291–393.
20. R. J. McIntyre, 'Distribution of gains in uniformly multiplying avalanche photodiodes', *IEEE Trans. Electron. Devices*, **ED-19**, 569–580, 1967.
21. C. A. Armiento, S. H. Groves, and C. E. Hurwitz, 'Ionization coefficients of electrons and holes in InP', *Appl. Phys. Lett.*, **35**, 333–335, 1979.
22. Y. Takanashi, M. Kawashima, H. Saito, and Y. Horikoshi, 'Temperature dependence of noise performance of InGaAsP avalanche photodiode', *Paper* OQE80-13, Technical Group Meeting of IECE Japan, Optics and Quantum Electronics, 1980.

23. I. Umebu, A. N. M. M. Choudhury, and P. N. Robson, 'Ionization coefficients measured in abrupt InP junctions', *Appl. Phys. Lett.*, **36**, 302–303, 1980.
24. H. D. Law, K. Nakano, and L. R. Tomasetta, 'III–V alloy heterostructure high-speed avalanche photodiodes', *IEEE J. Quantum Electron.*, **QE-15**, 549–558, 1979.
25. Y. Takanashi and Y. Horikoshi, 'Ionization coefficient of InGaAsP/InP APD', *Jpn. J. Appl. Phys.*, **18**, 2173–2174, 1979.
26. M. Ito, T. Kaneda, K. Nakajima, Y. Toyama, T. Yamaoka, and T. Kotani, 'Impact ionisation ratio in $In_{0.73}Ga_{0.27}As_{0.57}P_{0.43}$', *Electron. Lett.*, **14**, 418–419, 1978.
27. T. P. Pearsall, 'Impact ionization rates for electrons and holes in $Ga_{0.47}In_{0.53}As$', *Appl. Phys. Lett.*, **36**, 218–220, 1980.
28. C. W. Kao and C. R. Crowell, 'Impact ionization by electrons and holes in InP', *Solid-State Electron.*, **23**, 881–891, 1980.
29. T. P. Pearsall, R. E. Nahory, and C. R. Chelikowsky, 'Orientation dependence of free-carrier impact ionization in semiconductors: GaAs', *Phys. Rev. Lett.*, **39**, 295–298, 1977.
30. T. P. Pearsall, 'Threshold energies for impact ionization by electrons and holes in InP', *Appl. Phys. Lett.*, **35**, 168–170, 1979.
31. R. Yeats and S. H. Chiao, 'Leakage current in InGaAsP avalanche photodiodes', *Appl. Phys. Lett.*, **36**, 167–170, 1980.
32. Y. Matsushima, K. Sakai, S. Akiba, and T. Yamamoto, 'Dark-current of $In_{0.53}Ga_{0.47}As$/InP mesa-type avalanche photodetector', *Jpn. J. Appl. Phys.*, **19**, 573–574, 1980.
33. F. Capasso, R. A. Logan, P. W. Foy, and S. Sumski, 'Low leakage current and saturated reverse characteristic in broad-area InGaAsP diodes', *Electron. Lett.*, **16**, 241–242, 1980.
34. Y. Takanashi, M. Kawashima, and Y. Horikoshi, 'Required donor concentration of epitaxial layers for efficient InGaAsP avalanche photodiodes', *Jpn. J. Appl. Phys.*, **19**, 693–701, 1980.
35. S. R. Forrest, M. DiDomenico, Jr, R. G. Smith, and H. J. Stocker, 'Evidence for tunnelling in reverse-biased III–V photodetector diodes', *Appl. Phys. Lett.*, **36**, 580–582, 1980.
36. H. Ando, H. Kanbe, M. Ito, and T. Kaneda, 'Tunnelling current in InGaAs and optimum design for InGaAs/InP avalanche photodiode', *Jpn. J. Appl. Phys.*, **19**, L277–L280, 1980.
37. T. P. Pearsall, 'Band-to-band tunnelling current in $Ga_{0.47}In_{0.53}As$ p–n junctions', *Electron. Lett.*, **16**, 771–773, 1980.
38. S. Akiba, K. Sakai, Y. Matsushima, and T. Yamamoto, 'A proposal to semiconductor heterostructure APD', *Natl. Conv. Rec. of IECE Japan*, 1979, p. 859.
39. K. Nishida, K. Taguchi, and Y. Matsumoto, 'InGaAsP Heterostructure avalanche photodiodes with high avalanche gain', *Appl. Phys. Lett.*, **35**, 251–253, 1979.
40. H. Kanbe, N. Susa, H. Nakagome, and H. Ando, 'InGaAs avalanche photodiode with InP p–n junction', *Electron. Lett.*, **16**, 163–165, 1980.
41. S. Sakai, M. Umeno, and Y. Amemiya, 'InGaAsP/InP double-heterostructure photodiodes', *Jpn. J. Appl. Phys.*, **17**, 1701–1702, 1978.
42. H. H. Wieder, A. R. Clawson, and G. E. McWilliams, '$In_xGa_{1-x}As_yP_{1-y}$/InP heterojunction photodiodes', *Appl. Phys. Lett.*, **31**, 468–470, 1977.
43. C. E. Hurwitz and J. J. Hsieh, 'GaInAsP/InP avalanche photodiodes', *Appl. Phys. Lett.*, **32**, 487–489, 1978.
44. K. Taguchi, Y. Matsumoto, and K. Nishida, 'InP–InGaAsP planar avalanche photodiodes with self-guard-ring effect', *Electron. Lett.*, **15**, 453–455, 1979.
45. N. Susa, Y. Yamauchi, H. Ando, and H. Kanbe, 'Planar type vapor-phase epitaxial

$In_{0.53}Ga_{0.47}As$ photodiode', *IEEE Electron Device Lett.*, **EDL-1**, 55–57, 1980.
46. Y. Matsushima, K. Sakai, S. Akiba, and T. Yamamoto, 'Planar $In_{0.53}Ga_{0.47}As$ avalanche photodiodes with guard-ring structure', *Jpn. J. Appl. Phys.*, **19**, 1441–1442, 1980.
47. T. P. Pearsall and M. Papuchon, 'The $Ga_{0.47}In_{0.53}As$ homojunction photodiode—A new avalanche photodetector in the near infrared between 1.0 and 1.6 μm', *Appl. Phys. Lett.*, **33**, 640–642, 1978; and private communication from T. P. Pearsall.
48. Y. Takanashi and Y. Horikoshi, 'InGaAsP/InP avalanche photodiode', *Jpn. J. Appl. Phys.*, **17**, 2065–2066, 1978.
49. T. P. Lee, C. A. Burrus, and A. G. Dentai, 'InGaAsP/InP photodiodes: microplasma-limited avalanche multiplication at 1–1.3 μm wavelength', *IEEE J. Quantum Electron.*, **QE-15**, 30–35, 1979.
50. H. D. Law, L. R. Tomasetta, and K. Nakano, 'Ion-implanted InGaAsP avalanche photodiode', *Appl. Phys. Lett.*, **33**, 920–922, 1978.
51. C. A. Armiento, J. P. Donnelly, and S. H. Groves, 'P–n junction diodes in InP and $In_{1-x}Ga_xAs_yP_{1-y}$ fabricated by beryllium-ion implantation', *Appl. Phys. Lett.*, **34**, 229–231, 1979.
52. Y. Matsushima, K. Sakai, S. Akiba, and T. Yamamoto, 'Zn-diffused $In_{0.53}Ga_{0.47}As$/InP avalanche photodetector', *Appl. Phys. Lett.*, **35**, 466–468, 1979.
53. Y. Takanashi and Y. Horikoshi, 'Effect of impurity diffusion on the characteristics of avalanche photodiode', *Jpn. J. Appl. Phys.*, **19**, 687–691, 1980.
54. Y. Yamamoto and H. Kanbe, 'Zn diffusion in $In_xGa_{1-x}As$ with $ZnAs_2$ source', *Jpn. J. Appl. Phys.*, **19**, 121–128, 1980.
55. P. K. Tien and B. I. Miller, 'Diffusion of Cd acceptors in InP and a diffusion theory for III–V semiconductors', *Appl. Phys. Lett.*, **34**, 701–704, 1979.
56. Y. Matsushima, K. Sakai, S. Akiba, and T. Yamamoto, 'Guard-ring structure InGaAs planar APD', *Natl. Conv. Rec. of IECE Japan*, 1980, p. 746.
57. J. P. Donnelly, C. A. Armiento, V. Diadiuk, and S. H. Groves, 'Planar guarded avalanche diodes in InP fabricated by ion implantation', *Appl. Phys. Lett.*, **35**, 74–76, 1979.
58. F. Osaka, K. Nakajima, T. Kaneda, T. Sakurai, and N. Susa, 'InP/InGaAsP avalanche photodiodes with new guard ring structure', *Electron. Lett.*, **16**, 716–717, 1980.
59. G. H. Olsen and H. Kressel, 'Vapour-grown 1.3 μm InGaAsP/InP avalanche photodiodes', *Electron. Lett.*, **15**, 141–142, 1979.
60. N. Susa, Y. Yamauchi, and H. Kanbe, 'Punch-through type InGaAs photodetector fabricated by vapor-phase epitaxy', *IEEE J. Quantum Electron.*, **QE-16**, 542–545, 1980.
61. N. Susa, Y. Yamauchi, and H. Kanbe, 'Vapor phase epitaxially grown InGaAs photodiodes', *IEEE Trans. Electron Devices*, **ED-27**, 92–98, 1980.
62. Y. Matsushima, K. Sakai, and Y. Noda, 'New type InGaAs/InP heterostructure avalanche photodiode with buffer layer', *IEEE Electron Dev. Lett.*, **EDL-2**, 179–181, 1981.
63. H. Kanbe, N. Susa, and H. Ando, 'Structure of InGaAs avalanche photodiodes', *Tech. Digest of Integrated and Guided-Wave Optics*, Nevada, 1980, pp. WD 1.1–1.4.
64. T. P. Lee and C. A. Burrus, 'Dark current and breakdown characteristics of dislocation-free InP photodiodes', *Appl. Phys. Lett.*, **36**, 587–589, 1980.
65. F. Capasso, P. M. Petroff, W. B. Bonner, and S. Sumski, 'Investigation of microplasmas in InP avalanche photodiodes', *IEEE Electron Device Lett.*, **EDL-1**, 27–29, 1980.
66. C. A. Burrus, A. G. Dentai, and T. P. Lee, 'InGaAsP p–i–n photodiodes with low dark current and small capacitance', *Electron. Lett.*, **15**, 655–657, 1979.

67. R. F. Leheny, R. E. Nahory, and M. A. Pollack, '$In_{0.53}Ga_{0.47}As$ p–i–n photodiodes for long-wavelength fiber-optic systems', *Electron. Lett.*, **15**, 713–715, 1979.
68. D. R. Smith, R. C. Hooper, K. Ahmad, D. Jenkins, A. W. Mabbitt, and R. Nicklin, 'p–i–n/F. E. T hybrid optical receiver for longer-wavelength optical communication systems', *Electron. Lett.*, **16**, 69–71, 1980.
69. S. M. Sze, *Physics of Semiconductor Devices*, John Wiley & Sons, New York, 1969, p. 114.
70. Y. Matsushima, K. Sakai, S. Akiba, and T. Yamamoto, 'High multiplication gain $In_{0.53}Ga_{0.47}As$/InP heterostructure avalanche photodiode (HAPD) fabricated by diffusion technique', *6th European Conf. on Optical Communication*, York, 1980, pp. 226–229.
71. N. Susa, H. Nakagome, O. Mikami, H. Ando, and H. Kanbe, 'New InGaAs/InP avalanche photodiode structure for the 1–1.6 μm wavelength region', *IEEE J. Quantum Electron.*, **QE-16**, 864–870, 1980.
72. A. R. Clawson, W. Y. Lum, G. E. McWilliams, and H. H. Wieder, 'Quaternary alloy $In_xGa_{1-x}As_yP_{1-y}$/InP photodetectors', *Appl. Phys. Lett.*, **32**, 549–551, 1978.
73. Y. Matsushima and K. Sakai, 'Investigation of surface passivation films on InP for APD', *Natl. Conv. Rec. of Japan, Soc. of Appl. Phys.*, **1**, p.Hg, 1981.
74. V. Diadiuk, C. A. Armiento, S. H. Groves, and C. E. Hurwitz, 'Surface passivation techniques for InP and InGaAsP p–n junction structures', *IEEE Electron Device Lett.*, **EDL-1**, 177–178, 1980.
75. T. P. Pearsall and R. W. Hopson, Jr, 'Growth and characterization of lattice-matched epitaxial films of $Ga_xIn_{1-x}As$/InP by liquid-phase epitaxy', *J. Electron. Mater.* **7**, 133–146, 1978.
76. M. A. Washington, R. E. Nahory, M. A. Pollack, and E. D. Beebe, 'High-efficiency $In_xGa_{1-x}As_yP_{1-y}$/InP photodetectors with selective wavelength response between 0.9 and 1.7 μm', *Appl. Phys. Lett.*, **33**, 854–856, 1978.
77. S. R. Forrest, R. F. Leheny, R. E. Nahory, and M. A. Pollack, '$In_{0.53}Ga_{0.47}As$ photodiodes with dark current limited by generation–recombination and tunnelling', *Appl. Phys. Lett.*, **37**, 322–325, 1980.
78. M. Feng, J. D. Oberstar, T. H. Windhorn, L. W. Cook, G. E. Stillman, and B. G. Streetman, 'Be-implanted 1.3 μm InGaAsP avalanche photodetectors', *Appl. Phys. Lett.*, **34**, 591–593, 1979.
79. Y. Takanashi and Y. Horikoshi, 'InGaAsP/InP avalanche photodiode prepared by Zn-diffusion', *Jpn. J. Appl. Phys.*, **18**, 1615–1616, 1979.
80. T. P. Lee, C. A. Burrus, A. G. Dentai, A. A. Ballman, and W. A. Bonner, 'High avalanche gain in small-area InP photodiodes', *Appl. Phys. Lett.*, **35**, 511–513, 1979.
81. R. Yeats and S. H. Chiao, 'Long-wavelength InGaAsP avalanche photodiodes', *Appl. Phys. Lett.*, **34**, 581–583, 1979.
82. Y. Takanashi and Y. Horikoshi, 'Noise performance of 1.3 μm InGaAsP avalanche photodiode at -190°C', *Jpn. J. Appl. Phys.*, **19**, L163–L166, 1980.
83. V. Diadiuk, S. H. Groves, and C. E. Hurwitz, 'Avalanche multiplication and noise characteristics of low-dark-current GaInAsP/InP avalanche photodetectors', *Appl. Phys. Lett.*, **37**, 807–810, 1980.
84. D. R. Smith, A. K. Chatterjee, M. A. Z. Rejman, D. Wake, and B. R. White, 'P–i–n F.E.T. hybrid optical receiver for 1.1–1.6 μm optical communication systems', *Electron. Lett.*, **16**, 750–751, 1980.
85. R. C. Hooper, M. A. Z. Rejman, S. T. D. Ritchie, D. R. Smith, and B. R. White, 'PIN-FET hybrid optical receivers for longer wavelength optical communication system', *6th European Conf. on Optical Communication*, York, 1980, pp. 222–225.
86. R. F. Leheny, R. E. Nahory, M. A. Pollack, A. A. Ballman, E. D. Beebe, J. C. DeWinter,

and R. J. Martin, 'Integrated $In_{0.53}Ga_{0.47}As$ p–i–n F.E.T. photoreceiver', *Electron. Lett.*, **16**, 353–355, 1980.
87. J. Barnard, H. Ohno, C. E. C. Wood, and L. F. Eastman, 'Double heterostructure $Ga_{0.47}In_{0.53}As$ MESFETs with submicron gates', *IEEE Electron Device Lett.*, **EDL-1**, 174–176, 1980.
88. K. Iga and B. I. Miller, 'GaInAsP/InP laser with monolithically integrated monitoring detector', *Electron. Lett.*, **16**, 342–343, 1980.
89. J. A. Copeland, A. G. Dentai, and T. P. Lee, 'PNPN optical detectors and light-emitting diodes', *IEEE J. Quantum Electron.*, **QE-14**, 810–813, 1978.
90. J. C. Campbell, T. P. Lee, A. G. Dentai, and C. A. Burrus, 'Dual-wavelength demultiplexing InGaAsP photodiode', *Appl. Phys. Lett.*, **34**, 401–402, 1979.
91. M. Tobe, Y. Amemiya, S. Sakai, and M. Umeno, 'High-sensitivity InGaAsP/InP phototransistors', *Appl. Phys. Lett.*, **37**, 73–75, 1980.
92. J. C. Campbell, A. G. Dentai, C. A. Burrus, and J. F. Ferguson, 'High sensitivity InP/InGaAs heterojunction phototransistor', *Electron. Lett.*, **16**, 713–714, 1980.
93. T. Miya, Y. Terunuma, T. Hosaka, and T. Miyashita, 'Ultimate low-loss single-mode fiber at 1.55 μm', *Electron. Lett.*, **15**, 106–108, 1979.
94. T. P. Pearsall, '$Ga_{0.47}In_{0.53}As$: a ternary semiconductor for photodetector applications', *IEEE J. Quantum Electron.*, **QE-16**, 709–720, 1980.
95. J. Barnard, H. Ohno, C. E. C. Wood, and L. F. Eastman, 'Integrated double heterostructure $Ga_{0.47}In_{0.53}As$ photoreceiver with automatic gain control', *IEEE Electron Device Lett.*, **EDL-2**, 7–9, 1981.
96. Thomas P. Pearsall, M. Piskovski, A. Brochet, and J. Chevrier, "A $Ga_{0.47}In_{0.53}As$/InP heterodiode with reduced dark current", *IEEE J. Quant. Electron* **QE-17**, 255–258, 1981.

Note added in proof

Since this manuscript was written, recent important developments of photodetectors have been published in *IEEE J. Quantum. Electron.*, **QE-17**, 1981, 'Special issue on quaternary compound semiconductor materials and devices—sources and detectors'.

GaInAsP Alloy Semiconductors
Edited by T. P. Pearsall

Chapter 17

Field-effect Transistors

HIDEO OHNO* AND J. BARNARD†

* *Dept. of Electrical Engineering,*
Hokkaido University, Sapporo 060, Japan
† *School of Electrical Engineering, Cornell University, Ithaca, NY 14853, USA*

17.1 INTRODUCTION

The GaInAsP quaternary alloys possess material parameters which make the alloy highly suitable for high-frequency field-effect transistors (FET), although much interest in GaInAsP quaternary alloys has been focused on optical applications, especially on infrared lasers and detectors for optical-fibre communications. In addition, advantages should arise from integrating quaternary alloy lasers and detectors with amplifiers on a chip, which make quaternary alloy FETs indispensable for optical-fibre communication applications.

This chapter describes FETs made of GaInAsP quaternary alloys and GaInAs ternary alloys, the ternary limit, together with a summary of the relations between material parameters and device performance. Details on microwave FETs are left out of the chapter, since most of the microwave FETs are currently made of GaAs which is beyond the scope of this volume. InP FETs are mentioned briefly. Readers are referred elsewhere for a review article.[1]

In section 17.2, the relations between the FET performance and material parameters are reviewed to clarify the origin of interest in the quaternary alloys from the point of view of field-effect transistor applications. This is done with the emphasis on the comparison of the potential of GaInAsP alloys with that of GaAs. Since GaAs Schottky-barrier gate FETs (metal–semiconductor FETs; MESFETs) have been the most successful among microwave FETs with different types of structures, it is natural first to try to transfer GaAs MESFET technology to the alloy FETs. The single greatest difficulty in doing so is the very low Schottky-barrier heights of the material. Thus Schottky-barrier formation is dealt with in 17.3. The last section, 17.4, deals with the fabrication and performance of quaternary and ternary FETs. The high-frequency results of the devices are mentioned wherever they are available, otherwise only DC performance is described.

17.2 FIELD-EFFECT TRANSISTOR PERFORMANCE AND MATERIAL PARAMETERS

A schematic diagram of a field-effect transistor is shown in Figure 17.1. A FET consists of two ohmic contacts: the source and drain contacts. Between the two ohmic contacts is placed a gate, which has the ability to deplete or enhance the electron concentrations in the active layer under the gate by the gate potential thus controlling the conductivity between the two ohmic contacts. The gate contact is usually a metal–semiconductor contact, a p–n junction or a metal–insulator–semiconductor contact. The electrons flow into the active layer from the source contact, controlled by the gate, and are then taken out from the drain contact. Semi-insulating substrates are commonly used for isolation of the active layer from the substrate. As the lattice-matching of the active layer to the substrate material is a crucial requirement for high-performance quaternary FETs, the GaInAsP alloys lattice-matched to InP substrates are dealt with throughout the chapter except where otherwise stated (for the lattice-matching condition see Chapter 2).

The relations between material parameters and their effects on the FET performance are summarized in Table 17.1.

17.2.1 Transit time

For high-frequency operation the electron transit time under the gate, where the control of currents takes place, should be as short as possible. Thus velocities of electrons passing under the gate should be high. Depending on the electric field under the gate, which depends on the dimensions of the device as well as the operating conditions, three factors become important to determine the response. For relatively low electric field, the low-field electron mobility is the limiting

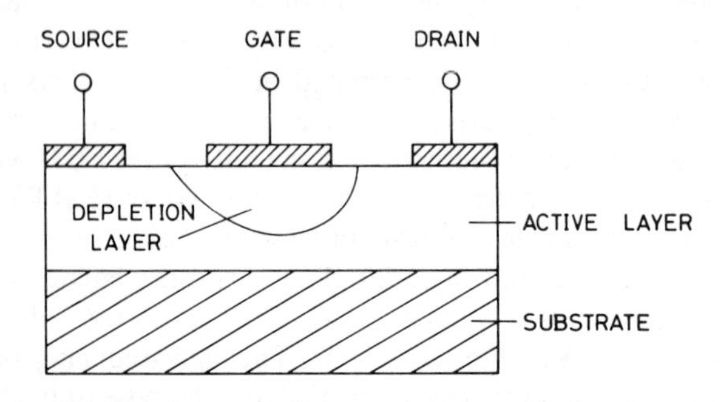

Figure 17.1 A schematic diagram of a field-effect transistor. Electrons flow from source to drain and are controlled by the depletion region under the gate. For an inversion-type FET, see Figure 17.4

Table 17.1 Material parameters and FET performance

High low-field mobility	Short transit time Small parasitic resistance
High peak velocity	Short transit time (depends on dimensions and operation conditions)
High saturation velocity	Short transit time (depends on dimensions and operation conditions)
Small effective mass	High low-field mobility Enhances overshoot
Small scattering rate	High low-field mobility Enhances overshoot
Presence of alloy scattering	Reduces low-field mobility Reduces overshoot May increase stability
Large Γ–L separation	Enhances overshoot Reduces excess noise
Large bandgap	Ensures room-temperature operation Reduces breakdown
Semi-insulating substrate	Reduces parasitic capacitance

factor.[2] This situation corresponds to a logic application of FETs where a small logic voltage swing is preferred to achieve a faster operation and a smaller power dissipation. A high electron mobility also reduces parasitic resistances such as source-gate resistance. The peak electron velocity together with the low-field electron mobility dominate the transit time in a medium strength field. In the high electric field region, high saturation velocities become increasingly important with high-field diffusion constants for a faster operation of FETs.[3] The operation conditions of microwave FETs are commonly in relatively high electric fields, making the entire shape and the absolute value of the velocity–field characteristics highly significant. Details of the velocity–field characteristics for quaternary alloys have already been reviewed in the earlier chapters of this volume.

Comparing GaAs and GaInAsP from the velocity–field characteristics point of view, two statements can be made:

(1) Low-field mobilities are similar to GaAs for the alloys with $E_g \sim 0.9$ eV and are more than 50% higher at the ternary limit, $Ga_{0.47}In_{0.53}As$ ($E_g \sim 0.75$ eV) than GaAs for the same doping levels.[4–8]

(2) Peak velocities of alloys are higher than GaAs near the two limits of the quaternary alloy, InP and $Ga_{0.47}In_{0.53}As$.[6]

Calculations made by Littlejohn *et al.*[9] showed the highest peak velocity at $y = 0.6$ in $Ga_xIn_{1-x}As_yP_{1-y}$. Experimental results, however, indicate that alloy scattering is dominant in the mid-range of the alloy composition, making the low-field mobility and the peak velocity lower in this range.[4–6, 10] Although the GaInAs ternary alloy suffers from alloy scattering, the highest mobility among the GaInAsP alloy, $13800 \, cm^2 \, V^{-1} \, s^{-1}$ at room temperature, has been obtained in GaInAs by Oliver and Eastman.[7] Peak velocities are also high ($\gtrsim 2.2 \times 10^7 \, cm \, s^{-1}$) for the ternary alloy, suggesting the effect of alloy scattering to be small at this limit.[6, 11, 12] The highest low-field electron mobilities among GaInAsP alloys and high peak velocities comparable to InP make GaInAs particularly attractive for FET applications. No measurements have been made on the saturation velocities and high-field non-isotropic diffusion constants of the quaternary alloys.

17.2.2 Transient effects

Since the frequency response of a FET is determined by the transit time of electrons passing under the gate, the gate length has to be reduced to realize a high-frequency operation of the device. In the limit of such small dimensions that the electrons cannot reach an equilibrium state, the transient phenomena may become one of the governing factors determining the frequency response. The transient phenomena can be divided into two regimes, which are called velocity overshoot[13] and ballistic electron motion.[14] The difference between the two is that while velocity overshoot is mainly due to the finite energy relaxation time, ballistic electron motion occurs in the collision-free region and is thus limited by the band structure of the material. For normal FETs having dopings around $10^{17} \, cm^{-3}$, collisions occur too frequently for electrons to move ballistically. Therefore only overshoot phenomena are discussed here. Results of Monte Carlo calculations give insights to what is happening under the short gate in the overshoot regime.

(1) The overshoot depends strongly on the electron effective mass m^* and the degree of alloy scattering.[15] The effect of alloy scattering on overshoot phenomena is rather weak when the electron effective mass is relatively small ($0.041 \, m_0$ in the case of reference 15).

(2) Materials with a high low-field electron mobility and a large energy minima difference $\Delta E_{\Gamma L}$ show a higher value of overshoot velocity.[16]

GaAs has a relatively small Γ–L energy minima difference, 0.3 eV, compared to the quaternary alloys. While $\Delta E_{\Gamma L}$ is not known precisely for the alloys, (see note added in proof, Ref. 53) it is estimated to be around 0.6 eV, more or less independent of the lattice-matched alloy composition. The electron effective mass is smaller for GaInAsP with $y \gtrsim 0.2$ than for GaAs ($0.068 \, m_0$) and the lowest value is observed in GaInAs ($0.041 \, m_0$).[17] While GaInAsP alloys suffer from alloy scattering, which is absent in GaAs, the small electron effective masses may make the contribution of alloy scattering insignificant as far as overshoot phenomena are concerned.[15] Among the quaternary alloys, GaInAs shows the highest mobility, has the smallest electron effective mass, and a large Γ–L energy minima difference compared to GaAs, and has been reported to suffer less from alloy scattering.[6] These facts together make the ternary FETs more capable of utilizing the overshoot phenomena than the quaternary alloys and even than GaAs.

17.2.3 Other parameters

Other parameters which influence the FET performance are as follows.

(1) The bandgap of the material: the bandgap should be large enough to ensure the room-temperature operation of the devices, i.e. the intrinsic carriers concentration n_i must be low enough not to disturb the extrinsic carrier concentration. A relatively wide bandgap is also important in preventing the device from various breakdown phenomena such as avalanche breakdown and Zener breakdown.

(2) The substrate material: in order to suppress parasitic capacitances, the substrate has to be either semi-insulating or insulating. Hence the quaternary alloy FETs are made on semi-insulating InP substrates.

(3) Low-noise capability of the material: the capability depends largely on the material parameters.[18] For example, phonon scattering rates which depend on the material are sources of noise, and another example is that materials having smaller energy band minima difference may suffer more from the inter-valley scattering noise.[19]

(4) Alloy scattering: this is one of the causes of the mobility reduction, and makes the mobility less temperature-dependent, and thereby may increase the thermal stability of the devices.[20]

17.3 SCHOTTKY BARRIER

17.3.1 Normal Schottky Barrier

For microwave FETs and logic FETs a large variety of structures (MESFET; JFET, junction gate FET; and IGFET, insulated gate FET) can be employed, depending on the structure of the gate. The MESFET approach has been the most successful for GaAs FETs owing to its simplicity of structure and fabrication and its high performance in the high-frequency region both in low-noise FETs and power FETs. The major obstacle for transferring the GaAs MESFET technology directly to GaInAsP alloys is the low Schottky-barrier height. As Greene *et al.*,[4] Leheny *et al.*,[5] and Marsh *et al.*[6] showed experimentally, the GaInAsP alloys have a higher mobility with decreasing energy bandgap. The highest mobility in the alloy system is obtained in GaInAs ($E_g \sim 0.75$ eV). On the other hand, it has been recognized that the barrier heights ϕ_b of the alloys are rather small compared to that of GaAs ($\phi_b = 0.8$ eV) ranging from $\phi_b = 0.5$ eV for InP to $\phi_b = 0.2$ eV for GaInAs at room temperature. This makes a conventional MESFET structure difficult because of the large leakage current, which increases both the drain output conductance and the pinch-off drain

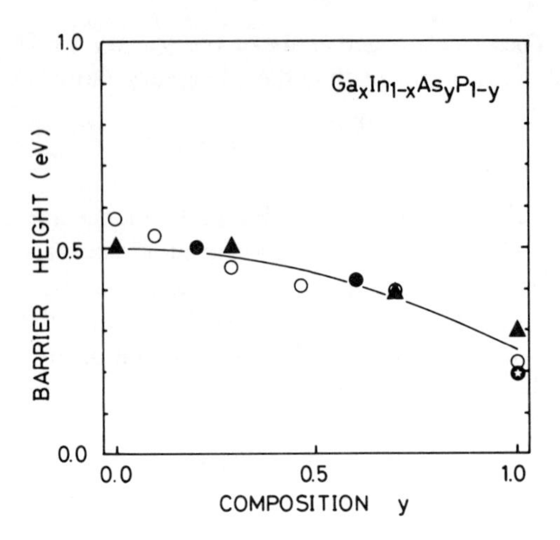

Figure 17.2 Schottky-barrier height versus alloy composition. Data are as follows: ○, ref. 21; ▲, ref. 22; ●, ref. 23; ○, ref. 24. The full curve represents $\phi_b = -0.248y^2 + 0.506$

current. This is serious especially at the small-bandgap end where the highest mobility has been obtained. Schottky-barrier heights without intentional interfacial layers are shown in Figure 17.2 versus the alloy composition.[21–23] The measured Schottky-barrier height shows a monotonic decrease with decreasing bandgap. Results on Au/p-GaInAsP by Escher *et al.*[24] were included in the figure assuming that the barrier heights for p- and n-type alloys sum up to the bandgap.

17.3.2 Quasi-Schottky barrier

To overcome difficulties associated with the low Schottky-barrier heights, various structures have been proposed and investigated, some of them with device results, some of them without. Most of the structures have a thin interfacial oxide layer between the metal and the semiconductor, essentially an MIS structure, to increase the effective barrier heights. Thus the barrier is called a quasi-Schottky barrier. The linear dependence of the effective barrier height on the thickness of the oxide layer was interpreted using the tunnelling model by Morgan *et al.*[25] The relation between the effective barrier height ϕ_b^* and the barrier height ϕ_b is,

$$\phi_b^* = \phi_b + Dd \tag{17.1}$$

using standard tunnelling theory, where D is a constant at a constant temperature and d is the thickness of the oxide layer. However, a different model exists for the role of the interfacial oxide layer on the effective barrier height.[26] Models should be able to explain both the linear dependence of ϕ_b^* on d and the temperature dependence of the barrier height. There is still some uncertainty in both models.

Table 17.2 Effective Schottky-barrier heights on GaInAsP

	E_g(eV)	ϕ_b^*(eV)	n	Metal	Interfacial layer†
GaInAsP[28]	0.99–1.1	0.57	1.04	Au	70 Å Al_2O_3 With SiO_2, similar results MgO also tried
GaInAsP[23]	0.77	0.51	1.14	Au	Oxidization in HNO_3
	0.95	0.72	2.0	Au	In air at room temperature for 24 h
	0.96	0.50	1.17	Au	Oxidization in HNO_3
	1.20	0.65	1.18	Au	In air at room temperature for 24 h
	1.20	0.70	1.19	Au	In air at room temperature for 24 h
GaInAsP[22]	0.9	0.6		Au, Al	SiO_2
GaInAs[22]	0.75	0.5		Au, Al	SiO_2
GaInAs[27]	0.75	0.49		Au	100 Å SiO_x

† For Schottky-barrier heights without interfacial layers, see Figure 17.1.
E_g = bandgap of the alloy.
ϕ_b^* = effective barrier height.
n = ideality factor of the Schottky barrier.

The experimental results on quasi-Schottky barriers are summarized in Table 17.2.[22, 23, 27, 28] The device results are discussed in the next section.

17.3.3 Schottky heterobarrier

An alternative approach to quasi-Schottky barriers is a Schottky heterostructure in which a thin insulating layer is replaced by a semiconductor layer having a higher barrier height. A similar structure was first applied to GaAs FETs with GaAlAs layers between the metal and GaAs.[29] For GaInAsP alloys, InP was first proposed for the intermediary layer.[30] This is because InP has the highest Schottky-barrier height in the alloy system and also because the alloys are lattice-matched to InP so that no interfacial traps due to a large lattice mismatch are expected. To realize a good barrier with reasonable rectification, the structure would have to be metal/p^+-InP/n-GaInAsP, since InP itself has a relatively low Schottky-barrier height even though it is the highest for the alloy system. To date, no experimental results have been reported on the structure using InP.

The Schottky heterobarrier was realized by the use of AlInAs for GaInAs.[31] $Al_{1-x}In_xAs$ can be lattice matched to InP at $x = 0.52$ with 1.46 eV direct bandgap at room temperature.[32] As the GaAs–AlAs system shows a trend towards higher Schottky-barrier height and smaller electron affinity with increasing AlAs mole fraction in the direct-bandgap region,[33] $Al_{0.48}In_{0.52}As$ shows considerably higher barrier height than that of $Ga_{0.47}In_{0.53}As$. The diffusion potential ϕ_d of aluminium metal deposited onto molecular-beam epitaxy (MBE) grown AlInAs measured by C–V method was $\phi_d = 0.80$ eV. The AlInAs was lattice matched to InP doped with germanium to 5.7×10^{17} cm^{-3}. Since aluminium was deposited *in vacuo* after growth, there was no interfacial

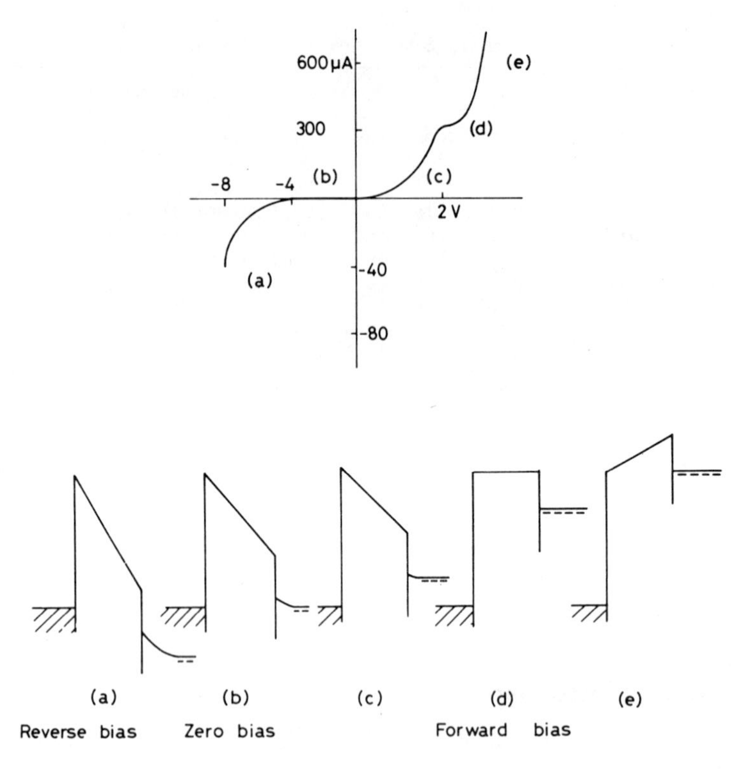

Figure 17.3 An I–V characteristic of a Schottky heterostrwcture with associated band diagrams. The structure investigated here is Al/AlInAs/GaInAs/AlInAs/InP. The terrace in the I–V characteristic is an indication of the presence of such Schottky heterostructure. The AlInAs layer was 1600 Å thick including 1000 Å buffer layer on n^+-InP substrate

layer between the metal and the semiconductor. The I–V characteristics of the Schottky heterostructures are shown in Figure 17.3 with associated band diagrams. The forward I–V characteristic shows a terrace which is an indication of the presence of the heterojunction right after the Schottky contact as shown in the diagram.[34] The effectiveness of the structure in reducing the reverse bias current is clear and devices using this structure were made, which will be discussed later.

17.4 FIELD-EFFECT TRANSISTORS

17.4.1 JFET

Conventional junction FETs were fabricated by Leheny *et al.*[35] The junction was formed by diffusion for 70 s at 750°C in a $ZnAs_2$ atmosphere. The device with approximately 20 μm gate length and 1 mm gate width showed a DC transconductance $g_m = 1\,\text{mS}\,\text{mm}^{-1}$ of gate width. The low DC transconductance reported here is due to immature processing technology and non-optimized device structures rather than that the material potential is poorer than expected. Using JFETs, Leheny *et al.* have integrated GaInAs p–i–n detectors with FETs which is an alternative to avalanche photodetectors (APD) for the front-end design of the optical-fibre communication system.[36] No microwave result has been reported so far for JFETs.

JFET is of potential interest because of its high built-in potential, which not only makes normally-off FETs easier to fabricate but also reduces the leakage current of the gate. The problem associated with JFETs is the way to form the junctions. The diffusion process is rather difficult to control especially when the gate length becomes submicrometre in dimension. It has been demonstrated that beryllium implantation could be done for p-layer formation in GaInAsP.[37] No results on the application of ion implantation to JFETs have been reported.

17.4.2 MISFET

Shinoda *et al.* succeeded in fabricating *n*-channel inversion-mode GaInAsP MISFETs.[38] The p-GaInAsP layer ($E_g \sim 1.2\,\text{eV}$) was doped with Cd to $5 \times 10^{17}\,\text{cm}^{-3}$. Al_2O_3 was used as the gate insulator. A cross-sectional view of the device is shown in Figure 17.4. The device with $9\,\mu\text{m} \times 100\,\mu\text{m}$ gate dimensions showed $2600\,\text{cm}^2\,\text{V}^{-1}\,\text{s}^{-1}$ field-effect mobility, which is much higher than that of InP MISFETs and close to the bulk mobility, suggesting the interface state density to be lower in quaternary alloy–oxide systems than in InP–oxide. This MISFET approach is particularly interesting for an application of quaternary alloys to high-speed low-power integrated circuits. Improvements in the threshold voltage drift similar to that of InP MISFETs, due partly to the quality of the gate insulator and partly to the complex nature of oxide–quaternary semiconductor interface, have to be made for stable and reliable operation of quaternary MISFETs.

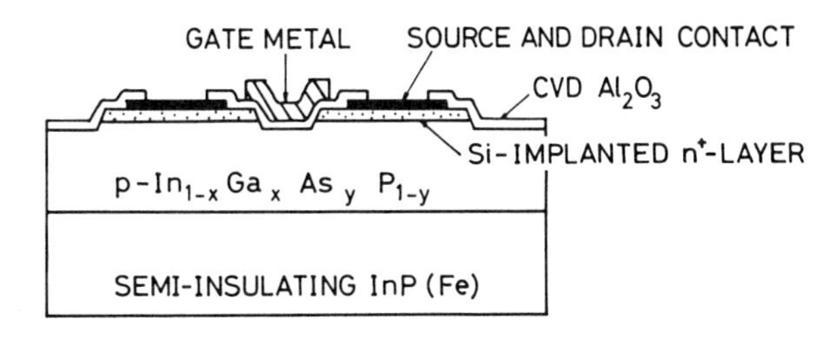

Figure 17.4 A schematic diagram of an *n*-channel inversion-mode MISFET reported by Shinoda *et al.*[38] The gate insulator was 1000 Å thick Al_2O_3. Reproduced by permission of *Japanese Journal of Applied Physics*

17.4.3 MESFET

Decker *et al.* fabricated GaInAs MESFETs with the InAs mole fraction of 0.03–0.045.[39] The layers were grown on semi-insulating GaAs, which is not a lattice-matched system. Since the composition of the layers was close to GaAs, Schottky barriers were high enough for MESFET gates. These devices showed no superior performance to that of similar GaAs MESFETs except that the drain conductance was somewhat reduced, which was attributed to the conduction band discontinuity between the active GaInAs layer and the GaAs substrate.

In spite of the difficulty of low Schottky-barrier heights on GaInAsP alloys lattice-matched to InP substrates, GaInAsP MESFETs were made with the alloys having E_g of 1.2 eV and 1.15 eV. Morkoç *et al.* have reported MESFETs having $1\,\mu m \times 200\,\mu m$ gate dimensions with $5\,\mu m$ channel length.[40] The device with 1.2 eV bandgap GaInAsP showed 8 mS DC transconductance which is $40\,mS\,mm^{-1}$. The device with 1.15 eV GaInAsP showed an improved DC transconductance, 10 mS ($50\,mS\,mm^{-1}$). Microwave measurements indicated the maximum available gain (MAG) of 5 dB for 1.2 eV GaInAsP and 7 dB for 1.15 eV GaInAsP at 10 GHz. Au–Ge–Ni alloy contacts were used for ohmic contacts. No details of the gate metal and the gate structure were included in the paper.

Schottky barrier and ohmic contact formation together with FET results were investigated by Morkoç *et al.*[22] By choosing the right processing parameters, contact resistances of $5.8 \times 10^{-6}\,\Omega cm^2$ and $5 \times 10^{-7}\,\Omega cm^2$ were obtained for GaInAsP with 1.15 eV bandgap and GaInAs, respectively, using Au–Ge–Ni alloy for the ohmic metal. Both had around $10^{17}\,cm^{-3}$ doping levels. Gate structures having a thin SiO_2 layer between the metal and the semiconductor were employed for Schottky-barrier formation to the alloy as described in section 17.4.2. The transconductance of the device with 1.15 eV GaInAsP was 7 mS ($35\,mS\,mm^{-1}$). Maximum available gain at 8 GHz was 9 dB and the noise figure was 5.6 dB with 7 dB associated gain at 7 GHz. For InP FETs with quasi-Schottky gate having thicker SiO_2, better Schottky diode characteristics were obtained with lower FET performances. This indicates that a good two-terminal characteristic is not a measure of good gate control. It is not clear whether this is due to immature processing technology or to an unavoidable interface problem in oxide–GaInAsP systems. In spite of the success in fabricating GaInAsP FETs with 1.15 eV bandgap, the quasi-Schottky approach taken here did not give satisfactory gate control for GaInAsP with 0.9 eV bandgap and GaInAs.

17.4.4 Double-Heterostructure MESFET

Using afore-mentioned Schottky heterostructure (section 17.4.3), double-heterostructure (DH) GaInAs MESFETs were fabricated.[31,41,42] Thin intermediary lattice-matched AlInAs layers were used not only to raise the effective Schottky-barrier height but also to form an electron confinement buffer layer. The band diagram of the structure is shown in Figure 17.5, with a cross-section of

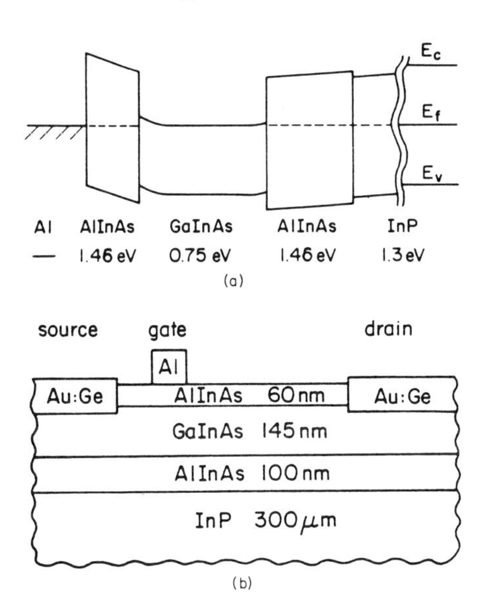

Figure 17.5 The band diagram and a cross-sectional view of the double-heterostructure GaInAs MESFET. Figures in the band diagram represent room-temperature bandgaps. Figures in the cross-sectional view show thicknesses of the layers

the device. The wafers were grown entirely by molecular-beam epitaxy including the aluminium gate metal without exposure to air. The aluminium gates were defined by optical image projection. An etch back process was used to avoid oxidizing the AlInAs under the gate by exposing it to the atmosphere. Other metals may be used instead of aluminium, if they can be etched back without damaging AlInAs and GaInAs, for example by dilute NaOH. Then source and drain windows were opened. Before depositing gold–germanium eutectic, the AlInAs Schottky assist layer was removed to minimize contact resistances. The AlInAs layer was etched away by H_3PO_4:H_2O_2:H_2O (1:1:38), which etches both GaInAs and AlInAs at a rate of 1000 Å min^{-1} at 21.5°C. After patterning the source and drain contacts by a lift-off technique, mesa etching was done to isolate individual devices by H_3PO_4:H_2O_2:H_2O (1:1:8), which has a rate of 5000 Å min^{-1} for GaInAs at room temperature. The wafer was then heated to 350°C for 30 s to reduce the contact resistances. No degradation of the aluminium Schottky gates was observed after the thermal cycle.

The devices had 0.6 μm gate lengths, 65 μm gate widths, 0.8 μm source–gate spacings, and 3.5 μm source–drain spacings. The undoped AlInAs layer underneath the gate was 600 Å thick while the undoped AlInAs buffer layer was 1000 Å. The GaInAs active layer was 1450 Å thick doped with germanium to 1.2×10^{17} cm^{-3}. A typical drain I–V characteristic is shown in Figure 17.6. The

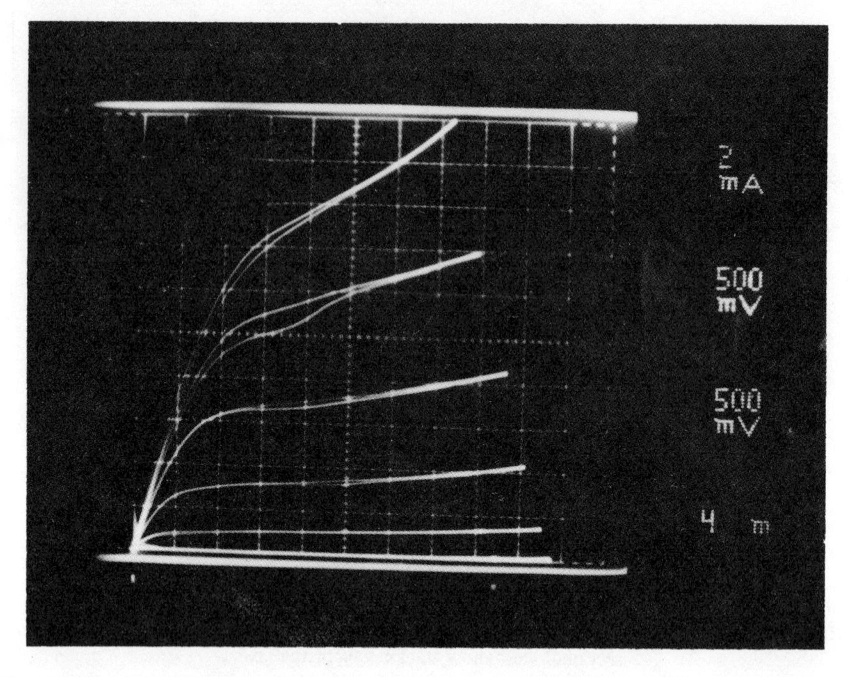

Figure 17.6 A typical drain I–V characteristic of a double-heterostructure GaInAs MESFET having gate width of 65 μm. The DC transconductance g_m was 135 mS mm^{-1} at 2.0 V drain-to-source bias

DC transconductance was 135 mS mm^{-1} at $V_{gs} = 0$ V and $V_{ds} = 2$ V, and 230 mS mm^{-1} at $V_{gs} = 0$ V and $V_{ds} = 4$ V where an anomalous increase of the drain current was noticeable. The pinch-off voltage was -2.20 V and the knee voltage was 0.95 V. At $V_{gs} = 0$ V and $V_{ds} = 2$ V, drain-to-source current was 177 mA mm^{-1} of gate width. Since the devices were made to demonstrate the DC capability of the material, the gate pad was placed over the active layer and connected to the gate by an aluminium connection on a thin mesa-etched active layer as shown in Figure 17.7. No back-gating effect was observed. This was demonstrated by observing that the channel current could not be modulated by a voltage applied to the gate pad when the aluminium connection between the contact pad and the Schottky gate was severed. The drain-to-source leakage current was less than 31 μA mm^{-1} of gate width for $V_{gs} = -2.50$ V and $V_{ds} < 5$ V. The gate-to-drain I–V characteristic is shown in Figure 17.8, which indicates the effectiveness of the intermediary AlInAs layer. The reverse-biased gate leakage to the drain was less than 62 nA μm^{-2} for V_{dg} less than 3 V from where it rose to 310 nA μm^{-2}.

Looping, which was found both in the drain I–V characteristic and in the gate-to-drain I–V characteristic, may be evidence of trapping. Since it is difficult to

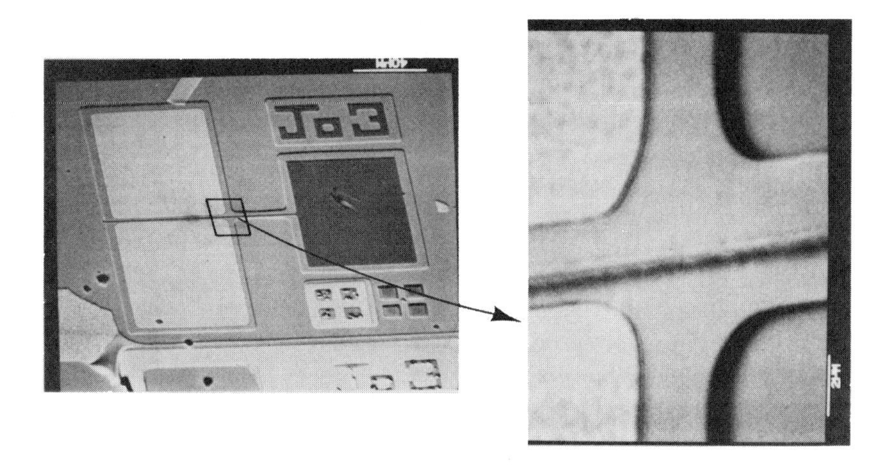

Figure 17.7 A scanning electron micrograph of a double-heterostructure GaInAs MESFET

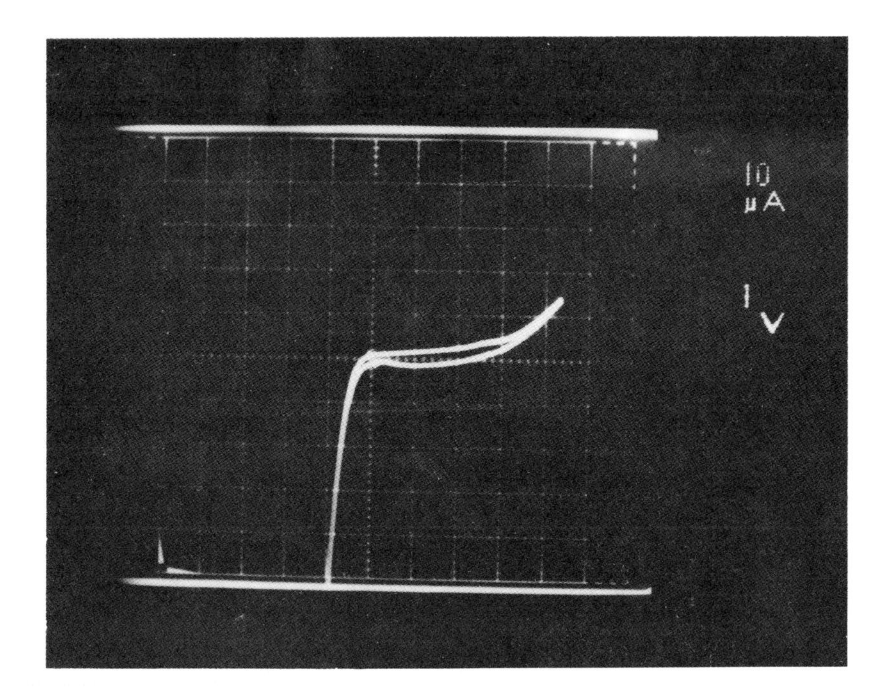

Figure 17.8 A gate-to-drain I–V characteristic of a double-heterostructure GaInAs MESFET. This shows the effectiveness of the AlInAs intermediary layer

grow a conducting AlInAs layer without intentional doping of over $10^{17}\,cm^{-3}$, it is evident that AlInAs layers used in the experiments contain a large amount of deep levels, which is probably the cuase of the looping.

Comparison can be made with GaAs FETs having similar structures, namely without a gate recess. Good GaAs FETs without a gate recess show $110\,mS\,mm^{-1}$ DC transconductance whereas the GaInAs MESFETs showed $135\,mS\,mm^{-1}$ at low drain-to-source bias of 2 V. This implies a higher average electron velocity in GaInAs than in GaAs. Also the presence of the AlInAs layer had lowered the g_m, since the potential drop across the AlInAs layer did not contribute to the modulation of the current. Although no microwave measurement has yet been reported on the device, this is the first device result which indicates the superiority of GaInAs to GaAs.

The integration of GaInAs dual-gate FETs with GaInAs high-speed photoconductive detectors was demonstrated by Barnard *et al.*,[43] which may also be an alternative design for the front end of the optical-fibre communication system.

A resistivity increase of GaInAs with initial resistivity of $3.3 \times 10^{-2}\,\Omega cm$ was observed after either proton, boron, oxygen, or nitrogen ion implantation. The highest resistivity obtained was $10\,\Omega\,cm$ by boron ion implantation at 100 kV for $10^{12}\,cm^{-2}$ with subsequent heating at 200°C.[44]

Modulation doping, reported by Dingle *et al.*[45] and made into FETs by Mimura *et al.*[46] for the GaAs–GaAlAs system, may not result in a significant improvement in the low-field electron mobility at low temperature for the present GaInAs–AlInAs system, since alloy scattering gives smaller temperature dependence to μ ($\mu \alpha T^{-0.5}$). This limits the 77 K mobility of GaInAs to about $90\,000\,cm^2\,V^{-1}\,s^{-1}$.[7] The modulation-doped structure, however, may enhance the overshoot phenomena according to the calculation by Hill *et al.*,[47] who calculated the effect of ionized-impurity scattering on overshoot. Overshoot phenomena are reduced by ionized-impurity scattering especially in low electric fields where GaInAs FETs are supposed to be used.

The anomalous increase of the drain current in GaInAs DH MESFETs in relatively high drain bias is possibly due to the high field in GaInAs at the end of the drain contact. Drain contacts and source contacts were made directly to GaInAs layers after removing AlInAs layers and a part of the active layer in the contact region to reduce contact resistances. This recess structure may cause a current crowding at the gate edge of the drain contact. Even without such structures, high electric fields may be present in the near-drain edge of the gate where a stationary Gunn domain is formed. The presence of such high fields may trigger a band-to-band tunnelling. Takanashi *et al.* showed that the soft breakdown in GaInAsP APDs was due to the band-to-band tunnelling.[48] Forrest *et al.* also reported that the dark-current behaviour in GaInAs photodiodes could be explained by taking the band-to-band tunnelling into account.[49] The band-to-band tunnelling is enhanced with small bandgap and small effective mass. These results suggest the anomaly in GaInAs MESFETs to be caused by Zener breakdown. In addition to 'the tunnelling, Takeda *et al.* observed current instabilities due to transferred electron oscillation followed by impact ionization in GaInAs.[50] This shows an inability of GaInAs FETs to form a large stationary Gunn domain to accommodate a large drain-to-source bias voltage.

Although further experiment is necessary to verify the origin of the increase in drain conductance, the high-field breakdown phenomena clearly restrict the range of application of GaInAs FETs. It is clear from the results obtained so far that GaInAs is not a suitable material for power FETs because of its small bandgap. However for low-noise FETs, both calculations[16,18] and experiments [41,42] show the potential of the material to be superior to GaAs.

ACKNOWLEDGMENTS

The authors wish to thank Dr C. E. C. Wood, Professor L. F. Eastman, and Professor H. Sakaki for useful discussions.

REFERENCES

1. C. A. Liechti, 'Microwave field-effect transistors—1976', *IEEE Trans. Microwave Theor. Tech.*, **MTT-24**, 279–300, 1976.
2. M. Ino and M. Ohmori, 'Intrinsic response time of normally off MESFET's of GaAs, Si and InP', *IEEE Trans. Microwave Theor. Tech.*, **MTT-28**, 456–459, 1980.
3. T. Wada and J. Frey, 'Physical basis of short-channel MESFET operation', *IEEE J. Solid-State Circuits*, **SC-14**, 398–412, 1979.
4. P. D. Greene, S. A. Wheeler, A. R. Adams, A. N. El-Sabbahy, and C. N. Ahmad, 'Background carrier concentration and electron mobility in LPE $In_{1-x}Ga_xAs_yP_{1-y}$ layers', *Appl. Phys. Lett.*, **35**, 78–80, 1979.
5. R. F. Leheny, A. A. Ballman, J. C. DeWinter, R. E. Nahory, and M. A. Pollack, 'Compositional dependence of the electron mobility in $In_{1-x}Ga_xAs_yP_{1-y}$', *J. Electron. Mater.*, **9**, 561–568, 1980.
6. J. H. Marsh, P. A. Houston, and P. N. Robson, 'Compositional dependence of the mobility, peak velocity and threshold field in $In_{1-x}Ga_xAs_yP_{1-y}$', *Gallium Arsenide and Related Compounds 1980*, Conf. Ser. no. 56, Institute of Physics, Bristol, 1981, pp. 613–620.
7. J. D. Oliver and L. F. Eastman, 'Liquid phase epitaxial growth and characterization of high purity lattice matched $Ga_xIn_{1-x}As$ on (111)B InP', *J. Electron. Mater.*, **9**, 693–712, 1980.
8. T. P. Pearsall, G. Beuchet, J. P. Hirtz, N. Visentin, and M. Bonnet, 'Electron and hole mobilities in $Ga_{0.47}In_{0.53}As$', *Gallium Arsenide and Related Compounds 1980*, Conf. Ser. no. 56, Institute of Physics, Bristol, 1981, pp. 639–649.
9. M. A. Littlejohn, J. R. Hauser, and T. H. Glisson, 'Velocity-field characteristics of $Ga_{1-x}In_xP_{1-y}As_y$ quaternary alloys', *Appl. Phys. Lett.*, **30**, 242–244, 1977.
10. P. K. Bhattacharya, J. W. Ku, S. J. T. Owen, G. H. Olsen, and S. H. Chiao, 'LPE and VPE $In_{1-x}Ga_xAs_yP_{1-y}$/InP: transport properties, defects and device considerations', *IEEE J. Quantum Electron.*, **QE-17**, 150–161, 1981.
11. A. Sasaki, Y. Takeda, N. Shikagawa, and T. Takagi, 'Liquid phase epitaxial growth, electron mobility and maximum drift velocity of $In_{1-x}Ga_xAs$ ($x \simeq 0.5$) for microwave

devices', *Jpn. J. Appl. Phys.*, **16**, *Suppl.* 16–1, 239–243, 1977.
12. J. C. Gammel, H. Ohno, and J. M. Ballantyne, 'High-speed photoconductive detectors using GaInAs', *IEEE J. Quantum Electron.*, **QE-17**, 269–272, 1981.
13. J. G. Ruch, 'Electron dynamics in short channel field-effect transistors', *IEEE Trans. Electron. Devices*, **ED-19**, 652–654, 1972.
14. M. S. Shur and L. F. Eastman, 'Ballistic transport in semiconductor at low temperature for low-power high-speed logic', *IEEE Trans. Electron. Devices*, **ED-26**, 1677–1683, 1979.
15. M. A. Littlejohn, L. A. Arledge, T. H. Glisson, and J. R. Hauser, 'Influence of central valley effective mass and alloy scattering on transient drift velocity in $Ga_{1-x}In_xP_{1-y}As_y$', *Electron. Lett.*, **15**, 586–588, 1979.
16. A. Cappy, B. Carney, R. Fauquemberques, G. Salmer, and E. Constant, 'Comparative potential performance of Si, GaAs, GaInAs, InAs submicrometer-gate FET's', *IEEE Trans. Electron. Devices*, **ED-27**, 2158–2162, 1980.
17. T. P. Pearsall, '$Ga_{0.47}In_{0.53}As$: a ternary semiconductor for photodetector applications', *IEEE J. Quantum Electron.*, **QE-16**, 709–720, 1980.
18. J. M. Golio and R. J. Trew, 'Compound semiconductors for low-noise microwave MESFET applications', *IEEE Trans. Electron. Devices*, **ED-27**, 1256–1262, 1980.
19. J. Frey, 'Effects of intervalley scattering on noise in GaAs and InP field-effect transistors', *IEEE Trans. Electron. Devices*, **ED-23**, 1298–1303, 1976.
20. A. R. Adams, 'Calculations of the intrinsically higher temperature stability of electron devices made from quaternary alloys', *Electron. Lett.*, **16**, 177–178, 1980.
21. K. Kajiyama, Y. Mizushima, and S. Sakata, 'Schottky barrier height of n-$In_xGa_{1-x}As$ diodes', *Appl. Phys. Lett.*, **23**, 458–459, 1973.
22. H. Morkoç, T. J. Drummond, and C. M. Stanchak, 'Schottky barriers and ohmic contacts on n-type InP based compound semiconductors for microwave FET's', *IEEE Trans. Electron. Devices*, **ED-28**, 1–5, 1981.
23. P. K. Bhattacharya and M. D. Yeaman, 'Enhanced barrier height of Au–$In_{1-x}Ga_xAs_yP_{1-y}$ Schottky diodes', *Solid-State Electron.* **24**, 297–300, 1981.
24. J. S. Escher, L. W. James, R. Sankaran, G. A. Antypas, R. L. Moon, and R. L. Bell, 'Schottky-barrier height of Au/p-InGaAsP alloys lattice matched to InP', *J. Vac. Sci. Technol.*, **13**, 874–875, 1976.
25. D. V. Morgan and J. Frey, 'Comments on the modification of Schottky barrier height by interfacial oxides', *Phys. Stat. Sol.*, a, **51**, K29–K33, 1979.
26. H. Sakaki, Y. Sekiguchi, D. C. Sun, M. Taniguchi, H. Ohno, and A. Tanaka, 'Schottky barrier properties of nearly-ideal ($n \simeq 1$) Al contacts on MBE- and heat cleaned-GaAs surfaces', *Jpn. J. Appl. Phys.*, **20**, L107–L110, 1981.
27. D. V. Morgan and J. Frey, 'Increasing the effective barrier height of Schottky contacts to n-$In_xGa_{1-x}As$', *Electron. Lett.*, **14**, 737–738, 1978.
28. D. V. Morgan, J. Frey, and W. J. Devlin, 'Rectifying and ohmic contacts to GaInAsP', *J. Electrochem. Soc.*, **127**, 1202–1205, 1980.
29. S. Umebachi, K. Asahi, M. Inoue, and G. Kano, 'A new heterojunction gate GaAs FET', *IEEE Trans. Electron. Devices*, **ED-22**, 613–614, 1975.
30. H. Morkoç, J. D. Oliver, and L. F. Eastman, 'High mobility in $In_{0.53}Ga_{0.47}As$ for high performance MESFET's', *Proc. 1979 Cornell Conf. on Active Microwave Semiconductor Devices and Circuits*, Cornell University, 1979, pp. 71–80.
31. H. Ohno, J. Barnard, C. E. C. Wood, and L. F. Eastman, 'Double heterostructure $Ga_{0.47}In_{0.53}As$ MESFETs by MBE', *IEEE Electron. Devices Lett.*, **EDL-1**, 154–155, 1980.
32. M. R. Lorentz and A. Onton, 'Electronic structure of III–V alloys from luminescence', *Proc. Int. Conf. on the Physics of Semiconductors*, US Atomic Energy Commission, Washington DC, 1970, pp. 444–449.

33. J. S. Best, 'The Schottky barrier height of Au on n-$Ga_{1-x}Al_xAs$ as a function of AlAs content', *Appl. Phys. Lett.*, **34**, 522–524, 1979.
34. H. Ohno, C. E. C. Wood, L. Rathbun, D. V. Morgan, and L. F. Eastman, 'GaInAs–AlInAs structures grown by molecular beam epitaxy', *J. Appl. Phys.*, **52**, 4033–**4037**, 1981.
35. R. F. Leheny, R. E. Nahory, M. A. Pollack, A. A. Ballman, E. D. Beebe, J. C. DeWinter, and R. J. Martin, 'An $In_{0.53}Ga_{0.47}As$ junction field-effect transistor', *IEEE Electron. Devices Lett.*, **EDL-1**, 110–111, 1980.
36. R. F. Leheny, R. E. Nahory, M. A. Pollack, A. A. Ballman, E. D. Beebe, J. C. DeWinter, and R. J. Martin, 'Integrated $In_{0.53}Ga_{0.47}As$ p–i–n F. E. T. photoreceiver', *Electron. Lett.*, **16**, 353–355, 1980.
37. C. A. Arimento, J. P. Donnelly, and S. H. Groves, 'p–n junction diodes in InP and $In_{1-x}Ga_xAs_yP_{1-y}$ fabricated by beryllium-ion implantation', *Appl. Phys. Lett.*, **34**, 229–231, 1979.
38. Y. Shinoda, M. Okamura, E. Yamaguchi, and T. Kobayashi, 'InGaAsP *n*-channel inversion-mode MISFET', *Jpn. J. Appl. Phys.*, **19**, 2301–2302, 1980.
39. D. Decker, R. Fairman, and C. Nishimoto, 'Microwave InGaAs Schottky-barrier-gate field-effect transitors—Preliminary results', *Proc. 1975 Cornell Conf. on Active Semiconductor Devices for Microwaves and Integrated Optics*, Cornell University, 1975, pp. 305–314.
40. H. Morkoç, J. T. Andrews, Y. H. Houng, R. Sankaran, S. G. Bandy, and G. A. Antypas, 'Microwave $In_xGa_{1-x}As_yP_{1-y}$/InP F. E. T.', *Electron. Lett.*, **14**, 448–449, 1978.
41. J. Barnard, H. Ohno, C. E. C. Wood, and L. F. Eastman, 'Double heterostructure $Ga_{0.47}In_{0.53}As$ MESFETs with submicron gates', *IEEE Electron. Devices Lett.*, **EDL-1**, 174–176, 1980.
42. H. Ohno, J. Barnard, L. Rathbun, C. E. C. Wood, and L. F. Eastman, 'Double heterostructure $Ga_{0.47}In_{0.53}As$ MESFETs by molecular beam epitaxy', *Gallium Arsenide and Related Compounds 1980*, Conf. Set. no. 56, Institute of Physics, Bristol, 1981, pp. 465–473.
43. J. Barnard, H. Ohno, C. E. C. Wood, and L. F. Eastman, 'Integrated double heterostructure $Ga_{0.47}In_{0.53}As$ photoreceiver with automatic gain control', *IEEE Electron. Devices, Lett.*, **EDL-2**, 7–9, 1981.
44. J. Barnard, C. E. C. Wood, and L. F. Eastman, 'Resistivity increase in MBE $Ga_{0.47}In_{0.53}As$ following ion bombardment', *IEEE Electron Device Letters*, **EDL-2**, 193–195, 1981.
45. R. Dingle, H. L. Stormer, A. C. Gossard, and W. Wiegmann, 'Electron mobilities in modulation-doped semiconductor heterojunction superlattices', *Appl. Phys. Lett.*, **33**, 665–667, 1978.
46. T. Mimura, S. Hiyamizu, T. Fujii, and K. Nanbu, 'A new field-effect transistor with selectively doped GaAs/n-$Al_xGa_{1-x}As$ heterojunctions', *Jpn. J. Appl. Phys.*, **19**, L225–L227, 1980.
47. G. Hill, P. N. Robson, A. Majerfield, and W. Fawcett, 'Effect of ionized impurity scattering on the electron transit time in GaAs and InP F.E.T.s', *Electron. Lett.*, **13**, 235–236, 1977.
48. Y. Takanashi, M. Kawashima, and Y. Horikoshi, 'Required donor concentration of epitaxial layers for efficient InGaAsP avalanche photodiodes', *Jpn. J. Appl. Phys.*, **19**, 693–701, 1980.
49. S. R. Forrest, R. F. Leheny, R. E. Nahory, and M. A. Pollack, '$In_{0.53}Ga_{0.47}As$ photodiodes with dark current limited by generation–recombination and tunnelling', *Appl. Phys. Lett.*, **37**, 322–325, 1980.
50. Y. Takeda, N. Shikagawa, and A. Sasaki, 'Transferred-electron oscillation in n-$In_{0.53}Ga_{0.47}As$', *Solid-State Electron.*, **23**, 1003–1006, 1980.

Note added in proof

The following articles have been published since this chapter was originally written. Readers are suggested to refer to them for the latest development in this field.

On the velocity-field characteristics:

51. J. Degani, R. F. Leheny, R. E. Nahory, and J. P. Heritage, "Velocity-field characteristics of minority carriers (electrons) in p-$In_{0.53}Ga_{0.47}As$", *Appl. Phys. Lett.*, **39**, 569–572, 1981.
52. T. H. Windhorn, L. W. Cook, and G. E. Stillman, "The electron velocity-field characteristic for n-$In_{0.53}Ga_{0.47}As$ at 300 K", *IEEE Electron Device Letters*, **EDL-3,** 18–20, 1981.

On the Γ–L separation in GaInAs:

53. K. Y. Cheng, A. Y. Cho, S. B. Christman, T. P. Pearsall, and J. E. Rowe, "Measurement of the Γ–L separation in $Ga_{0.47}In_{0.53}As$ by ultraviolet photoemission", *Appl. Phys. Lett.*, **40**, 423–425, 1982.

On the modulation doping in the GaInAsP system:

54. K. Y. Cheng, A. Y. Cho, T. J. Drummond, and H. Morkoc, "Electron mobilities in modulation doped $Ga_{0.47}In_{0.53}As/Al_{0.48}In_{0.52}As$ heterojunctions grown by molecular beam epitaxy", *Appl. Phys. Lett.*, **40**, 147–149, 1982.
55. M. Razeghi, M. A. Poisson, J. P. Larivan, B. de Cremoux, J. P. Dushman, and M. Voos, "TEG in LP-MO CVD $Ga_{0.47}In_{0.53}As$-InP superlattice", *Electron. Lett.*, **18**, 339–340, 1982.

On the modified Schottky barriers and related FET results:

56. S. Bandy, C. Nishimoto, S. Hyder, and C. Hooper, "Saturation velocity determination for $In_{0.53}Ga_{0.47}As$ field-effect-transistors", *Appl. Phys. Lett.*, **38**, 817–819, 1981.
57. C. Y. Chen, A. Y. Cho, P. A. Garbinski, and K. Y. Cheng, "Characteristics of an $In_{0.53}Ga_{0.47}As$ very shallow junction gate structure grown by molecular beam epitaxy", *IEEE Electron Device Letters*, **EDL-3**, 15–17, 1981.
58. C. Y. Chen, A. Y. Cho, K. Y. Cheng and A. Garbinski, "Quasi-Schottki barrier diode on n-$Ga_{0.47}In_{0.53}As$ using a fully depleted p^+-$Ga_{0.47}In_{0.53}As$ layer grown by molecular beam epitaxy", *Appl. Phys. Lett.*, **40**, 401–403, 1982.
59. T. Y. Chang, R. F. Leheny, R. E. Nahory, E. Silberg, A. A. Ballman, E. A. Caridi, and C. J. Harrold, "Junction field-effect transistors using $In_{0.53}Ga_{0.47}As$ material grown by molecular beam epitaxy", *IEEE Electron Device Letters*, **EDL-3**, 56–58, 1982.
60. P. O'Connor, T. P. Pearsall, K. Y. Cheng, A. Y. Cho, J. C. M. Hwang, and K. Alavi, "$In_{0.53}Ga_{0.47}As$ FET's with insulator-assisted Schottky gates", *IEEE Electron Device Letters*, **EDL-3**, 64–66, 1982.

On the MIS structures and MISFETS:

61. H. H. Wieder, A. R. Clawson, D. I. Elder, and D. A. Collins, "Inversion-mode insulated gate $Ga_{0.47}In_{0.53}As$ field-effect transistors", *IEEE Electron Device Letters*, **EDL-2**, 73–74, 1981.

62. Y. Shinoda and T. Kobayashi, "InGa AsP n-channel inversion-mode metal-insulator-semiconductor field-effect transistor with low interface state density", *J. Appl. Phys.*, **52**, 6386-6394, 1981.
63. A. S. H. Liao, B. Tell, R. F. Leheny, R. E. Nahory, J. C. De Winter, and R. J. Martin, "An $In_{0.53}Ga_{0.47}As$ p-Channel MOSFET with plasma-grown native oxide insulated gate.", *Electron Dev. Lett.*, **EDL-3**, 158–160.

GaInAsP Alloy Semiconductors
Edited by T. P. Pearsall

Appendix

Some Physical Properties of GaInAsP Alloys—$Ga_xIn_{1-x}As_yP_{1-y}$

						Reference
Composition	x	0	0.16	0.27	0.40	0.47	
	y	0	0.33	0.60	0.85	1.0	
Property							
Bandgap	300 K, eV	1.35	1.112	0.953	0.800	0.75	
	77 K, eV	1.414	1.175	1.203	0.864	0.789	1, 2, 3
	4 K, eV	1.425	1.188	1.018	0.881	0.812	
Absorption edge 300 K, μm		0.92	1.12	1.30	1.55	1.65	3
Electron effective mass		0.08	0.067	0.053	0.045	0.041	4
Light hole mass		0.120	0.091	0.072	0.062	0.051	5, 6
Heavy hole mass		0.56	—	0.50	—	0.50	5, 7, 8
Electron mobility; $T = 300/77$ K							
$N_D + N_A = 10^{16}$ cm^{-3}		4100/44 000	3800/8000	6300/13 000	7700/15 000	10 500/28 500	9, 10
$= 10^{17}$ cm^{-3}		3500/5800	3000/5000	3600/5000	6500/7500	9000/15 000	9, 10
$= 10^{18}$ cm^{-3}		2400/2400	2400/2800	2800/2800	5400/5000	5000/7000	9, 10